Terrigenous Mass Movements

Biswajeet Pradhan · Manfred Buchroithner
Editors

Terrigenous Mass Movements

Detection, Modelling, Early Warning and Mitigation Using Geoinformation Technology

Springer

Prof. Dr. Biswajeet Pradhan
Institute of Advanced Technology
Spatial & Numerical Modelling
Laboratory
University Putra Malaysia (UPM)
Serdang, Selangor Darul Ehsan
43400, Malaysia

Prof. Dr. Manfred Buchroithner
Institute of Cartography
Dresden University of
Technology
Helmholtzstr. 10
01062 Dresden
Germany

ISBN 978-3-642-25494-9 e-ISBN 978-3-642-25495-6
DOI 10.1007/978-3-642-25495-6
Springer Heidelberg New York Dordrecht London

Library of Congress Control Number: 2012930655

Cover design: eStudio Calamar

Printed on acid-free paper

Springer is part of Springer Science+Business Media (www.springer.com)

Preface

Mass movements are a serious geologic hazard common to almost every part of the world. Globally, the damage to property caused by all forms of mass movements has been running into billions of dollars annually and thousands of lives have also been lost. Geoinformation technology has revolutionised our understanding of the Earth as an integrated system, giving us a growing capability to forecast changes in weather and climate and allowing observations of changes in land cover and land use and also prediction of landslides.

Over the past twelve years the topics of terrigenous mass movements and the use of space-borne remote sensing and GIS has been instrumental for their monitoring, prediction and mitigation. This has been a subject of continuous cooperation between the two editors of this book. Their geological background and their intensive use of remote sensing data of various sensors over many years, combined with an eager interest in modelling, made different types of "geo" mass movements including glacier lake outbursts a major focus of their research. They brought their backgrounds in geology and spatial modeling, along with experiences in the use of remote sensing, to the task of editing a book on mass movements for spatial modelers, environmental scientists, and land planners. Its objective is to show how to make best use of the tools of remote sensing, numerical modelling and GIS and then employ them wisely for landslide assessment and predictions.

Thus it was just a logical step to compile the accumulated knowledge and expertise into a book. However, in order to make this volume more balanced and comprehensive, from the beginning on we decided to invite a series of internationally renowned experts to contribute their shares. As soon as the message of our undertaking spread within the community, we got a lot of enthusiastic feed-backs. In addition, we also placed a series of dedicated invitations to well-established colleagues in particular fields. With the typescript submission deadline coming closer, however, quite a number of potential authors apologised for not being able to submit, or simply did not respond at all. Others did deliver, after several reminders, with a significant delay, though.

From the outset of our work for this book, our idea was to compile a volume on different aspects of terrigenous mass movements. This starts from the GIS data base construction (inventory mapping) to the development of highly sophisticated modelling tools for landslide predictions. We have organizsed the book's chapters along the lines of overall landslide assessment. The initial chapters offer a foundation in the areas of remote sensing most relevant to landslide inventory mapping, along with assessments of the tools available and a better understanding of when and when not to use remote sensing. Other chapters introduce the well-established domains of statistical, heuristic and data mining techniques in landslide predictions. Throughout the text, there is an emphasis on applications of remote sensing and GIS tools aided with soft computing techniques for landslide mapping with examples and case studies showing the way.

As editors, we jointly and mutually made—to a certain extent–critical remarks with regard to both the articles' contents and the authors' phrasing, and in some cases even directly corrected minor "slips of the pen". Finally, after quite some of editorial work, the entire typescript was ready to be sent to Springer Verlag. There, Dr. Christian Witschel and Agata Oelschläger were appreciative and understanding partners. Their cooperative way of dealing with us and our special requests is thankfully acknowledged.

Further, the support by Alexander von Humboldt Foundation (AvH), Deutscher Akademischer Auslandsdienst (DAAD) as well as our colleagues at our home institutions, the University Putra Malaysia (UPM) in Kuala Lumpur and at the Technische Universtät Dresden (TUD), Germany, has to be thankfully mentioned.

The production of this book would not have been possible without the professional and formidable efforts of Steffi Sharma, Omar Althwaynee, Sheila Pradhan, and Anne Lange. We also extend our thanks and gratitude to them.

Kuala Lumpur and Dresden, August 2011

Biswajeet Pradhan
Manfred F. Buchroithner

Contents

Chapter 1
An Automated Approach for Detection of Shallow Landslides from LiDAR Derived DEM Using Geomorphological Indicators in a Tropical Forest

Ulrich Mann, Biswajeet Pradhan, Nikolas Prechtel and Manfred F. Buchroithner

Abstract Landslide inventories in the tropical dense forested areas are routinely compiled by means of a terrain model interpretation (e.g. using stereo-radargrammetry; stereo-aerial photographs; stereo-optical imagery), aided with field investigations. However, construction of the landslide inventories from aerial photographs and field based studies are excessively time consuming which involves relatively high cost. Moreover, these techniques are less effective when applied to dense tropical forest where landslide scars are difficult to map from the aerial photographs. This chapter attempts an automatic procedure for detection of rotational shallow landslides from airborne based light detection and ranging (LiDAR) derived high resolution digital elevation model (DEM) in a tropical forest in Cameron Highlands, Malaysia. For the extraction of landslides from DEM, we used various geomorphic indicators such as surface roughness index, vegetation index and breaklines. The entire landslide extraction process was implemented in ArcGIS platform and custom Python scripts was used for the implementation and model construction. For modeling purpose, the Python Imaging Library (PIL) was used. The terrain zone classification was tested for various DEM resolutions of 1.5 m, 2 m, 3 m, 4 m, 5 m and 8 m. For testing purposes, the resolutions with the best results were used for further processing. To automate the classification of the terrain zones, a rule based region growing threshold was defined depending on the resolution of the DEM. Finally, a statistical description was applied to rank the extracted terrain zones according to their compliance with the landslide signature. Subsequently, the landslide probability

U. Mann · B. Pradhan · N. Prechtel · M. F. Buchroithner
Institute for Cartography, Faculty of Forestry, Geo and Hydro Science, Dresden University of Technology, 01062 Dresden, Germany

B. Pradhan (✉)
Institute of Advanced Technology, Spatial and Numerical Modeling Laboratory, University Putra Malaysia, 43400 UPM, Serdang, Malaysia
e-mail: biswajeet@lycos.com; biswajeet24@gmail.com

B. Pradhan and M. Buchroithner (eds.), *Terrigenous Mass Movements*,
DOI: 10.1007/978-3-642-25495-6_1,

index (LPI) was calculated by performing zonal operation using each of the geomorphic parameters. Hence, the LIDAR-derived DEM provides adequate landslide factor maps to identify the landslide occurred areas, which could be used for further landslide assessment and site-planning purposes in the tropical regions.

Keywords LiDAR · Landslide · Inventory · DEM · GIS · Remote sensing · Malaysia

1.1 Introduction

Every year, thousands of humans are threatened by natural disasters such as earthquakes, floods, tropical storms, forest fires or mass movements. Specifically the topic mass movement is of major concern to many developing countries. Mass movements, or landslides, which is used synonymously in this chapter, cause approximately a thousand deaths as well as S$2 billion of property damage each year worldwide, as reported by the ESA's CEOS Disaster Management Support Group (Bishop and Shroder 2004; CCRS 2005; Rivard 2009). Especially, in the developing countries, where little open space for housing exists and therefore building construction is done on slopes which are sometimes unstable. Here, preparedness for the occurrence of this geohazard is a key issue. Landslides carry high economic and social loses to Malaysia. Occurrences of landslides are prevalent in hill complexes of Malaysia both in the highlands and lowlands. These have caused loss of lives and properties in recent years. While agriculture in landslide occurring areas has caused severe soil erosion downstream, hill construction projects for infrastructure and residential purposes were the main triggering factors of landslides. Planners have taken cognizance that developable areas in the gentle terrain have become scarce and encroachments into the more sensitive hilly areas particularly in the highlands are inevitable. Development planning in the hilly areas are of paramount importance for landslide management (Gokceoglu and Sezer 2009). In this context a national landslide hazard zoning map is a prerequisite to assist in decision making for approving developments in the prone areas. To further strengthen this initiative, landslide detection and monitoring system should also be developed for timely mitigation measures. However, currently there is not much work has been done for landslide hazard analysis and mitigation in Malaysia. Pradhan (2010b) stated that in Malaysia landslides occur frequently and damage could be mitigated, if an early warning system is available. For an example, he mentioned a landslide, which occurred in Kuala Lumpur in October 2002. It killed six members of a family and is still in the public's memory (Pradhan 2010a, c, d).

Over the years, geohazard susceptibility and and risk assessment have been conducted in an increasing numbers worldwide (Akgun and Bulut 2007; Akgun and Turk 2010; Akgun et al. 2008; Can et al. 2005; Duman et al. 2006; Ercanoglu et al. 2008; Iovine et al. 2003; Youssef et al. 2009; McKean and Roering 2003;

Nefeslioglu and Gokceoglu 2011; Guzzetti 2004; Lee and Jones 2004; Sassa et al. 2005). Furthermore, studies on the prediction of the occurrence of future landslides are carried out, using analytical and soft computing and statistical approaches (Ercanoglu 2005; Ercanoglu and Gokceoglu 2002, 2004; Lee 2005; Nefeslioglu et al. 2008, 2010; Sezer et al. 2011; Guzzetti et al. 2006; Pradhan 2011; Pradhan et al. 2010a, b, c; Oh and Pradhan 2010; Yilmaz 2009, 2010a, b). For these tasks, information about past and recent landslides is fundamental and most of the prediction approaches use landslide inventory maps of the study area (Pradhan 2010a, b; Pradhan and Lee 2010a, b, c). Compiling those maps is still conducted manually by experts interpreting aerial imagery and generally needs a supporting field study to verify the results (Pradhan et al. 2010a; Brardinoni et al. 2003). Pradhan et al. (2010a) asserted that "Landslides were detected from aerial photographs by interpreting breaks in the forest canopy, bare soil, and other typical geomorphological characteristics of landslide scars." Furthermore, it is a time consuming task, as McKean and Roering (2003) asserted: "Landslide inventory maps are tedious to compile, difficult to make in vegetated terrain using conventional techniques, and tend to be subjective."

In this chapter, an approach for the automated detection of rotational shallow landslides from a high-resolution digital elevation model (DEM), using geomorphological indicators is presented. The aim is to substitute or find an alternative way for outlining erosional features with more efficient techniques. This can provide the possibility of enhancing future statistical estimations about landslide occurrence, by using objective, automated generated inventory maps. Furthermore in this way, the question, if landslides can be recognized in tropical forested terrain could be answered to some extent.

Danneels et al. (2008) presented an approach on automated mapping of landslides and compilation of inventories in Tian Shan, Kyrgyz Republic. They used satellite imagery as well as various DEM derived supplementary for manual delineation of medium sized slope failure features. Primarily, they obtained a sliding likelihood value for each pixel and performed segmentation afterwards. Subsequently, they filtered classified landslides according to their shape and geomorphologic properties such as the slope angle (Clark et al. 2004; Razak et al. 2009, 2010; Schulz 2007).

In this chapter, a short characterisation of rotational landslides morphology is conducted and the geomorphologic parameter i.e. slope curvature is discussed. The limitation on rotational landslides is used as there exist too many types of landslides, each with different characteristics. Additionally, the classification approach is presented and a closer look to the classification steps is provided. Finally the results are presented and discussed.

1.2 Study Area

Many parts of middle and south-eastern Asian countries pose high risk due to mass movements. For example, Malaysia suffers from several landslide incidents each year, causing a severe numbers of fatalities and economic loss

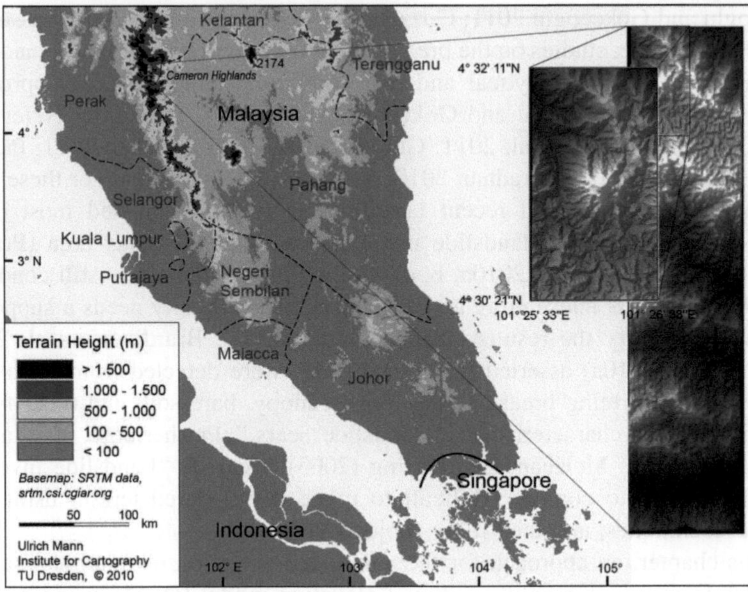

Fig. 1.1 Location map of the study area (Cameron Highlands in Malaysia)

(Lazacode 2010). They are commonly triggered by heavy rainfall, and anthropogenic activities such as improper slope management. For this research, the Cameron Highlands in Malaysia was chosen as the study area. This mountain range is situated in the north-westernmost part of the state of Pahang bordering to the state of Perak in the west and Kelantan in the north (Fig. 1.1). Here, landslides occur frequently because geomorphological and meteorological circumstances abet this kind of natural hazard.

The study area is located within 4° 30′N and 101° 20′E and covers an area of approximately 710 km^2. In the study area terrain heights range up to more than 1500 m above Mean Sea Level (msl), and slopes up to 60 degree are common, as identified from the elevation dataset.

Highest peaks of rainfall occur from March to May and September to December. Pradhan et al. (2010a) observed that most landslide incidents in this region occurred above a level of 208 mm of daily rainfall. The geology of the region basically consists of igneous and metamorphic bedrock (Pradhan et al. 2010a). Large parts consist of granite rocks that were formed in the late Triassic period (Fortuin 2006). The landscape is densely vegetated due to the tropical climate. Between the steep slopes there exist man-made plains, often established as terraces, used for tea and vegetable growing.

The available DEM covers an area of about 53.6 km^2. For the testing of geoprocessing techniques, a smaller region was used as application site, as the computing times are very time-consuming. The application site is located at 381533,5 to 383360,5 m East and 498562,65 to 501924,15 m North in local coordinates, which is approximately 4° 30′ 21″ to 4° 32′ 11″N and 101° 25′ 33″ to 101° 26′ 33″E in geographic coordinates. The used local coordinate system is Kertau RSO Malaya Meters. Heights in the dataset range between 1,230 to 1,650 m above msl and the steepest slopes reach more than 65 degree of inclination.

1.3 Geomorphological Features and Landslides

A rotational landslide, as defined by Abbott (2004) is the "downward and outward movement of a mass on top of a concave upward failure surface". Material is displaced, due to a triggering event, slumping downslope along a failure surface. At the top part of the slide, the crown, cracks occur in the original ground surface due to the shearing stress, which form the main scarp, a steep slope along the failure surface. While the material is moving downwards, due to internal rotation, several faults are formed cross-directionally along the movement of slope. So the material moves down in blocks, called the head and several units, separated by longitudinal faults and cracks.

At the foot of the landslide, displaced material moves beyond the depletion zone and accumulates above original undisturbed ground surface. At the end of the accumulation zone, this material is building the toe, which is the typically curved part of the accumulated material, most distant from the main scarp (Fig. 1.2). It can be assumed that the materials along the slide moves downwards, until it reaches a stable position (Lee and Jones 2004).

From the above description, the most important characteristic is the shape of the depletion zone which is typically formed concave in direction of the movement of the slide, which means in the direction of the steepest path on the slope. Typically, rotational landslides also show a concave curvature in the horizontal direction, formed by the declining slope at the flanks of the sliding zone, and lower heights in the centre of the displaced zone. The horizontal curvature also occurs, if the surface of the rupture is occurred on a planar sliding plane, but the flanks are similar to those of a rotational slide.

Landslides occur in hilly and mountainous terrain where the maximum slope angle exceeds the critical slope angle, which is dependent on the local soil characteristics and other geologic factors (Schuster and Wieczorek 2002). Takahashi (2009) defined a critical slope angle for debris-flow events for 15 degrees. Hylland and Lowe (1997) derived critical slope angles for slumps and debris slides in Utah, U.S., ranging between 9 and 27 degrees. Rivard (2009) specified that slow earth flows occur on slopes of 10 to 35 degree, whereas debris slides and avalanches occur on very steep slopes of more than 25 degrees.

Local topographic surface roughness is generally considered to be higher on the erosional areas of a landslide, on the depletion as well as on the accumulation zone, than in adjacent unfailed terrain. McKean and Roering (2003) used this characteristic to differentiate a large landslide complex from surrounding terrain, with reasonably good results.

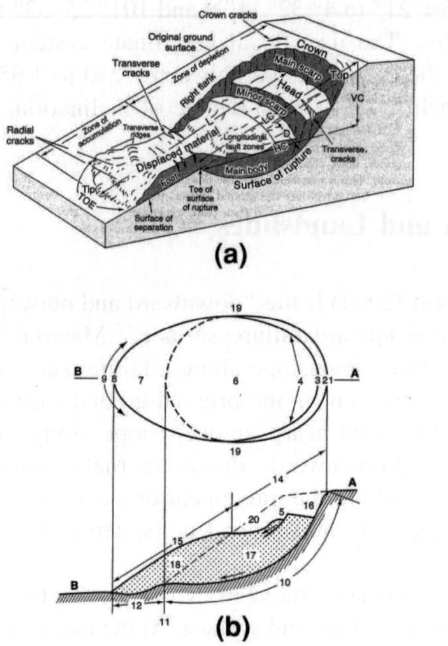

No.	Name	Definition
1	Crown	Practically undisplaced material adjacent to highest parts of main scarp
2	Main scarp	Steep surface on undisturbed ground at upper edge of landslide caused by movement of displaced material (13, stippled area) away from undisturbed ground, i.e. it is visible part of surface of rupture (10)
3	Top	Highest point of contact between displaced material (13) and main scarp (2)
4	Head	Upper parts of landslide along contact between displaced material and main scarp (2)
5	Minor scarp	Steep surface on the displaced material of landslide produced by differential movements within displaced material
6	Main body	Part of displaced material of landslide that overlies surface of rupture between main scarp (2) and toe of surface of rupture (11)
7	Foot	Portion of landslide that has moved beyond toe of surface of rupture (11) and overlies original ground surface
8	Tip	Point on toe (9) farthest from top (3) of landslide
9	Toe	Lower, usually curved margin of displaced material of a landslide, most distant from main scarp (2)
10	Surface of rupture	Surface that forms (or that has formed) lower boundary of displaced material (13) below original ground surface (20)
11	Toe of surface of rupture	Intersection (usually buried) between lower part of surface of rupture (10) of a landslide and original ground surface (20)
12	Surface of separation	Part of original ground surface (20) now overlain by foot (7) of landslide
13	Displaced material	Material displaced from its original position on slope by movements in landslide; forms both depleted mass (17) and accumulation (18): stippled on Fig. 2.6b
14	Zone of depletion	Area of landslide within which displaced material (13) lies below original ground surface (20)
15	Zone of accumulation	Area of landslide within which displaced material (13) lies above original ground surface (20)
16	Depletion	Volume bounded by main scarp (2), depleted mass (17) and original ground surface (20)
17	Depleted mass	Volume of displaced material (13) that overlies surface of rupture (10) but underlies original ground surface (20)
18	Accumulation	Volume of displaced material (13) that lies above original ground surface (20)
19	Flank	Undisplaced material adjacent to sides of surface of rupture (right and left is as viewed from crown)
20	Original ground surface	Surface of slope that existed before landslide took place

Fig. 1.2 Landslide morphology and features of an idealised rotational landslide (*Source* Lee and Jones 2004; after Cruden and Varnes 1996; IAEG Commission on Landslide 1990)

In areas with a dense vegetation canopy, covering will be less on failed slopes, due to unearthing of bare soil and translation of covering vegetation. Thus, a vegetation index is expected to be an indicator of probable landslide existence, in the study area.

1.4 The Parameter: Slope Curvature

The parameter curvature is a descriptive mathematical characteristic of a slope and is used to describe the shape of ground elements. Dikau (1988) defined relief entities in the sense of geomorphometric objects, as the smallest elements. The delineation of those entities is based on the homogeneity of the forms with respect to certain geomorphologic parameters. Dikau (1988) further broke down curvature into more precise relief parameters: profile and plan curvatures. Profile curvature is the curvature in vertical direction and plan curvature in horizontal direction, respectively. These characteristics are joined for each ground element and describe the character of the entities of the georelief. This approach of Dikau (1988) is utilized in this work, in order to derive relief-forms on slopes which are characterised by their profile and plan curvature (Kugler 1974).

For the derivation of the curvature value, the surface point is examined in relation to its vicinity. As a result, it becomes obvious, that curvature is a scale-dependant property, which means that it differs for one and the same point with a change in scale. This is crucial to notice, as it reveals the importance of the factor scale for the geomorphologic analysis. It is possible that a point on a surface owns a convex curvature in a large scale, but if we 'zoom out', the same point will be placed in a concave shaped form and thus have a concave curvature.

1.5 Data Used

In this chapter, existing elevation data of the Cameron Highlands, was compiled from a high-resolution airborne laser scanning (ALS) point cloud. The term LiDAR (Light Detection and Ranging) is often used synonymously for this data collection method and will be used throughout the chapter.

The advantages of a DEM from this source are obvious: data can be gathered in short time, investigations in a wide region are possible, smaller features can be detected and thus surface characteristics can be determined due to better resolution, and the extraction of a bare earth model is possible, making investigations more or less independent of vegetation effects. For example, Jakob and Hungr (2005) stated, that LiDAR has the ability to detect landslide scarps under a fairly dense tree canopy. Carter et al. (2001), were able to detect landslide scarps beneath a tree canopy and could identify features which otherwise would have been remained undetected. This makes airborne LiDAR a competitive data collecting technique (Sithole and Vosselman 2004).

For the calculation of a vegetation index, satellite images with red and infra-red bands are necessary. The images need to fit with the collection of the laser scanning point cloud in the temporal component, otherwise precise examination and reliable results cannot be guaranteed for the classification, as surface and vegetation characteristics are changing over time.

1.6 LiDAR Processing Steps

The entire process chain is based on three different stages, which are depicted in Fig. 1.3. The first is a layer of region polygons, derived from the elevation model. These polygons represent surface areas, which own a concave shape, in the horizontal as well as in the vertical direction (Glenn et al. 2006). Those regions are expected to represent potential landslide areas, based on the morphology of a rotational slide.

The second stage is a surface roughness index and the third one a vegetation index, each represented as a parameter layer. These indices will be used for the previously classified areas, to more precisely investigate their properties.

Fig. 1.3 Three stages of the landslide classification

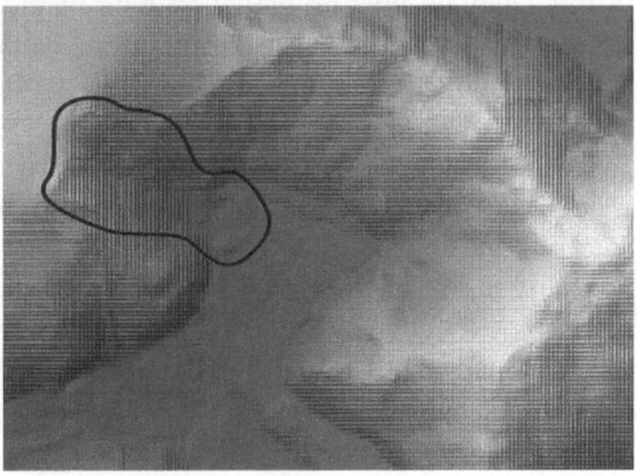

Fig. 1.4 Showing landslide feature location

It is expected, that good measures for defining a landslide can be derived with these indices, so they serve as indicators of eroded locations. Of course, each of them should not be taken individually, as different areas might show the same characteristics. But considering together in a decision making matrix, they can serve as a landslide signature from geomorphological and spectral properties. Considering these different aspects of a landslide, it is expected to define erosional areas by developing a probability index for their occurrence. In this way, each classified region polygon can be treated in respect to a certain probability of being a landslide. The classification idea follows the rule: Terrain regions of concave shape, which are situated on a slope, owning a high surface roughness and a low vegetation index are classified with a high landslide probability index.

In Fig. 1.4, a part of the study area is shown. It includes a presumed landslide area, which is outlined by the red contour. Clearly recognizable is a breakline at the upper part (left side in the image) and accumulated material, building a small pond surrounded by a levee, at the toe of the slide. This area was used to determine the suitable classification methods and to adjust the parameters such as filtering thresholds and DEM resolution (De Smith et al. 2007). In this way, an adjustment to the characteristics of the terrain in the study area could be conducted.

1.6.1 Potential Landslide Zone Classification

To classify zones from the DEM, which likely correspond to erosional areas. Curvature has been derived in both horizontal and vertical directions, which are referred to as plan curvature and profile curvature respectively (Wohl and Oguchi 2004). During the calculation for each ground element in the DEM, the curvature value is derived from a polynomial fitting in its neighbourhood. Each ground cell can then be grouped by its curvature values after following the classification approach of Dikau (1988) into convex, straight or concave. By doing so, it allows a geomorphological description of a slope at a certain scale (Dikau and Schmidt 1999).

Subsequently, straight or concave curvature (in both directions), were identified from the LiDAR derived DEM dataset. The classification of straight cells is also necessary, as there is a smooth transition between straight and concave and just classifying areas above a certain curvature would result in a too coarse distribution of classified pixels. That means, a binary raster dataset is produced, consisting of pixel values, where "1" indicates a concave shaped ground cell and "0" a convex shape, respectively. Furthermore, inclination at the current position is evaluated, to ensure, only pixels on a slope with a certain inclination upwards are taken into account, as plains should be excluded from the classification, as these areas are not expected to show eroded features. Thus, a slope inclination threshold was applied, by following different proposals as given by experts, defining a critical slope angle for the occurrence of landslides in their research. For the critical slope angle, 10 degree was chosen for this classification. This seems to be adequate, considering the propositions of critical angles and the terrain in the study area, which is, except of slopes also consisting of man-made plains, which should not be considered for the classification.

To derive more dense clusters, where the majority of ground cells have the same curvature, binary filtering algorithms were utilized. In this region growing operation, a 3-by-3 pixels kernel was tested with a FOCALSUM filter. That means, if in the kernel a certain number of pixels above a certain threshold are classified as concave, the kernel is evaluated to true, which classifies the current DEM cell as a concave pixel. This operation results in a densification of concave pixels in areas, where agglomeration of concave DEM cells exist.

In the next step, those densified areas were fused, so that the classified regions will have homogenous types of concave cells. Therefore, mathematical morphology operations are employed in a 3-by-3 pixel kernel. This will generate solid polygons from the clustered areas. In a successive application, an Opening operation is applied before a Closing operation. The opening removes pixels from the outer border of concave clusters. This means that solitary pixels and the outer parts of larger areas of classified pixels are removed and their core remains. In the Closing operation, existing clusters are enlarged in a way, that they get concave pixels accumulated at their outer border. This will resize them to the original width. So the areas, that showed the best initial clustering, are separated and classified as concave shaped terrain features. A slope evaluation during the processing prevents the closing operation from classifying pixels which are not situated on slopes. This was not possible in the original classification and hence, the potential landslide zones should not extend into flat terrain after following the processing steps.

At this stage of the analysis, a new raster-based dataset of concave curvature exists. For further use in GIS, it is important to have the data available in vector format. So, the last step is a raster-vector conversion, which gives a dataset of simple polygons that can now be the basis of a zonal evaluation of different landslide describing parameters.

1.6.2 Surface Roughness Index

The surface roughness index computation is based on the calculation of deviations between the elevation model surface and a trend-surface, fitted in the local range of a moving window. As the roughness of a surface in a distinct scale is characterised by the frequency and amplitude of its height values, the calculation of normalised heights can be used as a descriptor of the local surface roughness. To get those normalised heights, the deviations of the actual terrain height from a planar trend-surface were computed (Gray 1997; Penck 1894; Pollack 2002; Ritter 2010).

For each ground cell of the elevation model, a trend-surface of a first order polynomial is fitted in the range of a 5-by-5 pixel kernel. This medium sized kernel is expected to return the best interpretation of the surrounding terrain, whereas a 3-by-3 pixel kernel would include a relatively smaller neighbourhood and a 7-by-7 pixel kernel to generate a larger neighbourhood to give representative results. To compute a trend-surface, the current extent from the DEM has to be transformed into a point cloud dataset first. Then the deviation of the actual heights from the planar trend-surface over the entire kernel is computed and the maximum deviation is returned. This value will represent the deviation at the current kernel position in the roughness index dataset. In this way, a raster of maximum deviations is generated. It was decided to use the maximum value here, as it best outlines the maximum roughness at the current position. Minimum deviations would smooth the data too much and no meaningful conclusion could be derived.

1.6.3 Vegetation Index

The bare-earth of erosional areas in vegetated terrain shows a high difference in the vegetation index, in comparison to surrounding terrain (NASA 2011). The derivation of a vegetation index shall serve as a parameter for the differentiation between vegetation and bare soil. This classification can be derived in a continuous scale, which means that it can easily be integrated into a landslide probability indexation.

A vegetation index, frequently used in remote sensing, is the normalized difference vegetation index (ndvi), which is expected to give reasonable results for this work. It is computed from the returned intensities of the red and near infrared wavelengths by following Eq. 1.1:

$$NDVI = \frac{(NIR - R)}{(NIR + R)} \tag{1.1}$$

In this work, the influence of a vegetation index could not be tested due to the lack of satellite multi-spectral images for the test site. It is not facile to get the right data, due to time constraints in data collection. For reliable classification results, it is important that the data used for the processing of the vegetation index should be collected at the same time as the DEM raw data or little time afterwards. Data collected later influences the results of the processing, even more for a tropical landscape, because vegetation recovers bare spots comparatively fast here and will thus change the actual parameter index for a classified zone.

Nevertheless, the principle of operation of such an index would generally give reasonable results during data processing. However, further tests have to be conducted with respect to the question, how this index influences the results of a mass movement classification in vegetated terrain.

1.6.4 Breaklines

An important fact, that gives conclusions about the state of the terrain and eroded locations, is the occurrence of scarps and breaklines (Fara and Scheidegger 1963). From their nature, they show discontinuities of the terrain surface, whether they are of natural or man-made origin. In this way, they can correspond to the scarps of a landslide.

The suggestion here is, to use the breaklines for a more reliable determination of landslide features. However, it is not intended to use the breakline detection for the proposed landslide indexation. This is because the linear type of the breakline features is not comparable with the polygons of the zone layer. That means, an area-based value for an entire polygon cannot be derived directly. But, it is possible to use an auxiliary layer, containing this information, for the manual judgement of an expert about classified landslide locations. If the expert has the information about concave zones, classified with a certain landslide probability, and there exist further information, like the location of such a zone below a curved breakline on a slope, this may help to improve the classification of this area.

The breaklines can be automatically detected from the DEM without the need of any further data. For the detection, the same approach as for the curvature derivation of the areas is used. Deriving curvature in vertical direction, the profile curvature, it is possible to detect points of highest change in steepness. This has to be imagined as a vertical cross-section through the slope. Now, brims on slopes are usually convex features in vertical direction. Considering the cross-section profile, the curvature of its corresponding curve on the surface is greater, i.e. the higher the value of convexity at a certain point. If there exists a high convexity in the data, this may correspond to the section of the lateral cut with the brim. Now, if we make several cross-sections next to each other through the slope, the connection of all the points of highest convexity will approximate the breakline.

1.6.5 Landslide Probability Index

For the potential landslide zones, a value for each parameter has to be calculated. That means, with the zone layer and each parameter layer, a zonal operation is performed in orderto retrieve the parameter index for the current zone. This is accomplished with the Map Algebra function *ZONALMEAN* or *ZONALMAX*, respectively (ESRI 2010). With this calculation, all raster values from a parameter layer, that fall within a specific zone are used to determine the value for that zone. The mean or the maximum value is returned and assigned to the entire zone.

Problems with the correct assignment of a value may occur, if the zoning brings suboptimal results. That is, if the regions do not fit to the real extent of a landslide feature, being either too large or too small. In the first case this leads to a decreased probability index, as surrounding parts of the slide, owning different properties for the parameters, are included into the calculation. In the second case, not all valid cells, belonging to the landslide feature are taken into account. In general, it has to be said, that it is better to classify smaller regions instead of too large regions. This assures that a landslide does not get lost within a large region, which will be classified with a lower probability.

Having determined all the characteristics of the potential landslide zones, it is important to establish a ranking of all the classified zones in the dataset according to their probability of being a landslide. This is necessary to get a statistical evaluation of the zones for further use of the data.

The probability classification follows the rule:

$$\textit{high surface roughness AND low vegetation intensity} = \textit{high landslide probability}.$$

Basically, additional parameters, apart from the topographic surface roughness and the vegetation intensity can be applied to the calculation, depending on the terrain characteristics, as defined by an expert.

Figure 1.5 depicts the classification scheme: zones that own a high local topographic surface roughness and low vegetation intensity have a high probability of being a landslide. On the other hand, zones with a low value in surface roughness and a high vegetation index will get a low probability index for the landslide indexation.

The influence of those contrary effects of different parameters must be considered. For clear designation, they are called positive-effect and negative-effect parameters here. For a positive-effect parameter, landslide probability increases

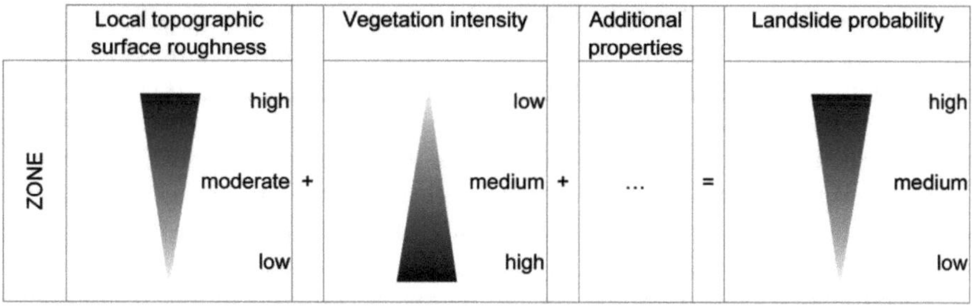

Fig. 1.5 LPI Classification scheme

with a higher index. For negative-effect parameters, the probability decreases with higher indices. Hence, surface roughness is a positive-effect parameter and the vegetation index a negative-effect parameter, respectively. This probability determination is quite a vague proposition, even more for all parameter values between their minimum and maximum. So, a formula is given to calculate the Landslide Probability Index (LPI):

$$LPI = \sum (PI_x w_x) \tag{1.2}$$

where PI_x is the index for a parameter and W_x the weight, given to this parameter. The total of weights in the calculation adds to one. The calculation of the parameter index is differentiated for positive-effect parameters PI_{pe} and negative-effect parameters PI_{ne}:

$$PI_{pe} = \frac{(p_x + |p_x \min|)}{(p_x \max - p_x \min)} \tag{1.3}$$

$$PI_{ne} = \frac{(p_x \max - p_x \min) - (p_x + |p_x \min|)}{(p_x \max - p_x \min)} \tag{1.4}$$

where p_x is the current parameter value for the zone. p_x maxand p_x minare the maximum and minimum bounds of the possible range of values for the particular parameter. The addition of the absolute value of the particular minimum value makes the calculation independent of the different range of values' that can occur for different parameters. The calculation of the LPI gives probability values in a common range from zero, for a very unlikely landslide area, up to one for the highest likelihood.

1.6.6 Implementation

In this work, ArcGIS and custom Python scripts were used for the implementation. For different tasks, the Python Imaging Library (PIL) was used. This allowed image processing capabilities, which would not have been possible with the ArcGIS standard tools alone. Thus it was possible to automate a moving window calculation and define exactly what to do with each image pixel (Gajski 2004).

The terrain zone classification has been tested for various types of DEM resolutions such as: 1.5 m, 2 m, 3 m, 4 m, 5 m and 8 m on the landslide locations. The clustering of concave shaped ground cells and the width of unclassified areas differs with respect to the DEM resolution. For testing purposes, the resolutions with the best results have been used for further processing.

To retrieve a valid threshold for the region growing operation, datasets with 2 m and 3 m resolution were tested with different thresholds varying between 2 and 6. That means, if 2 or more out of 9 (or 6 or more out of 9, respectively) pixels in the kernel have the value '1', then a value of '1' is also assigned to the pixel at the current kernel centre. A significant improvement in the zoning can be seen after the region growing. A better clustering is present in all datasets, distances between classified pixels are smaller and sparsely distributed pixels are removed, dependant on the threshold.

To automate the classification of the terrain zones, a rule for automatically calculating the region growing threshold has to be defined. It is dependant on the resolution of the DEM. The following rule was defined, based on the processing of the 2 m and 3 m resolution DEM:

$$\textit{if DEM}-\text{resolution} > 8 : \text{region growing threshold} = 8$$

$$\textit{otherwise} : \textit{region growing threshold} = \textit{floor}(\textit{DEM}-\text{resolution}) + 1,$$

where floor indicates a function which converts a floating point value to an integer by truncation. This works well for the 2 m and 3 m resolution but needs further enhancement for other resolutions.

For testing mathematical morphology operations, the dataset with the threshold of 3 is used. They are implemented by a Python script in Map Algebra syntax which again uses a moving window. For the *Opening* operation, the standard threshold was defined with 8. In a closer look, that means that only classified pixels, that are enclosed within a larger amount of concave pixels, also remain. If in the kernel three or more unclassified cells occur, the classified pixel is subtracted from the amount.

The threshold for the *Closing* operation was defined as two, which means, that a pixel is classified if it has at least two classified neighbouring pixels. In the present case, *Opening* and *Closing* operations have been applied twice each.

The computing time for the surface roughness index derivation was too time-consuming for the entire test site. Hence it was evaluated for two separate small areas only. The main part of the script running time is consumed by computing the trend-surface and the computation of the standard deviations. It also takes a lot of time, to convert a certain DEM extent to a point dataset. The run-time per pixel could be reduced after some optimizations in the code from around 19 to approximately 10 s. Around 38% of that is consumed by the trend-surface computation and 54% by the standard deviations computation.

1.7 Results and Discussion

In this chapter, a principal idea of compiling an automated procedure for landslide inventory mapping is suggested. The inventory map outlines zones in the dataset which may correspond to probable landsliding zones. Therefore, characteristics and properties of landslides in the examined countryside have to be taken into account. This work suggests using local topographic surface roughness and vegetation density as indicators.

Landslide inventory map examples have been produced on two small parts of the test site, as computation time for the entire data was proven to be too long. One of them is depicted in Fig. 1.6. Furthermore, it has to be stated that only one factor, namely local topographic surface roughness, could be used for testing the landslide probability categorization. That means that this work may not yet provide a reliable complete landslide inventory dataset of the entire study area. It is intended to provide ideas and suggestions for a technical approach to the automatic extraction of those features from a LiDAR derived DEM.

The used DEM had an original resolution of approximately 1 m. Even though it would have been good to make use of this resolution, problems arose from inconsistencies in the data. Presumably due to tiling effects, originating from the production, regular outliers in height values occurred. This resulted in a grid-shaped error over the entire dataset, where the pixels on the grid-lines had anomalous heights, in relation to their neighbouring pixels. This was not acceptable, as the necessary computation of curvature in the DEM did generate high deviations from an expected value for those grid locations. Therefore, a resampling of the DEM was necessary, which resulted in a lowered resolution. To keep lower resolution as small as possible, resampling has been done in small irregular intervals. Sample resolutions of 1.12 m, 1.18 m or 1.25 m were tested, but the errors could not be removed entirely. The use of two irregular resampling steps down to 1.35 m and 1.5 m finally and produced the best results.

In the process, concave shaped terrain forms could be extracted from the DEM. Those areas are assumed to correspond to erosional landslide features. Best results were achieved in the 2 m and 3 m resolution. These terrain zones build the basis of the landslide classification with different parameters. The employment of mathematical morphology in the process is the restricting criterion for the size of movements, which can be detected with this approach. The opening operation deletes outer pixels of clusters. Here, this operation was applied twice. This implies, that clusters of concave pixels which are smaller than two times the DEM resolution in width, are removed from the dataset. Accordingly, it is possible to detect landslide features which are larger than two times the DEM resolution in their smallest dimension.

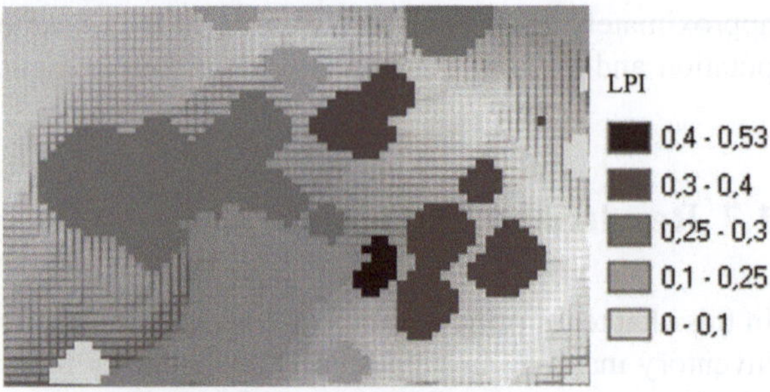

Fig. 1.6 LPI inventory image

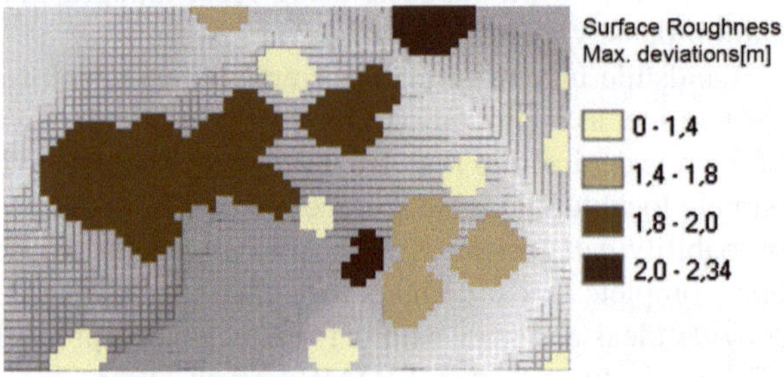

Fig. 1.7 Surface roughness index, zonalmax

In the Figs. 1.7 and 1.8, the classification of the roughness parameter is shown. In Fig. 1.7, it can be seen that the polygon containing this feature is classified with a dark brown, using the *ZONALMAX* derivation. This means that it is considered to have a high surface roughness. In the other Fig. 1.8, where a *ZONALMEAN* calculation was deployed, the same feature shows a lower surface roughness. In fact, even if the ZONALMAX shows the better classification here, it is expected to give more unreliable results. This is due to the fact, that one outlier in the zone with an abnormally large height will classify the entire region with a high surface roughness, influencing not only the classification for this factor heavily but also the entire probability value. The *ZONALMEAN* calculation instead gives averaged results, taking the entire roughness distribution in the zone into account. For that reason, it is generally preferred here. But issues do also appear with this type of calculation. As can be seen from the resulting classification, it is also not providing reliable results. The following problems are identified: firstly, a part of the landslide which shows the highest roughness values is near the main scarp. Unfortunately, this part is not classified correctly by the zoning algorithm. That means, important parts which would enhance the roughness index upwards are not considered by the calculation. Secondly, the polygon containing the landslide feature is larger than the landslide in the eastward direction. This results in an averaged roughness calculation for a zone, which likely consists of the landslide feature and

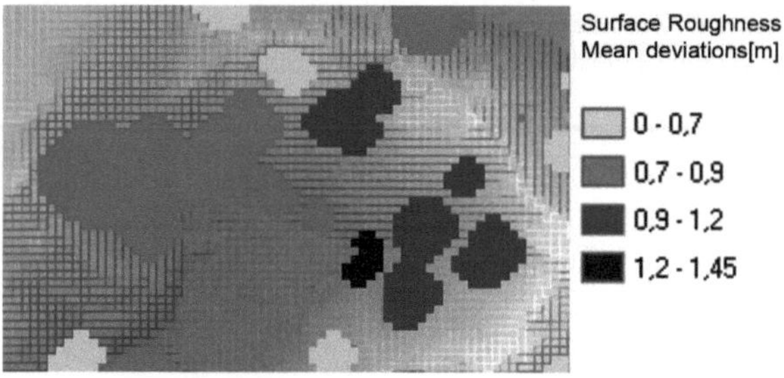

Fig. 1.8 Surface roughness index, zonalmean

uninvolved terrain. It is obvious, that the part which is not a landslide lowers the averaged roughness value.

This was tested on the dataset. The polygon was divided into two features, the polygon of the expected landslide extent and the leftover. On this dataset, the same probability classification was conducted. The result (Fig. 1.9) shows, that the value of the landslide part rises, whereas the second polygon is decreased in the roughness determination, which results in a lower landslide probability classification. Thus it is better to classify smaller areas to prevent this incorrect classification.

A problem that occurs with an automated classification is that landslides may not be detected correctly. But also false positives may appear, i.e., features, that are not landslides, could erroneously be detected. This should be avoided by employing additional properties of landslide features. To get more reliable results which can be treated as an inventory map, a vegetation dataset would be necessary to also produce the vegetation index layer.

1.8 Conclusions and Perspective

In this chapter, a methodology was developed to automate the process of classifying rotational landslides from a high-resolution LiDAR based DEM. Additionally, characteristics of landslides were investigated and described, in general as well as in more detail, for rotational landslides, to understand their behavior. However, it is clear that landslide is a complex system, which is difficult to restrict to a few parameters. Nevertheless it is necessary to establish a descriptive context for landslide features to differentiate them from unfailed surrounding terrain. This is intended by the term of a landslide signature from geomorphological and spectral properties.

Main factors of the classification, like using surface roughness for the evaluation of slides or the determination of homogeneous terrain form elements for a geomorphological description of terrain characteristics, are not new but have proven their functionality in previous research. These basically useful techniques have been aggregated and adapted to serve the classification needs. Standard

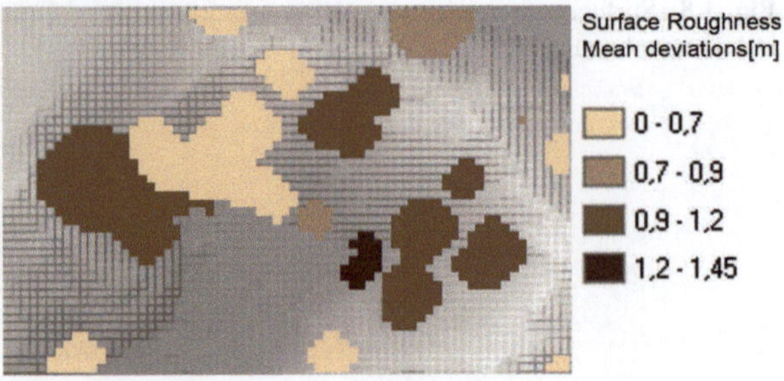

Fig. 1.9 Surface roughness on divided landslide feature

geoprocessing techniques and custom Python scripts have been applied, to make the automation of classification tasks possible.

Finally, a statistical description is applied to rank the extracted terrain zones according to their compliance with the aforementioned landslide signature. Therefore the Landslide Probability Index (LPI) is introduced. However, not included in the classification, is the accumulation area of the landslide. To also classify this part of a slide it would be necessary to test different algorithms that for instance detect convex shaped form elements.

Unfortunately, the results of the classification cannot yet be proven, as no reliable landslide inventory of the study area exists for the given scale that was produced under similar conditions. To support the ideas and results, detailed fieldwork would be necessary. Furthermore, thresholds used in the classification have to be investigated more in detail and need further adjustment for correct behavior in different scales.

It is expected that this approach can serve as a base of a semi-automatic derivation of a landslide inventory, which allows the geoscientist to adapt parameters and thresholds as is reasonable in the current study area. It would also make sense to establish a processing application which would decrease the programming labour and would be flexible enough to support the researchers during the classification.

Acknowledgments This research is fully supported by the Alexander von Humboldt Foundation (AvH), Germany by providing AvH fellowship and Georg Forester Return Fellowship to carry out research at Dresden University of Technology, Germany and University Putra Malaysia respectively.

References

Abbott PL (2004) Natural disasters. McGraw-Hill Companies Inc., New York, p 460, BBC (2010) Weather Cameron Highlands. http://www.bbc.co.uk/weather/world/city_guides/results.shtml ?tt=TT002580. Accessed 8 Nov 2010

Akgun A, Dag S, Bulut F (2008) Landslide susceptibility mapping for a landslide prone area (Findikli, NE of Turkey) by likelihood-frequency ratio and weighted linear combination models. Environ Geol 54:1127–1143

Akgun A, Bulut F (2007) GIS-based landslide susceptibility for Arsin-Yomra region (Trabzon, North Turkey). Environ Geol 51:1377–1387

Akgun A, Turk N (2010) Landslide susceptibility mapping for Ayvalik (Western Turkey) and its vicinity by multicriteria decision analysis. Environ Earth Sci 61(3):13595–13611

Brardinoni F, Slaymaker O, Hassan MA (2003) Landslide inventory in a rugged forested watershed: a comparison between air-photo and field survey data. Geomorphology 54: 179–196

Bishop MP, Shroder JF Jr (2004) Geographic information science and mountain geomorphology. Springer, Berlin

Can T, Nefeslioglu HA, Gokceoglu C, Sonmez H, Duman TY (2005) Susceptibility assessment of shallow earthflows triggered by heavy rainfall at three subcatchments by logistic regression analyses. Geomorphology 72:250–271

Carter W, Shrestha R, Tuell G, Bloomquist D, Sartori M (2001) Airborne laser swath mapping shines new light on earth's topography. EOS Trans, Am Geophys Union 82(46):549–555

CCRS (2005) Canada centre for remote sensing glossary. http://www.ccrs.nrcan.gc.ca/glossary/index_e.php. Accessed 29 Nov 2010

Clark ML, Clark DB, Roberts DA (2004) Small-footprint LiDAR estimation of sub-canopy elevation and tree height in a tropical rain forest landscape. Remote Sens Environ 91(1):68–89

Cruden DM, Varnes DJ (1996) Landslide types and processes. In: Turner AK, Shuster RL (eds) Landslides: investigation and mitigation. Transportation Research Board Special Report 247:36–75

Danneels G, Havenith H-B, Strom A, Pirard E (2008) Landslide detection methods, inventory analysis and susceptibility mapping applied to the Tien Shan, Kyrgyz Republic. In: Sassa K, Canuti P (eds) Landslides. Disaster risk reduction. Springer, Berlin

De Smith MJ, Goodchild MF, Longley PA (2007) Geospatial analysis: a comprehensive guide to principles, techniques and software tools, 3rd edn. Troubador, London

Dikau R (1988) Entwurf einer geomorphographisch: analytischen Systematik von Reliefeinheiten. In: Heidelberger geographische bausteine, vol 5. Heidelberg

Dikau R, Schmidt J (1999) Georeliefklassifikation. In: Schneider-Sliwa R, Schaub D, Gerold G (Hrsg.) Angewandte Landschaftsökologie – Grundlagen und Methoden, Heidelberg pp 217–244

Duman TY, Can T, Gokceoglu C, Nefeslioglu HA, Sonmez H (2006) Application of logistic regression for landslide susceptibility zoning of Cekmece Area, Istanbul, Turkey. Environ Geol 51:241–256

Ercanoglu M, Gokceoglu C (2002) Assessment of landslide susceptibility for a landslide prone area (north of Yenice, NW Turkey) by fuzzy approach. Environ Geol 41:720–730

Ercanoglu M, Gokceoglu C (2004) Use of fuzzy relations to produce landslide 18 Susceptibility map of a landslide prone area (West Black Sea Region, Turkey). Eng Geol 75:229–250

Ercanoglu M (2005) Landslide susceptibility assessment of SE Bartin (West Black Sea region, Turkey) by artificial neural networks. Nat Hazards Earth Syst Sci 5:979–992

Ercanoglu M, Kasmer O, Temiz N (2008) Adaptation and comparison of expert opinion to analytical hierarchy process for landslide susceptibility mapping. Bull Eng Geol Environ 67: 565–578

ESRI (2010) Performance of ArcObjects. http://resources.esri.com/help/9.3/arcgisengine/dotnet/26b7a578-7854-4ff9-b96c-948eded73626.htm. Accessed 01 Dec 2010

Fara HD, Scheidegger AE (1963) An eigenvalue method for the statistical evaluation of fault plane solutions of earthquakes. Seismol Soc Am Bull 53:811–816

Fortuin R (2006) Soil erosion in Cameron Highlands. Saxion University Deventer, Environmental Technology, The Netherlands. http://www.reach.org.my/index.php?option=com_content&view=article&id=80:soil-erosion-in-cameron-highlands&catid=68:land-slides-erosion-and-iltation&Itemid=13. Accessed 8 Nov 2010

Gajski D (2004) Rasterbasierte Geländeoberflächeanalysen. Dissertation, Wien

Glenn NF, Streutker DR, Chadwick J, Glenn DJ, Thackray GD, Dorsch SJ (2006) Analysis of Lidar-derived topographic information for characterizing and differentiating landslide morphology and activity. Geomorphology 73:131–148

Gokceoglu C, Sezer E (2009) A statistical assessment on international landslide literature (1945–2008). Landslides 6:345–351

Guzzetti F (2004) Landslide mapping, hazard assessment and risk avaluation: limits and potential. In: Proceedings international symposium on landslide and debris flow hazard assessment, National Center for Research on Earthquake Engineering, Taipei, 7–8 Oct 2004, pp C1–C17

Guzzetti F, Reichenbach P, Ardizzone F, Cardinali M, Galli M (2006) Estimating the quality of landslide susceptibility models. Geomorphology 81:166–184

Gray A (1997) Osculating circles to plane curves. In: Modern differential geometry of curves and surfaces with mathematica, 2nd edn. CRC Press, Boca Raton, pp 111–115

Hylland MD, Lowe M (1997) Regional landslide-hazard evaluation using landslide slopes, western Wasatch County, Utah. Environ Eng Geosci 3(1):31–43

IAEG Commission on Landslides (1990) Suggested nomenclature for landslides. Bull Int Assoc Eng Geol 41:13–16

Iovine G, Di Gregorio S, Lupiano V (2003) Debris-flow susceptibility assessment through cellular automata modelling: an example from 15–16 December 1999 disaster at Cervinara and San Martino Valle Caudina (Campania, southern Italy). Nat Hazards Earth Syst Sci 3:457–468

Jakob M, Hungr O (2005) Debris-flow hazards and related phenomena. Springer, Berlin

Kugler H (1974) Das Georelief und seine kartographische Modellierung. Dissertation B, Martin-Luther-Universität Halle, Wittenburg

Lazacode (2010) Cameron Highlands History. Lazacode Malaysia Education and Creative Information portal. http://lazacode.com/place-empire/cameron-highlands-history. Accessed 8 Jan 2011

Lee EM, Jones DKC (2004) Landslide risk assessment. Thomas Telford, London

Lee S (2005) Application of logistic regression model and its validation for landslide susceptibility mapping using GIS and remote sensing data. Int J Remote Sens 26:1477–1491

McKean J, Roering J (2003) Objective landslide detection and surface morphology mapping using high-resolution airborne laser altimetry. Geomorphology 57:331–351

NASA (2011) Measuring vegetation (NDVI and EVI). By Weier J, Herring D. NASA Earth Observatory. http://earthobservatory.nasa.gov/Features/MeasuringVegetation. Accessed 8 Jan 2011

Nefeslioglu HA, Gokceoglu C (2011) Probabilistic risk assessment in medium scale for rainfall induced earthflows: Catakli catchment area (Cayeli, Rize, Turkey). Math Probl Eng, Article ID 280431

Nefeslioglu HA, Duman TY, Durmaz S (2008) Landslide susceptibility mapping for a part of tectonic Kelkit Valley (Eastern Black Sea region of Turkey). Geomorphology 94(3–4):401–418

Nefeslioglu HA, Sezer E Gokceoglu C, Bozkir AS, Duman TY (2010) Assessment of landslide susceptibility by Decision Trees in the metropolitan area of Istanbul, Turkey. Math Probl Eng. pp 1–15. doi:10.1155/2010/901095,

Oh HJ, Pradhan B (2010). Application of neuro-fuzzy model to landslide susceptibility mapping for shallow landslides in a tropical hilly area. Comput Geosci (Article on-line first available). doi:10.1016/j.cageo.2010.10.012

Penck A (1894) Morphologie der Erdoberfläche. J. Engelhorn, Stuttgart

Pollack M (2002) Methodischer Beitrag zur GIS-basierten mittelmaßstäbigen geomorphologischen Klassifizierung und Kartierung des Zentral-Altai auf Grundlage von DGM und Geländeerhebungen. Diploma Thesis, TU Dresden, Institute for Cartography

Pradhan B (2010a) Remote sensing and GIS-based landslide hazard analysis and cross-validation using multivariate logistic regression model on three test areas in Malaysia. Adv Space Res 45(10): 1244–1256

Pradhan B (2010b) Manifestation of an advanced fuzzy logic model coupled with Geoinformation techniques for landslide susceptibility analysis. Environ Ecol Stat (article-on line first available). doi:10.1007/s10651-010-0147-7

Pradhan B (2010c) Application of an advanced fuzzy logic model for landslide susceptibility analysis. International Journal of Computational Intelligence Systems 3(3):370–381. doi:10.2991/ijcis.2010.3.3.12

Pradhan B (2010d) Landslide susceptibility mapping of a catchment area using frequency ratio, fuzzy logic and multivariate logistic regression approaches. J Indian Soc Remote Sens 38(2):301–320. doi:10.1007/s12524-010-0020-z

Pradhan B (2011) Use of GIS based fuzzy relations and its cross application to produce landslide susceptibility maps in three test areas in Malaysia. Environ Earth Sci 63(2):329–349. doi:10.1007/s12665-010-0705-1

Pradhan B, Lee S (2010a) Regional landslide susceptibility analysis using back-propagation neural network at Cameron Highlands, Malaysia. Landslides 7(1):13–30

Pradhan B, Lee S (2010b) Delineation of landslide hazard areas on Penang Island, Malaysia, by using frequency ratio, logistic regression, and artificial neural network models. Environ Earth Sci 60(5):1037–1054

Pradhan B, Lee S (2010c) Landslide susceptibility assessment and factor effect analysis: backpropagation artificial neural networks and their comparison with frequency ratio and bivariate logistic regression modeling. Environ Model Softw 25(6):747–759

Pradhan B, Lee S, Buchroithner MF (2010a) A GIS-based back-propagation neural network model and its cross-application and validation for landslide susceptibility analyses. Comput Environ Urban Syst 34:216–235

Pradhan B, Lee S, Buchroithner MF (2010b) Remote sensing and GIS-based landslide susceptibility analysis and its cross-validation in three test areas using a frequency ratio model. Photogrammetrie, Fernekundung, Geoinformation 1(1):17–32

Pradhan B, Sezer A, Gokceoglu C, Buchroithner MF (2010c) Landslide susceptibility mapping by neuro-fuzzy approach in a landslide prone area (Cameron Highland, Malaysia). IEEE Trans Geosci Remote Sens 48(12):4164–4177. doi:10.1109/TGRS.2010.2050328

Razak KA, Straatsma MW, Van Westen CJ, de Jong SM (2010). Airborne laser scanning of forested landslides: terrain model quality and visualization. Geomorphology (article on-line first available). doi:10.1016/j.geomorph.2010.11.003

Razak KA, Straatsma MW, Van Westen CJ, Malet JP (2009) Utilization of airborne LIDAR data for landslide mapping in forested terrain: status and challenges. Presented at the 10th South East Asian Survey Congress (SEASC) in conjunction with 16th UNGEGN Workshop—16th ABLOS Business Meeting and 4th Indonesian Geoinformation Technology Exhibition, 04–07 Aug 2009, Bali, Indonesia, p 10

Ritter ME (2010) The physical environment: an introduction to physical geography. http://www.uwsp.edu/geo/faculty/ritter/geog101/textbook/title_page.html. Accessed 9 Jan 2011

Rivard LA (2009) Geohazard-associated Geounits. Springer, Berlin

Sassa K, Fukuoka H, Wang F, Wang G (2005) Landslides risk analysis and sustainable disaster management. Springer, Berlin

Sezer EA, Pradhan B, Gokceoglu C (2011) Manifestation of an adaptive neuro-fuzzy model on landslide susceptibility mapping: Klang valley, Malaysia. Expert Syst Appl 38(7):8208–8219

Sithole G, Vosselman G (2004) Experimental comparison of filter algorithms for bare-earth extraction from airborne laser scanning points clouds. ISPRS J Photogramm Remote Sens 59(1–2):85–101

Schulz WH (2007) Landslide susceptibility revealed by LIDAR imagery and historical records, Seattle, Washington. Eng Geol 89:67–87

Schuster RL, Wieczorek GF (2002) Landslide triggers and types. In: Rybář J, Stemberk J, Wagner P (eds) Landslides. Swets and Zeitlinger B.V., Lisse

Takahashi T (2009) Mechanics-based approach toward the mitigation of debris flow disasters. In: Sassa K, Canuti P (eds) Landslides. Disaster risk reduction. Springer, Berlin

Wohl E, Oguchi T (2004) Geographic information systems and mountain hazards. In: Bishop MP, Shroder JF Jr (eds) Geographic information science and mountain geomorphology. Springer, Berlin, pp 309–341

Yilmaz I (2009) Landslide susceptibility mapping using frequency ratio, logistic regression, artificial neural networks and their comparison: a case study from Kat landslides (Tokat-Turkey). Comput Geosci 35(6):1125–1138

Yilmaz I (2010a) The effect of the sampling strategies on the landslide susceptibility mapping by conditional probability (CP) and artificial neural network (ANN). Environ Earth Sci 60:505–519

Yilmaz I (2010b) Comparison of landslide susceptibility mapping methodologies for Koyulhisar, Turkey: conditional probability, logistic regression, artificial neural networks, and support vector machine. Environ Earth Sci 61:821–836

Youssef AM, Pradhan B, Gaber AFD, Buchroithner MF (2009) Geomorphological hazard analysis along the Egyptian red sea coast between Safaga and Quseir. Nat Hazards Earth Syst Sci 9:751–766

Chapter 2
Landslide Susceptibility Mapping Using a Spatial Multi Criteria Evaluation Model at Haraz Watershed, Iran

H. R. Pourghasemi, Biswajeet Pradhan, Candan Gokceoglu and K. Deylami Moezzi

Abstract The purpose of this study is to prepare landslide susceptibility map using a spatial multi criteria evaluation approach (SMCE) in a landslide-prone area (Haraz) in Iran. In the first stage, landslide locations were identified in the study area from interpretation of aerial photographs, and field surveys. In the second stage, twelve data layers were used as landslide conditioning factors for susceptibility mapping. These factors are slope, aspect, altitude, lithology, land use, distance from rivers, distance from roads, distance from faults, topographic wetness index, stream power index, stream transport index, and plan curvature. Next, landslide-susceptible areas were analyzed using the SMCE approach and mapped using landslide conditioning factors. For verification, the results of the analyses was compared with the field-verified landslide locations. Additionally, the receiver operating characteristics (ROC) curves for all landslide susceptibility models were drawn and the area under curve values was calculated. Landslide locations were used to validate results of the landslide susceptibility map generated using the SMCE approach and the verification results showed a 76.84% accuracy. According to the results of the AUC evaluation, the produced map has exhibited good performance.

Keywords Landslide · Susceptibility · GIS · Remote sensing · Spatial multi criteria evaluation · AHP · Iran

H. R. Pourghasemi · K. Deylami Moezzi
Department of Watershed Management Engineering,
College of Natural Resources and Marine Sciences, Tarbiat Modares University,
International Campus, Noor, Iran

B. Pradhan (✉)
Institute of Advanced Technology, Spatial and Numerical Modeling Laboratory, University Putra Malaysia, 43400 Serdang, Selangor, Malaysia
e-mail: biswajeet24@gmail.com; biswajeet@lycos.com

C. Gokceoglu
Applied Geology Division, Department of Geological
Engineering, Engineering Faculty, Hacettepe University, Ankara, Turkey

B. Pradhan and M. Buchroithner (eds.), *Terrigenous Mass Movements*,
DOI: 10.1007/978-3-642-25495-6_2,

2.1 Introduction

Undesired effects on human life and economic activity resulting from landslides are observed throughout the world. During the 1990's, nearly nine percent of worldwide natural disasters constitutes of landslides (Gokceoglu et al. 2005). According to Schuster and Fleming (1986), in many countries, the economic losses and causalities due to landslides are greater than commonly recognized and generate a yearly loss of property larger than that from any other natural disaster, including earthquakes, floods and windstorms.

Over the last decade, it is possible to find many studies on landslide susceptibility assessment. The basic concept was first introduced by Radbruch (1970), Dobrovolny (1971), and Brabb and Pampeyan (1972) as the spatial distribution of factors related to the instability processes in order to determine zones of landslide-prone areas without any temporal implication. Guzzetti et al. (1999) summarized most of the landslide susceptibility mapping studies. More recently, probabilistic models have been proposed (Dai and Lee 2001; Gokceoglu et al. 2005; Akgun and Bulut 2007; Akgun et al. 2008; Lee and Pradhan 2007; Oh and Lee 2009). The logistic regression model has also been employed for landslide susceptibility mapping (Nefeslioglu et al. 2008; Pradhan 2010a; Chauhan et al. 2010; Bai et al. 2010; Akgun 2011). Shou and Wang (2003) and Zhou et al. (2003) have used the geotechnical and factor of safety parameter models to investigate the slope failure of the studied areas. Data mining using fuzzy logic, artificial neural network and decision tree models have also been applied in Geographical Information Systems (GIS) as a new landslide susceptibility assessment approach (e.g. Ercanoglu and Gokceoglu 2002, 2004; Ermini et al. 2005; Lee et al. 2006; Melchiorre et al. 2006; Castellanos and VanWesten 2007; Kanungo et al. 2006; Wang 2008; Tangestani 2009; Wan 2009; Saito et al. 2009; Pradhan 2010a, b, c, 2011; Pradhan et al. 2010; Pradhan and Buchroithner 2010, Pradhan and Lee 2009, 2010a, b, c; Akgun and Turk 2010; Nefeslioglu et al. 2010; Yeon et al. 2010; Sezer et al. 2011; Akgun 2011), multicriteria decision analysis (MCDA) approach (Ayalew et al. 2005; Komac 2006; Akgun and Balut 2007; Yalcin 2008; Akgun and Turk 2010; Akgun 2011). In this study, a GIS based spatial multi criteria model was used for landslide susceptibility mapping at Haraz watershed.

2.2 Study Area

The study area is located in the northern part of Iran, which is one of the most landslide prone areas in Iran (Pourghasemi 2008). The watershed area centered between the longitudes of 52° 06′ 02′ E to 52° 18′ 13″ E, and latitudes of 35° 49′ 05″ N to 35° 57′ 39″ N, is mountainous and is located in the Alborz Folded geological zone (Fig. 2.1). It covers two adjacent 1:50,000 topographic sheets of

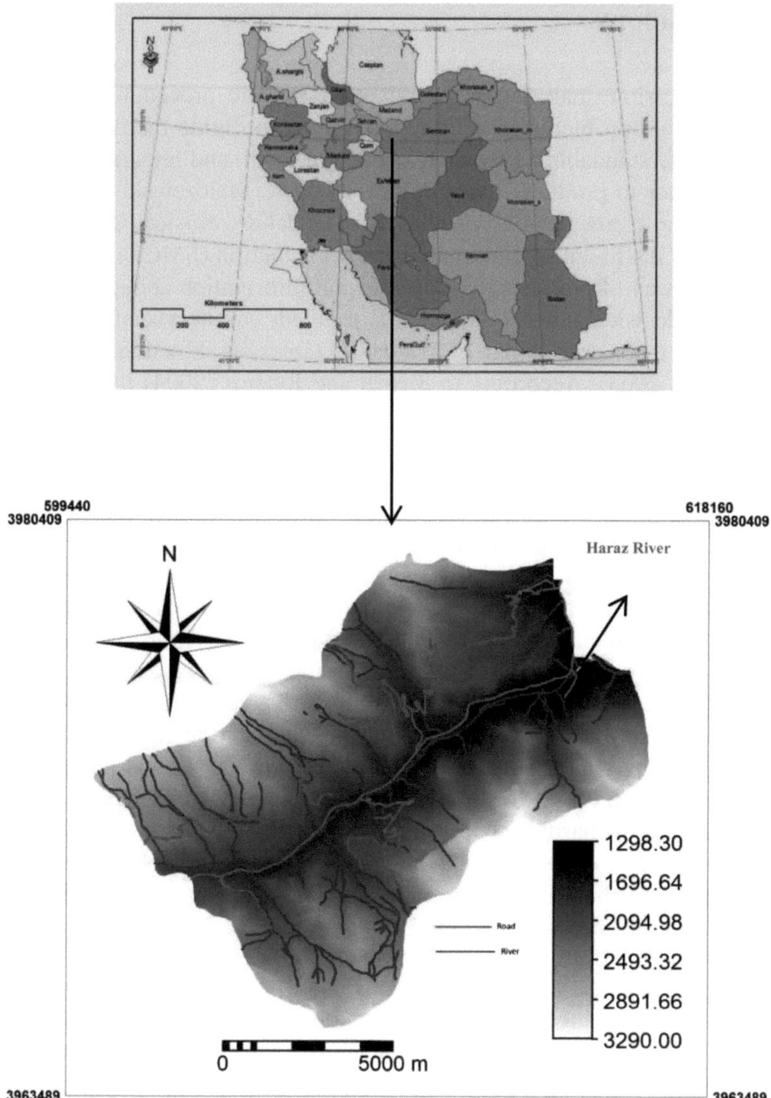

Fig. 2.1 Location of the study area showing Mazandaran province in Iran

the Army Geographic Institute of Iran and has an extent of about 114.5 km^2. In the study area, the main river is the Haraz. Based on the data from the Iranian Meteorological Department, the temperature in the study area varies between −25°C in winter and 36.5°C in summer. The mean annual rainfall is around 500 mm, while the maximum precipitation falls between November and January in general. Altitude values in the study area vary between 1200 to 3290 m.asl.

2.3 Methodology

In general, decision analysis uses a set of systematic procedures for analyzing complex decision problems. The basic strategy is to divide the decision problem into small, understandable parts, analyze each of them, and integrate these parts in a logical manner to produce a meaningful solution (Malczewski 1999). To solve spatial-based problems such as geo-hazards (landslide, erosion, earthquakes) and site selection, GIS-based spatial multi criteria evaluation (SMCE) have been used.

SMCE is a way of producing policy-relevant information about spatial decision problems for decision makers. An SMCE problem can be visualized as an evaluation table of maps or as a map of evaluation tables, as shown in Fig. 2.2 (Sharifi and Herwijnen 2003). According to Sharifi and Retsios (2004) if the objective of the evaluation is a ranking of the alternatives, then the evaluation table of maps has to be transformed into one final ranking of alternatives. Actually, the function has to aggregate not only the effects but also the spatial component (Sharifi and Herwijnen 2003). At times, defining such a function can be highly complicated. Therefore, it is required to simplify the function by dividing it into at least two operations. Those operations are: (i) aggregation of the spatial component, and (ii) aggregation of the criteria. These two operations can be carried out in different orders as visualized in Fig. 2.3 as Path 1 and Path 2. These two path features resembles the order of aggregation. If we consider a step wise analysis during the first path, then the first step is the aggregation across spatial units (spatial analysis is the principal tool); the second step is the aggregation across criteria (multi criteria analysis playing the main role). Similarly, in the second path, these steps are taken in reverse order. In the first case, the effect of one alternative for one criterion is a map (Sharifi and Retsios 2004; Sharifi and Herwijnen 2003). This case can be used when evaluating the spatial evaluation problem using the so-called 'Path 1'. In the second case, every location has its own zero-dimensional problem and can best be used when evaluating the spatial problem using the so-called 'Path 2' (Fig. 2.3). For implementing the whole semi-quantitative model the SMCE module of ILWIS-GIS (integrated land and water information system) was used (Castellanos, 2008). The SMCE application assists and guides users in doing multi-criteria evaluation in a spatial manner (ITC 2001). The model is built by making criteria tree, where the conditioning parameter maps are grouped, standardized and weighted. The landslide casual parameters are weighted by means of direct, pair-wise, and rank ordering comparison and the output is a composite index map (Castellanos and Van Westen 2007). Figure 2.4 presents an overview of the various components of the landslide susceptibility method.

In this study a pair-wise comparison based weighting was used. This method assumes that the users comparably evaluate the difference of magnitude among all unique pairs of factors qualitatively. Pair-wise comparison method was established by Saaty (1980) in the context of the analytical hierarchy process (AHP). In this process, the weights are defined by standardizing the eigenvector correlated with the highest eigenvalue of the ratio matrix. The AHP consists of three main steps;

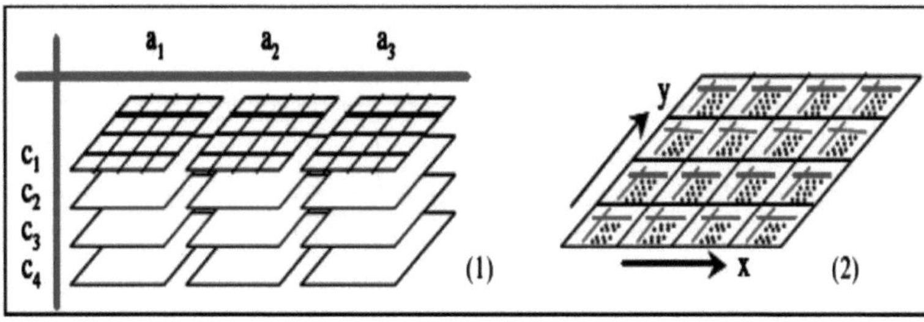

Fig. 2.2 Two interpretations of a two-dimensional decision problem (*1*: table of maps, *2*: map of tables); *Source* Sharifi and Retsios 2004)

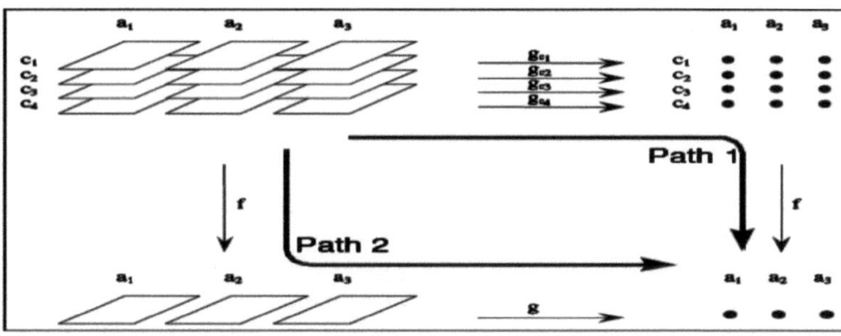

Fig. 2.3 Two paths of spatial multi criteria evaluation (adapted from Herwijnen 1999). The result of both path 1 and 2 is a ranking of alternatives a1, a2, and a3, with respect to their performance in terms of the four spatial effects (criteria c1, c2, c3, and c4) for which they are evaluated (functions f) and the spatial distribution of these effects, which is aggregated in functions g. (*Source* Sharifi and Retsios 2004)

1) generating the pair wise comparison matrix, 2) computing the weights of the criterion, and 3) estimating the consistency ratio (Malczewski 1999). In the development of comparison matrix, the method employs an underlying scale with values from 0 to 1 to rate the relative preferences for two criteria which can be seen in the Table 2.1.

This study used the combination between bivariate statistical analysis and pair-wise comparison. Firstly, to know the scored value for each class parameter, we calculated the density of landslides by using some steps in the bivariate statistical analysis (frequency ratio model). The second process is grouping the conditioning factors into four induced factors such as geomorphological, geological, hydro-logical and anthropogenic.

Next, the levels of weight values were used to standardize the input value by means of pair wise comparison resulting values from 0 to 1. After this process the steps in spatial multi criteria evaluation were followed again by means of pair-wise comparison method. The difference of this improved method was located on

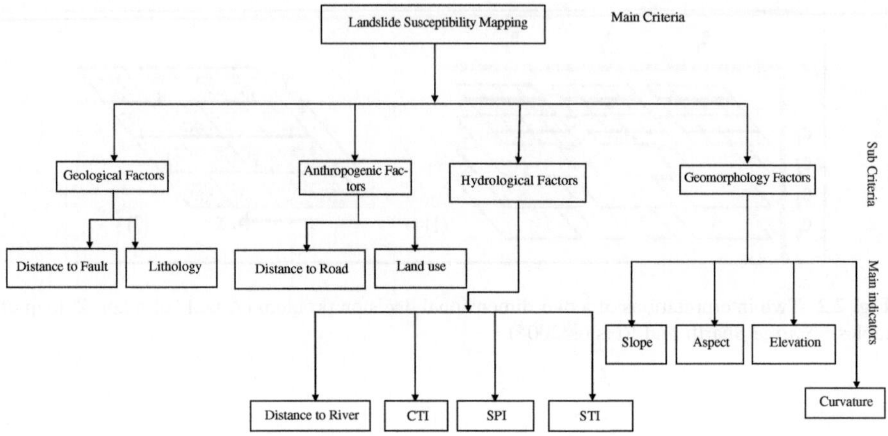

Fig. 2.4 The methodological flow chart showing the step wise processes for landslide susceptibility mapping in Haraz watershed

Table 2.1 Scale of relative importance suggested by Saaty (1997)

Inensity of importance	Definition	Explanation
1	Equal importance	Two activities contribute equally to objective
3	Weak importance of one over another	Experience and judgment slightly favor one activity over another
5	Essential or strong importance	Experience and judgment strongly favor one activity over another
7	Demonstrated importance	An activity is strongly favored and its dominance demonstrated in practice
9	Absolute importance	The evidence favoring one activity over another is the highest possible order of affirmation
2, 4, 6, 8	Intermediate values between the two adjacent judgments	When compromise is needed

1/9 1/8 1/7 1/6 1/5 1/4 1/3 1/2 1 2 3 4 5 6 7 8 9

Less Important ← → **More Important**

giving the weighting value of each parameter. The weighting value of this method was given by calculation process of analytical hierarchy process (AHP). The values were extracted based on the level of influences. Expert opinion which depends on observed physical characteristic of landslide sites determined the levels of the influencing factors.

Fig. 2.5 Landslide distribution map of the study area

2.4 Construction of Spatial Database

2.4.1 Landslide Inventory Map

The existing landslide inventory map is very essential for studying the relationship between the landslide distribution and the conditioning factors. To produce a detailed and reliable landslide inventory map, extensive field surveys and observations were performed in the study area. A total of 78 landslides were identified and mapped in the study area by evaluating aerial photos at 1:25,000 scale and by multiple field studies (Fig. 2.5). The modes of failure for the landslides identified in the study area were determined according to the landslide classification system proposed by Varnes (1978).

2.4.2 Slope

The most important parameter in the slope stability analysis is the slope angle (Lee and Min 2001). Because the slope angle is directly related to the landslides and it is frequently used in preparing landslide susceptibility maps (Clerici et al. 2002; Saha et al. 2005; Cevik and Topal 2003; Ercanoglu and Gokceoglu 2004; Lee et al. 2004a; Lee 2005; Yalcin 2005). For this reason, the slope map of the

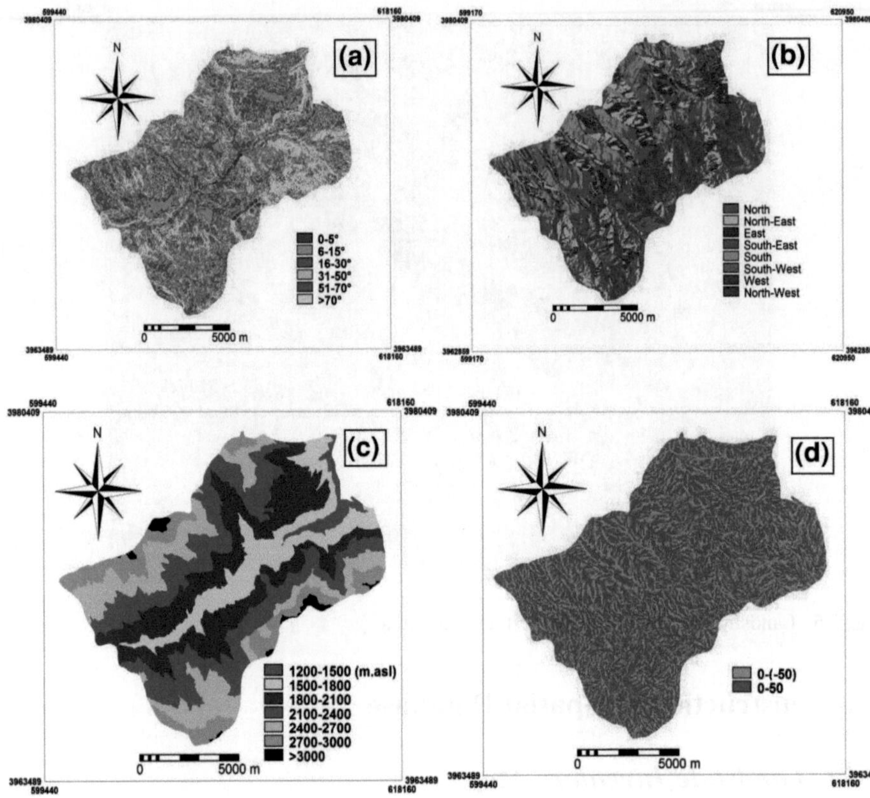

Fig. 2.6 Topographical parameter maps of the study area; **a** slope (degree), **b** aspect, **c** altitude (m.asl), **d** plan curvature

study area is prepared from the DEM, and divided into six slope categories (Fig. 2.6a). The spatial relationship between slope and landslide are presented in Table 2.2.

2.4.3 Aspect

Aspect is also considered as a landslide conditioning factor, and this parameter has been considered in several other studies (Van Westen and Bonilla 1990; Gokceoglu and Aksoy 1996; Saha et al. 2005; Ercanoglu and Gokceoglu 2004; Lee et al. 2004a, b; Yalcin 2005). Some of the meteorological events such as the direction of the rainfall, amount of sunshine, the morphologic structure of the area affect the slope stability (Mohammadi 2008). The hillsides receiving dense rainfall reach saturation faster, however this is also related to infiltration capacity of the slope controlled by various parameters such as topographic slope, type of soil,

Table 2.2 Spatial relationship between landslides and landslide conditioning factors

Factor	Class	No. of pixels in domain	Percentage of domain	No. of landslide	Percentage of landslide	Frequency ratio	a	b	c	d
Slope (Deg)	0–5	13851	1.21	1	1.82	1.50	0.9	0.325	0.181	0.053
	6–15	64268	5.62	2	3.64	0.65	0.1	0.325	0.181	0.006
	16–30	155602	13.59	10	18.18	1.34	0.75	0.325	0.181	0.044
	31–50	343634	30.03	19	34.55	1.15	0.57	0.325	0.181	0.034
	51–70	262117	22.91	10	18.18	0.79	0.23	0.325	0.181	0.014
	>70	304809	26.64	13	23.64	0.89	0.33	0.325	0.181	0.019
Aspect	North	149997	13.12	5	9.09	0.69	0.28	0.067	0.181	0.003
	Northeast	195301	17.07	9	16.36	0.96	0.41	0.067	0.181	0.005
	East	129167	11.29	2	3.64	0.32	0.1	0.067	0.181	0.001
	Southeast	171144	14.95	16	29.09	1.95	0.9	0.067	0.181	0.011
	South	135677	11.85	3	5.46	0.46	0.17	0.067	0.181	0.002
	Southwest	131718	11.51	9	16.36	1.42	0.64	0.067	0.181	0.008
	West	79979	6.99	7	12.73	1.82	0.84	0.067	0.181	0.01
	Northwest	151298	13.22	4	7.27	0.55	0.21	0.067	0.181	0.003
Altitude (m)	1200–1500	28463	2.49	0	0	0	0.1	0.107	0.181	0.002
	1500–1800	157018	13.72	19	34.54	2.52	0.9	0.107	0.181	0.017
	1800–2100	303058	26.48	22	40	1.51	0.58	0.107	0.181	0.011
	2100–2400	305844	26.73	7	12.73	0.48	0.25	0.107	0.181	0.005
	2400–2700	208321	18.20	6	10.91	0.60	0.29	0.107	0.181	0.006
	2700–3000	125384	10.96	1	1.82	0.17	0.15	0.107	0.181	0.003
	>3000	16193	1.42	0	0	0	0.1	0.107	0.181	0.002
Lithology	A	459914	40.19	30	54.55	1.36	0.9	0.833	0.267	0.2
	B	153621	13.43	3	5.45	0.41	0.34	0.833	0.267	0.076
	C	147386	12.88	2	3.64	0.28	0.26	0.833	0.267	0.058
	D	19655	1.72	0	0	0	0.1	0.833	0.267	0.022
	E	363705	31.78	20	36.36	1.14	0.77	0.833	0.267	0.171

(continued)

Table 2.2 (continued)

Factor	Class	No. of pixels in domain	Percentage of domain	No. of landslide	Percentage of landslide	Frequency ratio	a	b	c	d
Land use	Best Range	246601	21.55	12	21.82	1.01	0.16	0.667	0.490	0.052
	Mixing orchard and agriculture	152518	13.33	20	36.36	2.73	0.41	0.667	0.490	0.134
	Residential	3450	0.30	1	1.82	6.07	0.9	0.667	0.490	0.294
	Moderate range	741712	64.82	22	40	0.62	0.1	0.667	0.490	0.033
Distance to faults	0–100 m	44942	3.93	3	5.45	1.39	0.31	0.167	0.267	0.014
	101–200 m	43132	3.77	4	7.27	1.93	0.52	0.167	0.267	0.023
	201–300 m	43144	3.77	6	10.91	2.89	0.9	0.167	0.267	0.04
	301–400 m	44914	3.92	2	3.64	0.93	0.13	0.167	0.267	0.006
	>400 m	968149	84.61	40	72.73	0.86	0.1	0.167	0.267	0.004
Distance to rivers	0–100 m	263584	23.03	33	60	2.61	0.9	0.132	0.062	0.007
	101–200 m	205759	17.98	5	9.09	0.51	0.15	0.132	0.062	0.0012
	201–300 m	159801	13.97	7	12.73	0.91	0.29	0.132	0.062	0.002
	301–400 m	131420	11.49	3	5.45	0.47	0.13	0.132	0.062	0.001
	>400 m	383717	33.53	7	12.73	0.38	0.1	0.132	0.062	0.0008
Distance to roads	0–100 m	136228	11.90	23	41.82	3.51	0.9	0.333	0.490	0.147
	101–200 m	110283	9.64	4	7.27	0.75	0.16	0.333	0.490	0.026
	201–300 m	93440	8.17	5	9.10	1.11	0.25	0.333	0.490	0.041
	301–400 m	83876	7.33	3	5.45	0.74	0.15	0.333	0.490	0.024
	400–500 m	74626	6.52	3	5.45	0.84	0.18	0.333	0.490	0.029
	>500 m	645828	56.44	17	30.91	0.55	0.1	0.333	0.490	0.016
TWI	0–4	144529	12.63	50	90.91	7.20	0.9	0.377	0.062	0.021
	4–8	983621	85.96	4	7.27	0.08	0.11	0.377	0.062	0.003
	8–12	16077	1.40	1	1.82	1.3	0.24	0.377	0.062	0.006
	>12	54	0.005	0	0	0	0.1	0.377	0.062	0.002

(continued)

Table 2.2 (continued)

Factor	Class	No. of pixels in domain	Percentage of domain	No. of landslide	Percentage of landslide	Frequency ratio	a	b	c	d
SPI	0–20	266962	23.33	15	27.27	1.17	0.9	0.422	0.062	0.024
	20–40	267926	23.42	12	21.82	0.93	0.26	0.422	0.062	0.007
	40–60	191325	16.72	8	14.55	0.87	0.1	0.422	0.062	0.003
	60–80	130680	11.42	6	10.91	0.96	0.34	0.422	0.062	0.009
	80–100	87780	7.67	4	7.27	0.95	0.31	0.422	0.062	0.008
	>100	199608	17.44	10	18.18	1.04	0.55	0.422	0.062	0.014
STI	0–10	271966	23.77	16	29.09	1.22	0.73	0.070	0.062	0.003
	10–20	362255	31.66	17	30.91	0.98	0.53	0.070	0.062	0.002
	20–30	267619	23.39	12	21.82	0.93	0.49	0.070	0.062	0.002
	30–40	139582	12.20	5	9.09	0.75	0.34	0.070	0.062	0.001
	40–50	58732	5.13	4	7.27	1.42	0.9	0.070	0.062	0.004
	>50	44127	3.85	1	1.82	0.47	0.1	0.070	0.062	0.0004
Plan curvature	Concave	553227	48.35	21	38.18	0.79	0.1	0.501	0.181	0.009
	Convex	591054	51.65	34	61.82	1.20	0.9	0.501	0.181	0.082

Domain: pixels in study area, domain (%): (domain/total pixels in study area)* 100, landslide: number of landslide occurrences, landslide (%): (landslide/total number of landslide occurrences) *100 and frequency ratio: landslide (%)/domain (%).

$A = Q^{sc}, Q_2^t$ and Q_1^t, $B = Q^{ag}, Q^{ta}, Q^{tu}$ and Q^b, $C = K_k^{tv}$ and E_k^{gy}, $D = PE_z$ and PE_f, $E = K_2, K_t, J_1, J_d, J_S, TR_{el}$ and P_d.

a: Normalized Value, b: parameter's Value, c: Group's Value, d: Final Weight

permeability, porosity, humidity, the organic ingredients, land cover, and the climatic season. As a result, pore water pressure of the slope-forming material increases. Consequently, in this study, the aspect map of the study area is produced to show the relationship between aspect and landslide (Fig. 2.6b).

2.4.4 Altitude

Altitude is also a significant landslide conditioning factor because it is controlled by several geologic and geomorphological processes (Gritzner et al. 2001; Dai and Lee 2002; Ayalew et al. 2005; Pourghasemi 2008). To assess altitude as an input parameter for the landslide susceptibility map, an altitude map is prepared from the 10m × 10mdigital elevation model (Fig. 2.6c).

2.4.5 Plan Curvature

The term curvature is theoretically defined as the rate of change of slope gradient or aspect, usually in a particular direction (Wilson and Gallant 2000). The curvature value can be evaluated by calculating the reciprocal value of the radius of curvature of that particular direction. Hence, while the curvature values of broad curves are small, the tight ones have higher values. Plan curvature is described as the curvature of a contour line formed by intersecting a horizontal plane with the surface. The influence of plan curvature on the slope erosion processes is the convergence or divergence of water during downhill flow. For this reason, this parameter constitutes one of the conditioning factors controlling landslide occurrence (Nefeslioglu et al. 2008). The plan curvature map was produced using the script written by Hengl et al. (2003) and run in ILWIS 3.3 software (Fig. 2.6d).

2.4.6 Lithology

Landslides are greatly controlled by the lithology properties of the land surface. Since different lithologic units have different landslide susceptibility values, they are very important in providing data for susceptibility mapping. For this reason, it is essential to group the lithologic properties properly (Carrara et al. 1991; Anbalagan 1992; Mejia-Navarro and Wohl 1994; Mejia-Navarro and Garcia 1996; Pachauri et al. 1998; Luzi and Pergalani 1999; Dai et al. 2001; Yalcin 2005; Duman et al. 2006).

Therefore, a lithology map of the study area is digitized from the existing geology map (sheet number 6461) at the scale of 1:100,000 from the Geological

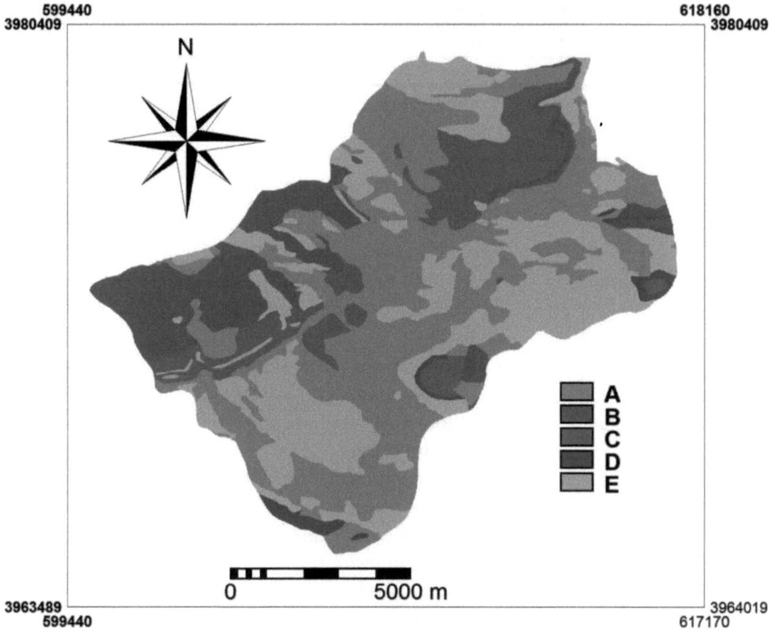

Fig. 2.7 The lithology map of the study area

Survey of Iran (GSI). The study area is covered with various types of lithologic formations. The general geological setting of the area is shown in Fig. 2.7, and the lithological properties are summarized in Tables 2.2 and 2.3.

2.4.7 Land Use

In this study, a land use map was prepared from the LANDSAT satellite image by applying a supervised classification scheme and field surveys. There are four types of landuse are identified in the study area: best range, moderate range, mixed orchard and agriculture and residential areas (Fig. 2.8). Most of the study area is covered by moderate range (64.32%). Several researchers (Maharaj 1993; Fernandez et al. 1999; Jakob 2000; Ocakoglu et al. 2002) emphasized on the importance of vegetation cover or land use characteristics on the stability of slopes, and they considered vegetation cover to assess the conditioning factors of landslides. For the study area, the spatial relationship between land use factor and landslide occurrence are presented in Table 2.2.

Table 2.3 Geology formation of research area

Code	Class	Formation	Lithology	Geological age
Q^{sc}	A	–	Scree	Quaternary
Q_2^t		–	Young terraces	Quaternary
Q_1^t		–	Old terraces	Quaternary
Q^{ag}	B	–	Agglomerate	Quaternary
Q^{ta}		–	Trachy andesitic lava flows	Quaternary
Q^{tu}		–	Ash tuff, lapilli tuff	Quaternary
Q^b		–	Olivine basalt	Quaternary
K_k^{tv}	C	Karaj	Green tuff, basaltic and limestone with gypsum and conglomerate	Eocene
E_k^{gy}		Karaj	Gypsum	Eocene
PE_z	D	Ziarat	Limestone bearing nummulites and alveolina, conglomerate	Paleocene
PE_f		Fajan	Conglomerate, agglomerate, some marl and limestone	Paleocene
K_2	E	–	Biogenic and cherty limestone	Late Cretaceous
K_t		Tizkuh	Orbitoline bearing limestone	Late Cretaceous
J_1		Lar	Massive to well bedded, cherty limestone	Late Jurassic
J_d		Dalichai	Well bedded, partly oolitic-detritic limestone, marly limestone	Late Jurassic
J_S		Shemshak	Dark shale and sandstone with plant remains, coal	Late Jurassic
TR_{eL}		Elika	Thin bedded limestone	Early Triassic
P_d		Dorud	Cross bedded, quartzitic sandstone	Early Permian

2.4.8 Distance from Rivers

An important parameter that controls the stability of a slope is the saturation degree of the material on the slope. The closeness of the slope to drainage structures is another important factor in terms of stability. Streams may adversely affect stability by eroding the slopes or by saturating the lower part of material until resulting in water level increases (Gokceoglu and Aksoy 1996). Five different buffer zones are created within the study area to determine the degree to which the streams affected the slopes (Fig. 2.9).

2.4.9 Distance from Roads

Similar to the effect of the distance to streams, landslides may occur on the road and on the side of the slopes affected by roads (Pachauri and Pant 1992; Pachauri et al. 1998; Ayalew and Yamagishi 2005; Yalcin 2005). A road constructed beside slopes causes a decrease in the load on both the topography and on the toe of slope.

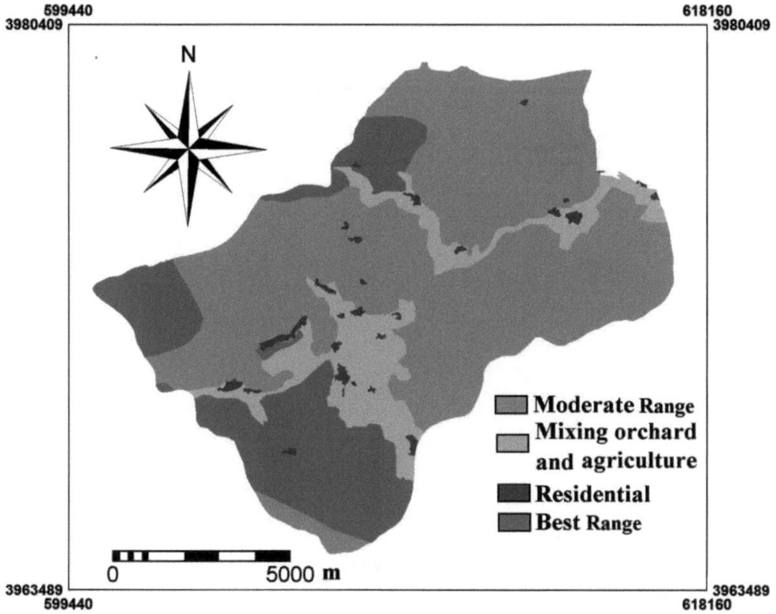

Fig. 2.8 The land use map of the study area

As a result of an increase in stress on the back of the slope, because of changes in topography and decrease of load on toe, some tension cracks may develop. Although a slope is balanced before the road construction, some instability may be observed because of negative effects of excavation. In fact, during the field works, some landslides were recorded whose origin can be attributed to road construction. For this reason, five different buffer areas are created on the path of the road to determine the effect of the road on the stability of slope (Fig. 2.10). The landslide percentage distribution and its frequency ratio are determined considering the distance classes to the road achieved by comparing the map of the distance to the road and the landslide inventory (Table 2.2).

2.4.10 Distance from Faults

The distance from fault is calculated at 100 m intervals using the lithology map (Fig. 2.11). Faults form a line or zone of weakness characterized by heavily fractured rocks. Selective erosion and movement of water along fault planes promote such phenomena. Besides the major thrusts and faults on the geological maps complementary information regarding possible faults and structural dislocations were recognized as lineaments by means of image enhancement (filtering) of satellite imagery. The recognition of lineaments as possible faults is performed

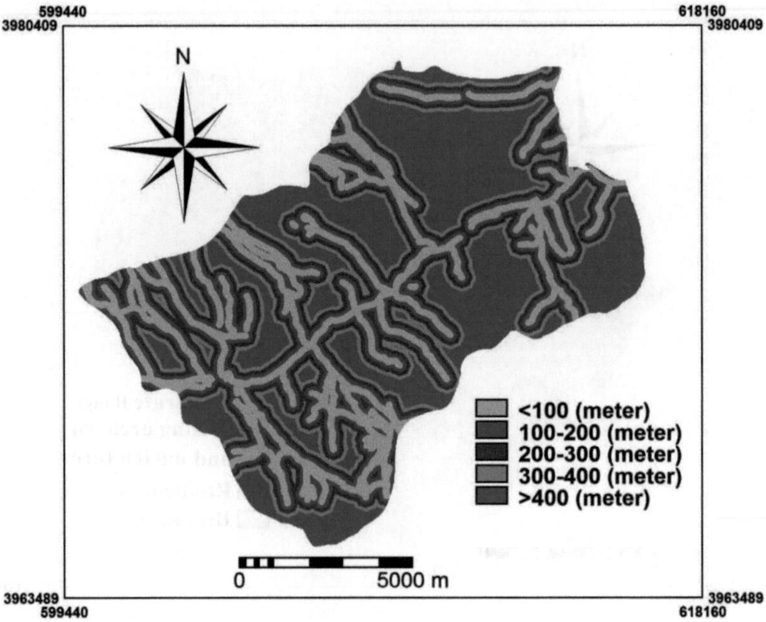

Fig. 2.9 The distance from rivers map of the study area

step-by-step form large to smaller scales allowing the generalization of many neighboring small order lineaments taking into account the spatial scale of the study. The spatial relationship between distance from faults and landslide are presented in Table 2.2.

2.4.11 Topographic Wetness Index (TWI)

The topographic wetness index (TWI) has been used extensively to describe the effect of topography on the location and size of saturated source areas of runoff generation. Moore et al. (1991) proposed Eq. (2.1) for the calculation of TWI under the assumption of steady state conditions and uniform soil properties (i.e. transmisivity is constant throughout the catchment and equal to unity).

$$TWI = \ln(A_S / \tan\beta) \quad (2.1)$$

where A_S is the specific catchment's area (m^2/m), and β is slope gradient (in degrees).

According to Wood et al. (1990), the variation in the topographical components is often far greater than the local variability in soil transmisivity, and Eq. (2.1) can be used to calculate TWI. The TWI map was produced using the script written by Hengl et al. (2003) and run in ILWIS 3.3 software (Fig. 2.12a).

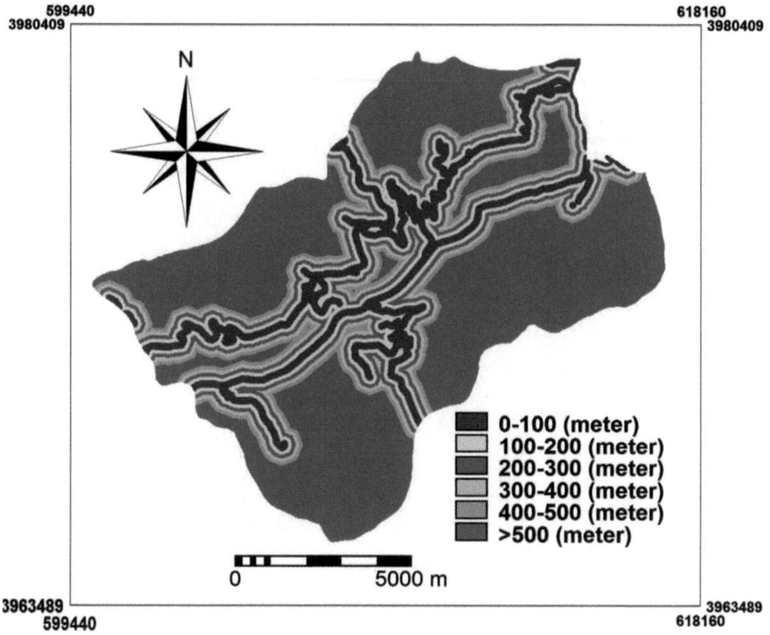

Fig. 2.10 The distance from roads map of the study area

2.4.12 Stream Power Index (SPI)

The stream power index (SPI) is a measure of the erosive power of water flow based on the assumption that discharge (q) is proportional to specific catchment area (As) (Eq. 2.2) (Moore et al. 1991).

$$SPI = A_S \times \tan\beta \tag{2.2}$$

where A_S is the specific catchment's area (m^2/m), and β the slope gradient in degrees. As the specific catchment's area and gradient increase, the amount of water contributed by upslope areas and the velocity of water flow increase; hence, the SPI and slope-erosion risk increase (Moore et al. 1991). Moore et al. (1993) stated that the SPI controls the potential erosive power of overland flow. Therefore, these processes can be considered as one of the components of landslide occurrence (Lee and Min 2001; Gokceoglu et al. 2005; Nefeslioglu et al. 2008; Yilmaz 2009; Akgun and Turk 2010). The SPI map was produced using the script written by Hengl et al. (2003) and run in ILWIS 3.3 software (Fig. 2.12b).

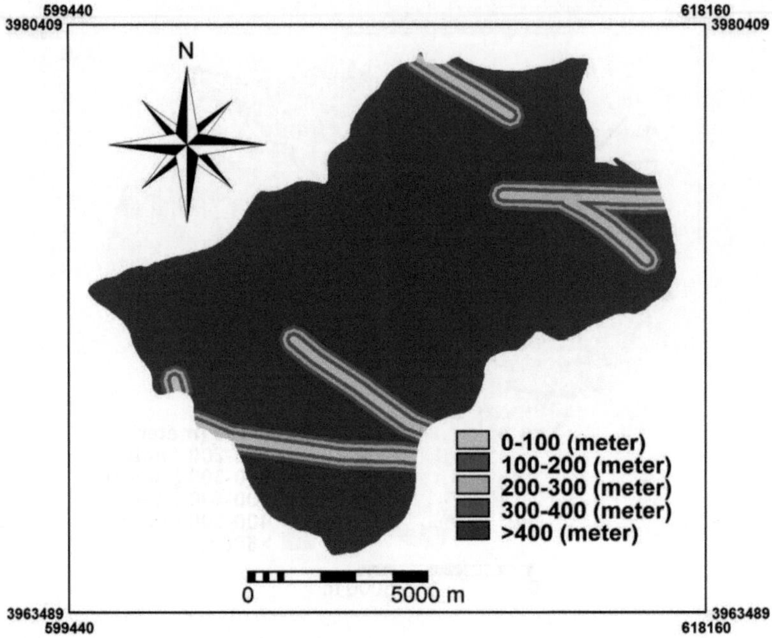

Fig. 2.11 The distance from faults map of the study area

2.4.13 Stream Transport Index (STI)

Another index often used to reflects the erosive power of the overland flow is the sediment transport index (Moore et al. 1993). The STI is calculated from the following formula:

$$STI = \left(\frac{A_S}{22.13}\right)^{0.6}\left(\frac{\sin\beta}{0.0896}\right)^{1.3} \tag{2.3}$$

where A_S is the specific catchment's area (m^2/m), and β the slope gradient.

This empirical formula resembles the Universal Soil Loss Equation and can thus be used to depict locations of potential erosion risk (Moore and Burch 1986). If a close inspection on Eq. (2.3) is performed, it is revealed that the physical meaning of this factor is the capability of sediment transportation controlled a specific catchment area and slope gradient. For that reason, the main causes for this phenomenon may be the disturbed drainage system and the low slope gradient trend on landslide bodies. Therefore, this distinct anomaly can be considered as a good indicator of landslide occurrence (Nefeslioglu et al. 2008). The STI map was produced using the script written by Hengl et al. (2003) and run in ILWIS 3.3 software (Fig. 2.12c).

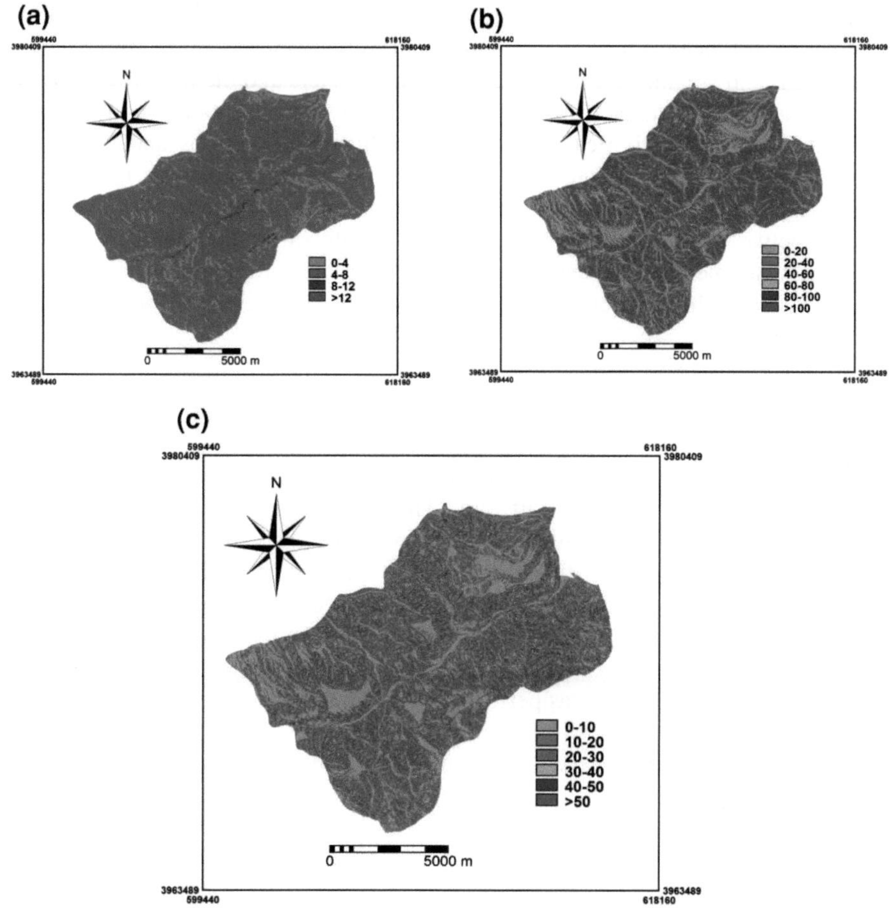

Fig. 2.12 Hydrological based terrain maps of the study area; **a** TWI, **b** SPI, **c** STI

The spatial relationship between TWI, SPI, STI, and landslide occurrence are presented in Table 2.2.

2.5 Landslide Susceptibility Mapping

In this study, the employed SMCE method was built based on analyzing the weight value in bivariate statistical analysis (Table 2.2). All comparisons are based on pair-wise method published by Saaty (1980) in terms of analytical hierarchy process. In this method, all factors were classified into a few groups. The first group consists of slope, aspect, altitude and plan curvature parameters; the second one includes lithology and distance from faults parameters which were extracted from geological map. The next groups presenting the hydrological condition contains of distance from rivers, topographical wetness index (TWI), stream power index (SPI) and stream transport index (STI) parameters and the last group consists of land use and distance from roads parameters, because both of these were induced by human activities. The level of influence for groups and parameters were determined by the range of weighting and were determined by the range of weighting values between a spectrum from minimum to maximum. The range value is between the minimum and maximum weight value. The standardization of each class parameter is compared to each other in order to determine the level of influence. Normalized priority value for each class parameter had been extracted by following Eq. (2.4):

$$N_V = 0.8\left(\frac{X_i - X_{Min}}{X_{Max} - X_{Min}}\right) + 0.1 \tag{2.4}$$

The final weight values were automatically calculated by means of spatial multi criteria evaluation in ILWIS software. The final weight value for each class parameter is produced by multiplying the group weight value, parameter weight value and normalized priority value of class parameter (Table 2.2). Based on weighting values in AHP, the levels of the influence of parameters were generated. The anthropogenic factor has the most influence and the hydrological factor which has the less influence and was categorized in the lowest level. Pair-wise comparison method (Table 2.1) was performed to extract the weight value as presented in Table 2.4.

Based on total weight value, the susceptibility map for Haraz watershed was constructed (Fig. 2.13).

2.6 Validation of the Landslide Susceptibility Map

Validation is a fundamental step in the development of a susceptibility and determination of its prediction ability. The prediction capability of a landslide susceptibility model is usually estimated by using independent information that is not available for building the model. An alternative way to the above statistics is the threshold (cut-off value) calculations, is the receiver operating characteristic (ROC) value and the area under the ROC curve (AUC) (Zweig and Campbell 1993). This method has been widely used as a measure of performance of a predictive rule (Yesilnacar and Topal 2005; Van Den Eeckhaut et al. 2006; Baeza et al. 2010). ROC plots the different accuracy values obtained against the whole range of possible threshold values of the functions, and the AUC serves as a global accuracy statistic for the model, regardless of a specific discriminate threshold. This curve is obtained by plotting all combinations of sensitivities and proportions of false negatives (1-specificity) which may be obtained by varying the decision threshold. The range of values of the ROC curve area is 0.5–1 for a good-fit, while

Table 2.4 The weight value for each group and parameter using pair-wise comparison for SMCE

Number	Groups and parameters	The weight value	Inconsistency ratio
1	Geomorphologic factor	0.181	
	Slope	0.325	
	Aspect	0.067	0.09
	Altitude	0.107	
	Plan curvature	0.501	
2	Geological factor	0.267	
	Lithology	0.833	0.00
	Distance to fault	0.167	
3	Hydrological factor	0.062	
	Distance to river	0.132	
	Topographic wetness index	0.377	0.09
	Stream power index	0.422	
	Stream Transport Index	0.070	
4	Human induced	0.490	
	Land use	0.667	0.00
	Distance to road	0.333	

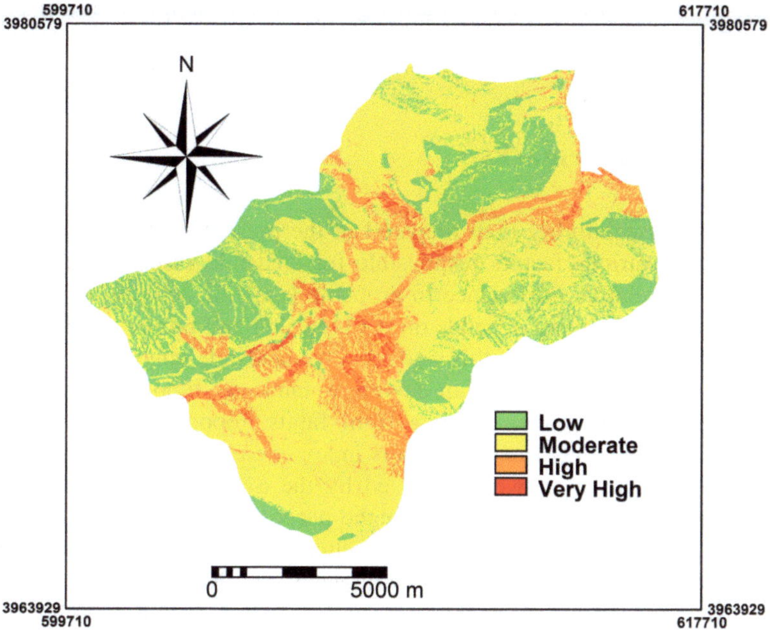

Fig. 2.13 Landslide susceptibility map produced by spatial multi criteria evaluation

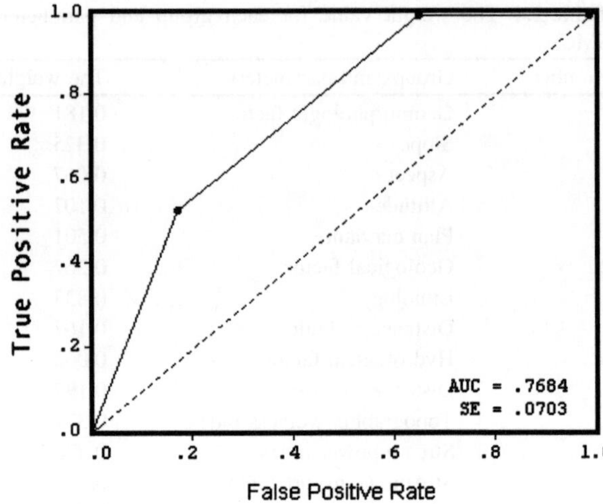

Fig. 2.14 ROC curve for the SMCE

values below 0.5 represent a random fit (Hanley and McNeil 1983). Figure 2.14 shows the ROC curve of the spatial multi criteria evaluation model for the training sample. The AUC value is 0.7684, indicate the good ability of function to correctly discriminate between failed and unfailed groups in the sample used for building the model.

2.7 Concluding Remarks

The landslide susceptibility map prepared in the present study is the result of a combination of various factors responsible for landslide susceptibility, in which each factor has relative importance to probable landslide activity. A reliable and accurate susceptibility map depends on the inclusion and proper determination of the role of these parameters. In this study, twelve landslide-controlling parameters, namely slope, aspect, altitude, plan curvature, land use, lithology, distance from rivers, distance from roads, distance from faults, topographic wetness index, stream power index and stream transport index, were considered. Subsequently, landslide-susceptible areas were analyzed by the SMCE approach and mapped using landslide conditioning factors. For the purpose of verification, the learning set of landslides was randomly sampling choose from a total of 78 landslides population disregarding the temporal component. The ROC curve of block entry SMCE was produced based on the test data set, which was randomly collected from landslide bodies and safe zones. The results showed a 76.84% accuracy with standard error of 0.0703. According to the results of the AUC evaluation, the produced map has exhibited promising results.

References

Akgun A (2011) A comparision of landslide susceptibility maps produced by logistic regression, multi-criteria decision, and likelihood ratio methods: a case study at Izmir, Turkey. Landslides doi:10.1007/s10346-011-0283-7

Akgun A, Bulut F (2007) GIS-based landslide susceptibility for Arsin-Yomra (Trabzon, 381 North Turkey) region. Environ Geol 51:1377–1387

Akgun A, Dag S, Bulut F (2008) Landslide susceptibility mapping for a landslide-prone area (Findikli, NE of Turkey) by likelihood frequency ratio and weighted linear combination models. Environ Geol 54(6):1127–1143

Akgun A, Turk N (2010) Landslide susceptibility mapping for Ayvalik (Western Turkey) 379 and its vicinity by multicriteria decision analysis. Env Earth Sci 61(3):595–611

Anbalgan R (1992) Land hazard evaluation and zonation mapping in mountainous terrain. Eng Geol 32:269–277

Ayalew L, Yamagishi H (2005) The application of GIS-based logistic regression for landslide susceptibility mapping in the Kakuda-Yahiko Mountains, Central Japan. Geomorphology 65(1/2):15–31

Ayalew L, Yamagishi H, Marui H, Kanno T (2005) Landslide in Sado Island of Japan: Part II. GIS-based susceptibility mapping with comparison of results from two methods and verifications. Eng Geol 81:432–445

Baeza C, Lantada N, Moya J (2010) Validation and evaluation of two multivariate statistical models for predictive shallow landslide susceptibility mapping of the Eastern Pyrenees (Spain). Environ Earth Sci 61:507–523

Bai S, Lü G, Wang J, Zhou P, Ding L (2010) GIS-based rare events logistic regression for landslide-susceptibility mapping of Lianyungang, China. Environ Earth Sci 62(1):139–149

Brabb EE, Pampeyan EH (1972) Preliminary map of landslide deposits in San Mateo County, California. US Geological Survey Miscellaneous Field Studies, Map MF-360, scale 1:62.500 (reprinted in 1978)

Carrara A, Cardinali M, Detti R, Guzzetti F, Pasqui V, Reichenbach P (1991) GIS techniques and statistical models in evaluating landslide hazard. Earth Surf Proc Land 16:427–445

Castellanos AEA (2008) Multi-Scale landslide risk assessment in Cuba. PhD thesis, ITC, ITC.Nl

Castellanos E, Van Westen CJ (2007) Generation of a landslide risk index map for Cuba using spatial multi-criteria evaluation. Landslide 4:311–325

Cevik E, Topal T (2003) GIS-based landslide susceptibility mapping for a problematic segment of the natural gas pipeline, Hendek (Turkey). Environ Geol 44:949–962

Chauhan S, Sharma M, Arora MK, Gupta NK (2010) Landslide susceptibility zonation through ratings derived from artificial neural network. Int J Appl Earth Observ Geoinf 12:340–350

Clerici A, Perego S, Tellini C, Vescovi P (2002) A procedure for landslide susceptibility zonation by the conditional analysis method. Geomorphology 48:349–364

Dai FC, Lee CF, Xu ZW (2001) Assessment of landslide susceptibility on the natural terrain of Lantau Island, Hong Kong. Environ Geol 40(3):381–391

Dai FC, Lee CF (2002) Landslide characteristics and slope instability modeling using GIS, Lantau Island, Hong Kong. Geomorphology 42:213–228

Dai FC, Lee CF (2001) Terrain-based mapping of landslide susceptibility using a geographical information systems: a case study. Can Geotech J 38:911–923

Dobrovolny E (1971) Landslide susceptibility in and near anchorage as interpreted from topographic and geologic maps, in the great Alaska earthquake of 1964-Geology volume. Publication 1603. U.S. Geological survey open file report 86-329, National Research Council, Committee on the Alaska Earthquake, National Academy of Sciences, USA, pp 735–745

Duman TY, Can T, Gokceoglu C, Nefeslioglu HA, Sonmez H (2006) Application of logistic regression for landslide susceptibility zoning of Cekmece Area, Istanbul, Turkey. Env Geol 51:241–256

Ercanoglu M, Gokceoglu C (2004) Use of fuzzy relations to produce landslide susceptibility map of a landslide prone area (West Black Sea Region, Turkey). Eng Geol 75:229–250
Ercanoglu M, Gokceoglu C (2002) Assessment of landslide susceptibility for a landslide-prone area (north of Yenice, NW Turkey) by fuzzy approach. Environ Geol 41:720–730
Ermini L, Catani F, Casagli N (2005) Artificial neural networks applied to landslide susceptibility assessment. Geomorphology 66:327–343
Fernández CI, Castillo TF, Hamdouni RE, Montero JC (1999) Verification of landslide susceptibility mapping: a case study. Earth Surf Proc Land 24(6):537–544
Gokceoglu C, Sonmez H, Nefeslioglu HA, Duman TY, Can T (2005) The 17 March 2005 Kuzulu landslide (Sivas, Turkey) and landslide-susceptibility map of its near vicinity. Eng Geol 81:65–83
Gokceoglu C, Aksoy H (1996) Landslide susceptibility mapping of the slopes in the residual soils of the Mengen region (Turkey) by deterministic stability analyses and image processing techniques. Eng Geol 44:147–161
Gritzner ML, Marcus WA, Aspinall R, Custer SG (2001) Assessing landslide potential using GIS, soil wetness modeling and topographic attributes, Payette River, Idaho. Geomorphology 37:149–165
Guzzetti F, Carrarra A, Cardinali M, Reichenbach P (1999) Landslide hazard evaluation: are view ofcurrent techniques and their application in a multi-scale study, Central Italy. Geomorphology 31:181–216
Hanley JA, McNeil BJ (1983) A method of comparing the areas under receiver operating characteristic curves derived from the same cases. Radiology 148:839–843
Hengl T, Gruber S, Shrestha DP (2003) Digital terrain analysis in ILWIS. International Institute for Geo-Information Science and Earth Oservation Enschede, The Netherlands, p 62
Herwijnen MV (1999) Spatial Decision Support for Environmental Management. Vrije Universiteit, Amsterdam 274
ITC (2001) ILWIS 3.0 academic-user's guide. ITC, Enschede, p 520
Jakob M (2000) The impacts of logging on landslide activity at Clayoquot Sound, British Columbia. Catena 38:279–300
Komac M (2006) A landslide susceptibility model using analytical hierarchy process method and multivariate statistics in perialpine-Slovenia. Geomorphology 74:17–28
Kanungo DP, Arora MK, Sarkar S, Gupta RP (2006) A comparative study of conventional, ANN black box, fuzzy and combined neural and fuzzy weighting procedures for landslide susceptibility zonation in Darjeeling Himalayas. Eng Geol 85(3–4):347–366
Lee S (2005) Application of logistic regression model and its validation for landslide susceptibility mapping using GIS and remote sensing data. Int J Remote Sens 26:1477–1491
Lee S, Choi J, Min K (2004a) Probabilistic landslide hazard mapping using GIS and remote sensing data at Boun, Korea. Int J Remote Sens 25:2037–2052
Lee S, Min K (2001) Statistical analysis of landslide susceptibility at Yongin, Korea. Environ Geol 40:1095–1113
Lee S, Pradhan B (2007) Landslide hazard mapping at Selangor, Malaysia using frequency ratio and logistic regression models. Landslides 4:33–41
Lee S, Pradhan B (2006) Probabilistic landslide risk mapping at Penang Island. Malays J Earth Syst Sci 115(6):1–12
Lee S, Ryu JH, Won JS, Park H (2004b) Determination and application of the weights for landslide susceptibility mapping using an artificial neural network. Eng Geol 71:289–302
Lee S, Ryu JH, Lee MJ, Won JS (2006) The application of artificial neural networks to landslide susceptibility mapping at Janghung, Korea. Math Geol 38(2):199–219
Luzi L, Pergalani F (1999) Slope instability in static and dynamic conditions for urban planning: the ''Oltre Po Pavese'' case history (Region Lombardia-Italy). Nat Hazards 20:57–82
Maharaj RJ (1993) Landslide processes and landslide susceptibility analysis from an upland watershed: a case study from St. Andrew, Jamaica, West Indies. Eng Geol 34: 53–79
Malczewski J (1999) GIS and multi criteria decision analysis. Wiley, New York. ISBN: 978-0-471-32944-2, p 408

Mejia-Navarro M, Garcia LA (1996) Natural hazard and risk assessment using decision support systems, application: Glenwood Springs, Colorado. Environ Eng Geosci 2(3):299–324

Mejia-Navarro M, Wohl EE (1994) Geological hazard and risk evaluation using GIS: methodology and model applied to Medellin, Colombia. Bull Assoc Eng Geol 31:459–481

Melchiorre C, Matteucci M, Remondo J (2006) Artificial neural networks and robustness analysis in landslide susceptibility zonation, IEEE. 2006 International Joint Conference on Neural Networks Sheraton Vancouver Wall Centre Hotel, Vancouver, 16–21 July, pp 4375–4381

Mohammadi M (2008) Mass movement hazard analysis and presentation of suitable regional model using GIS (Case Study: A part of Haraz Watershed), M.Sc. Thesis, Tarbiat Modarres University International Campus, Iran, p 80

Moore ID, Burch GJ (1986) Sediment transport capacity of sheet and rill flow: application of unit stream power theory. Water Res Res 22:1350–1360

Moore ID, Gessler PE, Nielsen GA, Peterson GA (1993) Soil attribute prediction using terrain analysis. Soil Sci Soc Am J 57:443–452

Moore ID, Grayson RB, Ladson AR (1991) Digital terrain modeling: a review of hydrological, geomorphological, and biological applications. Hydrol Process 5:3–30

Nefeslioglu HA, Sezer E, Gokceoglu C, Bozkir AS, Duman TY (2010) Assessment of landslide susceptibility by decision trees in the metropolitan area of Istanbul, Turkey, Mathematical Problems in Engineering, Article ID 901095, p 15, doi:10.1155/2010/901095

Nefeslioglu HA, Duman TY, Durmaz S (2008) Landslide susceptibility mapping for a part of tectonic Kelkit Valley (Eastern Black Sea region of Turkey). Geomorphology 94:401–418

Ocakoglu F, Gokceoglu C, Ercanoglu M (2002) Dynamics of a complex mass movement triggered by heavy rainfall: a case study from NW Turkey. Geomorphology 42(3):329–341

Oh HJ, Lee S, Chotikasathien W, Kim CH, Kwon JH (2009) Predictive landslide susceptibility mapping using spatial information in the Pechabun area of Thailand. Environ Geol 57: 641–651

Pachauri AK, Pant M (1992) Landslide hazard mapping based on geological attributes. EngGeol 32:81–100

Pachauri AK, Gupta PV, Chander R (1998) Landslide zoning in a part of the Garhwal Himalayas. Environ Geol 36(3–4):325–334

Pourghasemi HR (2008) Landslide hazard assessment using fuzzy logic (Case Study: A part of Haraz Watershed), M.Sc. Thesis, Tarbiat Modarres University International Campus, Iran, pp 92

Pradhan B (2010a) Remote sensing and GIS-based landslide hazard analysis and cross-validation using multivariate logistic regression model on three test areas in Malaysia. Adv Space Res 45(10):1244–1256

Pradhan B (2010b) Manifestation of an advanced fuzzy logic model coupled with Geoinformation techniques for landslide susceptibility analysis. Environ Ecol Stat 18(3):471–493

Pradhan B (2010c) Application of an advanced fuzzy logic model for landslide susceptibility analysis. Int J Comput Intell Sys 3(3):370–381

Pradhan B (2011) Use of GIS-based fuzzy logic relations and its cross application to produce landslide susceptibility maps in three test areas in Malaysia. Environ Earth Sci 63:329–349

Pradhan B, Buchroithner MF (2010) Comparison and validation of landslide susceptibility maps using an artificial neural network model for three test areas in Malaysia. Environ Eng Geosci 16(2):107–126

Pradhan B, Lee S (2009) Landslide risk analysis using artificial neural network model focusing on different training sites. Int J Phys Sci 3(11):1–15

Pradhan B, Lee S (2010a) Regional landslide susceptibility analysis using back-propagation neural network model at Cameron Highland, Malaysia. Landslides 7(1):13–30

Pradhan B, Lee S (2010b) Landslide susceptibility assessment and factor effect analysis: backpropagation artificial neural networks and their comparison with frequency ratio and bivariate logistic regression modeling. Environ Modell Softw 25:747–759

Pradhan B, Lee S (2010c) Delineation of landslide hazard areas using frequency ratio, logistic regression and artificial neural network model at Penang Island, Malaysia. Environ Earth Sci 60:1037–1054

Pradhan B, Sezer E, Gokceoglu C, Buchroithner MF (2010) Landslide susceptibility mapping by neuro-fuzzy approach in a landslide prone area (Cameron Highland, Malaysia). IEEE T Geosci Remote 48(12):4164–4177

Radbruch DH (1970) Map of relative amounts of landslides in California. US Geological Survey Open-File Report 70-1485, pp 36, map scale 1:500.000. US Geological Survey Open-File Report, pp 85–585

Saaty T (1980) The analytical hierarchy Process. McGraw-Hill, New York

Saaty TL (1997) A scaling method for priorities in hierarchical structures. J Math Psychol 15:234–281

Saha AK, Gupta RP, Sarkar I, Arora MK, Csaplovics E (2005) An approach for GIS-based statistical landslide susceptibility zonation with a case study in the Himalayas. Landslides 2:61–69

Saito H, Nakayama D, Matsuyama H (2009) Comparison of landslide susceptibility based on a decision-tree model and actual landslide occurrence: the Akaishi mountains, Japan. Geomorphology 109:108–121

Schuster RL, Fleming RW (1986) Economic losses and fatalities due to landslides. Bull Assoc Eng Geol 23:11–28

Sezer EA, Pradhan B, Gokceoglu C (2011) Manifestation of an adaptive neuro-fuzzy model on landslide susceptibility mapping: Klang valley, Malaysia. Expert Syst App 38(7):8208–8219

Sharifi MA, Herwijnen MV (2003) Spatial decision support systems. International Instiute for Geo-Information Science and Earth Observation (ITC). p 201

Sharifi MA, Retsios V (2004) Site selection for waste disposal through spatial multiple criteria decision analysis. J Telecommun Inf Technol 3:1–11

Shou KJ, Wang CF (2003) Analysis of the Chiufengershan landslide triggered by the 1999 Chi–Chi earthquake in Taiwan. Eng Geol 68:237–250

Tangestani MH (2009) A comparative study of dempster–shafer and fuzzy models for landslide susceptibility mapping using a GIS: an experience from zagros Mountains, SW Iran. J Asian Earth Sci 35:66–73

Van Den Eeckhaut M, Vanwalleghem T, Poesen J, Govers G, Verstraeten G, Vandekerckhove L (2006) Prediction of landslide susceptibility using rare events logistic regression: a case-study in the Flemish Ardennes (Belgium). Geomorphology 76:392–410

Van Westen CJ, Bonilla JBA (1990) Mountain hazard analysis using PC-based GIS. 6th IAEG Congress, vol 1. Balkema, Rotterdam, pp 265–271

Varnes DJ (1978) Slope movement types and processes. In: Schuster RL, Krizek RJ (eds) Landslides analysis and control. Special Report, Transportation Research Board. vol 176 National Academy of Sciences, New York, pp 12–33

Wan S (2009) A spatial decision support system for extracting the core factors and thresholds for landslide susceptibility map. Eng Geol 108:237–251

Wang R (2008) An expert knowledge-based approach to landslide susceptibility mapping using GIS and fuzzy logic, A dissertation submitted in partial fulfillment of the requirement for the degree of Docto of Philosophy, University of Wisconsin-Madison, p 175

Wilson JP, Gallant JC (2000) Terrain analysis principles and applications. Wiley, New York

Wood EF, Sivapalan M, Beven KJ (1990) Similarity and scale catchment storm response. Rev Geophisics 28:1–18

Yalcin A (2005) An investigation on Ardesen (Rize) region on the basis of landslide susceptibility, KTU, PhD Thesis (in Turkish)

Yalcin A (2008) GIS-based landslide susceptibility mapping using analytical hierarchy process and bivariate statistics in Anderson (Turkey): comparision of results and confirmations. Catena 1:1–12

Yeon YK, Han JG, Ryu KH (2010) Landslide susceptibility mapping in Injae, Korea, using a decision tree. Eng Geol 116:274–283

Yesilnacar E, Topal T (2005) Landslide susceptibility mapping: a comparison of logistic regression and neural networks methods in a medium scale study, Hendek region (Turkey). Eng Geol 79(3–4):251–266

Yilmaz I (2009) Landslide susceptibility mapping using frequency ratio, logistic regression, artificial neural networks and their comparison: a case study from kat landslides (Tokat-Turkey). Comp Geosci 35(6):1125–1138

Zhou G, Esaki T, Mitani Y, Xie M, Mori J (2003) Spatial probabilistic modeling of slope failure using an integrated GIS Monte Carlo simulation approach. Eng Geol 68:373–386

Zweig MH, Campbell G (1993) Receiver-operating characteristics (ROC) plots. Clin Chem 39: 561–577

Chapter 3
Soft Computing Modeling in Landslide Susceptibility Assessment

C. Gokceoglu and E. Sezer

Abstract In the landslide literature, the most frequently encountered term is "landslide susceptibility". The main reasons include: (i) developments in computer techniques and Geographical Information Systems; (ii) increasing awareness of the socio-economic significance of landslides and (iii) the complex nature of landslides. Owing to the complex nature of landslides, various approaches have been considered for the assessment of landslide susceptibility at regional scale. The assessment approaches can be classified into three main groups such as heuristic, statistical methods and soft computing approaches. In this chapter, the soft computing methods used for landslide susceptibility are considered, and the related international literature is assessed.

Keywords Landslide · Landslide susceptibility · Fuzzy approach · Artificial neural networks

3.1 Introduction

The complexity of the Earth System's behavior makes it extremely difficult to accurately forecast the future of the Earth System, and presents a major challenge to the global change research community (Pielke et al. 2003). Landslides are one of the components of the earth surface processes. Earth surface processes are governed by

C. Gokceoglu (✉)
Department of Geological Engineering, Hacettepe University,
06800 Beytepe, Ankara, Turkey
e-mail: cgokce@hacettepe.edu.tr

E. Sezer
Department of Computer Engineering, Hacettepe University,
06800 Beytepe, Ankara, Turkey

B. Pradhan and M. Buchroithner (eds.), *Terrigenous Mass Movements*,
DOI: 10.1007/978-3-642-25495-6_3,

various geological, hydrological and environmental conditions, and climatologic events. Human activities have been the additional landslide causative factor for hundred years. Geological complexity and heterogeneity is the major problem when modeling the landslide occurrence and predicting the future landslide activity.

It is possible to group the uncertainties under four classes such as conceptual, model, parameter and data. These uncertainties are encountered frequently when assessing regional landslide susceptibility and hazard. The conceptual uncertainty refers to the difficulties in problem formulation. Due to the complexity of the landslide phenomenon, to formulate the regional landslide assessment is too difficult, is sometimes impossible. In fact, similar problem is faced when creating a regional landslide assessment model. The other uncertainty is sourced from parameters used in the assessments. Generally, the selection of representative inputs is highly difficult for prediction of possible landslide areas because the regional assessments require a generalization. The last problem is related to the data. The data should be representative and reliable. Sometimes, to find the data having a sufficient quality and quantity is difficult. All these uncertainties or some combinations of them arise in modeling landslide susceptibility, hazard and risk assessments. If these uncertainties are eliminated or minimized, the quality of a regional landslide assessment increases. Up to now, various techniques have been proposed and applied for the regional landslide susceptibility, hazard and risk assessments. It is possible to produce a landslide susceptibility map employing various indirect mapping techniques such as combination of index maps (i.e. Pachauri and Pant 1992; Gokceoglu and Aksoy 1996; Turrini and Visintainer 1998; Donati and Turrini 2002), bivariate and multivariate statistical analyses (i.e. Carrara et al. 1991; Atkinson and Massari 1998; Guzzetti et al. 1999; Baeza and Corominas 2001; Lee and Min 2001; Clerici et al. 2002; Lee 2005; Can et al. 2005; Gokceoglu et al. 2005; Wang and Sassa 2006; Duman et al. 2006; Guzzetti et al. 2006; Nefeslioglu et al. 2008a; Gorum et al. 2008), neural networks (i.e. Lee 2007a; Gomez and Kavzoglu 2005; Kanungo et al. 2006; Nefeslioglu et al. 2008b) and fuzzy approach (i.e. Juang et al. 1992; Binaghi et al. 1998; Ercanoglu and Gokceoglu 2002 and 2004).

In recent years, the assessment of landslide hazard and risk has become a topic of major interest for both geoscientists and engineering professionals as well as for the community and the local administrations in many parts of the world (Aleotti and Chowdury 1999). Thc reasons for the international interest in landslides are twofold: firstly an increasing awareness of the socio-economic significance of landslides and secondly, the increased pressure of development and urbanization on the environment (Aleotti and Chowdury 1999). Especially, the urbanization has a crucial role on the landslide development and the increase in landslide losses. According to Helmore (1996), a hundred years ago, the world population totaled 1.1 billion, and about 5% of people lived in cities. Today, the population has risen to 5.3 billion and approximately 45% of it is concentrated in urban areas. The most explosive growth has been in the developing world, where urban populations have tripled in the last 30 years (Guzzetti et al. 1999). Between 1950 and 1995, the number of cities with population of more than one million increased sixfold in the third world (Helmore 1996). This change will continue depending on industrialization throughout the world. As a result of this change, the landslide threat on the cities and the population may increase. Therefore, the landslide hazard mitigation efforts should be considered. The first stage of the landslide mitigation efforts is accepted as mapping of landslide inventory, landslide susceptibility and landslide hazard. After this stage, the possible landslide risk should be considered and assessed. In this chapter, the general concepts on landslide susceptibility, hazard and risk are given firstly. Then, the general principles of the soft computing techniques and application of the soft computing techniques to landslide susceptibility assessments are defined in detail. Finally, further recommendations are discussed.

3.2 General Concepts

Recently, some guidelines on the landslide susceptibility, hazard and risk zoning have been published by the Joint Technical Committee on Landslides and Engineered Slopes (Fell et al. 2008). In literature, it is possible to encounter some misconceptions on landslide susceptibility, hazard and risk assessments. Landslide susceptibility can be defined as a quantitative or qualitative assessment of the classification, volume (or area), and spatial distribution of landslides which exist or potentially may occur in an area (Fell et al. 2008). The term "hazard" is defined as a condition with the potential for causing an undesirable consequence. The description of landslide hazard should include the location, volume (or area), classification and velocity of the potential landslides and any resultant detached material, and the probability of their occurrence within a given period of time (Fell et al. 2008). According to International Strategy for Disaster Reduction (ISDR), the term "risk" can be defined as a measure of "the probability of harmful consequences, or expected losses (deaths, injuries, property, livelihoods, economic activity disrupted or environmental damaged) resulting from interactions between natural or human-induced hazards and vulnerable conditions" (Hufschmidt and Crozier 2008). When considering this description for landslides, the landslide risk can be defined as the potential for adverse consequences, loss, harm, or detriment to human population and things that human beings value due to landslide occurrences (Kanungo et al. 2008). As can seen from these definitions, there are important differences among the landslide susceptibility, hazard and risk concepts. In the following subtopics, these concepts and their assessments are explained in detail.

3.2.1 Landslide Susceptibility

Susceptibility is the propensity of an area to generate landslides. In mathematical form, landslide susceptibility is the probability of spatial occurrence of known slope failures, given a set of geoenvironmental conditions (Guzzetti et al. 2005). The first extensive papers on the use of spatial information in a digital context for landslide mapping date back to the late seventies and early eighties of the last century. Among the pioneers in this field were Brabb et al. (1978) in California and Carrara et al.

(1977) in Italy (van Westen et al. 2008). However, in recent years, a huge literature on the assessment of landslide susceptibility has been published in the major international journals. According to Aleotti and Chowdhury (1999), an ideal susceptibility assessment should answer the questions of "where will the landslide occur?", "what type of failures will occur?" and "how will the landslides occur?". In the first stage, to answer these questions, a well-documented landslide inventory map of the area to be studied should exist, because a landslide inventory map of an area shows the location, classification, volume, activity, date of occurrence and other characteristics of landslide. Also, assuming landslides will occur in the future because of the same conditions that produced them in the past (Guzzetti et al. 1999). In fact, a landslide susceptibility assessment is to predict the future landslide locations considering the existing landslides. For this reason, when producing landslide susceptibility maps, various indirect methods such as statistical models and soft computing have been employed. A recent review on the landslide susceptibility modeling approaches was published by Brenning (2005). According to Brenning (2005), logistic regression and discriminant analysis are the most frequently chosen models. However, in this chapter, use of the soft computing techniques for the landslide susceptibility assessments will be given. In the recent years, the artificial neural networks (ANN) and fuzzy approach have been extensively used for the landslide susceptibility assessment. The ANN has been applied by Lee et al. (2003a, b, 2004, 2006, 2007a, b), Gomez and Kavzoglu (2005), Ermini et al. (2005), Melchiorre et al. (2008), Yao et al. (2008), Yesilnacar and Topal (2005), Kanungo et al. (2006), Nefeslioglu et al. (2008b), Lee and Evangelista (2006), Wang and Sassa (2006), Lu and Rosenbaum (2003), Caniani et al. (2008), Lee (2007a). The other soft computing technique, the fuzzy approach, has also been employed by various researchers (Juang et al. 1992; Binaghi et al. 1998; Ercanoglu and Gokceoglu 2002 and 2002; Champati ray et al. 2007; Tangestani 2004 and 2006; Gorsevski and Jankowski 2008). However, when compared with the ANN and the statistical approaches, the use of the fuzzy approach for the assessment of the landslide susceptibility is limited in the world literature. The application of the fuzzy approaches is more difficult and time consuming than the application of the black-box methods such as the ANN and the statistical models. When preparing a regional landslide susceptibility map, a huge dataset should be processed. Especially, construction of rule-based fuzzy systems for this purpose requires much time and effort although the fuzzy systems produce satisfactory results. For this reason, the researchers prefer the black-box approaches. Among the soft computing techniques, a hybrid approach, the neuro-fuzzy method, exists. However, in the literature, an application of the neuro-fuzzy method for the assessment of the landslide susceptibility has not been encountered yet.

3.2.2 Landslide Hazard

Landslide hazard is described as the probability of occurrence of a potentially damaging phenomenon (landslide) within a given area and in a given period of time by Varnes (1978). When considering this description, a regional landslide hazard assessment requires the probability of occurrence of future landslides. Unfortunately, this is not the case of most published regional landslide "hazard" assessments, which generally only provide a ranking of terrain units in terms of susceptibility, not considering the temporal component of the hazard (Zezere et al. 2004). In literature, many susceptibility assessments have been presented as hazard assessment. However, some authors have attempted to produce landslide hazard map including period of time (Zezere et al. 2004; Catani et al. 2005; Chau et al. 2004). In fact, landslides are localized processes, which normally do not happen with different frequency and magnitude at the same location (van Westen et al. 2006). This is the main problem for the preparation of landslide hazard maps based on the landslide inventory maps. However, the landslide inventories used in literature traditionally reflect the actual terrain conditions like failed slopes. However, the question of what the terrain parameters, especially slope degree were at the failure time should be answered because one of the main targets of the landslide susceptibility maps is to determine the susceptibility degree of the slopes which has not failed yet (Gorum et al. 2008). Similarly, according to van Westen et al. (2006), for most types of landslides, once the movement has occurred, the slope conditions are changed and a repetition of a similar event in the same location is not likely to happen. Due to this problem, although it is possible to produce some significant landslide susceptibility maps, assessment of regional landslide hazard is highly difficult. However, it is possible to elaborate frequency–magnitude relationships for landslide occurrences over a larger area, such as an entire watershed, by mapping the landslides taking place due to particular triggering events and relate the spatial frequency to the return period (van Westen et al. 2006). Finally, if a representative landslide susceptibility assessment of an area exists and the return period of the triggering factor such as rainfall or earthquake is known, a realistic landslide hazard assessment can be carried out.

3.2.3 Landslide Risk

The landslide risk is defined as a measure of the probability and severity of an adverse effect to health, property or the environment. Risk is often estimated by the product of probability of a phenomenon of a given magnitude times the consequences (Fell et al. 2008). The risk can be formulated as follows Eq. 3.1 (Varnes 1984; Fell 1994; Leroi 1996)

$$Risk = \sum HVA \tag{3.1}$$

Where, "H" (hazard) is expressed as probability of occurrence within a reference period: "V" (physical vulnerability) is expressed as a particular type of element at risk for a specific type of hazard and for a specific element at risk; "A" is the amount or cost of the particular elements at risk.

In literature, the regional landslide risk assessments are limited because a landslide risk assessment requires landslide susceptibility and hazard assessments.

The difficulties about landslide hazard assessments are described previously. These difficulties prevent the production of high quality regional landslide risk assessments. However, the available approaches such as statistical or soft computing methods are able to produce satisfactory results. The main problem is sourced from high quality data availability. Similar idea was pointed out by van Westen et al. (2006). The absence or incompleteness of landslide records is one of the major drawbacks in the assessment of landslide hazard risk. For this reason, it has been impossible in most parts of the world to establish the quantitative relationship of the occurrence of landslides with important triggering factors, such as earthquakes and rainfall, of which magnitude-frequency functions are known (van Westen et al. 2006). Although some regional landslide risk assessments have been published (Rautelal and Lahhera 2000; Lee and Pradhan 2006; Remondo et al. 2005; Catani et al. 2005; Corominas et al. 2005; Romeo et al. 2006), none of these is based on soft computing methods.

3.3 Fuzzy Modeling and Relations

In recent years, Fuzzy Inference System (FIS) and Fuzzy Relations have been used extensively for the production landslide susceptibility maps. Because, unlike classicial set theory; fuzzy set theory is flexible, and focuses on the degree of being a member of a set (Berkan and Trubatch 1997). Additionally, the advantages of the fuzzy logic approach to solve engineering geological problems can be summarized as (Alvarez Grima 2000); (a) it allows express explicit of the knowledge of the system via fuzzy if-then rules, (b) it deals with subjective uncertainty (fuzziness, vagueness, imprecision) inherent to the way experts approach their problems, (c) numerical and categorical data can be combined, and (d) it provides a sound mathematical basis. For this reason, in this section the FIS and fuzzy relations are introduced to help the earth scientists to produce landslide susceptibility maps by fuzzy approach. Before the explanation of FIS and fuzzy relations, some basic fuzzy concepts should be introduced.

3.3.1 Basic Fuzzy Concepts

All concepts in fuzzy domain are based on fuzzy set theory. The fuzzy set theory is an extension of crisp set theory especially on the boundary of the set. In fuzzy sets, the transition from belonging to a set to the not belonging is gradual, and the degree of transition is denoted by the membership functions. If A is a fuzzy set in universe of discourse U, A is presented as follows Eq.3.2:

$$A = \{(x, \mu_A(x)) \,|x \in U\} \tag{3.2}$$

where, $\mu_A(x)$ means membership function and its result gives the membership degree of "x" to "A" and its values varies between 0 and 1 closed interval. If integers is "U", "integers near to 0" is the set called "A", then elements of "A"

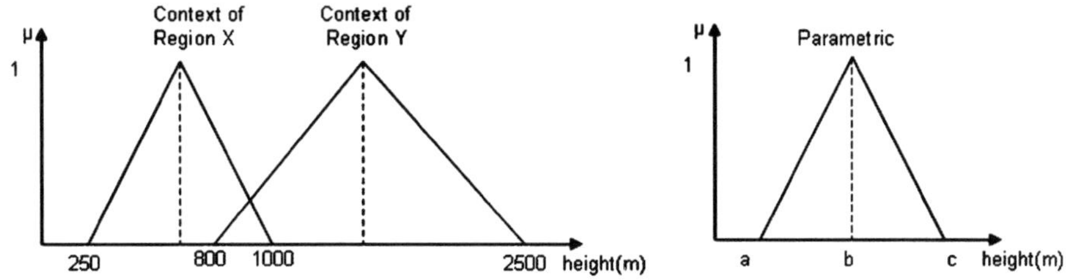

Fig. 3.1 Advantage of using parametric membership function

and their membership degrees presented like this: A = {(1.0/0), (0.8/1),(0.6/2), (0.4/3), (0.2/4), (0.0/5)}. Membership degrees can be mapped by point to point or they are resulted from a fuzzy function. In fact, the use of fuzzy functions is supported because of their parametric structure to support context knowledge modeling. For example, the determination of average height of mountainous regions differs from region to region. So membership functions can be parameterized according to region context. In Fig. 3.1, this context modeling by using parametric triangular membership function is illustrated.

There are many types of fuzzy membership functions and some of them are triangular, trapezoid, sigmoid, Gauss, etc. Because of usage and calculation simplicity, triangular and trapezoid functions are used more frequently than others and symmetric functions usages are enforced according to Yen and Langari (1998). Symmetric functions means fuzzy function is symmetric around "a" certain point "x = c".

Fuzzy sets are applicable for fuzzy operations: union, intersection and complement. If "A" and "B" are fuzzy sets on universe of discourse of U, fuzzy set operations are given in Eqs. 3.3–3.5

$$\text{Union}: \mu_{(A\cup B)}(x) = \max(\mu_A(x),\ \mu_B(x)) \quad \textit{where} \quad x \in U \tag{3.3}$$

$$\text{Intersection}: \mu_{(A\cap B)}(x) = \min(\mu_A(x),\ \mu_B(x)) \quad \textit{where} \quad x \in U \tag{3.4}$$

$$\text{Complement}: \mu_{(\neg A)}(x) = 1 - \mu_A(x) \quad \textit{where} \quad x \in U \tag{3.5}$$

Max and min are fuzzy mathematic operators. Max means selection of the element which has the maximum membership degree from any series and min means reverse selection. In this situation, the minimum membership degree is selected.

Linguistic variable means the variable takes its value in linguistic form. For example, "slope degree of a region" can be linguistic variable and its values can be "high", "low", etc. If all concepts are collected together, a linguistic variable can be the input or output of the problem. Fuzzy sets are the values of the linguistic variables and membership functions are tools which enable to give assigning crisp values to the fuzzy sets with membership degrees.

Fuzzy sets and the other concepts based on it, take their power from two sources: ability of membership functions and linguistic variables. They are very useful to express human thinking behavior in the rule based form. Because the rules used by any experts in any domain do not use crisp values too much.

Fig. 3.2 An example for the relation matrix

		6	7	8	9
	3	0.33	0.40	0.45	0.50
R=	4	0.20	0.27	0.33	0.38
	5	0.09	0.17	0.23	0.29

For example, one earth scientist expresses the volume of a debris flow with the "low" or "moderate", and qualifies it with "very", "almost", "nearly" terms. This type of thinking can be simulated easily with usage of linguistic variables and values of them. Collected crisp values from any region supports some rules with some degrees. In addition, some inputs are placed on the boundaries of the sets, fire multiple rules and results multiple conclusions. This scheme of rule evaluation is very close to scheme of expert thinking.

3.3.2 Fuzzy Relations

Fuzzy relations play an important part in fuzzy modeling, and fuzzy relations are based on philosophy that everything is related to some extend or unrelated (Dubois and Prade 1980). Fuzzy relation is an extension of the crisp relation with the same approach while extending fuzzy sets from crisp sets. Fuzzy relations can be built between two or more fuzzy sets (binary and n-ary respectively). If "A" and "B" is the universe of discourse, then the fuzzy relation between them denoted by R is described in Eq. 3.6.

$$R = \{((a,b),\ \mu_R(a,b))|(a,b) \in AxB\} \tag{3.6}$$

As a result of the Eq. 3.6, if "A" has "n" and "B" has "m" elements, the resulted relation matrix must be "*n x m*" dimensions and cell(i, j) shows the relation degree between i^{th} element of A (in line) and j^{th} element of B (in column). In addition, this degree denoted by μ_R(a,b) should be in a close interval of [0.1]. To calculate the membership degree by using membership function, new membership functions can be designed illustrated in Eq. 3.7 or minimum operator of fuzzy mathematics can be used and given in Eq. 3.8.

$$\mu_R(a,b) = (b-a)/(b+a) \tag{3.7}$$

$$\mu_R(a,b) = \min(\mu_A(a), \mu_B(b)) \tag{3.8}$$

The relation matrix using Eq. 3.7 for A = {3, 4, 5} and B = {3, 6, 7, 8} is illustrated in Fig. 3.2.

The relation matrix can be helpful to answer the question like that "what can be b where *a* ?" and the answer is given by Eq. 3.9 (Yen and Langari 1998).

$$\forall w_j = \left[Possible - A(w_j) \leftrightarrow \bigcup_{h_i} Possible - B(h_i) \cap R(h_i, w_j) \right] \tag{3.9}$$

By using given binary relations, some other relations can be built with the max–min composition. Let's assume R_1 and R_2 are fuzzy relations. R_1 is built on *AxB* and R_2 is built on *BxC*. By using max–min composition, the membership degree of *AxC* tuples like (a, c) can be calculated as presented in Eq. 3.10.

$$R_1 \circ R_2 = \left\{ \left[(a,c),\ \max_b\ \min\left(\mu_{R_1}(a,b),\ \mu_{R_2}(b,c)\right) \right] | a \in A, b \in B, c \in C \right\} \tag{3.10}$$

As can be seen in Eq. 3.6, the operations on relations are similar to matrix multiplication with the minor difference that arithmetic operations are replaced with max–min operations ("x" and "+" are replaced with "min" and "max" respectively).

3.3.3 Fuzzy Inference Systems

Fuzzy Inference System (FIS) has many synonyms in literature and most widely used forms are "fuzzy model", "fuzzy rule based system" or only "fuzzy system". FIS and fuzzy relations have many successful applications in landslide literature (Ercanoglu and Gokceoglu 2002 and 2004; Lee 2007b; Tangestani 2004 and 2009; Kanungo et al., 2008; Muthu et al., 2008; Muthu and Petrou, 2007). As its name reflects, FIS produces results (in crisp or fuzzy form) according to included fuzzy rules by using inputs. A simple fuzzy rule can be simulated as "if x is high then y is low". In this figurative rule "high" and "low" are the values of linguistic variables "x" and "y". FIS uses the collection of fuzzy rules which can be populated by using boolean operators: and, or, not. This collection is organized and stored in a *rule base* with the purpose of expert thinking modeling. As can be seen, fuzzy rules are the most important part of the FIS because of their purposes.

The FIS has an input space and it is divided into fuzzy sets so it can be said that rules of the FIS are the subset models of the input space. In other words, each rule or rule group is modeling a local behavior of the system. Some inputs may fire more than one rule or rule group, because they belong more than one fuzzy set. Normally, if the input takes place on the boundary of any fuzzy set, it may be the member of another fuzzy set with another degree. At that time, two different rules (groups) are fired and two different results representing two different local behaviors of the system are produced. This situation requires aggregating local behaviors and producing one result reflecting each local behavior according to degrees. As a result, the other part of the FIS is called *aggregator*.

The remaining parts of the FIS are *fuzzificator* for convert crisp inputs to linguistic values with some membership degrees by using membership functions. In other words, fuzzification is the assignment of the crisp input values coming from real world to the fuzzy sets with some degrees. *Fuzzy matcher* is used to calculate matching degree of rules antecedent parts and produce rules evaluation results. The final part is called *defuzzifer* and usage of it is optional. The mission of defuzzifier is conversion of fuzzy result coming from aggregator to crisp value.

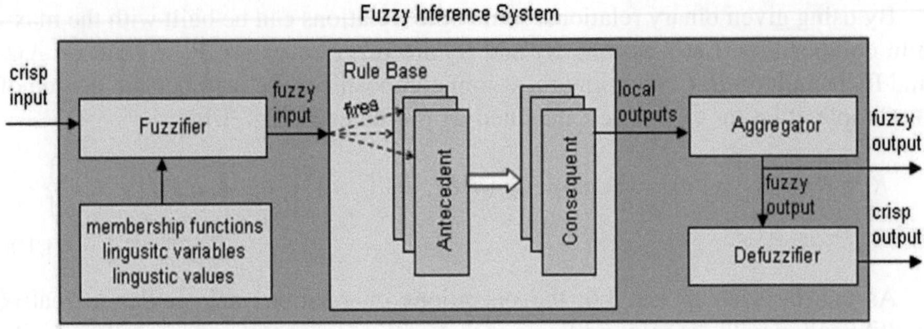

Fig. 3.3 FIS conceptual model

If crisp value is required to go on operation, defuzzifier is used. In Fig. 3.3, the conceptual model of FIS is illustrated.

All FIS models run accordingly to the block diagram in Fig. 3.3 at conceptual level. When crisp input is applied, firstly it is fuzzified *(fuzzifier)*. Secondly, rules having corresponding antecedent parts are fired and consequences of the rules are produced *(fuzzy matcher)*. Thirdly, the consequences coming from different rules are aggregated and single fuzzy result is produced *(aggregator)*. Finally and optionally, fuzzy result is converted to the crisp result *(defuzzifier)*. It is evident that fuzzy inputs of the FIS fire the rules without fuzzification.

There are three types of fuzzy inference system that have been widely employed: the Mamdani Model, the Sugeno Model and the Tusokomoto Model. The divergent parts of the models are consequent parts of the rules and normally production of consequences, aggregation and defuzzification. The Mamdani fuzzy model is perhaps the most appealing fuzzy method employed in engineering geological problems (Alvarez Grima 2000). Because of this reason it is mentioned more detailed than the others in continued parts. In addition "fuzzy model" is used instead of FIS in the where types of FIS mentioned so fuzzy model is used in the later section.

Constructing a fuzzy inference system can be carried out as follows:

1. Define the problem.
2. Choose type of fuzzy approach (inference system or relations).
3. Determine input(s) and output(s) of the system.
4. Define linguistic variables (if a FIS is constructing).
5. Determine fuzzy sets representing linguistic values for each input and output.
6. Choose membership functions.
7. Adjust parameters of membership functions with the help of experts.
8. Design fuzzy rules by using domain knowledge with the help of experts.
9. Evaluate the selected fuzzy approach in step 2.
10. Tune system if it is needed.

These steps are common for all fuzzy models mentioned above, but the step numbered 8 need to be done accordingly selected fuzzy model in step 2 because of the different types of the rules are used in different models.

3.3.3.1 Mamdani Fuzzy Model

Mamdani Fuzzy Model (Mamdani and Assilian 1975) was the first application of controlling a steam engine and boiler by using linguistic rules designed by experts. The Mamdani model uses the rules can be exampled as

$$\text{If a is } A_1 \text{ and b is } B_1 \text{ then c is } C_1$$
$$\text{If a is } A_2 \text{ and b is } B_2 \text{ then c is } C_2$$

In these exemplary rules, "A1","A2", "B_1", "B_2", "C_1" and "C_2" are fuzzy sets (linguistic terms). As can be seen, these examples uses two inputs connected with Boolean operators and single output. Normally, many variations may be organized with different Boolean operators, different input and output numbers. In fact, the attentions should be focused on the consequent part of the rule. It says that what is the linguistic value of linguistic variable should be if the antecedent part of the rule is supplied. The reason of the fact that fuzzy uses the membership degrees, the antecedent part of the rule can be supplied with some degree in [0,1]. Accordingly consequent part of the rule can be achieved with some degree reflecting the degrees of the antecedent part.

In the Mamdani Fuzzy Model, *min* operator is used for production of local results. In other words fuzzy matcher uses *min* operator to conclude achievement degree of consequent. To aggregate local consequents *max* operator is used and inferred fuzzy result is produced as illustrated in Fig. 3.4.

There are mostly four different methods for defuzzification of Mamdani model: Center of Area (COA), Mean of Maximum (MOM), Smallest of Maximum (SOM) and Largest of Maximum (LOM) and obtaining a crisp output from fuzzy result via COA and MOM is illustrated in Fig. 3.4. Continuity, disambiguity, plausibility, computational complexity and weight counting are the main criteria for selecting a defuzzification method (Hellendoorn and Thomas 1993). However, none of the defuzzification methods commonly used satisfies all these criteria (Alvarez Grima 2000). It is evident that, especially the COA method requires more computational effort than other types.

3.3.3.2 Sugeno Fuzzy Model

The Sugeno Fuzzy Model, also known as TSK Fuzzy Model, was proposed by Takagi, Sugeno and Kang (Sugeno and Kang 1988; Takagi and Sugeno 1985). The Sugeno model uses the fuzzy rule exampled as

$$\textit{If a is A and b is B then } z = f(a, b)$$

where, "A" and "B" are fuzzy sets. As like in Mamdani different variations can be done in the antecedent part of the rule. In the exemplary rule, "z" is the crisp output produced with the function of "f(a, b)". As can be seen, the Sugeno model does not use defuzzificaion process; its consequent part is designed to produce

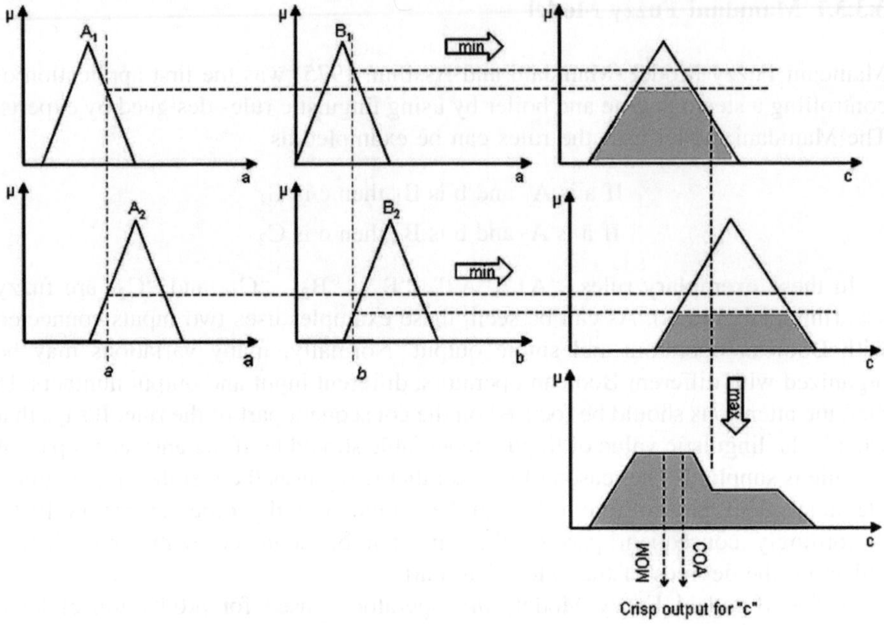

Fig. 3.4 Schematic illustration of the Mamdani model

crisp result directly. The function used in consequent part is often the polynomial which uses input variables, but this is not an obligation. New function forms can be designed to reflect the matching degree of antecedent parts of the rule, for example "z = f(a,b)" may be "z = a − b + 1". In the Sugeno model, after production of all local consequences; the aggregation process is implemented with either weighted average or weighted sum of them.

3.3.3.3 Tsukamoto Fuzzy Model

Tsukomoto fuzzy model was proposed by Tsukomoto (1979) and it uses the rules which can be exampled as

If a is A then b is B

where, "A" and "B" are fuzzy sets. It is similar with the rule structure of the Mamdani model. Accordingly, the Mamdani and the Tsukomoto models are similar in the processes from fuzzification to production of local consequents. However, the Tsokomoto model uses weighted averages of local consequents instead of the defuzzification process.

3.3.4 Landslide Susceptibility Applications of Fuzzy Approach

In literature, the first fuzzy approach application on landslide susceptibility assessment has been performed by Juang et al. (1992). The Mamdani, Sugeno and Tsukamoto models have not been applied on the landslide susceptibility assessment directly. However, the fuzzy functions and expert systems including fuzzy "if-then" rules are used extensively in the last decade. Binaghi et al. (1998) used certainty factor and fuzzy Dempster-Shafer approaches to produce landslide susceptibility maps. According to Binaghi et al. (1998), the apparatus of fuzzy logic serves as a natural frame for modeling the gradual transition from membership to nonmembership in intrinsically vague classes. The Dempster-Shafer theory based on the concept of belief function may be used to model and quantify the subjective credibility induced by partial evidence (Swets 1988). Identification of data layers governing landslides, definition of linguistic terms qualitatively describing each data layer class, definition of membership functions associated to the corresponding fuzzy set, definition of susceptibility classes, and elicitation and generation of the decision rules by examples selected by the expert are the complete fuzzy Dempster-Shafer modeling. Applying the fuzzy framework, the data layers involved in the classification process have been qualified by linguistic labels according to the expert description. The labeling process produced crisp and fuzzy labels, quantified with crisp membership functions, for data layer classes having nominal values, and standard quadratic membership functions, for data layer classes having numerical values (Binaghi et al. 1998). Binaghi et al. (1998) applied fuzzy Dempster-Shafer approaches to Fabriano (Marche, Italy) region and they obtained acceptable results. According to these authors (Binaghi et al. 1998), this approach has great potential in dealing with landslide susceptibility zonation. Similarly, a study on the landslide susceptibility mapping by the Dempster-Shafer and fuzzy models was performed by Tangestani (2009). Working in GIS with map layers, generally the attribute of interest is measured over discrete intervals, and the membership function can be expressed as a table expressing as a table relating map classes to membership values. Fuzzy logic is attractive because it is straightforward to understand and implement (Tangestani 2009). The fuzzy membership values are selected based on subjective judgment about the relative importance of the map classes by Tangestani (2009). For this purpose, Anbalagan's (1992) classification scheme was used. According to Tangestani (2009), the gamma fuzzy logic approach provides a flexible method with which to include an expert's opinion in developing an inference network, while the Dempster-Shafer belief model uses the combination rules in which the expert only can assign the belief functions for each factor map class. Tangestani (2009) pointed out that comparison of fuzzy output susceptibility map and the plausibility image of the Dempster-Shafer model with the field criteria collected in the known landslides map revealed that the fuzzy gamma approach with a value of 0.94 for λ gives more reasonable results.

Ercanoglu and Gokceoglu (2002) used a rule-based fuzzy approach to produce landslide susceptibility map of a landslide-prone area from Turkey. Ercanoglu and

Gokceoglu (2002) constructed the "if-then" rules and fuzzy sets representing each landslide conditioning factor from the data. Each landslide conditioning factor was divided into five linguistic variables by Ercanoglu and Gokceoglu (2002). In this fuzzy system, some fuzzy sets are continuous form while some others are in the discrete form. These different types of fuzzy sets were combined in a FIS by if-then rules. The prediction performance of the produced landslide susceptibility maps is very high. In other words, 93% of the landslides are located in high and very high zones. According to Ercanoglu and Gokceoglu (2002), the results of the procedure followed by the authors produce acceptable maps for a medium-scaled landslide susceptibility assessments.

Tangestani (2004) employed the fuzzy gamma approach to assess regional landslide susceptibility assessment. According to Ercanoglu and Gokceoglu (2002) and Tangestani (2004), all the available methods for regional landslide assessment have some uncertainties arising from lack of knowledge and variability. This is because regional landslide assessments require some generalizations and simplifications, although these assessments are complex. For this reason, a perfect assessment method for landslide susceptibility does not exist. The fuzzy logic introduced by Zadeh (1965) is one of the tools used to solve these complex problems. Performing the fuzzy gamma approach, Tangestani (2004) produced landslide susceptibility map of Kakan catchment from Iran. After assigning each fuzzy membership values for the conditioning factors, he used the fuzzy gamma operators and performed a combination each factor to produce the landslide susceptibility map. In fact, the methodology followed by Tangestani (2004) is very useful and applicable for such type problems. The approach employed by Tangestani (2004) provides a flexiable and transparent method by which to include and expert opinion in developing an inference network. Also, integration of fuzzy combination operators and GIS facilitates enables the expert to examine different scenarios and produce and intermediate map, or to add any new data layer to the model and test its effect on the final possibility map (Tangestani 2004). In this methodology, the most critical point is the gamma values to be assigned for each gamma combination operation because this has a crucial importance on the output fuzzy membership functions. The other important study on the fuzzy algebraic operators for producing landslide susceptibility map was performed by Lee (2007b). Lee (2007b) used similarly fuzzy gamma operators with Tangestani (2004) to produce landslide susceptibility map of the Gangneung area from South Korea. Lee (2007b) extracted the spatial relationships between each conditioning factor and fuzzy membership values. In addition, the different λ values were assigned and the factors were combined by the gamma operators. Tangestani (2004) used directly expert's opinion when assigning fuzzy membership values while Lee (2007b) extracted from the fuzzy membership values from the actual data. This is the main difference between the studies performed by Tangestani (2004) and Lee (2007b). This also show that the flexibility of the use of the fuzzy approach when producing landslide susceptibility maps. According to the verification process performed by Lee (2007b), the case of applying the gamma operator ($\lambda = 0.975$), showed the best accuracy (84.68%), whereas the fuzzy and (66.79%)

and fuzzy or (66.50) operators showed the worst accuracy. However, an interesting conclusion on the performance of the landslide susceptibility map methodology was carried out by Lee (2007b). According to Lee (2007b), in his study, the data-derived model (frequency ratio) and the knowledge-derived model (fuzzy logic) were combined. As a result, the combined data and knowledge derived model is useful for landslide susceptibility mapping considering the prediction accuracy (Lee 2007b). Another study considering fuzzy relations were performed by Ercanoglu and Gokceoglu (2004). Ercanoglu and Gokceoglu (2004) employed cosine amplitude method. The cosine amplitude method is one of the commonly used similarity methods, is used to evaluate the relationship between the landslide occurrence and the parameters considered (Ercanoglu and Gokceoglu 2004). In this approach, the membership degrees of elements for each fuzzy set are calculated by the strength of relationship between the landslide inventory and the conditioning parameters. During this calculation process, Ercanoglu and Gokceoglu (2004) considered Cartesian product and fuzzy max operator. Finally, the data produced from the Cartesian product are combined by the fuzzy max operator and the landslide susceptibility map was produced for a landslide prone area from Turkey. The results showed that 86% of the areal extent of the actual landslides exists in the high or very high susceptibility zones of the final susceptibility map. Another study performed by Kanungo et al. (2008) considers the cosine amplitude approach. According to Kanungo et al. (2008) the fuzzy set based approach using cosine amplitude similarity procedure could bring out the relative importance (ratings) of different categories of causative factors in terms of landslide occurrences in an unbiased manner. Kanungo et al. (2008) also performed the landslide risk assessment by fuzzy linguistic rules. The linguistic rules proposed by Anbalagan and Singh (1996) were considered. However, these rules are qualitative such as very low, low, moderate, high, very high. The qualitative risk matrices produced by Kanungo et al. (2008) were quantified in terms of fuzzy membership values. As mentioned by Ross (1995), in the fuzzy set theory, membership values of elements are computed in [0, 1] closed interval depending upon varying degrees of support or confidence of a phenomenon. There are several ways of computing membership values including Cartesian product, closed-form expression, linguistic rules of knowledge and similarity methods in data manipulation. However, the membership values to various categories of landslide susceptibility zones and resource elements were determined on the basis of a linguistic scale derived from expert knowledge (Kanungo et al. 2008). Kanungo et al. (2008) produced four different landslide susceptibility maps by conventional weighting approach, ANN black-box approach, fuzzy set based approach and combined neural and fuzzy approach. The most reliable methods were obtained from the combined neural and fuzzy approach.

Muthu and Petrou (2007) constructed an expert system for landslide susceptibility assessment of Caramanico Terme from Italy. The fuzzy "if-then" rules used by Muthu and Petrou (2007) were extracted from the existing topographical and remote sensing data. In fact, Muthu and Petrou (2007) developed an alert system using a rule-based expert system. They used 16 landslides to the test system. Muthu and Petrou (2007) considered the most relevant combination of maps in accordance with the historic landslide activity period to produce the outputs. When comparing the other fuzzy-based studies, the study is different because the results of this study are concrete for the everyday human life. In fact, the results of the other landslide susceptibility maps will contribute to decrease the losses sourced from the landslides. However, this contribution is indirect. After preparation of reliable landslide susceptibility maps, it is possible to produce some alert maps based on fuzzy expert systems. The results of the study performed by Muthu and Petrou (2007) show that the fuzzy expert systems are the useful tool for these purposes. In the same region (Caramanico Terme, Italy), Muthu et al. (2008) performed a study to produce landslide susceptibility map by fuzzy approaches such as the fuzzy expert system and the fuzzy neural networks. The rules were extracted by the expert opinion. Muthu et al. (2008) fuzzified all conditioning parameters in a form that takes into consideration the membership function with which each factor belongs to the class governing the landslides. According to Muthu et al. (2008), the proposed method is meant to operate at the regional, rather than at the local level, where accurate ground measurements are not possible because they are labor intensive. However, although some uncertainties sourced from the physical conditions and high complexity, the fuzzy expert system constructed by Muthu et al. (2008) exhibits a very good performance (75%). This means that the fuzzy expert system produces acceptable and useful map.

3.4 Artificial Neural Networks (ANN)

ANN is the most popular branch in the machine learning domain. Machine learning aims to train computers, or in other words computers can learn from examples and experiences if machine learning mechanisms are used. ANN can be defined as structures comprised of densely interconnected adaptive simple processing elements that are capable of performing massively parallel computations for data processing and knowledge representation (Basheer and Hajmeer 2000; Hetch-Nielsen 1990; Schalkof 1997) and it has been applied many real world problems in many different areas (Basheer and Hajmeer 2000). The attractiveness of the ANN comes from the remarkable information processing characteristics of the biological system such as nonlinearity, high parallelism, robustness, fault and failure tolerance, learning, ability to handle imprecise and fuzzy information and their capability to generalize (Jain et al. 1996).

3.4.1 Understanding Biological Neural Networks and ANN Relation

Biological nervous system consists of neurons with different types. Each neuron is divided into three main parts: axon-synapse, cell body and dendrites. Dendrites receive electrical signal from other neurons and pass it to the cell body, namely

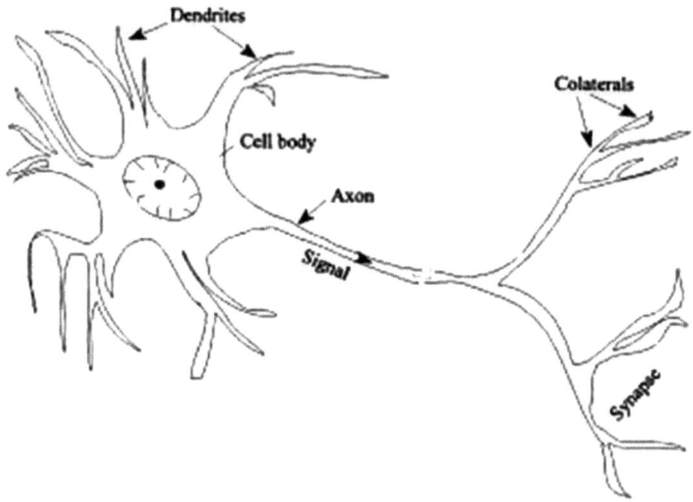

Fig. 3.5 Details of the biological neurons (Basheer and Hajmeer 2000)

dendrites are the receivers of the neurons. The cell body has a nucleus that contain information about heredity traits and a plasma that holds the molecular equipment used for producing the material needed by the neuron (Jain et al. 1996). The axon carries electrical signals to the dendrites of other neurons through synapses; in other words they transmit the output of the neuron to the other neuron. Figure 3.5 (Basheer and Hajmeer 2000) illustrates the details of the biological neurons. The biological neuron has large number of input receivers (dendrites) and output transmitters (synapses), it can process many signals at the same time. These signals can be suppressed or reproduced. It can be considered as weights are assigned to the axons and dendrites.

This type of signal transition between neurons and collaboration of the parts of the neuron constitute the basics of the ANN model. Both the biological nervous network and ANN learn by incrementally, adjusting the magnitudes of the weights or synapses' strengths (Zupan and Gasteiger 1993).

3.4.2 Understanding ANN principles

An ANN consists of neurons which are similar to biological neurons in the brain. In other words, a neuron is the simplest processing element of an ANN. It has one or more receivers and takes the inputs via them in the weighted form and one or more outgoing connection. In fact, ANN means many of interconnected neurons which take inputs from external environment or other neurons' outputs. In Fig. 3.6, the typical architecture of an ANN consisting of three layers can be seen: input layer, hidden layer and output layer. Input layer consists of the neurons

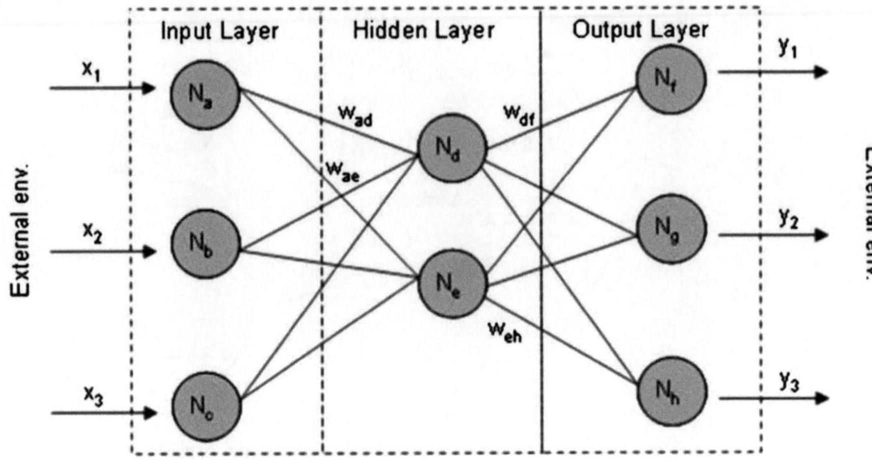

Fig. 3.6 The typical architecture of an ANN consisting of three layers

taking the input values from external environment and transmitting the output values to the other neurons. Hidden layer consists of the neurons taking the inputs from previous neurons and transmits the outputs to the succeeding neurons again. Finally, output layer takes inputs from hidden layer neurons and gives the output values to the external environment. Input and output layers of the ANN model can be used as one hidden layer. However, an ANN model can include one or more hidden layers. For example, an ANN can have one input, one output and two hidden layers. As the hidden layers are not visible from the external environment, in fact, they are hidden at the same time.

As can be seen in Fig. 3.6, the receiver and outgoing connections of the neurons enables them to be interconnected and all connections can be called as link. The link has tree component: source, destination and weight. In other words links are directed and weighted. The weight of the link shows the importance of the input on the link for the destination (neuron). In fact, the learning mechanism of the ANN is based on the adjustments of the weights to produce required outputs and actually this process is repetitive (iterative).

Up to now, the general concepts and components of the ANN are introduced and it can be concluded from introduction that the construction of the ANN model starts with the selection of the architecture, in other words, number of the neurons, layers, connections and the learning mechanism of the neurons should be determined.

3.4.3 Perceptron

Perceptron was introduced by Rosenblatt (1958) and the perceptron learning rule (Rosenblatt 1960) was proposed by the same researcher. The perceptron reflects atomic characteristics of the ANN. It employs single neuron with weighted inputs, linear combiner mechanism and a threshold value. In fact, the mission of the perceptron is classification of inputs by using the threshold value. Typical perceptron divides the inputs into the two classes according to be above or below of the threshold. In some resources (Negnevitsky 2002) threshold is used synonym of the hard limiter. The linear combiner mechanism of the perceptron which has "n" weighted input is illustrated in Eq. 3.11 (Basheer and Hajmeer 2000). It can be named as activation or transfer function (Negnevitsky 2002).

$$y = \begin{cases} 1 & \sum_{i=1}^{n} x_i * w_i \geq treshold, \\ 0 & \sum_{i=1}^{n} x_i * w_i \prec treshold \end{cases} \tag{3.11}$$

where 0 and 1 indicate "off" and "on"or "to be in class A_1" and "to be in class A_2".

The learning rule of the perceptron is based on weight adjustment policy applied for reduction of the difference between produced output and desired output. If y_m is the output in the mth iteration, "p" indicates produced output and "d" indicates desired output, the error in m^{th} iteration is calculated by Eq. 3.12 (Negnevitsky 2002).

$$e_m = y_m(d) - y_m(p) \tag{3.12}$$

According to Negnevitsky (2002), initial weights of the perceptron are randomly selected from the range of [−0.5, 0.5]. Inputs of the perceptron are externally supplied and unchangeable. As a result, the perceptron can adjust the only weights to reach desired output with acceptable error with iterative approach. If error in m^{th} iteration is negative then produced output should be decreased or if it is positive then produced output should be increased by using w_i parameters in Eq. 3.11. Thus, the perceptron learning rule can be illustrated as like in Eq. 3.13 (Negnevitsky 2002).

$$w_i(m+1) = w_i(m) + (\alpha * x_i(m) * e(m)) \tag{3.13}$$

where, "m" is the number of iteration, x_i is the ith input of the perceptron, "e" denotes error and "α" is learning rate and positive constant less than 1.

It is evident from the Eqs. 3.11–3.13 that the perceptron is suitable for the input spaces which can be divisible with a line because of its learning rule, namely linear combiner. This situation causes the strictly restriction to solve all problems with the perceptron, because most of the problems are not suitable for only linear classification. To overcome the limitation of linear separation, multilayer perceptron (MLP) is introduced. MLP is synonym of the multilayer neural network. As its name reflects, MLP uses more than one layer, in fact, three layers at least: one input layer, one hidden layer and one output layer. Figure 3.6 can be an example for a typical MLP and as mentioned before, hidden layer may be more than one. In fact, hidden is the term that can qualify node, neuron and layer and means the invisible part of the ANN from external viewpoint. The aim of the MLP is to give a solution to the nonlinear problems and it tries to characterize the nonlinearity by using multiple layers at the same time.

3.4.4 ANN Types and Terminology

Basheer and Hajmeer (2000) lists the criteria for classifications of ANNs as follows:

- The function that the ANN is designed to serve (pattern classification, clustering, etc.)
- The degree (partial/full) of connectivity of the neurons in the network
- The direction of flow of information with in the network (recurrent/non recurrent)
- The type of learning algorithm
- The learning rule
- Degree of learning supervision needed for ANN training (supervised/unsupervised)

Full/partial connectivity: If each neuron in one layer is connected to the all neurons in the next adjacent layer then connectivity is named as full. In counter situation, the connectivity is named as partial.

Recurrent/non recurrent: In a recurrent network, the outputs of some neurons fed back to the same neurons or neurons in preceding layers (Basheer and Hajmeer 2000). As the information can flow in both sides of forward and backward; system intelligence and learning capability are increased.

Supervised/unsupervised learning: In supervised learning correct outputs or in other word target outputs are used in training process. Weight adjustment is applied according to error between produced and target outputs. In unsupervised learning, synonym with self organization; instead of target outputs, structure of data exploration is employed. Different cases in data and their internal relations are used to organize clusters according to their similarity. Both of learning scheme can be used in hybrid models.

Learning algorithm and learning rule: Learning algorithm represents a set of systematic equations that utilize the outputs from the network along with an arbitrary performance measure to update the internal structure of the ANN (Basheer and Hajmeer 2000). Learning rule is the heart of the learning algorithm. Because it defines the weight adjustment policy while passing between iterations. There are four basic types of learning rules (Haykin, 1994; Hassoun, 1995).

Error correction learning (ECL): It is used in supervised learning in which the arithmetic difference, namely error, between the produce output from ANN at any iteration during training and the corresponding target output is used to modify the connection weights so as to gradually reduce the overall network error (Basheer and Hajmeer 2000).

Boltzmann learning (BL): It is similar to ECL; however, each neuron produces the output based on Boltzmann statistical distribution (Jain et al. 1996).

Hebian learning (HL): It suggests (Hebb 1949) that if neurons on both sides of synapse are activated synchronously and repeatedly the synapse's strength is selectively increased (Basheer and Hajmeer 2000).

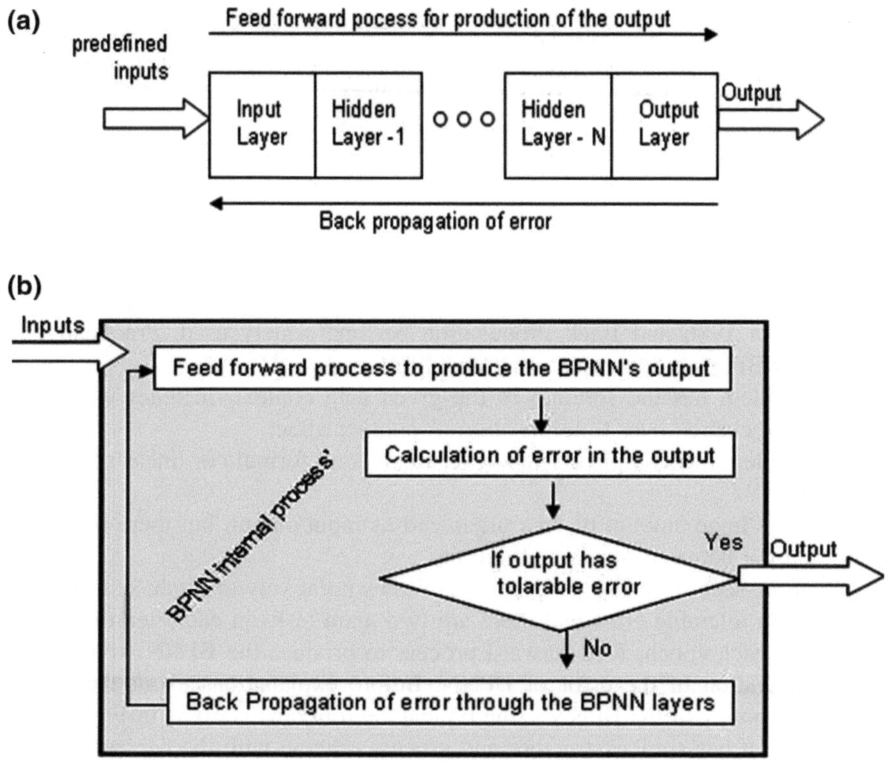

Fig. 3.7 The general flow of the BPNN's running model

Competitive learning (CL): It forces neurons to compete among themselves to be one activated neuron in a given iteration with all weights attached to it adjusted (Jain et al. 1996)

As can be seen, ECL and BL rules use all neurons with weights; however, HL rules use locally neurons and CL rules select one neuron.

3.4.5 Back Propagation Neural Network (BPNN)

To give a detailed example from ANNs, BPNN is selected because of its popularity, flexibility and adaptability in modeling a wide spectrum of problems in many application areas (Basheer and Hajmeer 2000). BPNN is the type of ANN with the characteristics of full connectivity, at least three layers (typical MLP) and supervised learning with ECL rule. Back propagation means error calculated at the output level of the BPNN in the any iteration is backed to the previous layer, namely hidden layer, and it continues to the input layer. The general flows of the BPNN's running model can be illustrated as like in Fig. 3.7(a) and alternatively in

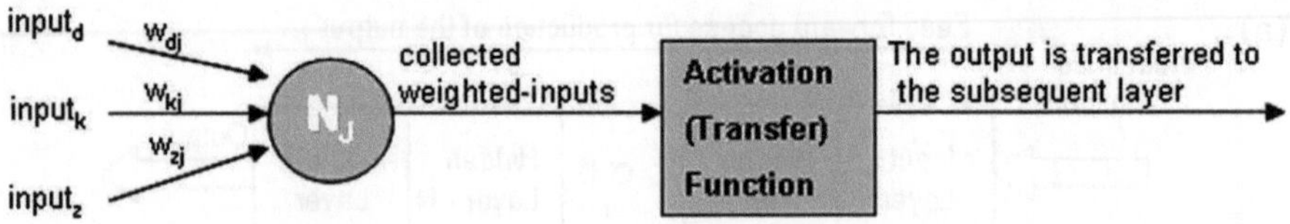

Fig. 3.8 The typical neuron view of the BPNN

Fig. 3.7(b). Paul Werbos proposed the Back Propagation algorithm in the first half of 1970's; however Rumelhart and McClelland (Rumelhart et al. 1986) rediscovered it in 1986 and Back Propagation became widely used. Practically the reasons for BPNN employment may be as follows:

The problem has the solution in the given data context. In other words, the solution can change over time, location or another affect.

The problem has complexity; however there is no formula or linear solution to model it.

There are huge amount of data organized as input/output, but there is no idea about relation between inputs and outputs.

As can be seen from Fig. 3.7, the BPNN runs iteratively to produce acceptable outputs with tolerable error and there are two main tasks in each iteration, alternatively in each epoch, feed forward process to produce the BPNN's output and back propagation of the error of BPNN. Before explanations about the weight adjustment policy of the BPNN's, the typical neuron view is given in Fig. 3.8. As can be seen it has similar structure and process scheme with the perceptron.

The BPNN should be initialized before the first iteration by assigning the initial values to the weights and thresholds of the each neuron. These values are randomly selected from a small range with the consideration of uniformly distribution (Haykin 1994). The complete one iteration of the BPNN is explained after this point with the formulas of weight adjustment which are taken from Negnevitsky (2002).

3.4.5.1 Feed Forward Processes of the BPNN

After supplying the external inputs to the input layer, all subsequent layers with the no matter of they are hidden or output layer behave in the same way. However, only the outputs of the output layer are visible and the error calculation starts at this level. For each neuron in the one layer; firstly collected weighted inputs is calculated by using Eq. 3.14, secondly the activation function of the neuron produces the real output which is input for the subsequent layer. The activation function constraints the output to be in the closed interval of [0, 1] and sigmoid activation function (Eq. 3.15) is suitable for BPNNs because of its derivation is easily computed.

$$X = \sum_{i=1}^{n} x_i * w_i - \Theta \tag{3.14}$$

$$Y^{sigmoid} = \frac{1}{1 + e^{-x}} \tag{3.15}$$

where "X" is collected weighted-inputs of the neuron; "θ" is the threshold, "n" is the number of inputs coming to the neuron, "w" denotes the weight and "Y" is the output of the neuron. After completing the each layer neuron by neuron, then in fact, last layer, namely the output layer, gives the BPNNs output to the external environment.

3.4.5.2 Calculation of error

The error calculation (Eq. 3.16) is carried out for each neuron in output layer and each neuron propagates its own error.

$$e_o = Y_{d,o}(m) - Y_o(m) \tag{3.16}$$

where "e" is the error, "m" is the iteration number, "o" is the oth neuron of the output layer, "Y" is the output and "Y_d" denotes desired (target) output.

3.4.5.3 Back Propagation Processes of the BPNN

Back propagation has the skeleton consisting of three items: "calculate error gradient", "calculate weight correction" and "produce new weight". As can be seen, it is very focused on "weights", and so objects of this step are the links sourced from one neuron to another. In Eq. 3.13, the weight adjustment of the perceptron has been given. In fact, the same philosophy lies behind the weight adjustment policy of the BPNN's neurons, but same equation is not suitable for the network. Because, there are non-predefined inputs for the layers except for the input layer and the connection of the neuron to the all of the neurons in subsequent layer makes the error concept to be more complex.

In BPNNs weight adjustment is done similarly with the perceptron but the weight correction (Δw) is calculated differently in the way of using error gradient (δ) instead of error (e). In all formulas after that; "w" has two subscribing symbol: first is the source and second is the target of the link which has the "w", "Y" (Eq. 3.15) is the output of the neuron which subscribes it and X (Eq. 3.14) is the collected weighted inputs of the neuron which subscribes it. The Eq. 3.17 gives the weight learning rule and the Eq. 3.18 gives the weight correction of the BPNN.

$$w_{ij}(m+1) = w_{ij}(m) + \Delta w_{ij}(m) \tag{3.17}$$

$$\Delta w_{ij}(m) = \alpha * Y_i(m) * \delta_j(m) \tag{3.18}$$

As can be seen from Eq. 3.18; learning rate (α), input of the jth neuron (Y_i) and the error gradient of jth neuron (δ_j) are used. Error gradient is calculated with the derivation of the activation function multiplied by the error at the neuron output. In Eq. 3.19, error gradient formula is given.

$$\delta_j(m) = \frac{\partial Y_j(m)}{\partial X_j(m)} * e_j(m) \tag{3.19}$$

If the sigmoid activation function Eq. 3.15 is used, Eq. 3.19 can be as like in Eq. 3.20

$$\delta_j(m) = \frac{\partial\left\{\frac{1}{1+\exp\left[-X_j(m)\right]}\right\}}{\partial X_j(m)} * e_j(m) = \frac{\exp\left[-X_j(m)\right]}{\left\{1+\exp\left[-X_j(m)\right]\right\}^2} * e_j(m) \tag{3.20}$$

While the Eq. 3.20 is evaluated, Eq. 3.21 is obtained with the consideration of the Eq. 3.22.

$$\delta_j(m) = Y_j(m) * \left[1 - Y_j(m)\right] * e_j(m) \tag{3.21}$$

$$Y_j(m) = \frac{1}{1+\exp\left[-X_j(m)\right]}. \tag{3.22}$$

Up to now, the weight correction skeleton of the BPNNs based on the formulas is given. In fact, all formulas are same for all neurons in the all layers in basic form. However, there are minor differences between layers since input layer has predefined inputs but others have not. In addition the output layer has simple calculation for error but others have not because of their fully connection to the other neurons in the subsequent layer. For this reason, additional formulas reflecting these differences are given below. The error gradient of the jth neuron in the output layer is calculated with Eq. 3.19 and the error gradient of the jth neuron in the preceding layers is given in Eq. 3.23.

$$\delta_j(m) = Y_j(m) * \left[1 - Y_j(m)\right] * \sum_{k=1}^{l} \delta_k(m) * w_{jk}(m) \tag{3.23}$$

The weight calculation formula of the jth neuron in the hidden layer which has the input layer previously is given in Eq. 3.12 and it is replaced in Eq. 3.23 again.

$$\Delta w_{ij}(m) = \alpha * x_i(m) * \delta_j(m) \tag{3.24}$$

It is evident that the differences can be seen when making a comparison between Eqs. 3.21 and 3.23, and Eqs. 3.18 and 3.24.

3.4.6 Constructing an ANN

In fact, it is too hard for the researchers creating and running an ANN firstly, but this process is aided by some software tools (for ex. MATLAB) to encapsulate researchers from too many details of the construction of the ANN. Accordingly, there are two way to explain this process: based on a tool or not. In this section, no tool is placed to the object and the construction process is explained, generally; however, in practically researchers use a tool. Even if the tool is used, accurate and efficient usage requires general knowledge about the construction of the ANN. The construction steps of an ANN are explained in subsequent parts.

Problem analysis: Input and output parameters of the problem should be determined and dependency relation between them should be examined. Input and/or output parameters of the problem can be distinct and/or continuous and they can be used together in the ANN architecture.

Data preprocessing: The most important steps of the data preprocessing are the data balancing and the normalization of the data. Data balancing means samples of data, representing the different variations, should have nearly uniform distribution. In other words, over-representation of any variation should be prevented. Data normalization means values should be scaled within a uniform range, for example [0–1]. It enables small values to be prevented from being suppressed by larger values.

Data set partitioning: There must be three distinct data sets for the ANN construction: training, test and validation. There are two important properties about sets: characteristics and size of them. Their characteristics can be notable from their definitions. Training data is responsible from enabling to the ANN to have correct weights. Hence, it should cover the all cases about the problem. Test data does test the performance of the ANN using untrained data. So, the accordance of the test and the training data sets in the perspective of cases is very important. According the test results, namely ANN performance, architectural decisions are reconsidered and if it is needed more training cycles are applied. Validation data set is supplied to the ANN finally, before it is completed. Validation enables to ensure from that it is the suitable and adjusted ANN. Therefore, it should cover different cases from train and test data sets.

In fact, the determination of the set size for each of them is not a mathematical problem and there are many advises and experiences about the size concept. Sonmez et al. (2006) have given a brief overview on this subject. Kavzoglu (2001) pointed out that a sufficient number of training samples being available is important to estimate ANN parameters accurately and same researcher suggested that the optimal number of training samples must be between [30 × numbers of input nodes × (numbers of input nodes + 1)] and [60 × numbers of input nodes × (numbers of input nodes + 1)]. Klimasauskas (1993) and Messer and Kittler (1998) proposed that at least 5–10 times the number of training samples as free parameters (weights) should be used. Swingler (1996) and Looney (1996) propose 20 and 25% of the data for testing, respectively Nelson and Illingworth (1990) recommend between 20 and 30% of data for testing.

ANN Design and Training: This step starts with selection of the ANN type and learning rule which have the best accordance with the problem and goes on with assignment of the values to the parameters of the ANN architecture. Parameter values have influence on the ANN performance and they can be listed as follows:

Initial weights: Initial weights are the first values assigned to the weights randomly from a small range. Basheer and Hajmeer (2000) pointed out that too small a range can cause to small error gradients and this situation may slow down the initial learning process. Also, Fahlman (1988) emphasized that weight initialization has an insignificant effect on the final ANN architecture. There are proposed different ranges in the literature and some of them are listed here: [−0.5; 0.5] by Sietsma and Dow (1991), [−0.3; 0.3] by Rumelhart et al. (1986), [−0.25; 0.25] by Gallahger and Downs (1997) and [−0.1; 0.1] by Paola (1994).

Back propagation learning rate (α): Learning rate controls the weight update step size. If the learning rate is selected as small, the training rate can be slow, however; the training phase can cause oscillations when learning rate is selected too large (Sonmez et al. 2006). A constant learning rate may be utilized throughout the training process (Basheer and Hajmeer 2000). There are some recommendations in literature about learning rate such as $\alpha = 0.1$–10 by Wyhthoff (1993), $\alpha = 0.3$–0.6 by Zupan and Gasteiger (1993) and $\alpha = 0.0$–1.0 by Fu (1995). In addition Sonmez et al. (2006) selected α as 0.01 and 0.1.

Back propagation momentum coefficient: A momentum term is commonly used in weight updating to help the search escape local minima and reduce likelihood of search instability (Haykin 1994). The momentum coefficient has a stabilizing effect in the back-propagation algorithm (Negnevitsky 2002). Some momentum coefficient values which are advised and experienced in literature are in the between 0.4–0.9 by Wyhthoff (1993), 0.0–1.0 by Hassoun (1995) and Fu (1995), and Sonmez et al. (2006) selected as 0.95.

Number of training cycles: System should be refrained from overlearning caused by application of too many training cycles, because the system looses the generalization ability. The error on the test data may not follow a smooth path, the onset of a major increase in the error is considered to represent the optimal number of cycles for the ANN architecture (Basheer and Hajmeer 2000). Alternatively, the point which the prediction performance graphs of train and test data sets should be closed to each other, is the optimal number of cycles for the ANN architecture.

Training modes: There are two training modes available for ANNs: example by example training (EET) and batch training (BT). The usage of them together is possible. EET means, as its name imply, network starts to learn in the first example, namely it starts to weight adjustment in the first example and goes on with succeeding examples. BT means, all examples are used in one batch and learning starts with the first iteration on the whole batch. In the second iteration, again the same batch is used and goes on in this way. The effectiveness of the EET or BT can be problem specific (Haykin 1994)

Hidden layer size: Hidden layer size means the number of layers and the number of neurons in each layer. Commercial ANNs incorporate three and sometimes four layers, including one or two hidden layers and each layer can contain from 10 to 1000 neurons (Negnevitsky 2002). Of course more layers with more and more neurons can be used in an ANN theoretically, but each layer consumes additional computational effort practically. As a result, the imagination of the ANN's is limited with the power of computational environment and number of data available.

Hecht-Nielsen (1987) point out that one hidden layer may be sufficient to solve most problems. Also, Rumelhart et al. (1986) stated that there is rarely an

Table 3.1 The heuristic proposed for the number of neuron to be used in hidden layer(s) (Ni: Number of input neuron, No: number of output neuron) (Sonmez et al. 2006)

Heuristic	Reference
$\leq 2 * N_i + 1$	Hecht-Nielsen (1987)
$3N_i$	Hush (1989)
$(N_i + N_o)/2$	Ripley (1993)
$\frac{2 + N_o * N_i + 0.5 * (N_o^2 + N_i) - 3}{N_i + N_o}$	Paola (1994)
$2N_i/3$	Wang (1994)
$\sqrt{N_i * N_o}$	Masters (1994)
	Kaastra and Boyd (1996)
$2N_i$	Kannellopoulas and Wilkinson (1997)

Table 3.2 Effect of extreme values of design parameters on training convergence and network generalization (Basheer and Hajmeer 2000)

Design parameter	Too high or too large	Too low or too small
Number of hidden nodes (NHN)	Overfitting ANN (no generalization)	Under fitting (ANN unable to obtain the underlying rules embedded in the data)
Learning rate	Unstable ANN (weights) that oscillates about the optimal solution	Slow training
Momentum coefficient	Reduces risk of local minima. Speeds up training. Increased risk of overshooting the solution (instability)	Suppresses effect of momentum leading to increased risk of potential entrapment in local minima. Slows training
Number of training cycles	Good recalling ANN (i.e., ANN memorization of data) and bad generalization to untrained data	Produces ANN that is incapable of representing the data
Size of training subset	ANN with good recalling and generalization	ANN unable to fully explain the problem. ANN with limited or bad generalization
Size of test subset	Ability to confirm ANN generalization capability	Inadequate confirmation of ANN generalization capability

advantage in using more than one hidden layers, while Masters (1994) emphasized that two hidden layers may be necessary for a learning rule with discontinuities. Sonmez et al. (2006) states that the number of neurons is the most critical task in the ANN structure and same researchers give the Table 3.1 listing heuristics proposed for this purpose.

Basheer and Hajmeer (2000) said that parameters of the ANN listed in Table 3.2 should not be set too high or too low and they are optimized. The Table 3.2 lists the effect of selection parameter's values from extreme points.

When the construction of the ANN is completed, different test and train data sets should be supplied to the model, and the results are compared by various performance indices.

3.4.7 Use of Artificial Neural Networks in Susceptibility Assessments

In the last decade, the artificial neural networks (ANN) became a popular method for producing landslide susceptibility maps. The ANN has many advantages compared with statistical methods. Firstly, the ANN method is independent of the statistical distribution of the data and there is no need of specific statistical variables. Compared with the statistical methods, ANN allows the target classes to be defined with much consideration to their distribution in the corresponding domain of each data source (Zhou 1999; Lee et al. 2004). In addition, integration of GIS data is suitable. Moreover, the continuous and categorical data can be used together. For this reason, several researchers have used the ANN to produce the landslide susceptibility maps. In this section, a brief overview is carried out on the landslide susceptibility maps produced by the ANN. One of the first ANN applications encountered in the literature was performed by Lu and Rosenbaum (2003). Lu and Rosenbaum (2003) employed the ANN to predict the stability of slopes. According to the authors (Lu and Rosenbaum 2003), the ANN appears to be appropriate where significant quantitites of data are available. In this study, the limit equilibrium method (LEM) was adopted in practice for the analysis and prediction of circular failures. The results obtained by Lu and Rosenbaum (2003) showed that using ANN appeared to be in agreement with the observed condition of the slopes employed, encouraging the adoption of the ANN as a basis for predicting the state of slope instability. Additionally, the need to evaluate potentially unstable slopes across an area would tend to favour the spatial framework offered by a geographic information system (Lu and Rosenbaum 2003). This increases the advantage of using ANN over limit equilibrium method and maximum likelihood estimation, not least because of the reasoning outputs that can be generated as raster images (Lu and Rosenbaum 2003). These images provide the landslide susceptibility zonation and contribute to landslide hazard mitigation efforts. Lee et al. (2003a) used the ANN for analysis of landslide susceptibility of a landslide-prone area in South Korea. Researchers (Lee et al. 2003a) combined the GIS and the ANN to produce the landslide susceptibility map. Various ANNs exist in the literature. Lee et al. (2003a) considered the back propogation neural network (BPNN) because the BPNN is the most frequently used ANN algorithm as mentioned in the previous sections. One of the advantages of the ANN is that accurate analysis is possible when there are few training datasets, because of the use of pixel-based calculations (Lee et al. 2003a). In the study of Lee et al. (2003a), the weights were calculated three times, using all 14 landslide conditioning factors in the first instance, then recalculating after removal of those 6 factors that had the smallest weights, and thirdly after removal of the remaining 4 least influential factors. The ANN structure included 14 input variables, 30 hidden neurons and 2 outputs. The results obtained by Lee et al. (2003a) demonstrated that the GIS and the ANN can be used to produce a landslide susceptibility index and, consequently, to manage landslide hazards effectively. Moreover, the weighting given to the various factors that are significant in the landslide susceptibility analysis provides a ranking of their relative significance (Lee et al., 2003a). Lee et al. (2003b) mentioned that the ANN program written in MATLAB can allow analysis of landslide susceptibility, but it is inconvenient for the management of spatial data, and modification of its input data is difficult. Therefore, it is necessary to integrate the GIS and the ANN to reduce the restrictions of using the two applications separately. Lee et al. (2003b) selected the learning rate of 0.01 and the random initial weights between 0.1 and 0.3. To test whether the variation of weights is dependent on initial weight or not, the weights which were calculated from many cases, were compared, and the results revealed that the initial weight does not have an influence on weight in the condition (Lee et al. 2003b). Lee et al. (2003b) constructed an ANN having a structure of $7 \times 15 \times 2$. The result of verification performed by Lee et al. (2003b) was a satisfactory agreement between the susceptibility map and the landslide location data. As a final conclusion carried out by Lee et al. (2003b), in the ANN method, it is difficult to follow the internal processes of the procedure, and the method entails a long execution time and a heavy computing load. There is a need to convert the database to another format such as ASCII; the method requires that data be converted to ASCII for use in the ANN program and later reconverted to incorporate it into a GIS layer. Moreover, the large amount of data in the numerous layers in the target area cannot be processed in ANN programs quickly and easily. Lee et al. (2004) determined the weights of the conditioning parameters for landslide susceptibility maps by the ANN. Lee et al. (2004) used the backpropagation algorithm proposed by Zhou (1999) when determining the weights. For comparison, the number of landslide was classified into the landslide susceptibility index. The range of index value is different in without weighting and weighting, so the index value needs to set the same interval for comparison (Lee et al. 2004). According to Lee et al. (2004), the objective and scientific weighting and rating are essential to landslide susceptibility mapping and the weights can be applied to an area that needs weighting and rating, such as groundwater pollution assessment and soil loss assessment.

Gomez and Kavzoglu (2005) assessed the shallow landslide susceptibility using the ANN in a landslide-prone area in Venezuela. Gomez and Kavzoglu (2005) emphasized that ANNs have the ability to handle imprecise and fuzzy data, so they can work with continuous, categorical and binary data without violating any assumptions. For this reason, the ANN can be applied to several problems such as prediction problems. Owing to this ability of the ANN, as assessment of probability for landsliding is performed through the forecast of future events from experience of past landslides, it may be considered as an ideal application for ANNs (Gomez and Kavzoglu 2005). The ANN, specifically multilayer perceptron, was applied to the area studied by Gomez and Kavzoglu (2005). Gomez and Kavzoglu (2005) constructed an ANN model including $9 \times 28 \times 1$ structure. According to Gomez and Kavzoglu (2005), it was found that ANNs have several distinct advantages for landslide susceptibility mapping. First, ANNs are nonlinear and therefore capable of handling complex data patterns. Second, they can identify subtle patterns in the input training data, which is probably ignored when employing conventional statistical methods. Third, they are able to take a specific set of input data and generalize a solution set, which allows solving problems with unknown patterns present in the input data. Fourth, ANNs have great potential when used with ground truth, as the ANN may extract the information content of the data automatically. Finally, continuous, near-continuous and categorical data can be used in ANNs without violating any model assumptions. On the other hand, it should be pointed out that ANNs have some drawbacks reported in the literature, some of which are the inconsistent results due to random initialization of the weights, slow convergence and difficulties in the design of the network structure and in the determination of optimum learning parameter, they require some degree of expertise (Gomez and Kavzoglu 2005). Ermini et al. (2005) applied ANN to produce landslide susceptibility map of Riomaggiore river basin in Italy. Some ANN techniques were applied to the landslide susceptibility mappings. In the application to landslide susceptibility of multilayer perceptron and probabilistic neural network, both techniques can be classified as "black box models", and furthermore, several ANNs have been developed on a statistical basis (Bishop 1995; Petterson 1996; Ermini et al., 2005). For this reason, it is quite difficult to find a widely accepted definition that classifies the differences between ANNs and statistical models (Ermini et al. 2005). As a general rule, the multilayer perceptron training phase is aimed at representing the multidimensional nature of the investigated process. Each data set record can be considered as a vector whose number of dimensions corresponds to the number N of variables taken into account in the dataset itself. Network complexity usually increases with the increasing number of variables and, consequently, the number of weights and hidden layers (Ermini et al. 2005). Ermini et al. (2005) produced two landslide susceptibility maps of the study area using multilayer perceptron and probabilistic neural networks. According to the results obtained by Ermini et al. (2005), the networks predictions are, in both cases, quite poor if considered in absolute terms, with a slightly better performance by the multilayer perceptron, but fairly good when suitably reclassified. It is interesting to note that the multilayer perceptron net performs better than the probabilistic neural network does and that a number of locations having high output values have been mapped as being unaffected by mass movements in the inventory. This may be due either to model overestimation errors or to possible cases of existent landslides, which have not been correctly recognized by the aerial photographs. The advantage of the multilayer perceptron model over the probabilistic neural network is even more understandable when rating the cumulative frequency distribution (Ermini et al. 2005). Although the poor predictions of the both models, Ermini et al. (2005) stated that geographic representations of the model predictions show satisfactory results for both models. The other interesting statement described by Ermini et al. (2005) is the "wait and see" for the validation of the landslide susceptibility models. According to Ermini et al. (2005), the estimation errors can be sourced from problems in the network construction, wrong or insufficient variables and noisy data. The output value, in particular, could be affected by two different kinds of noise: the first, which is relevant also to input variables, is essentially connected to survey mistakes. For example, errors might have been made during the geomorphological survey of landslides, which could lead to a misleading classification of landslide-free areas (Ermini et al. 2005). As a final conclusion of Ermini et al. (2005), improvements to the model could be obtained by analyzing the state of activity of the landslides and setting the different output values to the observed degree of activity, e.g., by using fuzzy set variables.

Yesilnacar and Topal (2005) produced landslide susceptibility maps of a landslide-prone area in Turkey by the logistic regression and the ANN. When producing landslide susceptibility map, Yesilnacar and Topal (2005) used the feed-forward back-propogation ANN. The feed-forward back-propogation learning algorithm is a well recognized procedure for training neural networks (multilayer perceptron - MLP) (Yesilnacar and Topal 2005). According to Yesilnacar and Topal (2005), it has been shown that a MLP with one hidden layer has the capability to approximate any function with an acceptable degree of accuracy if there are enough hidden nodes. In cases where the optimum number of hidden nodes on a single hidden layer is large, two hidden layers with a small number of nodes on each layer could be more appropriate (Yesilnacar and Topal 2005). In the case studied by Yesilnacar and Topal (2005), several iterations were tested to find the best number of hidden nodes for the problem. According to the results of Yesilnacar and Topal (2005), 13 hidden nodes showed the best increasing trend in different learning-rate momentum term couples. Then 36 trials were carried out to find the best learning rate-momentum term couple. Finally, 13 hidden nodes, learning rate of 0.3 and momentum term of 0.8 with 40,000 iterations were found to be best structure by Yesilnacar and Topal (2005). By using this ANN structure, a landslide susceptibility map having an overall accuracy of 82.12% for the test dataset was produced. As a result of the study performed by Yesilnacar and Topal (2005), the feed-forward back-propagation learning algorithm performed better than the logistic regression model. For this reason, the susceptibility map produced using the ANN method is found to be more realistic (Yesilnacar and Topal 2005). Lee et al. (2006) produced the landslide susceptibility map of a landslide prone area in Korea by the ANN method. According to Lee et al. (2006), the back-propagation training algorithm is the most frequently used network method and hence, they used this ANN algorithm. The back-propagation training algorithm is trained using a set of examples of associated input and output values (Lee et al. 2006). Lee et al. (2006) constructed an ANN model having $13 \times 30 \times 2$ structure and they set the learning rate of 0.01 and the momentum parameter of 0.01. Lee et al. (2006) tried 5 different cases. Case 4, when likelihood ratio on prone training sites and with likelihood ratio on nonprone sites, were better than the other cases (Lee et al. 2006). As a conclusion of the study of Lee et al. (2006), in addition to using multi-faceted approach to a solution, they enable the extraction of reliable results for a complex problem, and for continuous and discrete data processing. Wang and Sassa (2006) produced a rainfall-induced landslide susceptibility map of a landslide-prone area in Japan by the ANN. Wang and Sassa (2006) also used back-propagation neural network algorithm. ANNs are a powerful tool for the prediction of non-linearities. They can learn some target values (desired output) from a set of chosen input data that have been introduced to a network under both supervised and self-adjusted or unsupervised learning algorithm (Wang and Sassa 2006). Wang and Sassa (2006) constructed a five layer ANN model with 49 input nodes and one output node. Hidden nodes and the number of hidden layers were decided by reasonable errors from repeated trainings. Consequently, hidden nodes were 36, 10 and 16 in three hidden layers respectively (Wang and Sassa 2006). In the training model developed by Wang and Sassa (2006), the learning ratio was 0.9, the learning step was 0.7, the individual error was 0.01 and the collective error was 0.0001. Compared to statistical analysis methods commonly used, the BPNN model is more suitable for producing a map of landslide hazard. Since the BPNN model can improve the accuracy of prediction and classification of landslides after learning, much in the same way as human might (Wang and Sassa 2006). As a final conclusion of Wang and Sassa (2006), whatever neural networks are applied to predict the probability of landslide occurrence, it currently appears that ANNs provide a very promising avenue for spatial analysis and landslide hazard modeling, in spite of their limitations. Lee and Evangelista (2006) produced the earthquake-induced landslide susceptibility map of a landslide-prone area in Philippinnes by the ANN method. They constructed an ANN model having a structure of $8 \times 16 \times 2$. Lee and Evangelista (2006) determined the weight of each factor by the ANN method (BPNN algorithm) for the detected landslide locations. In the study of Lee and Evangelista (2006), the input data were normalized in the range of 0.1–0.9. In addition, the learning rate was set to 0.01, and the initial weights were randomly selected to values between 0.1 and 0.3 (Lee and Evangelista 2006). Although Lee and Evangelista (2006) produced a satisfactory landslide susceptibility map, they also emphasized some difficulties of the application of the BPNN. According to Lee and Evangelista (2006), the BPNN involves a long execution time, has a heavy computing load, and there is a requirement to convert the database to another format. Kanungo et al. (2006) produced the landslide susceptibility maps of a region in Darjeeling Himalayas by different approaches. One of these approaches According to Kanungo et al. (2006), ANN, which is a useful technique for regression and classification problems, has been successfully applied in other fields, and promises to be suitable for the delineation of areas prone to landslide activity. It has been found that the ANNs have several advantages for landslide susceptibility zoning, as these are non-linear and thus have the capability to analyze complex data patterns (Kanungo et al. 2006). The most widely used BPNN algorithm are gradient descent and gradient descent with momentum. These are often too slow for the solution of practical problems. The faster algorithms use standard numerical optimizer such as conjugate gradient, quasi Newton and Levenberg–Marquardt approach (Kanungo et al. 2006). For this reason, Kanungo et al. (2006) used Levenberg–Marquardt approach. Unlike gradient descent algorithms, it does not consider learning rate and momentum factor as its parameters (Kanungo et al. 2006). Kanungo et al. (2006) constructed a multilayer ANN model with one input layer, two hidden layers and one output layer. A total of 39 ANN architectures were created by varying number of neurons in the hidden layers (Kanungo et al. 2006). As a result of these trials, the network architecture 6 × 13 × 7 × 1 with training data accuracy (correlation coefficient of 0.918, RMSE 0.112 and 74.4% correct) and testing data accuracy (correlation coefficient of 0.896, RMSE 0.126 and 72.6 correct) was the most appropriate one (Kanungo et al. 2006). As a final conclusion of the study of Kanungo et al. (2006), it represents an objective approach where weights for factors are determined through the ANN connection weight approach and ratings of the categories of factors are determined through cosine amplitude similarity method based on fuzzy relation concept. Lee et al. (2007) produced landslide susceptibility maps of a landslide-prone area in Korea by likelihood ratio, logistic regression and the ANN. In the study of Lee et al. (2007), the structure of 14 × 30 × 2, input, hidden and output layers, respectively, was selected for the networks. In addition, the learning rate and the momentum parameter were selected same as 0.01. According to the results obtained by Lee et al. (2007), the success rates for the first two classes (90–100%) are better than those for the ANN model. For remainder of the classes (65–90%) and the middle classes (40–65%), the ANN model produced better results than the likelihood ratio method. The logistic regression is better than the other two models below the 35% classes (Lee et al. 2007). Lee (2007a) also prepared a landslide susceptibility map by the ANN method. In the study, Lee (2007a) used topography, geology, lineament, soil, forest and land cover data as the landslide related factors. According to Paola and Schowengerdt (1995), there are two stages involved in using ANN for multiscore classification: the training stage, in which the internal weights are adjusted; and the classifying stage. Typically, the BPNN algorithm trains the network until some targeted minimal error is achiheved between the desired and actual output values of the network. Once training is complete, the network is used as a feed-forward structure to produce a classification for the entire data. By following this procedure, Lee (2007a) constructed some ANN models for five different cases. The constructed ANN models have 16 × 30 × 2 layer structure. As a result of comparison of the landslide susceptibility maps prepared by Lee (2007a), using results from logistic regression as training sites prone and not prone to landslides (Case 5, 86.10%) showed the best prediction accuracy.

Melchiorre et al. (2008) produced landslide susceptibility map of a landslide prone area in Italy. The procedure, which involves the use of the ANNs and cluster analysis, demonstrates that an accurate sampling strategy improves the model results and increases the landslide occurrence prediction (Melchiorre et al. 2008). According to Melchiorre et al. (2008), the use of the ANNs can be a valid alternative in the indirect hazard mapping, when the conditioning factors are not approximable by a normal distribution and are strongly correlated. Moreover, the ANNs are able to give a good prediction even though trained with noisy and uncertain data. In the literature, different approaches to classification have been proposed ranging from the early decision tree presented by Quinlan (1986) to recent developments in support vector machines (Burges 1998) and ANNs (Bishop 1995). However, all these methods require the use of a set of labeled data for each class (Melchiorre et al. 2008). The main idea of the study performed by Melchiorre et al. (2008) is to use an unsupervised technique to find out pattern distribution in the dataset, in order to capture aspects (presence/absence of landslides) in the data structure and devise a sampling procedure able to improve the performance of the final classifier. One of the common problems encountered in the training stages of the ANNs is overlearning. For this reason, Melchiorre et al. (2008) used the early stopping technique proposed by Caruana et al. (2000) to prevent the overlearning problem and losing the generalization capacity of the model. The ANN was trained with six network inputs scaled in the range 0–1 and the network output was also defined in the range 0–1 by setting the output value to 1 for landslide presence and 0 for landslide absence (Melchiorre et al. 2008). The analysis was performed using the MLP network with the Levenberg–Marquardt training algorithm by Melchiorre et al. (2008). Melchiorre et al. (2008) selected the ANN structure with 14 hidden neurons as it ensures the best generalization without excessively increasing the network complexity. Finally, the most robust susceptibility map was obtained after the cluster sampling, since the cluster sampling-ANN model is able to distinguish and separate the unstable areas and thus to identify more reliable susceptibility classes. Although the results are encouraging, the model output and the discrimination of unstable areas can be improved (Melchiorre et al. 2008). Caniani et al. (2008) used the ANN to produce landslide susceptibility map of an urban area in Italy. According to Caniani et al. (2008), statistical methods guarantee a lower subjectivity, but they are more easily applied to areas characterized by a unique type of mass movements. Caniani et al. (2008) used the BPNN algorithm in their study. The structure of the ANN model developed by Caniani et al. (2008) includes 7 inputs, 5 hidden neurons and one output. According to the results obtained by Caniani et al. (2008), the landslide pixels were the sites falling in the medium and high susceptibility classes. Caniani et al. (2008) can observe that 80% of landslide pixels were correctly classified. The susceptibility map produced by Caniani et al. (2008) shows high performance with respect to the similarity approaches. Nefeslioglu et al. (2008b) applied the logistic regression and the ANN methods to produce the landslide susceptibility maps of a landslide-prone area in Turkey. When developing an ANN, the data is commonly partitioned into as least two subsets such as training and test data. It is expected that the training data include all the data belonging to the problem domain. Certainly, this subset is used in the training stage of the model development to update the weights of the network. On the other hand, the test data should be different from those used in the training stage. The main purpose of this subset is to check the network performance using untrained data, and to confirm its accuracy (Nefeslioglu et al. 2008b). No exact mathematical rule to determine the required minimum size of these subsets exists. However, some suggestions for the potions of these samplings are encountered in the literature (Basheer and Hajmeer 2000). Nefeslioglu et al. (2008b) selected 80% of the shole data as the training dataset. In the study of Nefeslioglu et al. (2008b), an ANN having a structure of $44 \times 7 \times 1$ was constructed. As the results of the study of Nefeslioglu et al. (2008b), when considering the accuracy and the precision evaluations, the BPNNs represent considerably high prediction capacities when comparing with the logistic regression analyses. However, the generalization capacities of the ANN algorithms were abruptly decreased for the random samplings in which the presence (3.1) data were taken from the landslide masses. Highly disturbed geomorphometric attributes on the recent landslides might be the possible reason for this pecularity. The susceptibility maps produced mapping the outputs of the BPNNs could be interpreted as highly optimistic, while of those generated using the resultant probabilities of the logistic regression equations might be considered as pessimistic (Nefeslioglu et al. 2008b).

3.5 Further Recommendations

In this chapter, the basic concepts o soft computing techniques such as FIS, fuzzy relations and ANN are given. Also, the constructing principles of the soft computing methods are introduced. Additionally, a literature overview is carried out on the landslide susceptibility maps produced by fuzzy approaches and ANNs. It is evident that the quality of a landslide susceptibility map depends on the quality of the data and the method employed. As can be seen from the literature review, the soft computing methods have been applied successfully to the production of landslide susceptibility maps.

In the near future, the other innovative researches about the landslides may be expected. These research are the assessment of landslide hazard and risk, prediction of runout and time of landslides and early warning systems by applying sofy computing methods. To reach such type purposes, the anatomy and mechanisms of the landslides should be clearly known and modelled. However, the landslide is perhaps one of the most complex natural pheomena in the nature. Due to this complexity, use of the soft computing methods including hybrid systems has a curcial importance. Considering these facts, it is possible to say that there is still a long way to understand this natural process. To reach the target and to minimize the losses caused by the landslides, the researchers should make much more investigation and publish their results.

References

Aleotti P, Chowdury R (1999) Landslide hazard assessment: summary review and new persoectives. Bull Eng Geol Environ 58:21–44

Alvarez Grima M (2000) Neuro-fuzzy modeling in engineering geology. Balkema AA, Rotterdam (eds), p 244

Anbalagan R (1992) Landslide hazard evaluation and zonation mapping in mountainous terrain. Eng Geol 32:269–277

Anbalagan R, Singh B (1996) Landslide hazard evaluation and zonation mapping in mountainous terrains–a case study from Kumaun Himalaya, India. Eng Geol 43:237–246

Atkinson PM, Massari R (1998) Generalised linear modelling of susceptibility to landsliding in the central Appenines, Italy. Comput Geosci 24:373–385

Baeza C, Corominas J (2001) Assessment of shallow landslide susceptibility by means of multivariate statistical techniques. Earth Surf Proc Land 26:1251–1263

Basheer IA, Hajmeer M (2000) Artificial neural networks: fundamentals, computing, design and application. J Microbiol Meth 43:3–31
Berkan RC, Trubatch SL (1997) Fuzzy system design principles, building fuzzy if-then rule bases. The Institute of Electrical and Electronics Engineers, New York, p 496
Binaghi E, Luzi L, Madella P, Pergalani F, Rampini A (1998) Slope instability zonation: a comparison between certainty factor and fuzzy Dempster-Shafer approaches. Nat Hazards 17:77–97
Bishop C (1995) Neural networks for pattern recognition. Oxford University Press, Oxford
Brabb EE, Pampeyan EH, Bonilla MG (1978) Landslide susceptibility in San Mateo county, California. U.S. Geological Survey Miscellaneous Field Studies Maps, MF-360, Map at 1:62.500 scale
Brenning A (2005) Spatial prediction models for landslide hazards: review, comparison and evaluation. Nat Hazards Earth Syst Sci 5:853–862
Burges CJC (ed) (1998) A tutorial on support vector machines for pattern recognition, data mining and knowledge discovery, vol 2. Kluwer Academic Publishers, The Netherlands, pp 121–167
Can T, Nefeslioglu HA, Gokceoglu C, Sonmez H, Duman TY (2005) Susceptibility assessments of shallow earthflows triggered by heavy rainfall at three catchments by logistic regression analyses. Geomorphology 72:250–271
Caniani D, Pascale S, Sdao F, Sole A (2008) Neural networks and landslide susceptibility: a case study of the urban area of Potenza. Nat Hazards 45:55–72
Carrara A, Carratelli EP, Merenda L (1977) Computer-based data bank and statistical analysis of slope instability phenomena. Zeitschhrift Für Geomorphologie 21:187–222
Carrara A, Cardinali M, Detti R, Guzzetti F, Pasqui V, Reichenbach P (1991) GIS techniques and statistical models in evaluating landslide hazard. Earth Surf Proc Land 16:427–445
Caruana R, Lawrence S, Giles CL (2000) Overfitting in neural nets: backpropagation, conjugate gradient and early stopping. Proc Neural Inf Process Syst, Denver, pp 402–408
Catani F, Casagli N, Ermini L, Righini G, Menduni G (2005) Landslide hazard mapping at catchment scale in the Arno River basin. Landslides 2:329–342
Champati ray PK, Dimri S, Lakhera RC, Sati S (2007) Fuzzy-based method for landslide hazard assessment in active seismic zone of Himalaya. Landslides 4:101–111
Chau KT, Sze YL, Fung MK, Wong WY, Fong EL, Chan LCP (2004) Landslide hazard analysis for Hong Kong using landslide inventory and GIS. Comput Geosci 30:429–443
Clerici A, Perego S, Tellini C, Vescovi P (2002) A procedure for landslide susceptibility zonation by the conditional analysis method. Geomorphology 48:349–364
Corominas J, Copons R, Moya J, Vilaplana JM, Altimir J, Amigo J (2005) Quantitative assessment of the residual risk in a rockfall protected area. Landslides 2:343–357
Donati L, Turrini MC (2002) An objective method to rank the importance of the factors predisposing to landslides with the GIS methodology: appliaction to an area of the Appenines (Valneria; Perugia, Italy). Eng Geol 63:277–289
Dubois D, Prade H (1980) Fuzzy sets and systems: theory and applications. Academic, New York
Duman TY, Can T, Gokceoglu C, Nefeslioglu HA, Sonmez H (2006) Application of logistic regression for landslide susceptibility zoning of Cekmece Area, Istanbul, Turkey. Env Geol 51:241–256
Ercanoglu M, Gokceoglu C (2002) Assessment of landslide susceptibility for a landslide-prone area (nortrh of Yenice, NW Turkey) by fuzzy approach. Env Geol 41:720–730
Ercanoglu M, Gokceoglu C (2004) Use of fuzzy relations to produce landslide susceptibility map of a landslide prone area (West Black Sea Region, Turkey). Eng Geol 75:229–250
Ermini L, Catani F, Casagli N (2005) Artificial neural networks applied to landslide susceptibility assessment. Geomorphology 66:327–343
Fahlman SE (1988) An empirical study of learning speed in backpropagation. Technical Report CMU-CS-88-162, Carnegie-Mellon University, Pittsburgh
Fell R (1994) Landslide risk assessment and acceptable risk. Can Geotech J 31:261–272

Fell R, Corominas J, Bonnard C, Cascini L, Leori E, Savage WZ (2008) Guidelines for landslide susceptibility, hazard and risk zoning for land use planning. Eng Geol 102:85–98
Fu L (1995) Neural networks in computer intelligence. McGraw-Hill, New York
Gallahger M, Downs T (1997) Visualisation of learning in neural networks using principal component analysis. In: Varma B, Yao X (eds) Proceedings of the international conference on computational intelligence and multimedia applications, Australia, pp 327–31
Gokceoglu C, Aksoy H (1996) Landslide susceptibility mapping of the slopes in the residual soils of the Mengen region (Turkey) by deterministic stability analyses and image processing techniques. Eng Geol 44:147–161
Gokceoglu C, Sonmez H, Nefeslioglu HA, Duman TY, Can T (2005) The march 17, 2005 Kuzulu landslide (Sivas, Turkey) and landslide susceptibility map of its close vicinity. Eng Geol 81:65–83
Gomez H, Kavzoglu T (2005) Assessment of shallow landslide susceptibility using artificial neural networks in Jabonosa River Basin, Venezuela. Eng Geol 78:11–27
Gorsevski PV, Jankowski P (2008) Discrening landslide susceptibility using rough sets. Comput, Environ Urban Syst 32:53–65
Gorum T, Gonencgil B, Gokceoglu C, Nefeslioglu HA (2008) Implementation of reconstructed geomorphologic units in landslide susceptibility mapping: the Melen Gorge (NW Turkey). Nat Hazards 46(3):323–351
Guzzetti F, Carrara A, Cardinali M, Reichenbach P (1999) Landslide hazard evaluation: a review of current techniques and tehir application in a multi-scale study, Central Italy. Geomorphology 31:181–216
Guzzetti F, Reichenbach P, Cardinali M, Galli M, Ardizzone F (2005) Probabilistic landslide hazard assessment at the basin scale. Geomorphology 72:272–299
Guzzetti F, Reichenbach P, Ardizzone F, Cardinali M, Galli M (2006) Estimating the quality of landslide susceptibility models. Geomorphology 81:166–184
Hassoun MH (1995) Fundamentals of artificial neural networks. MIT Press, New York
Haykin S (1994) Neural networks: a comprehensive foundation. Macmillan College Publishing Company, New York
Hebb DO (1949) The organization of behaviors. Wiley, New York
Hecht-Nielsen R (1987) Kolmogorov's mapping neural network existence theorem. In: Proceedings of the 1st IEEE international conference on neural networks, San Diego, pp 11–4
Hellendoorn H, Thomas C (1993) Defuzzification if fuzzy controllers. J Intell Fuzzy Syst 1:109–123
Helmore K (1996) Cities on the brink try new tactics. Choices 5:18–21
Hetch-Nielsen R (1990) Neurocomputing. Addison Wesley, Reading
Hufschmidt G, Crozier MJ (2008) Evolution of natural risk: analysis changing landslide hazard in Wellington, Aotearoa/New Zealand. Nat Hazards 45:255–276
Hush DR (1989) Classification with neural networks: a performance analysis. In: Proceedings of the IEEE international conference on systems Engineering Dayton Ohia, USA, pp 277–80
Jain AK, Mao J, Mohiuddin KM (1996) Artificial neural networks: tutorial. Comput IEEE 29:31–44
Juang CH, Lee DH, Sheu C (1992) Mapping slope failure potential using fuzzy sets. J Geotech Eng ASCE 118:475–493
Kaastra I, Boyd M (1996) Designing a neural network for forecasting financial and economic time series. Neurocomputing 10(3):215–236
Kanellopoulas I, Wilkinson GG (1997) Strategies and best practice for neural network image classification. Int J Remote Sens 18:711–725
Kanungo DP, Arora MK, Sarkar S, Gupta RP (2006) A comparative study of conventional, ANN black box, fuzzy and combined neural and fuzzy weighting procedures for landslide susceptibility zonation in Darjeeling Himalayas. Eng Geol 85:347–366
Kanungo DP, Arora MK, Gupta RP, Sarkar S (2008) Landslide risk assessment using concepts of danger pixels and fuzzy set theory in Darjeeling Himalayas. Landslides 5:407–416

Kavzoglu T (2001) An investigation of the design and use of feed forward artificial neural networks in classification of remotely sensed images. PhD thesis, The University of Nottingham, UK

Klimasauskas CC (1993) Applying neural networks. In: Trippi RR, Turban E (eds) Neural networks in finance and investigating. Probus, Cambridge

Lee S (2005) Application of logistic regression model and its validation for landslide susceptibility mapping using GIS and remote sensing data. Int J Remote Sens 26:1477–1491

Lee S (2007a) Landslide susceptibility mapping using an artificial neural network in the Gangneung area, Korea. Int J Remote Sens 21:4763–4783

Lee S (2007b) Application and verification of fuzzy algebreaic operators to landslide susceptibility mapping. Env Geol 52:615–623

Lee S, Evangelista DG (2006) Earthquake-induced landslide-susceptibility mapping using an artificial neural network. Nat Hazards Earth Syst Sci 6:687–695

Lee S, Min K (2001) Statistical analysis of landslide susceptibility at Yongin, Korea. Env Geol 40:1095–1113

Lee S, Pradhan B (2006) Probabilistic landslide hazard and risk mapping on Penang Island, Malaysia. J Earth Syst Sci 115:661–672

Lee S, Rhu J-H, Lee M-J, Won J-S (2003a) Use an artificial neural network for analysis of the susceptibility to landslides at Boun, Korea. Env Geol 44:820–833

Lee S, Ryu J-H, Min K, Won J-S (2003b) Landslide susceptibility analysis using GIS and artificial neural network. Earth Surf Proc Land 28:1361–1376

Lee S, Ryu J-H, Won L-S, Park H-J (2004) Determination and application of the weights for landslide susceptibility mapping using an artificial neural network. Eng Geol 71:289–302

Lee S, Ryu J-H, Lee M-J, Won J-S (2006) The application of artificial neural networks to landslide susceptibility mapping at Janghung, Korea. Math Geol 38:199–220

Lee S, Ryu J-H, Kim I-S (2007) Landslide susceptibility analysis and its verification using likelihood ratio, logistic regression and artificial neural network models: case study of Youngin, Korea. Landslides 4:327–338

Leroi E (1996) Landslide hazard-risk maps at different scales: objectives, tools and developments. In: Proceedings of 7th International symposium on Landslides, Trondheim, vol 1, pp 35–52

Looney CG (1996) Advances in feed-forward neural networks: demystifying knowledge acquiring black boxes. IEEE Trans Knowledge Data Eng 8(2):211–226

Lu P, Rosenbaum MS (2003) Artificial neural networks and grey systems for the prediction of slope stability. Nat Hazards 30:383–398

Mamdani EH, Assilian S (1975) An experiment in linguistic synthesis with a fuzzy logic controller. Int J Man-Mach Stud 7(1):1–13

Masters T (1994) Practical neural network recipes in C++. Academic, Boston

Melchiorre C, Matteucci M, Azzoni A, Zanchi A (2008) Artificial neural networks and cluster analysis in landslide susceptibility zonation. Geomorphology 94:379–400

Messer K, Kittler J (1998) Choosing an optimal neural network size to aid search through a large image database. Proceedings of the 9th British machine vision conference (BMVC98), University of Southamton, UK, p 235–44

Muthu K, Petrou M (2007) Landslide-hazard mapping using an expert system and a GIS. IEEE Trans Geosci Remote Sens 45:522–531

Muthu K, Petrou M, Tarantino C, Blonda P (2008) Landslide possibility mapping using fuzzy approaches. IEEE Trans Geosci Remote Sens 46:1253–1265

Nefeslioglu HA, Duman TY, Durmaz S (2008a) Landslide susceptibility mapping for a part of tectonic Kelkit Valley (Eastern Black Sea region of Turkey). Geomorphology 94:401–418

Nefeslioglu HA, Gokceoglu C, Sonmez H (2008b) An assessment on the use of logistic regression and artificial neural networks with different sampling strategies for the preparation of landslide susceptibility maps. Eng Geol 97:171–191

Negnevitsky M (2002) Artificial intelligence a guide to intelligent systems. Addison-Wesley, Reading

Nelson M, Illingworth WT (1990) A practical guide to neural nets. Addisin-Wesley, Reading
Pachauri AK, Pant M (1992) Landslide hazard mapping based on geological attributes. Eng Geol 32:81–100
Paola JD (1994) Neural network classification of multispectral imagery. MSc thesis, The University of Arizona, USA
Paola JD, Schowengerdt RA (1995) A review and analysis of backpropogation meural networks for classification of remotely sensed multi-spectral imagery. Int J Remote Sens 16:3033–3058
Petterson D (1996) Artificial neural networks. Prentince Hall, Singapore
Pielke RA, Schellnhuber HJ, Sahagian D (2003) Non-linearities in the earth system. Global Change News Lett 55:11–15
Quinlan JR (1986) Introduction of decision trees. Mach Learn 1:81–106
Rautelal P, Lahhera RC (2000) Landslide risk analysis between Giri and Tons rivers in Himachal Himalaya (India). Int J Appl Earth Obs Geoinf 2:153–160
Remondo J, Bonachea J, Cendrero A (2005) A statistical approach to landslide risk modelling at basin scale; from landslide susceptibility to quantitative risk assessment. Landslides 2:321–328
Ripley BD (1993) Statistical aspects of neural networks. In: Barndoff-Neilsen OE, Jensen JL, Kendall WS (eds) Networks and chaos-statistical and probabilistic aspects. Chapman & Hall, London, pp 40–123
Romeo RW, Floris M, Veneri F (2006) Area-scale landslide hazard and risk assessment. Env Geol 51:1–13
Rosenblatt F (1958) The perception: a probabilistic model for information storage and organization in the brain. Psychol Rev 65:386–408
Rosenblatt F (1960) Perceptron simulation experiments. Proc Ins Radio Eng 48:301–309
Ross TJ (1995) Fuzzy logic with engineering applications. McGraw Hill, New York
Rumelhart DE, Hinton GE, Williams RJ (1986) Parallel distributed processing. In: Rumelhart DE, McClelland JL (eds) Explorations in the microstructure of cognition, vol 1. MIT Press, Cambridge
Schalkof RJ (1997) Artificial neural networks. McGraw-Hill, New York
Sietsma J, Dow RJF (1991) Creating artificial neural network that generalize. Neural Networks 4:67–69
Sonmez H, Gokceoglu C, Nefeslioglu HA, Kayabasi A (2006) Estimation of rock modulus: for intact rocks with an artificial neural network and for rock masses with a new empirical equation. Int J Rock Mech Min Sci 43:224–235
Sugeno M, Kang GT (1988) Structure identification of fuzzy model. Fuzzy Sets Syst 28:15–33
Swets JA (1988) Measuring the accuracy of diagnostic systems. Science 240:1285–1293
Swingler K (1996) Applying neural networks: a practical guide. Academic, New York
Takagi T, Sugeno M (1985) Fuzzy identification of systems and its applications to modeling and control, IEEE Transactions on Systems. Man Cybern 15:116–132
Tangestani MH (2004) Landslide susceptibility mapping using fuzzy gamma approach in a GIS, Kaka catchments area, SW Iran. Aust J Earth Sci 51:439–450
Tangestani MH (2009) A comparative study of Demster-Shafer and fuzzy models for landslide susceptibility mapping using a GIS: an experience from Zagros Mountains, SW Iran. J Asian Earth Sci 35:66–73
Tsukomoto Y (1979) An approach to fuzzy reasoning method, Advanced Fuzzy Set Theory. Amsterdam, pp 137–149
Turrini MC, Visintainer P (1998) Proposal of a method to define areas of landslide hazard and application to an area of the Dolomites, Italy. Eng Geol 50:255–265
van Westen CJ, van Asch TWJ, Soeters R (2006) Landslide hazard and risk zonation–why is it so difficult? Bull Eng Geol Environ 65:167–184
van Westen CJ, Castellanos E, Kuriakose SL (2008) Spatial data for landslide susceptibility, hazard, and vulnerability assessment: an overview. Eng Geol 102:112–131

Varnes DJ (1978) Slope movement types and processes. In: Schuster RL, Krizek RJ (eds) Landslides analysis and control. Special report, vol 176. Transportation Research Board, National Academy of Sciences, New York, pp 12–33
Varnes DJ (1984) Landslide hazard zonation: a review of principles and practice. UNESCO, Paris
Wang C (1994) A theory of generalization in learning machines with neural application. PhD thesis, The University of Pennsylvania, USA
Wang HB, Sassa K (2006) Rainfall-induced landslide hazard assessment using artificial neural networks. Earth Surf Proc Land 31:235–247
Wyhthoff BJ (1993) Backpropagation neural networks: a tutorial. Chemometr Intell Lab Syst 18:115–155
Yao X, Tham LG, Dai FC (2008) Landslide susceptibility mapping based on support vector machine: a case study on natural slopes of Hong Kong, China. Geomorphology 101:572–582
Yen J, Langari R (1998) Fuzzy logic intelligence, control, and information. Prentice Hall, New Jersey
Yesilnacar E, Topal T (2005) Landslide susceptibility mapping: a comparison of logistic regression and neural networks methods in a medium scale study, Hendek region (Turkey). Eng Geol 79:251–266
Zadeh LA (1965) Fuzzy sets. IEEE inf control 8:338–353
Zezere JL, Reis E, Garcia R, Oliviera S, Rodrigues ML, Vieira G, Ferreira AB (2004) Integration of spatial and temporal data for the definition of different landslide hazard scenarios in the area north of Lisbon (Protugal). Nat Hazards Earth Syst Sci 4:133–146
Zhou W (1999) Verification of the nonparametric characteristics of backpropogation neural networks for image classification. IEEE Trans Geosci Rem Sens 38:771–779
Zupan J, Gasteiger J (1993) Neural networks for chemists: an introduction. VCH, New York

Chapter 4
Application and Verification of Fractal Approach to Landslide Susceptibility Mapping

Changjiang Li, Tuhua Ma, Leling Sun, Wei Li and Aiping Zheng

Abstract The existing methods for landslide susceptibility mapping, whether statistic method or physics-based method, require many data such as lithology, topography, soil properties, land use and so on. However, in many regions of the world, the abundance of data is not available, but the need for landslide susceptibility maps is great. For these regions, how should reliable susceptibility maps be produced from the limited data? We addressed the issue of the problem and developed a new landslide susceptibility analysis technique using historical landslide inventories and fractal statistics on a GIS platform. The aim of this article is to apply and verify the use of this new technique to landslide susceptibility mapping in the Zhejiang Province (101,800 km^2 in area), China.

Keywords Landslide · Susceptibility mapping · Historical landslide inventories · Fractal method · GIS

This chapter is an updated version of an earlier paper published under the title "Application and Verification of Fractal Approach to Landslide Susceptibility Mapping" in Natural Hazards (Published online: 8 April 2011. Doi: 10.1007/s11069-011-9804-x).

C. Li (✉) · T. Ma
Zhejiang Information Center of Land and Resources,
310007 Hangzhou, China
e-mail: zjigmr@mail.hz.zj.cn

L. Sun · W. Li
Zhejiang Provincial Department of Land and Mineral Resources,
310007 Hangzhou, China

A. Zheng
Department of Earth Sciences, Zhejiang University,
310027 Hangzhou, China

B. Pradhan and M. Buchroithner (eds.), *Terrigenous Mass Movements*,
DOI: 10.1007/978-3-642-25495-6_4,

4.1 Introduction

Landslides constitute one of the major natural hazards that could cause significant losses of life and property of various landslide disasters, the rainfall-induced landslides have the widest spatial and temporal distribution and the highest frequency of occurrence. Large and small landslides occur almost every year in nearly all hilly and mountainous regions of the world. One of the ways to reduce the damage impact of those hazards to life and economy is to increase preparedness through landslide forecasting systems (e.g., Keefer et al. 1987; Aleotti 2004; Li et al. 2010). A forecasting system can provide a timely warning for those individuals at high landslide risk areas during a developing storm for a region. Another way is to map or delineate areas prone to landsliding, which is essential for land-use activities and management decision making in hilly or mountainous regions.

Engineers, earth scientists, and planners are interested in assessment of landslide susceptibility for two purposes (Tangestani 2003): (1) The landslide susceptibility maps identify and delineate unstable hazard-prone areas, so that environmental regeneration programs can be initiated adopting suitable mitigation measures; (2) These maps help planners to choose favorable locations for siting development schemes, such as building and road construction. Even if the potential hazardous areas cannot be avoided entirely, their recognition in the initial stages of planning may help to take suitable precautionary measures. This study mainly deals with the landslide susceptibility zoning and mapping.

A number of researchers used various statistical techniques incorporating many types of data such as lithology, topography, soil properties, and land cover in Geographic Information Systems (GIS) to assess landslide susceptibility (e.g. Gupta and Joshi 1990; Chung and Fabbri 1999; Gritzer et al. 2001; Lineback et al. 2001; Santacana et al. 2003; Lee 2007; Lee et al. 2007). A few authors attempted to derive landslide susceptibility by means of physically-based methods (e.g. Montgomery and Dietrich 1994; Jibson et al. 2000; Savage et al. 2003). The physically-based methods rely on physical properties of hillslope materials, topographic information from a digital elevation model (DEM) and hydrologic and hydrogeological parameters in slope-stability models. However, the abundance of data such as that described above is not available in many hilly and mountainous regions of the world. For these regions, it is very challenge to produce reliable susceptibility maps from limited data. Coe et al. (2004) produced a landslide susceptibility map for a 980 km^2 area in eastern-central Guatemala by applying a moving-count circle approach using two types of data, a landslide inventory map and a DEM. Moreover, a semi-empirical method for analysing landslide susceptibility was given by the studies carried out by Iovine et al. (2003a, b) in Campania (Southern Italy), based on a cellular-automata modelling approach in which topographic data (1:5000 scale) plus a map of the soil cover were used as input matrices, and by further developments in which calibration was performed through genetic algorithms (Iovine et al. 2005; D'Ambrosio et al. 2006; Iovine 2008).

Zhejiang Province (101,800 km^2 in area) is one of the rainiest in the southeast coastal region of China. Due to its geologic, geomorphologic and climatic settings, every year more than 100 landslides are triggered by rainfall events. These landslides mainly belong to shallow earth slides and debris flows (Li 2004; Li et al. 2010).

In this study, we developed a fractal-based technique to map landslide susceptibility using historical landslide inventories only. To our knowledge, our approach differs from previous work in that we first measure the spatial, fractal clustering distribution of the existing landslides in a region, and then apply the derived fractal clustering relation to produce a landslide susceptibility map by means of GIS-supported spatial analysis. The method is illustrated using two examples. In the first example, based on the 3,285 landslides occurred from 1990 to 2003 in Zhejiang region, a landslide susceptibility map for the region is produced; the effectiveness of the susceptibility map is examined by the 459 landslides occurred from 2004 to 2007 in the region. In the second example, the method is applied to a relatively small area. i.e., Yueqing County (1,174 km^2 in area) located in the southeast costal region of Zhejiang. A landslide susceptibility map for the Yueqing County is produced by the procedure using the 156 landslides occurred from 2000 to 2003 in the area and the effectiveness of the susceptibility map is examined by the 149 landslides occurred from 2004 to 2007 in the same area.

4.2 Study Area

Zhejiang is located on the southeast coastal region of China (Fig. 4.1), with a population of about 70 millions. The province lies between latitudes 27°02′ N and 31°11′ N, and longitudes 118°01′ E and 123°25′ E, and has continental area of 101,800 km^2. This province is one of the most concentrated regions in terms of population and one of the most economical-developed areas in China. This region is complex in landform and greatly different in relief, and the hilly and mountainous terrain with elevation above 300 m accounts for 70.6% of its total area. The entire terrain is inclined from southwest towards northeast and drops in a step form.

This region belongs to a subtropical monsoon zone and has mean annual precipitation ranging from 1,000 to 2,000 mm. Due to the impact of the monsoon troughs and tropical cyclones, the distribution of rainfall over a year is not uniform usually, the amount of rainfall from May to September accounts for ca. 69% of the annual rainfall. Rainstorms occur frequently in summer because of the impact of tropical cyclones (Typhoon).

The rocks in Zhejiang are usually cut by faults and vary greatly in composition, degree of consolidation, amount of deformation, and depth of weathering. The superficial deposits in hills and mountainous areas are mainly colluvium and alluvium. Much of the colluvium with great-variable thickness and composition was deposited from the late Tertiary through the Holocene.

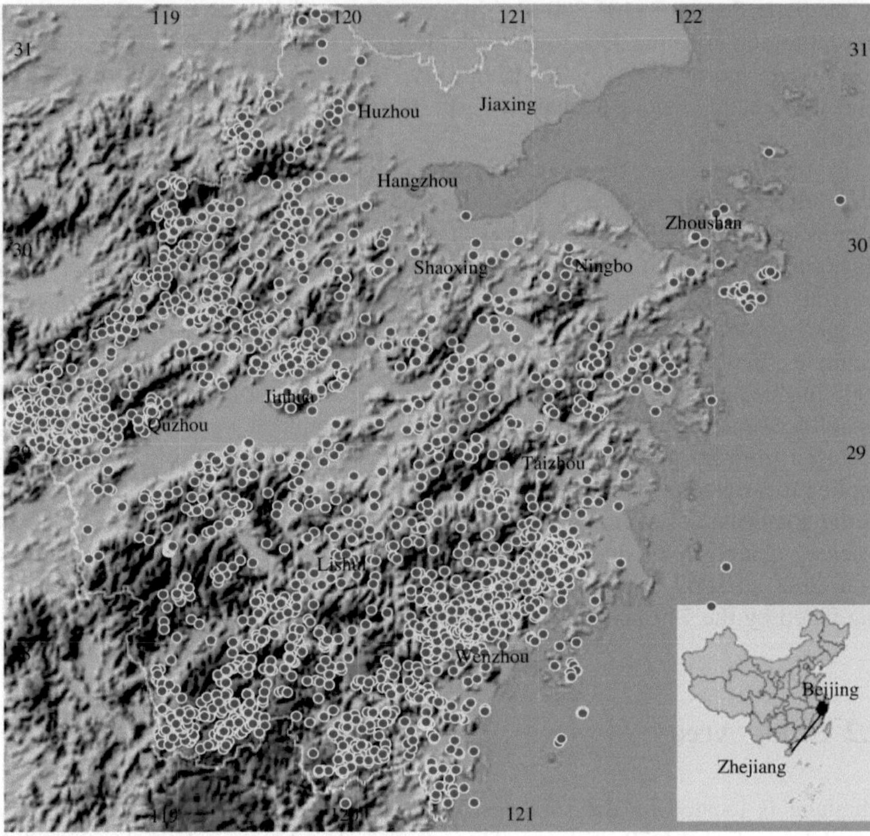

Fig. 4.1 Topographic relief map of Zhejiang Province showing the spatial distribution of rainfall-induced landslides. *Red dots* are the locations of 3,285 landslides occurred in the 1990–2003 period. Inset shows the location of Zhejiang in southeastern China

Owing to its geologic, geomorphologic and climatic settings, the region is highly prone to the occurrence of landslides. In the rainy season, a large number of shallow landslides are often triggered by heavy rainfall events and a heavy rainstorm may induce tens to over hundreds of landslides, which expose the population and economy to serious risks (Li 2004; Li et al. 2010).

Since 1990, after each serious landslide event, Zhejiang Provincial Department of Geology and Mineral Resources immediately sent experienced professionals to conduct surveys in the landslide site and recorded the occurrence date, location, failure types, the volume of the landslides, and the resulted casualties and economic losses. Especially from 2000 to 2003, a landslide-mapping project on a scale of 1:10,000 was carried out in 45 hilly and mountainous counties that cover all the landslide-prone areas in Zhejiang and account for ca. 70% of its total area. After completion of the project, a landslide database was compiled. Totally 4,454 landslides occurred from 1990 to 2007 were included in the compilation, about 75% are shallow soil slides, 8% are debris flows, and 17% are rock/soil falls. Among the 4,454 landslides, 3,744 landslides were recorded with locations (i.e., longitude and latitude). In this study, these 3,744 landslides data are used, of which 3,285 landslides that occurred from 1990 to 2003 are selected as the samples used to create a susceptibility map using the fractal-based approach for the Zhejiang region and the remaining 459 landslides which occurred from 2004 to 2007 are taken as the test samples to examine the effectiveness of the susceptibility map. The same method is also applied to a relatively small area (Yueqing County), where about 156 landslides occurred during 2000–2003. These data are used to create a landslide susceptibility map for Yueqing area using the fractal-based approach. We then examine the validity of the susceptibility map by considering 149 landslides occurred from 2004 to 2007 in the same area.

4.3 Description of Methodology

The spatial distribution of landslides is generally not uniform, but instead clustered at many different scales. The degree of clustering of landslides can be quantified using Mandelbrot's fractal concepts. A natural way to measure the degree of spatial clustering of landslides is to determine the fractal dimension (D). In a two-dimensional space, a fractal dimension of 2 describes a random distribution, while a fractal dimension of 0 describes a single point (Mandelbrot 1983). If the spatial clustering of landslides is fractal, the value of the fractal dimension generally lies in the range $0 < D < 2$. The fractal clustering distribution is different from random distributions (e.g. Poisson distribution) and also different from the uniform distribution. The differences between the fractal clustering distribution and the random distribution or the uniform distribution are illustrated in Fig. 4.2. The essential feature of the fractal distribution is its *self-similarity or scale invariance*. In a fractal clustering point pattern, there are no characteristic mean distances, and points are spaced in a hierarchy of clusters across scales. However, for any physical application there are upper and lower limits on applicability of the fractal distribution, i.e., the *scale invariance* exists only within a finite range of scale.

The fractal analysis method here used is the box-counting method popularized by Mandelbrot (1985). We consider the landslides in a region as a set of points in the two-dimensional space. For a region with a planar area of $L \times L$, we discretize it into square grid cells of size r and count the number $N(r)$ of cells that contain at least one landslide, as illustrated in Fig. 4.3. The process is repeated by reducing the cell size until a prescribed minimum size is reached. If $N(r)$ has the following relationship with the cell size r:

$$N(r) \propto r^{-D} \tag{4.1}$$

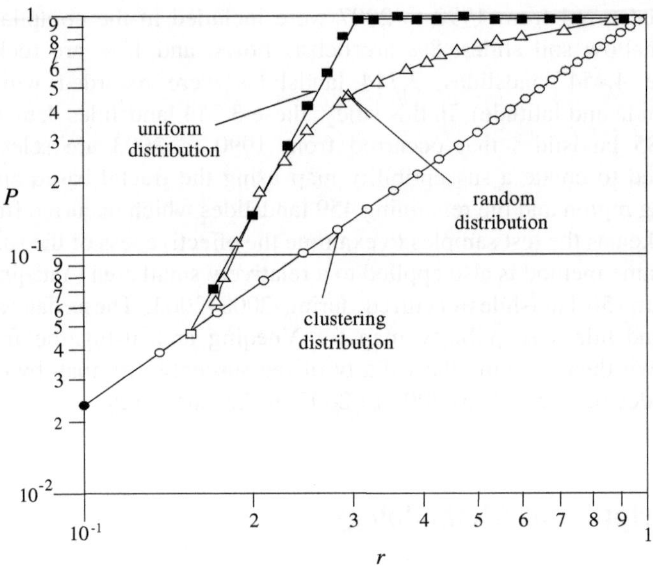

Fig. 4.2 Comparison of fractal clustering distribution with random and even distributions. P on the ordinate shows the probability that events occur in an observing scale with r

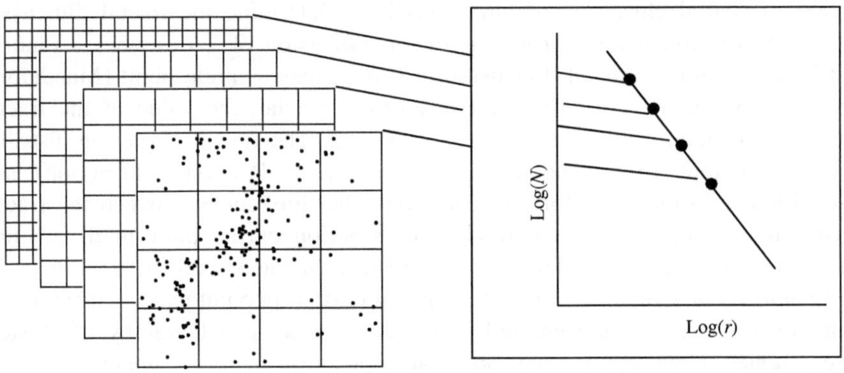

Fig. 4.3 Schematic diagram illustrating the fractal box-counting method in which the study area is divided into square grid cells of size r and the number of cells that contain one or more landslides is counted. The grid-cell size r is change and the counting procedure repeated. The number of occupied grid cells versus the grid-cell size r is plotted on log–log axes. The points were best fit by a single straight line. The slope of the straight line is the fractal dimension D

one says the spatial distribution of these landslides is fractal, characterized by a fractal dimension D. The D value is the slope of the straight-line fitted to Eq. (4.1).

Here, as an extension to application of Eq. (4.1), let $N(r_{\min})$ be the number of cells occupied by landslides when a minimum cell size $r_{\min}$ is larger than the typical distance between landslides, and let $P_i = N(r_i)/N(r_{\min})$, $r_{\min} \leq r_i \leq L$. We then have

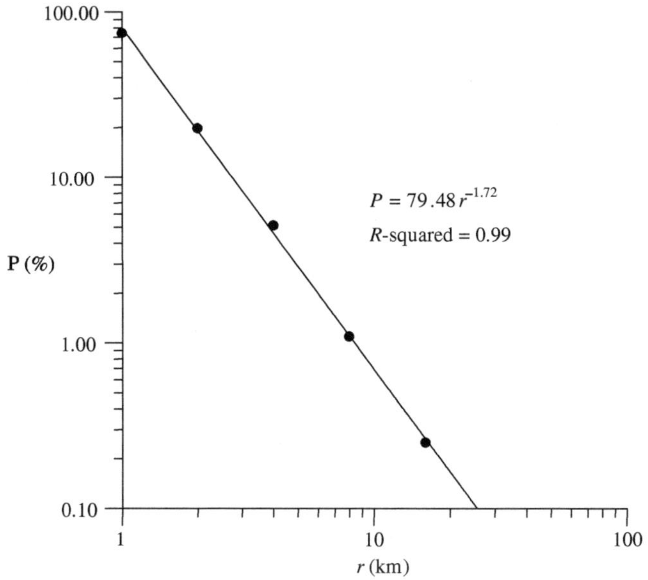

Fig. 4.4 Variation in P with r for 3,285 landslides during the period 1990–2003, Zhejiang region, China

$$P = Cr^{-D} \tag{4.2}$$

which means that the ratio of the landslides occurring at a given location over the total number of landslides can be defined as function of distance r from a known landslide. If the spatial distribution of landslides is characterized by a fractal clustering, then Eq. (4.2) can be utilized for landslide susceptibility zoning and mapping.

4.4 Application and Validation of Fractal Statistics to Landslide Susceptibility Mapping

4.4.1 Landslide Susceptibility Mapping

In this section, we apply the proposed method to produce landslide susceptibility maps for the Zhejiang region and for the Yueqing County, a relatively small area located in the southeast costal region of Zhejiang.

The 3,285 landslides occurred from 1990 to 2003 are used to produce a landslide susceptibility map of the Zhejiang region. Figure 4.4 is the result of the application of the box-counting method to 3,285 landslides occurred from 1990 to 2003. In the calculation, a 1 × 1 km grid cell is used as the minimum cell; landslides with distances smaller than 1 km are considered as a single landslide.

It can be seen from Fig. 4.4, the ratio of the landslides occurring at a given location over the total number of landslides (*P*) within a range of 1–15 km as function of distance (*r*) from a known landslide follows quite closely a straight line. From Eq. (4.2), the relationship given in Fig. 4.4 can be fitted by

$$P = 79.48r^{-1.72}, \tag{4.3}$$

where the slope $D = 1.72$ is obtained using a least square fitting method, with the coefficient of determination $R^2 = 0.99$. This result shows that these landslides are fractal clustering within a range of 1–15 km.

A distribution map can be made using the 3,285 landslides from 1990 to 2003. Using the buffer function of GIS and from Eq. (4.3), where *r* is radius, a buffer zone (area) around any known landslide can be built within a range of 2–15 km (when $r = 1$ km, the landslide itself is taken as a center). With each of 3,285 landslides serving in turn as the center, we can produce a series of buffer areas corresponding to different *r*, which are denoted as A1, A2,...,A*n*, where *n* is the number of the buffer areas and is related to the step length of *r* in the calculation. We take the maximum of each area *P*(*r*) = max (A1, A2, A3,...,A*n*), and then a *P*(*r*) distribution map is obtained. After the *P*(*r*) map is smoothed (here in 7 × 7 grids), we use the *W* function (Anderberg 1973; OAS 1991) to classify the landslide susceptibility zones for the *P*(*r*) map. A *k*-order *W* function is defined as follows

$$W = \sum_{i=1}^{k}\sum_{j=1}^{n_i}(x_{ij} - x_i)^2 = W_1 + W_2 + \cdots + W_k, \tag{4.4}$$

where, *x* denotes a set of observations [in our case, *x* is (*P*(*r*)], x_{ij} is the *j*th observation in the *i*th group, n_i is the number of observations in the *i*th group, and x_i is the mean value for the *i*th group. The objective is to find an optimal division of *k* groups such that the value of *W* is minimized. We use the principle of least squares, a common statistical approach, to this one-dimensional problem through minimizing the sum of squared deviations about the around means. Once the smallest *W* value is determined, the best grouping of the proportional values can be achieved.

For the previously mentioned *P*(*r*) map (smoothed), the W_i values ($W_1 = 38{,}540.97$, $W_2 = 16{,}985.69$, $W_3 = 15{,}591.37$, $W_4 = 10{,}706.27$, and $W_5 = 2{,}530.12$) can be determined from Eq. (4.4) for $k = 5$. From this, the range of *x* values is divided into five groups ($x \leq 3.37$, $3.37 < x \leq 6.75$, $6.75 < x \leq 10.13$, $10.13 < x \leq 13.51$, and, $13.51 < x \leq 16.88$), which classified the landslide susceptibility into 5 levels: very low zone (I), low zone (II), moderate zone (III), high zone (IV) and very high zone (V). The flow chart for producing a landslide susceptibility zoning map is shown in Fig. 4.5, and the resulting landslide susceptibility zoning map is illustrated in Fig. 4.6.

In the previous example, our method has been tested on the entire continental region of Zhejiang Province. In the following we apply the method to a relatively small area, i.e., the Yueqing County (see the shaded area in Fig. 4.6) in Wenzhou City.

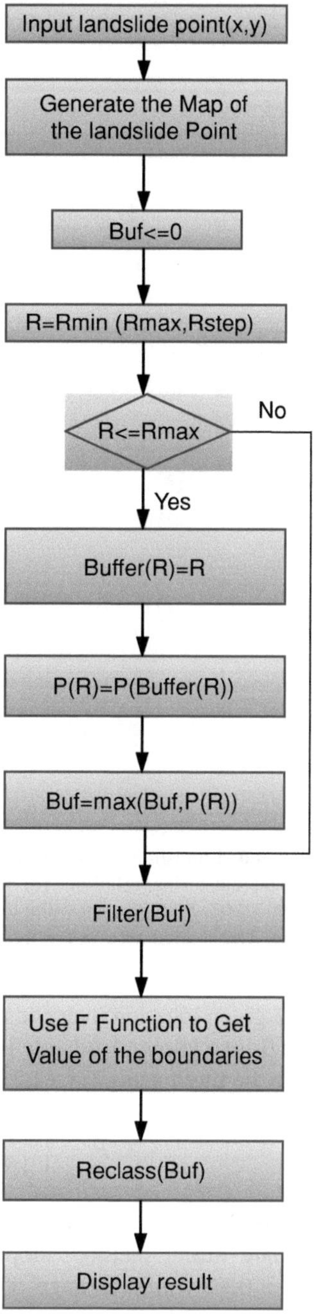

Fig. 4.5 Algorithm flow chart for producing landslide susceptibility zoning map

In the Yueqing area, about 156 landslides occurred during period 2000–2003. When the box-counting method is applied to the 156 landslides, a 25 × 25 m grid cell is taken as the minimum cell covering the area. The ratio of the landslides occurring

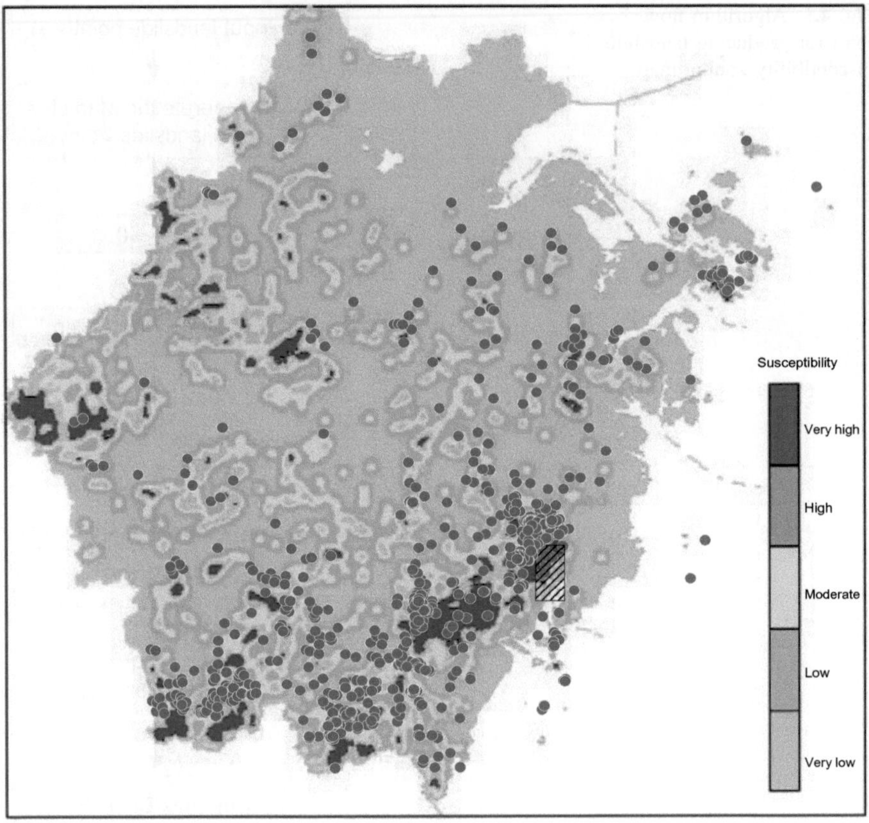

Fig. 4.6 Landslide susceptibility map of Zhejiang Province. The map is produced using the fractal clustering relations for the spatial distribution of 3,285 landslides during the period 1990–2003. *Pink dots* show the locations of 459 test landslides (2004–2007) on the susceptibility map. The shaded area in the figure shows the range of Yueqing County

at a given location over the total number of landslides (P) as a function of distance (r) from a known landslide is plotted in Fig. 4.7 and two line segments in the figure may be fitted using Eq. (4.2), i.e., the relationship between P and r is as follows

$$P = \begin{cases} 128.5r^{-0.07} & (25\,\text{m} < r \leq 800\,\text{m}) \\ 15789.0r^{-0.79} & (800\,\text{m} < r \leq 7000\,\text{m}) \end{cases} \tag{4.5}$$

The slopes of the two line segments in Fig. 4.7 are calculated using the least square fitting method; slope $D = 0.07$ within a range of 25–800 m and slope $D = 0.79$ within a range of 800–7,000 m. The fractal clustering distribution has different scaling exponents at two different scaling lengths, a result which is different from that obtained for the entire region of Zhejiang (see Fig. 4.4). Because the types of landslides in the Yueqing area are similar to those of landslides in other region of Zhejiang, and mainly belong to shallow landslides and are

Fig. 4.7 Variation in P with r for 156 landslides occurred during the period 2000–2003, Yueqing County, Wenzhou City, Zhejiang Province

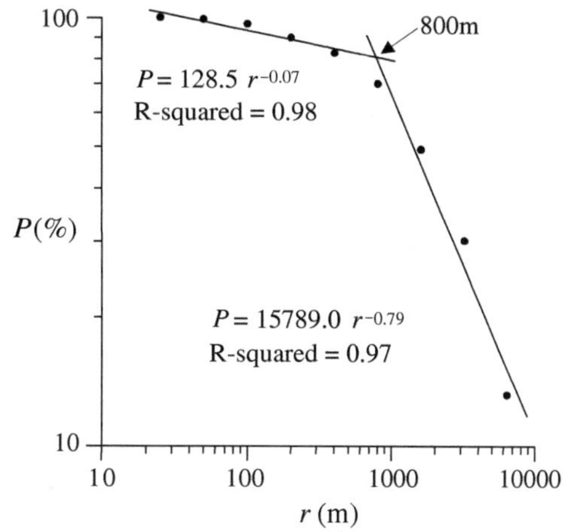

triggered by rainfall, we attribute this to the fact that, in the box-counting method, the minimum cell needed for covering the Yueqing County is much smaller than that for the entire region of Zhejiang where landslides with distances smaller than 1 km are considered as a single landslide.

Following to the same methodology used for the entire region of Zhejiang, a landslide distribution map for the Yueqing County is made using the 156 landslides occurred from 2000 to 2003. Then, using the buffer function of GIS and from Eq. (4.5), with r as radius, a buffer zone (area) around a known landslide is built in the range of 25 m $< r \leq$ 800 m and the range of 800 m $< r \leq$ 7,000 m, respectively, and the landslide susceptibility zoning map for the Yueqing County is obtained, as shown in Fig. 4.8. In the figure, the landslide susceptibility is divided into 4 levels ($k = 4$): very low zone (I), low zone (II), moderate zone (III), and high zone (IV).

4.4.2 Verification of the Landslide Susceptibility Maps

Before a landslide susceptibility map is applied to making predictions, its reliability must be examined. However, up to date there is no a generally-accepted method to verify the landslide susceptibility map. Casadei et al. (2003) defined an "optimal" model as the one that is able to identify the maximum number of landslides with the minimal of area predicted to be unstable. In our opinions, the effectiveness of a landslide susceptibility map can be examined in following two ways:

1. Compare a landslide susceptibility map with actual landslides (test samples) that occurred after the completion of the map and calculate the percentage of the test samples falling within areas of preferred susceptibility and the percentage of the

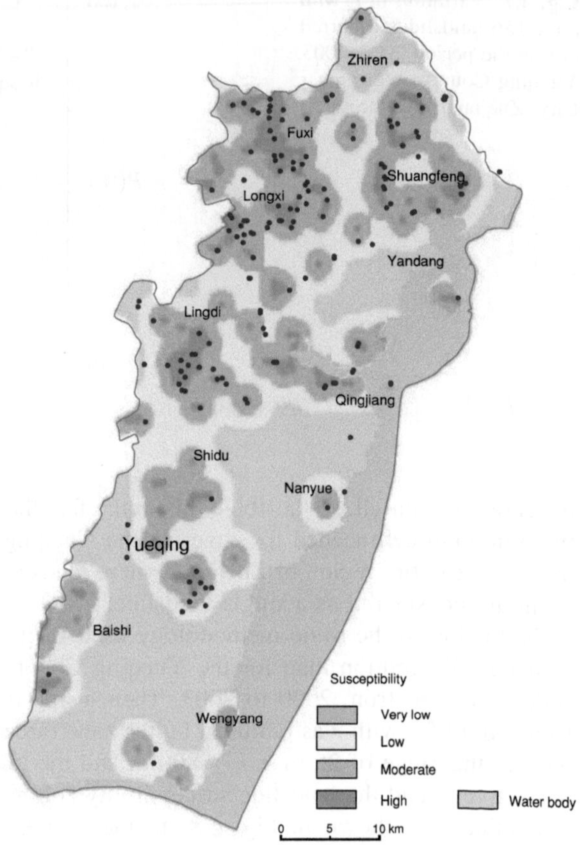

Fig. 4.8 Landslide susceptibility map of Yueqing County (see the shaded area of Fig. 4.6), Wenzhou City, Zhejiang Province. The map is produced using the fractal clustering relations for the spatial distribution of 156 landslides occurred during the period 2000–2003. *Black dots* show the locations of 149 test landslides 2004–2007) on the susceptibility map.

area that has no susceptibility (very low susceptibility zone). A primary goal of susceptibility mapping is to maximize both values (Coe et al. 2004).

2. Inspect the ratio between the percentage of the test samples that fall in class area and the percentage of the class area in total area. For a successful landslide susceptibility zoning, this ratio should gradually increase from very low susceptibility zone (or non- susceptibility zone) to high susceptibility zone.

Let $A(p_i)$ denote the percentage of area of the ith susceptibility zone over the total area of the study region, and $L(p_i)$ the percentage of landslides that fall in the ith susceptibility zone over the total landslides as the test group, according to the above procedure, the ratio

$$R_i = L(p_i)\,/A(p_i), \quad i = \mathrm{I}, \mathrm{II}, \ldots, \mathrm{V}, \tag{4.6}$$

can be used to evaluate the effectiveness of the landslide susceptibility mapping. As long as these ratios for different susceptibility zones satisfy

$$R_{\mathrm{I}} < R_{\mathrm{II}} < R_{\mathrm{III}} < R_{\mathrm{IV}} < R_{\mathrm{V}}, \tag{4.7}$$

Table 4.1 Results of verification for the landslide susceptibility map of Zhejiang Province using test samples

Susceptibility level	Percentage of the class area accounts for total area ($A(p_i)$)	Landslides occurring in each susceptibility level account for percentage of total samples ($L(p_i)$)	$R_i = L(p_i) / A(p_i)$
Very low zone (I)	52.01	4.18	0.08
Low zone (II)	16.10	2.0	0.12
Moderate zone (III)	16.66	10.16	0.61
High zone (IV)	10.79	47.54	4.41
Very high zone(V)	4.44	36.12	8.14

The test samples are 459 landslides occurred during the period 2004–2007

Table 4.2 Results of verification for the landslide susceptibility map of Zhejiang Province using test samples

Susceptibility level	Percentage of the class area accounts for total area ($A(p_i)$)	Landslides occurring in each susceptibility level account for percentage of total samples ($L(p_i)$)	$R_i = L(p_i) / A(p_i)$
Very low zone (I)	33.78	6.04	0.18
Low zone (II)	29.11	8.72	0.30
Moderate zone (III)	30.29	28.19	0.93
High zone (IV)	6.82	57.05	8.37

The test samples are 149 landslides occurred during the period 2004–2007

the landslide susceptibility zoning is reasonable.

In Fig. 4.6, the percentage of the area of each susceptibility level over the area of the total study region is, respectively, as follows: very low zone (I) 52.01%, low zone (II) 16.10%, moderate zone (III) 16.66%, high zone (IV) 10.79%, and very high zone (V) 4.44%. All 459 landslides that occurred during the period 2004–2007 are taken as test samples and used to evaluate the effectiveness of the susceptibility map. The test results (Table 4.1) show that the percentage of these landslides that fall in each zone over the total test samples is as follows: very low zone (I) 4.18%, low zone (II) 2.0%, moderate zone (III) 10.16%, high zone (IV) 47.54% and very high zone (V) 36.12%. The area of IV and V zones accounts for 15.23% of the total area, but the landslides that fall in these two zones account for 83.66% of the total test samples.

In Fig. 4.8 the susceptibility level is presented as four categories (very low, low, moderate, and high). The area of each category accounting for percentage of the total study area is, respectively, as follows: very low zone (I) 33.78%, low zone (II) 29.11%, moderate zone (III) 30.29%, and high zone (IV) 6.82%. All 149 landslides occurred during the period from 2004 to 2007 are taken as test samples and used to examine the effectiveness of the susceptibility map. As those shown in Table 4.2, the percentage of these landslides that fall in each zone over the total test samples is as follows: very low zone (I) 6.04%, low zone (II) 8.72%, moderate zone (III) 28.19%, and high zone (IV) 57.05%.

As shown in Tables 4.1 and 4.2, even though the two examples given here involve different regional scales, the distributions of R_i value can well satisfy Eq. (4.7), which shows that the landslide susceptibility zoning obtained from the fractal method is reasonable. However, it should be noted that our predictive model is a statistical model, and therefore, it is inevitable that two types of error may occur. Type I error (also known as a false positive) occurs if the model falsely predicts occurrence of landslides in an area while there is no landslides in the area at the given time. Type II error occurs if the model predicts that there is no landslide in an area at certain time when landslides did occur in the area at the time. As seen in Fig. 4.8, some high susceptibility areas are landslides free (the landslides used as test sets), on the contrary very low susceptibility areas contain few landslides. These errors may be reduced and our model improved if more landslide data are collected in the future.

4.5 Discussion

The existing methods for landslide susceptibility mapping, either statistic methods (e.g. Gupta and Joshi 1990; Binaghi et al. 1998; Chung and Fabbri 1999; Gritzer et al. 2001; Lineback et al. 2001; Sakellariou and Ferentinou 2001; Santacana et al. 2003; Lee 2007; Lee et al. 2007) or physically-based methods (e.g. Montgomery and Dietrich 1994; Jibson et al. 2000; Savage et al. 2003), require many types of data such as geology, topography, soil deposits, land cover and so on. However, since the data obtained from existing observations are always incomplete, any assessment of landslide susceptibility based on the data contains a considerable amount of uncertainties that would likely increase with the increase of the categories of the used data.

Though only the known landslide data are used in the fractal clustering distribution method described herein, if these landslides belong to a same type and are triggered by a same factor (e.g. rainfall) under the same geologic, geomorphologic and climatic settings, the involved uncertainties will be relatively low. On the other hand, though the mechanism of landslides is extremely complicated, as many complex phenomena in nature, rules underlain the landslide phenomena might be simple. If the spatial distribution of landslides is a fractal clustering, then within the statistically identified range of self-similarity, the property of fractal clustering of landslides can be used to assess the probability of landslide occurrences at a given distance from a known landslide and to delineate landslide susceptibility areas. Therefore, for many regions that are prone to rainfall-triggered landslides, where landslide inventory data are available and the landslides were recorded with occurrence locations (coordinates), this fractal approach based on GIS can provide an effective way to estimate landslide susceptibility.

In addition, landslide susceptibility zoning maps should be prepared at a scale appropriate for displaying the information needed at a particular zoning level. In using this method, the scale of mapping depends on the level of the available landslide inventory in a region. This technique is appropriate to landslide susceptibility mapping for any given region where the existing landslides spatially follow the fractal clustering distribution.

4.6 Conclusions

In this study, we propose a fractal-based method to map landslide susceptibility using historical landslide inventories only. The key points in this method are first to measure the spatial fractal clustering distribution of the existing landslides in a region, and then apply the derived fractal clustering relation to produce a landslide susceptibility map by means of the buffer function of GIS. The method has been demonstrated using two examples at different regional scales, based on the landslide inventory data from Zhejiang Province, where the landslides are mainly triggered by rainfall. In the first example, 3,285 landslides occurred during the period 1990–2003 in Zhejiang region are used to produce a landslide susceptibility map for the region, and then 459 landslides that occurred in the same region during the period 2004–2007 are used to examine the effectiveness of the susceptibility map. In the second case, the landslide susceptibility map of the Yueqing County located in the southeast costal region of Zhejiang is produced using 156 landslides occurred in the area during the period 2000–2003 and examined by 149 landslidesoccurred in the area during the period 2004–2007. The verification results show that the landslide susceptibility maps are reliable.

For many regions that are prone to rainfall-triggered landslides but with very limited data availability (e.g. landslide inventories only), using the technique described in this paper, a satisfactory landslide susceptibility mapping from the landslide inventories can be obtained.

Acknowledgments This study has been partially funded by the Science and Technology, Department of Zhejiang Province (No. 2006C13024). We would like to thank the Natural Hazards reviewers and the Terrigenous Mass Movements editors for their valuable comments, which have improved the paper. We also particularly thank Dr. Zhiming Lu of Los Alamos National Laboratory for his thorough and careful correction of an early draft of the manuscript.

References

Aleotti P (2004) A warning system for rainfall-induced shallow failures. Eng Geol 73:247–265

Anderberg MR (1973) Cluster analysis for applications. Academic Press, New York

Binaghi E, Luzi L, Madella P, Rampini A (1998) Slope instability zonation: a comparison between certainty factor and fuzzy Dempster-Shafer approaches. Nat Hazards 17:77–97

Casadei M, Dietrich WE, Miller L (2003) Testing a model for predicting the timing and location of shallow landslide initiation in soil-mantled landscapes. Earth Surface Process Landforms 28:925–950

Chung CF, Fabbri AG (1999) Probabilistic prediction models for landslide hazard mapping. Photogrammetric Eng Remote Sens 65:1389–1399

Coe JA, Godt JW, Baum RL, Bucknam RC, Michael JA (2004) Landslide susceptibility from topography in Guatemala. In: Lacerda WA, Ehrlich M, Fontura SAB, Sayão ASF (eds) Landslides: Evaluation and Stabilization, vol 1, pp 69–78

D'Ambrosio D, Spataro W, Iovine G (2006) Parallel genetic algorithms for optimising cellular automata models of natural complex phenomena: an application to debris-flows. In: Iovine G, Sheridan M, Di Gregorio S (eds) Computer simulation of natural phenomena for hazard assessment, Computers and Geosciences, 32(7): 861–875

Gritzer ML, Marcus WA, Aspinall R, Custer SG (2001) Assessing landslide potential using GIS, soil wetness modelling and topographic attributes, Payetti River, Idaho. Geomorphology 37:149–165

Gupta RP, Joshi BC (1990) Landslide hazard zoning using the GIS approach- A case study from the Ramganga catchment, Himalayas. Eng Geol 28:119–131

Iovine G, Di Gregorio S, Lupiano V (2003a) Debris-flow susceptibility assessment through cellular automata modeling: an example from the 15–16 December 1999 disaster at Cervinara and San Martino Valle Caudina (Campania, Southern Italy). Nat Hazards Earth Sys Sci 3:457–468

Iovine G, Di Gregorio S, Lupiano V (2003b) Assessing debris-flow susceptibility through cellular automata modelling: an example from the May 1998 disaster at Pizzo d'Alvano (Campania, southern Italy). In: Rickenmann D, Chen C L (eds) Debris-Flow Hazards Mitigation: Mechanics, Prediction and Assessment. In Proceedings of 3rd DFHM International Conference, Davos, Switzerland, September 10–12, 2003, Millpress Science Publishers, Rotterdam, 1:623–634

Iovine G, D'Ambrosio D, Di Gregorio S (2005) Applying genetic algorithms for calibrating a hexagonal cellular automata model for the simulation of debris flows characterised by strong inertial effects. Geomorphology 66(1–4):287–303

Iovine G (2008) Mud-flow and lava-flow susceptibility and hazard mapping through numerical modelling, GIS techniques, historical and geoenvironmental analyses. In: Sànchez-Marrè M, Béjar J, Comas J, Rizzoli A E, Guariso G (eds) Proceedings of the iEMSs Fourth Biennial Meeting: International Congress on Environmental Modelling and Software (iEMSs 2008). International Environmental Modelling and Software Society, Barcelona, Catalonia, July 2008. ISBN: 978-84-7653-074-0. 3:1447–1460

Jibson RW, Harp EL, Michael JA (2000) A method for producing digital probabilistic seismic landslide hazard maps. Eng Geol 58:271–289

Keefer DK, Wilson RC, Mark RK, Brabb EE, Brown WM, Ellen SD, Harp EL, Wieczorek GF, Alger CS, Zatkin RS (1987) Realtime landslide warning during heavy rainfall. Science 238: 921–925

Lee S (2007) Application and verification of fuzzy algebraic operators to landslide susceptibility mapping. Env Geol 52(4):615–623

Lee S, Ryu JH, Ii-Soo Kim IS (2007) Landslide susceptibility analysis and its inspection using likelihood ratio, logistic regression and artificial neural network models: case study of Youngin, Korea. Landslide 4:327–338

Li CJ (ed) (2004) Remote sensing survey and integrated investigation of land and resources in Zhejiang Province, Geological Publishing House, pp 39–82 (in Chinese with abstract in English)

Li CJ, Ma TH, Zhu XS (2010) aiNet–and GIS-based regional prediction system for the spatial and temporal probability of rainfall-triggered landslides. Nat Hazards 52:57–78

Lineback GM, Marcus WA, Aspinall R, Custer SG (2001) Assessing landslide potential using GIS, soil wet-ness modeling and topographic attributes, Payette River, Idaho. Geomorphology 37:149–165

Mandelbrot BB (1983) The fractal geometry of nature. W.H. Freeman and Company, New York, p 468

Mandelbrot BB (1985) Self-affine fractals and fractal dimension. Phys Scr 32:257–260

Montgomery DR, Dietrich WE (1994) A physically based model for the topographic control on shallow land-sliding. Water Resour Res 30:1153–1171
Organization of American States (1991) Landslide hazard assessment. In: Primer on Natural Hazard Management in Integrated Regional Development Planning. Washington, D.C.: Natural Hazards Project of the Department of Regional Development and Environment, Executive Secretariat for Economic and Social Affairs, Organization of American States. http://www.oas.org/dsd/publications/Unit/oea66e/begin.html. Cited 05 Jun 2008
Sakellariou MG, Ferentinou MD (2001) GIS-based estimation of slope stability. Nat Hazards Rev 2(1):12–21
Santacana N, Baeza B, Corominas J, De Paz A, Marturiá J (2003) A GIS-based multivariate statistical analysis for shallow landslide susceptibility mapping in La Po-bla de Lillet area (Eastern Pyrenees, Spain). Nat Hazards 30:281–295
Savage WZ, Godt JW, Baum RL (2003) A model for spatially and temporally distributed shallow landslide initiation by rainfall infiltration. In: Rickenmann D, Chen C (eds) Debris flow hazards mitigation: mechanics, prediction, and assessment. Mill-press, Rotterdam, pp 179 -187
Tangestani MH (2003) Landslide susceptibility mapping using the fuzzy gamma operation in a GIS, Kakan catchment area, Iran. Proceedings of 6th Annual International Conference Map India 2002, New Delhi, India

Chapter 5
Preliminary Slope Mass Movement Susceptibility Mapping Using DEM and LiDAR DEM

M. Jaboyedoff, M. Choffet, M.-H. Derron, P. Horton, A. Loye, C. Longchamp, B. Mazotti, C. Michoud and A. Pedrazzini

Abstract Hazard mapping in mountainous areas at the regional scale has greatly changed since the 1990s thanks to improved digital elevation models (DEM). It is now possible to model slope mass movement and floods with a high level of detail in order to improve geomorphologic mapping. We present examples of regional multi-hazard susceptibility mapping through two Swiss case studies, including landslides, rockfall, debris flows, snow avalanches and floods, in addition to several original methods and software tools. The aim of these recent developments is to take advantage of the availability of high resolution DEM (HRDEM) for better mass movement modeling. Our results indicate a good correspondence between inventories of hazardous zones based on historical events and model predictions. This paper demonstrates that by adapting tools and methods issued from modern technologies, it is possible to obtain reliable documents for land planning purposes over large areas.

Keywords DEM · Lidar · Rockfall · Debris-flow · Floods · Snow avalanches · Regional hazard mapping · Models · Flow-R · RAS · Conefall · HISTOFIT

5.1 Introduction

Slope mass movement hazard mapping has been a major concern since the 1970s. In Varnes (1984), the authors made an overview of the principal mapping practices that were mainly linked to field investigations and aerial photo interpretations

M. Jaboyedoff (✉) · M. Choffet · M.-H. Derron · P. Horton ·
A. Loye · C. Longchamp · B. Mazotti · C. Michoud · A. Pedrazzini
Institute of Risk Analysis-University of Lausanne, Lausanne, Switzerland
e-mail: michel.jaboyedoff@unil.ch

B. Pradhan and M. Buchroithner (eds.), *Terrigenous Mass Movements*,
DOI: 10.1007/978-3-642-25495-6_5,

(Cruden and Thomson 1987). Since the appraisal of geographical information systems (GIS) (Carrara and Guzzetti 1995) and the production of digital elevation models (DEM), hazard mapping of slope mass movements has increased significantly. In addition over the past decade, new techniques such as airborne laser scanning by Lidar (Light Detection and Ranging) provide to the earth sciences community high resolution DEM (HRDEM) with resolutions higher than 0.5 pts/m^2 (Shan and Toth 2008).

This computerization of hazard mapping has significantly supported susceptibility mapping at regional scale. This type of mapping is designed to provide a fast overview of area that is affected by potentially dangerous events. This is of primary importance for regional authorities and municipalities because of the responsibilities linked to risk management. The first step of a rational risk assessment is to have an overview of the area potentially endangered by slope movements such as deep-seated landslides, shallow landslides, debris flows, rockfall, flooding and erosion. This is most commonly done by producing preliminary susceptibility maps over the entire territory considered. According to Swiss guidelines, (Lateltin 1997; Loat and Petraschek 1997), this kind of maps is the first step of the process leading to detailed so-called "danger" maps used for local planning in communities.

However, such maps do not give any detailed information on the intensity or the frequency of occurrence of the slope movements. They only indicate the hazardous zones for instance at a 1:25,000 scale. The methods of mapping are various and numerous and some excellent recent overviews and recommendations published about landslides mapping can be found in Aleotti and Chowdhury 1999, Van Westen et al. 2006, Fell et al. 2008a, b, Cascini 2008. For floods, there is also a large variety of approaches (Merz et al. 2007). For snow avalanches, regional mapping was proposed by Toppe (1987) and more detailed mapping is also well established (Ancey et al. 2006; PPR 2011).

The present paper focuses on two examples in Western Switzerland of multi-hazards regional mapping. These examples show the variety of situations depending on the geological and geographical conditions but also on the number and types of data currently available. It is thus not possible to follow a unique method in all situations but it must be adapted to each circumstance (Jaboyedoff and Derron 2005). The proposed approach is based on deterministic simple modelling using DEM, interpretation of data produced from HRDEM like hillshades and fast field surveys. It demonstrates that up to a certain level, a simple relative hazard scale can be included in susceptibility maps. The use of all available modern documents HRDEM, vectorized topographic maps, orthophotos permits to obtain quite rapidly reliable results at regional scale. The limitations of such approach are usually: (1) the lack of data to calibrate the models, (2) the HRDEM permits to improve the quality of the results but they also induce some problems such as handling very large datasets or introducing some artefacts in the overly detailed topography.

5.2 Problem Identification/Conceptual Background

GIS has greatly improved the capacity of mass movement and flood hazard mapping and currently most such studies use a GIS framework (Carrara and Guzzetti 1995; Chacón et al. 2006). Hence, one of the principal improvements of these last ten years is the use of DEM and HRDEM, because it permits first to perform modeling and second to examine the details of the morphology. It must be observed that HRDEM often possess a resolution too high for modeling at regional scale, necessitating that its resolution be decreased. Therefore, regional scale mapping is now possible by relying on modeling, based only on geomorphologic approaches using HRDEM. Of course regional mapping benefits from old documents such as geological maps, which indicate some landslide locations, topographic maps, etc. However, the availability of new digital documents (vectorized geological maps, high resolution satellite images, series of orthophotos, etc.) is making it possible to improve all mapping methodologies, which have become highly dynamic. As a consequence, methods are no longer set permanently. This is an issue when working with regional/national authorities who would like to have a definitive methodology/product. It is possible that if the process of mapping takes a long time, the product is already outdated when it is issued, i.e., because a new HRDEM has been released in the mean time. Moreover at present, the coverage of regions by one type of document is frequently not homogeneous, which makes the creations of hazard maps more difficult.

One of the solutions for regional mapping purpose is to make cross validations of simple models using DEM with other documents and especially field investigations. Such an approach is one of the most efficient ways to obtain results making use of modeling and new documents. It is a heuristic type of approach mixed with a deterministic approach (Soeters and Van Westen 1996; Van Westen et al. 2006). The limits of such an approach are linked to the quality of data and the availability of inventories of events that are very important for calibrating the methods.

5.3 Review of Literature

Einstein (1988) presented one of the first reviews on landslide risk analysis including also the description of the maps needed for such purposes. Such risk approaches have underlined the necessity to use GIS to make further analysis for both mass movements, floods and erosion (Carrara and Guzzetti 1995; Consuegra et al. 1995). Nowadays publications on this topic abound and frequently papers related to GIS hazard assessments present several different methods. These methods have been classified by Soeters and Van Westen (1996), Van Westen et al. (2006) (see also Chacón et al. 2006) as:

1. Inventory based approach
2. Heuristic approach
3. Statistical and probabilistic approach
4. Deterministic approach.

The great difference between hazard assessment methods for landslides versus floods, erosion or snow avalanches is that the last tree are mainly based on inventories, because the events are relatively frequent and repeated in an area (Marco 1994; Gilard and Gendreau 1998; PPR 2011). While landslide mapping is often based on poor inventories in numbers and as a consequence the potentialities of events are often difficult to locate precisely. Furthermore the prevention of floods and snow avalanches is more based on alerts and risk management (McClung and Schaerer 1993; Directive 2007/60/EC), versus landslides. Nevertheless the objectives of regional hazard maps are the same; they must point out the areas of conflict between hazardous zone and human activities (Lateltin 1997; Loat and Petraschek 1997).

Considering the amount of existing literature, we will focus this short summary on regional hazard mapping using DEM. For landslides, one of the first attempts to use GIS and DEM was proposed by Carrara et al. (1978, 1991) using the concept of slope units (Guzzetti et al. 1999); this approach is based on inventory and statistical methods. By multivariate analysis, using a map of known landslides, a detailed hazard zoning is then produced. Other statistical methods have been developed that are mainly based on several multivariate regressions (Chung et al. 1995). DEM are also used in a lot of new techniques for hazard mapping, referred to as neural networks, i.e., (Zeng-wang 2001; Pradhan and Lee 2010), fuzzy logic (Ercanoglu and Gokceoglu 2002, 2004; Chung and Fabbri 2008) and also logistic regression (Bai et al. 2009; Dominguez-Cuesta et al. 2009). However, these methods are mainly applied at a regional scale, but do not introduce expert knowledge in a simple way, except during the training step, within the method, which is based on inventories. More simple approaches use relative ratings of several parameters including those that derivate from DEM (Gupta et al. 1999). Similar methods have been also applied to detect rockfall sources areas (Baillifard et al. 2003, 2004).

For flood hazard mapping and prediction, the main recent advance is TOPMODEL, which makes a simple and complete simulation of the hydrological processes that determines the discharge of rivers using a DEM (Beven and Kirkby 1979; Beven et al. 1995). Hazard assessments using TOPMODEL are closely linked to the return periods of high discharge, flood depth and velocity (Marco 1994; De Moel et al. 2009; Van Alphen et al. 2009; Loat and Petraschek 1997; PPR 1999). Flood depth can be modelled by shallow water approximation (Gilard and Gendrau 1998; De Moel et al. 2009), but the heuristic approach is recommended by both French and Swiss authorities (Loat and Petraschek 1997; PPR 1999), at least at the regional scale. USGS proposes a simplified physical model TrimR2D that uses Lidar-DEM (Jones 2004). Other models using HRDEM are based on stochastic approaches (Metzger 2003).

The philosophy of TOPMODEL (Beven and Kirkby 1979) led to the development of several deterministic (introducing random variables if necessary) models for shallow landslides using a "pixel" stability analysis: SHALSTAB (Montgomery and Dietrich 1994), SINMAP (Pack et al. 1998) and TRIGRS (Baum et al. 2002).

For modeling purposes rockfall, snow avalanches and debris flows have similar procedures; first we need to detect the source areas and second to estimate the propagations. At regional scale, the detection of source areas is often based on threshold angles: (1) for snow avalanches above 30° and less than 60° (McClung and Schaerer 1993; Salm 1983) and also on the slope orientation (McClung and Schaerer 1993); (2) for rockfall 37° was used as an example for a 10 m grid size DEM in Dolomites (Italy) (Frattini et al. 2008) and Guzzetti et al. (2003) used 60° with a 10 m grid size DEM in the special case of extreme glacial valley type of Yosemite (USA). For debris-flow, the zone of initiations is more complex to detect because it needs to estimate sediments availability, water input and slope gradient (Rickenmann and Zimmermann 1993; Takahashi 1981). The use of flow accumulation (Burrough and McDonnel 1998) permits to link slope angle and triggering conditions (Rickenmann and Zimmermann 1993; Heinimann et al. 1998).

The detection of rockfall sources can be refined using structural data that makes it possible to define the potential source areas that are affected by defined mechanisms by performing kinematic tests using the DEM (Willye and Mah 2004). This can be performed using a statistical approach (Jaboyedoff et al. 2004) that count the number of discontinuities per unit of topographic surface, or using stability analysis for each DEM pixel (Gokceoglu et al. 2000; Günther 2003; Günther et al. 2004). Fuzzy logic has also been tested manly based on DEM deduced parameters (Aksoy and Ercanoglu 2006).

For snow avalanches and rockfall, Toppe (1987) used the simplest evaluation, based on the concept of shadow angle (Heim 1932; Lied 1977; Perla et al. 1980; Evans and Hungr 1993). This states that the propagations are most probably restricted to an area that is defined by the intersection of the DEM and a cone centred on the source possessing an angle $\phi°$ equivalent to a friction angle (Jaboyedoff and Labiouse 2003; Evans and Hungr 1993). For snow avalanches $\phi°$ can be adapted to the morphology of the valley flanks (Lied and Bakkehoi 1980).

More advanced techniques for rockfall are using 3D trajectory simulations that lead to regional assessment, but they require good information on the ground (Guzzetti et al. 2002, 2003; Agliardi and Crosta 2003; Dorren et al. 2003; Frattini et al. 2008). In order to obtain a continuous zoning Lan et al. (2007) interpolate the trajectories results using geostatitics.

Several applications for debris-flows propagation have been proposed but very few have been used at the regional scale (Van Westen et al. 2006). They are mostly based on multiple flow direction (Huggel et al. 2003; Heinimann et al. 1998) or on random walk (Gamma 2000). The runout distance of the debris-flow is either assessed using a friction model or using a limiting angle slope (Heinimann et al. 1998).

In addition, the geomorphologic analysis has been greatly improved in the past ten years by the introduction of airborne Lidar-DEM, because it permits to recognize in detail landslide features, and deep gravitational deformations (Crosta and Agliardi 2002). The limitations and the advantages of these mapping techniques are now well known, but has clearly shown its efficacy in creating and correcting inventories (Haugerud et al. 2003; Ardizzone et al. 2007; Schultz 2004, 2007). Some attempts have been made to automatically detect zones of landslides using roughness or dispersion of the orientation of the topography (McKean and Roering 2004; Roering et al. 2005; Glenn et al. 2006). Morphological characteristics can be also easily extracted from HRDEM (Chigira et al. 2004).

The potential information that can be extracted from HRDEM is probably not yet fully used, especially for flood geomorphic analysis. The above review shows that a combination of methods is the best way to fully use all the potentiality offered by new techniques and data.

5.4 Study Area/Experimental Site

All the examples of susceptibility mapping presented in this paper has been taken from work conducted by the Institute of Geomatics and Risk Analysis of the University of Lausanne on two Swiss study areas: (1) the County of Vaud and (2) the Bagnes Valley. This chapter introduces the main relevant features for geohazards of these two areas.

5.4.1 Vaud County

The susceptibility mapping of potential slope movements (1:25,000) was performed on the entire territory of the county of Vaud (2800 km^2), western Switzerland, for the following processes: rockfalls, shallow and deep seated landslides, mud and debris flows (Jaboyedoff et al. 2008). The geology of the county can be divided into three main regions (Trümpy 1980) (Fig. 5.1): (1) The northwestern region is located within the limestones of the Jura chain. Its elevation ranges from 400 m to approx. 2000 m a.s.l.. This area is composed of folded and thrusted Mesozoic and Tertiary carbonates platform series, in a thin skin tectonic style. (2) The middle part of the Vaud county belongs to the Swiss Molasse Plateau. It corresponds to a foreland basin of Oligo-Miocene age. The rocks are mostly poorly consolidated sandstones with some layers of shales and conglomerates. The topography is gently hilly with few cliffs made of competent sandstones and some steep slopes resulting from fluvial erosion. (3) The southeast area belongs to the Prealpine units of the Alps. Here the steep and rugged topography ranges from 400 to more than 3000 m. The main rocks are: massive limestone, dolomites, marls, evaporites and shales of Mesozoic and Cenozoic ages.

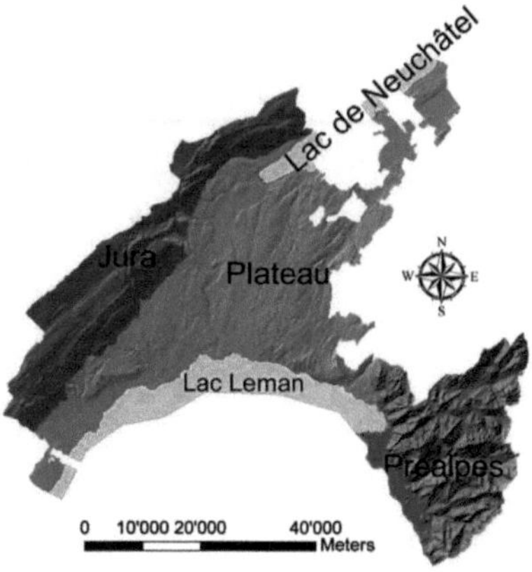

Fig. 5.1 The canton Vaud could be divided in three distinct geological areas, characterized by a different lithology and a different tectonic history (hillshade, swisstopo© 2005 SIT)

The morphology of the valleys has been strongly shaped by glacial erosion and then reworked by fluvial erosion and slope mass movements. This third domain is by the far the most susceptible to mass movements.

5.4.2 *Bagnes Valley*

The Bagnes Valley (Valais county, Switzerland) has an area of 300 km^2 with an elevation between 600 m and 4200 m a.s.l. The development of the valley is rapid because of the fast growing ski resort of Verbier. Susceptibility maps at 1:25,000 have been provided to the local authorities for the following processes: landslides, shallow landslides, rockfall, debris flows, snow avalanches, flooding and river overflowing. Similar methods as for the County of Vaud were used. In addition, as the area is smaller, some methods were improved, historical event were included, field checks were conducted and feedback from local geologists and specialists was considered.

The Bagnes Valley is one of the only alpine valleys where the three main paleogeographical domains of the Alps are present, i.e. Helvetic, Penninic and Austro-Alpine (Trümpy 1980). An extremely wide variety of rocks is then present, from some Cambrian polycyclic basements to Mesozoic-Cenozoic sedimentary covers (Sartori et al. 2006). In the lower part of the valley (Fig. 5.2), the Helvetic domain is mainly composed of massive limestone that can form high fractured cliffs. The Lower Penninic unit is dominated by various schists, as on the catchment area of the Merdenson where the quantity of mobilizable fine material mobilized by debris-flow is very important (Jaboyedoff et al. 2010).

The Middle Penninic unit is mainly constituted of a complex succession of many different types of silicate rocks, from the old polymetamorphic basements to Permo-Trias sediments. It includes quartzites, quartzitic micaschists, various volcano-detritic sediments, granites and several felsic intrusions, meta-gabbro and

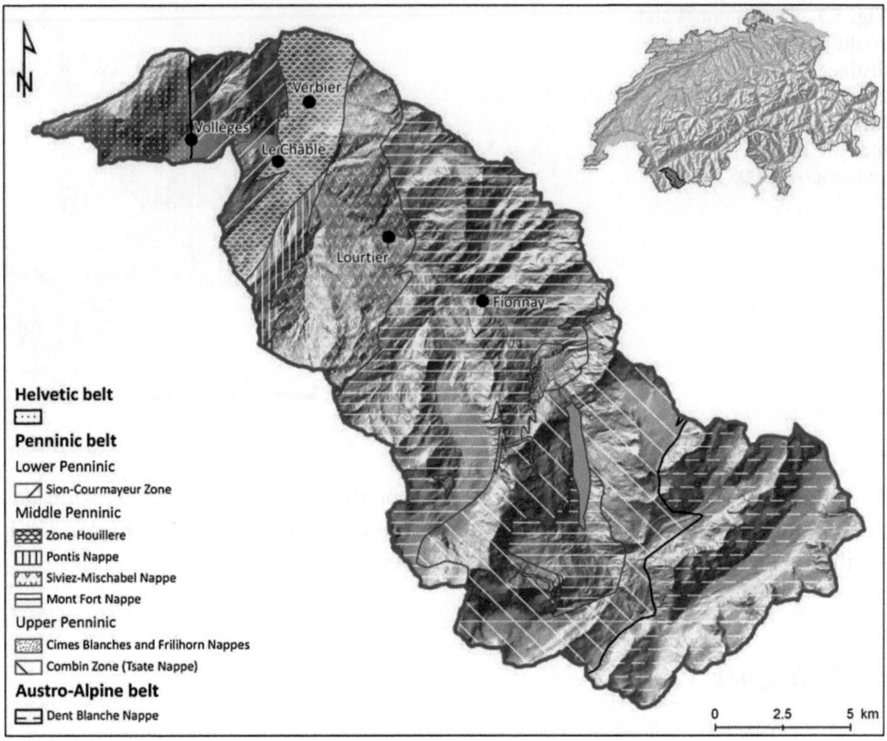

Fig. 5.2 Tectonic Units of the Val de Bagnes, from the geological national map. The valley cuts the three main belts of the Alps based on the Mesozoic paleogeography, the Helvetic, the Penninic and the Austro-Alpine belts. (modified from Thélin et al. 1994; geological national map and hillshade: ©swisstopo)

various types of amphibolitic gneisses. These lithologies are mainly outcropping between Le Châble and Mauvoisin and over all the width of the valley, it is the most important unit of the study area. All these rocks have been quite intensively deformed and metamorphised. Most of them are quite massive except some smicaschists and remnants gypsum and dolomites. It is in this tectonic unit that the most cases of rockfall occur along the road between Lourtier and Fionnay, or in Plamproz too (Jaboyedoff et al. 2010).

The Upper Penninic Unit in the Val de Bagnes is represented by a thick succession of calc-micashists and metamorphised ophiolites. The Borne du Diable cliff, on the left shore of the Mauvoisin Lake, is mostly composed of schists and it concentrates in a small areas two debris flows channels, a major rockslide, several shallow landslides and repeated rockfalls (Jaboyedoff et al. 2010). Finally, the Dent Blanche Nappe from the Austro-Alpine domain is mainly composed by massive orthogneisses forming the highest peak of the area and abrupt cliffs at the Southern end of the valley.

5.5 Data and Tools

First we briefly introduce datasets used for the Vaud County and Valley of Bagnes projects. In addition, four original software programs were developed by the authors to draw these susceptibility maps. For three of them we only provide a short description of their principles and functionalities. The fourth one, the numerical model FLOW-R, is described more in detail as it has been extensively used in these projects, as well for debris flows, rockfall, snow avalanches and flooding.

5.5.1 Data

For the County of Vaud, a 1 m resolution digital elevation model (HRDEM) derived from aerial laser scanning was available. With a mean point density of 2 point per m^2, it has an altimetric accuracy of 30 cm, with ±5 cm corresponding to one standard deviation. According to Van Den Eeckhaut et al. (2007), two different hillshade maps were created from this DEM, one with a sun elevation angle of 30° and a sun azimuth angle of 315° and another with a sun elevation angle of 30° and a sun azimuth angle of 45°. High resolution orthophotos with a ground resolution of 0.25 or 0.5 m were also used to complete and verify the DEM observations. National topographic maps 1:25,000 (Swisstopo), in both vector and raster format were been used to characterize the present-day land use.

For the Bagnes Valley, a 2 m resolution DEM performed by aerial laser scanning was used for the altitudes lower than 2000 m. As for the County of Vaud dataset, the elevation accuracy is 30 ± 5 cm (Swisstopo 2005). For the altitudes higher than 2000 m, a 25 m DEM (MNT25, swisstopo) derived from the national maps at 1:25,000 (CN25, swisstopo) was used. The altimetric accuracy is between 2 and 5 m in the study area, with a horizontal accuracy from 2.5 to 7.5 m (Swisstopo 2004). National topographic maps 1:25,000 (Swisstopo), in both vector and raster format were also used to characterize the present-day land use, in addition to maps of the geological atlas of Switzerland at 1:25,000 in the raster format (edited by Swisstopo), as well as geological and tectonic vector maps at 1:500,000. Finally, the whole Val de Bagnes is covered by orthophotos (0.5 m resolution) taken in 1999 from Swisstopo.

5.5.2 Tools

5.5.2.1 HISTOFIT

HISTOFIT is an Excel©-based application that computes the most-likely Gaussian curves in an iterative way, so that the sum of those gaussian curves fits the slope angle frequency distribution of a topography. The fitting process is done by

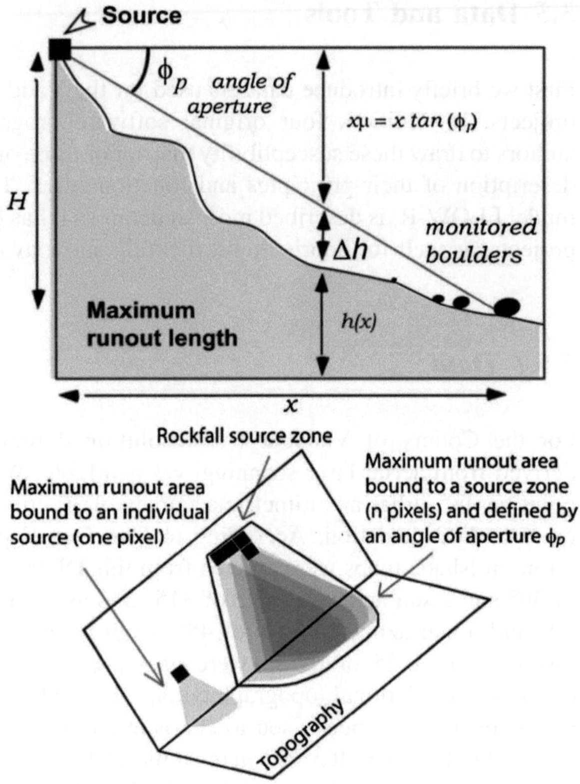

Fig. 5.3 **a** Longitudinal cross-section of the idealized shadow angle method showing the relationship between ϕ_π and the maximum runout distance (modified from Jaboyedoff 2003). **b** Scheme of the shadow angle method in 3D implemented in the GIS-based freeware CONEFALL (modified from Jaboyedoff and Labiouse 2003)

minimizing the standard error using optimisation procedures of the Excel solver. This tool enables to fit the sum of up to 5 gaussian curves to a target function represented here by a slope angle histogram. HISTOFIT is a freeware available at www.unil.ch/igar.

5.5.2.2 CONEFALL

CONEFALL is a freeware that enables to model the runout area of rock fall. The method used to simulate rockfall runout surface generalizes the shadow angle (Fahrböschung) theory (Heim 1932; Scheidegger 1973; Evans and Hungr 1993) in a 3D GIS environment. Using this approach, rockfall is considered as a sliding and rolling process going down a slope with a certain average friction angle. The model considers thus that an individual block can reach any place in the area situated inside a cone of given aperture 90° – φ_π (Fig. 5.3). The shadow angle method is empirical and does not require detailed input parameters, such as coefficient of friction and restitution coefficients.

The program requires a DEM and sources areas in a grid format. Beside the computation and display of the runout areas, the program can compute amongst other elements the number of contributing source pixels, the velocities of the blocs and their kinematic energy. CONEFALL can be downloaded from www.quanterra.org (Jaboyedoff 2003; Jaboyedoff and Labiouse 2011).

5.5.2.3 RAS

RAS is a software in development (previously at the Geological Survey of Norway and now at the University of Lausanne) to obtain a rapid assessment of snow avalanches and rockfall propagations over very large areas (up to more than one billion cells DEM). It uses the same shadow angle principle than Conefall, except that the angle of propagation is not kept constant. The angle of propagation is estimated for each source cell in function of the topography using the alpha/beta method of Lied and Bakkehoi (1980); (see also McClung and Schaerer 1993; Ancey et al. 2006). RAS has been used to map the snow avalanches potential propagation in the Bagnes Valley.

5.5.2.4 FLOW-R

The numerical model FLOW-R (Flow assessment at a Regional scale) has been developed for regional susceptibility mapping of gravitational processes (Horton et al. 2008). One of its strengths is to propose a choice of algorithms and an easy customization of the method in a graphical user interface (Fig. 5.4). The model, originally developed for debris flows, has proved to be relevant for other processes (rockfall, floodings and avalanches).

The procedure used has two steps. First, the sources are identified on the basis of various layers of data (e.g. DEM, land use map), and then these sources are propagated using a probabilistic and energetic approach (Horton et al. 2008). The volumes of the phenomena (debris-flow, rockfall etc.) are not taken into account in this model. Both the sources identification and the propagation area assessment are based on a regularly gridded DEM.

Source Area Identification

The source area identification is processed by combining various layers of data. In each layer, the cells are classified according to user-defined criteria into three possible values: possible source—excluded—ignored. The possible source option means that according to the selected criterion, the cell is a potential source area. The ignored option means that there is no evidence if the cell is a source or not, so sno decision is fixed. The excluded option means that the cell cannot be a source area. In combining the grids established for the different criteria, a cell is selected

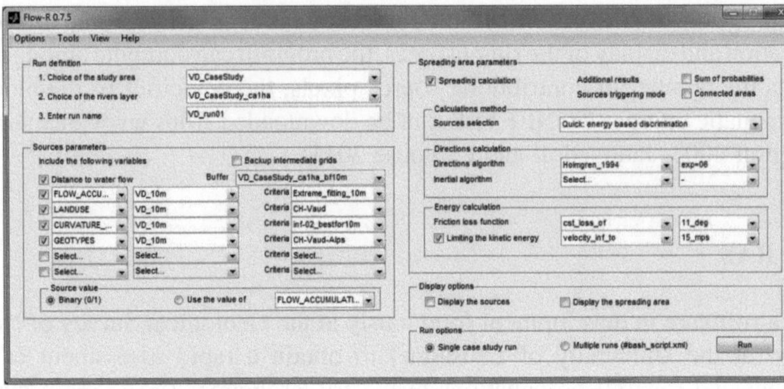

Fig. 5.4 Graphical user interface of the Flow-R model

as a source area if it was at least once identified as a possible source but never classified as excluded (Horton et al. 2008). Data can be either of a continuous or a discrete nature. In the first case, the selection is based on ranges, and in the second, on values.

Spreading Area Assessment

The spreading can be mathematically estimated by two types of algorithms: the first ones are called flow direction algorithms and determine the path that the debris flow will follow; the second ones determine mainly the runout distance (Horton et al. 2008). The propagation is calculated from each source cell. To calculate the propagation, only a grid with the source cells and a DEM are required.

1. Flow direction algorithms

First the flow direction algorithm apportions the flow from one cell to its eight neighbours in a way that there is always at least one cell in which the flow can run. The probability of spreading is a function of the slope angle and the persistence, which is a weighting of the directions according to the previous direction, allowing an integration of the notion of inertia (Gamma 2000). It is, however, not a probability in a strict mathematical sense, but it has to be interpreted in a qualitative way (Huggel et al. 2003).

The slope has a leading effect on the debris flow path. Various flow direction algorithms have been integrated and evaluated. All these algorithms are implemented in FLOW-R (Horton et al. 2008):

- D8: assigns the flow to only one adjacent cell. It is limited to directions of 45° and is very sensitive to small errors (Desmet and Govers 1996; Tarboton 1997; Erskine et al. 2006; Huggel et al. 2003; Endreny and Wood 2003).
- D∞: assigns the flow to one or two adjacent cells (Tarboton 1997).

- $\rho 8$: stochastic method which gives a probability to every cell having an altitude inferior to the central cell. The path is randomly determined afterwards, producing a single flow direction (Fairfield and Leymarie 1991).
- Multiple flow direction: based on the previous method and considers the spreading over every non-zero cell in a continuous, and not random, way (Quinn et al. 1991).
- Multiple flow direction modified: an exponent of 1.1 was added by Freeman (1991).
- Parametric flow direction: a variable exponent was added to control the spreading (Holmgren 1994). The higher is the exponent, the more convergent the flow becomes (Eq. 5.1).

$$f_{Si} = \frac{(\tan \beta_i)^x}{\sum_{j=1}^{8} (\tan \beta_i)^x} \quad \text{for all} \quad \tan \beta > 0 \tag{5.1}$$

where i, j = flow directions (1 to 8), f_{si} = flow proportion in direction i, tan β_i = slope gradient indirection as defined above and x = variable exponent.

A weighting of the directions is included to take into account the persistence of the debris flow. Based on Gamma (2000), the weight is a function of the change in angle from the last flow direction (Eq. 5.2).

$$\begin{cases} f_{pi} = W_0 & \text{if} \quad \alpha_i = 0^\circ \\ f_{pi} = W_{45} & \text{if} \quad \alpha_i = 45^\circ \\ f_{pi} = W_{90} & \text{if} \quad \alpha_i = 90^\circ \\ f_{pi} = W_{135} & \text{if} \quad \alpha_i = 135^\circ \\ f_{pi} = 0 & \text{if} \quad \alpha_i = 180^\circ \end{cases} \tag{5.2}$$

where i = flow directions (1 to 8), f_{pi} = flow proportion in direction i, α_i = angle between the previous direction and the direction from the central cell to cell i, $w_{0,45,90,135}$ = weights for the corresponding change in direction.

Resulting probabilities are the combination of the slope-related algorithm and the persistence (Eq. 5.3).

$$f_i = \frac{f_{si} \cdot f_{pi}}{\sum_{j=1}^{8} f_{sj} \cdot f_{pj}} \cdot f_0 \tag{5.3}$$

where i, j = flow directions (1 to 8), f_i = total flow proportion in direction i, f_{si} = flow proportion from the slope-related algorithm, f_{pi} = flow proportion from the persistence, f_0 = previously determined flow proportion of the central cell.

Each cell with a probability more than a minimal threshold is then included in the path. For the spreading assessment of a source cell, the calculation thus integrates different paths or divergences in one run (Fig. 5.5). There is no need for random multiple runs as the field of all probabilities is covered (Horton et al. 2008).

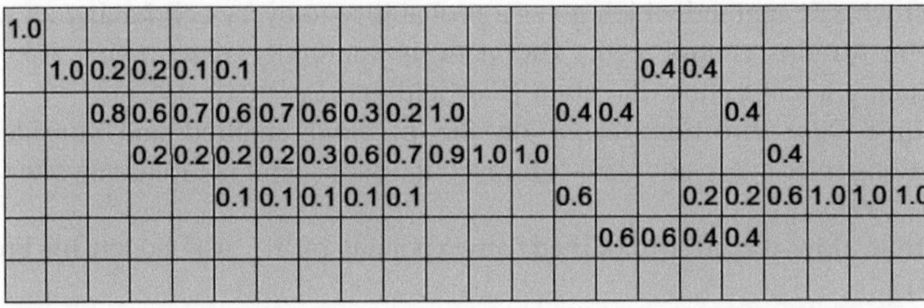

Fig. 5.5 Illustration of the conservative spreading. After Horton et al. (2008)

2. Runout distance calculation

The runout distance algorithms are energy-based calculations that define if a part of the flow can potentially reach the next cell of the DEM. Thus, they control the distance reached by the debris flow and in addition reduce the divergence. Therefore, the energy-based algorithms also influence the flow direction, as each cell that cannot be reached has a probability set to zero.

In a first regional assessment, the source mass is unknown. Thus, runout distance calculation is based on a unit energy balance (Eq. 5.4), a loss function and eventually a maximum threshold. This approach does not aim to represent exact physical processes, but to remain realistic (Horton et al. 2008).

$$E^{i}_{kin} = E^{i-1}_{kin} + \Delta E^{i}_{pot} - E^{i}_{loss} \tag{5.4}$$

where i = time step, E_{kin} = kinetic energy, ΔE_{pot} = change in potential energy and E_{loss} = loss.

The energy loss can be of two different kinds. The first case is a two parameters friction model (Perla et al. 1980) and the second is a constant loss characterized by an average slope angle along the path. The maximum threshold aims to limit the energy to reasonable values, mostly for the constant loss approach (Horton et al. 2008).

Figure 5.6 illustrates the runout distance calculation principles.

(a) At the start, a source has a certain unit of potential energy (without considering the volume) regarding its adjacent cells downhill
(b) During propagation, part of this energy is lost in friction
(c) The kinetic energy is increasing and may reach the maximum threshold, leading to an energy line having the same shape as the terrain
(d) The debris flow stops when the energy becomes null.

Results

The spreading areas of all sources are combined by keeping the maximum or the sum of the probability values. There are three outputs of the model: the sources, the propagation probabilities and the propagation kinetic energy (Fig. 5.7).

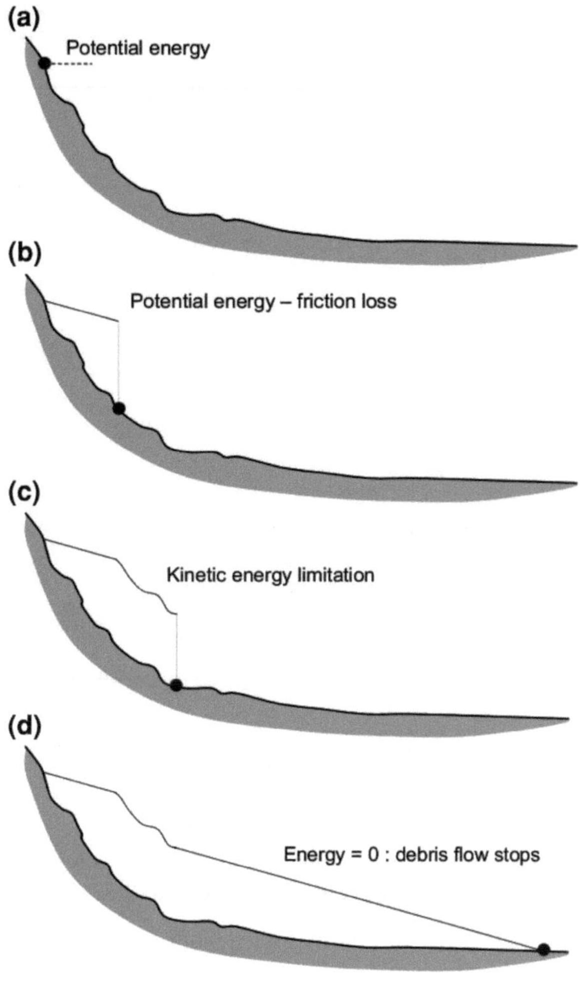

Fig. 5.6 Illustration of the runout distance calculation principles. After Horton et al. (2008)

5.6 Methodology

5.6.1 Landslide Inventory and Susceptibility

A landslide inventory is frequently the first step in landslide hazard identification. Several authors have outlined the importance of a uniform and well documented database to better define the potential unstable areas (Guzzetti et al. 2000; Malamud et al. 2004; Galli et al. 2008). Nowadays, the availability of high resolution digital data such as Aerial Laser Scanning digital elevation model, orthophotos and land use maps has made detailed mapping of geomorphological features possible (Chigira et al. 2004; Van Den Eeckhaut et al. 2007; Kasai et al. 2009).

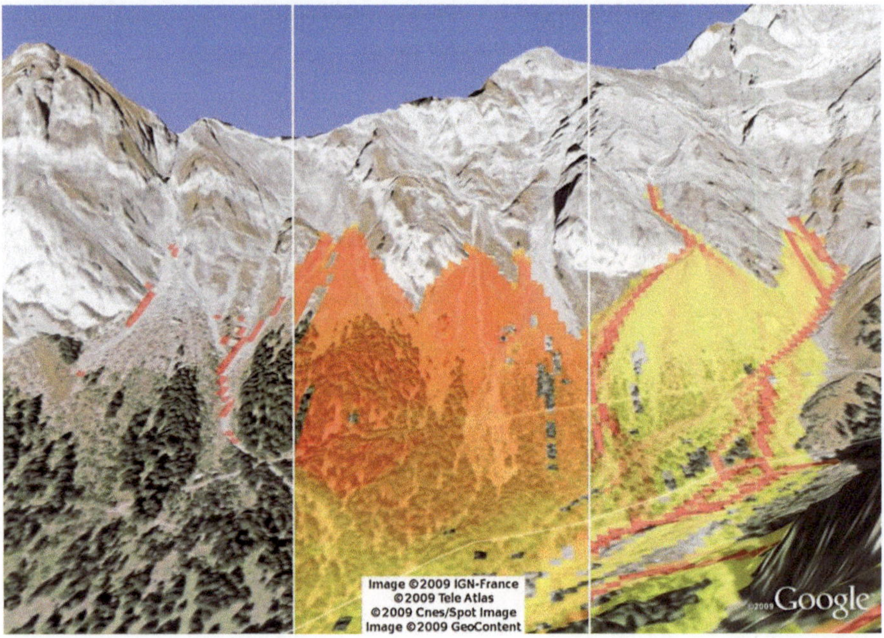

Fig. 5.7 Model outputs illustrations: sources, kinetic energy and probabilities (photos from GoogleEarth™)

5.6.1.1 Method

In both cases, Vaud County and Bagnes valley, the following method (Fig. 5.8) has been used to develop landslide (shallow and deep) inventories:

(1) A visual analysis of high resolution DEM (hillshade and 3D viewing) and orthophotos to provide the main relevant geomorphological features (trench, scars, slope deposits, undulation, etc.)
(2) An analysis of the 1:25,000 geological maps (www.swisstopo.ch) to include the sensitivity of lithologies to landslide and structural elements such as fault systems and tectonic lines
(3) An analysis of the slope angle map to identify active erosion areas, morphological changes and asses the activity of some mass movements
(4) The integration of registered historical events and fieldwork observations.

All the information related to the landslide inventory was stored in a GIS database. In order to improve the objectivity of the methodology, the database includes geomorphological criteria that have allowed the identification of the unstable area. The database contains five main attributes describing the characteristic of each detected landslide area:

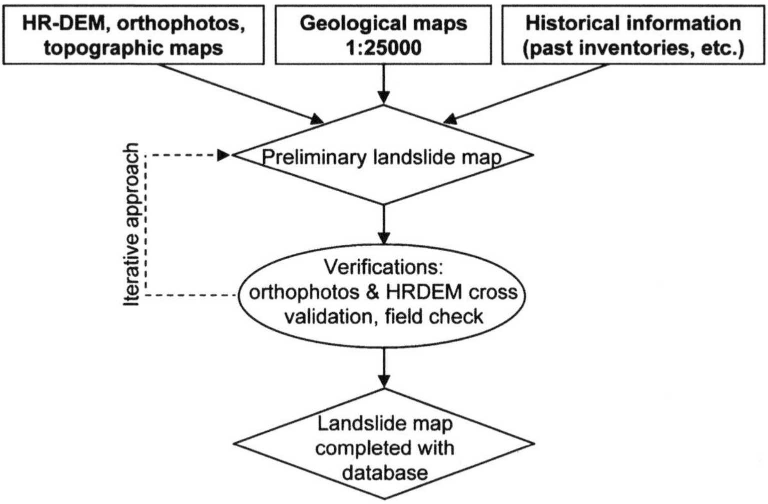

Fig. 5.8 Flowchart of the landslide inventory map creation

(1) The document(s) used for its detection
(2) Relevant geomorphological features
(3) Main processes leading to the instability (pure gravity-driven, fluvial erosion of the toe, etc.)
(4) Relative age ("old" when the geomorphological feature is smoothed and "recent" where the geomorphological feature of the landslide could be clearly differentiated compared of the surrounding topography)
(5) Depth (shallow: 0–4 m, medium: 4–10 m, deep >10 m or unknown).

The uncertainty about the real extension and the present-day activity was also qualitatively assessed using descriptive terms (verified, probable and undetermined). When at least two geomorphological evidences are detected and reliable (landslide scar and deposit or landslide scar and morphological depression, etc.) the landslide polygon is considered as "verified or proved". When morphological evidences are less distinct, with blur limits, the landslide polygon is considered as "probable or suspected". If the landslide area has been delimited in previous inventory maps or in the geological map but the HRDEM and the orthophoto observations do not point out any geomorphological evidence, the attribute "undetermined" was employed.

The limiting factor in interpreting HRDEM hillshade 2D and its 3D visualization depends principally on the data artefacts and the application of intense human reworked areas. Artefacts are mostly related to the occurrence of locally very dense vegetation cover or occurrence of clouds during the data acquisition as well as some steep rugged topography that truncates the laser signal during the data acquisition. Human activities (urban areas, roads or agricultures) disturb the original surface and

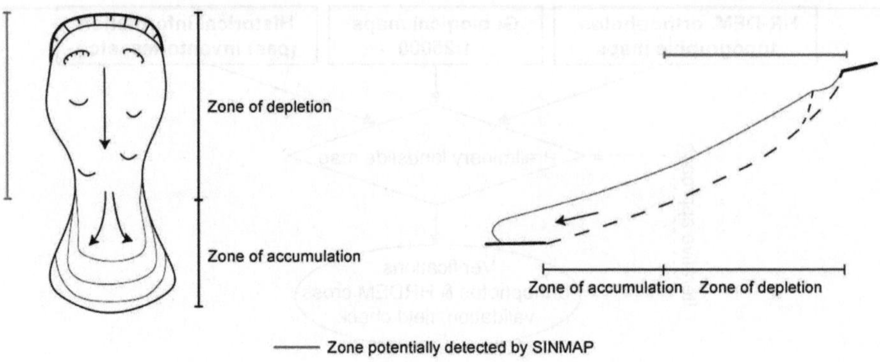

Fig. 5.9 Schematic representation of areas detected by the model SINMAP. Only the initiation zones are identified, while zones of accumulations are not detected (modified after Jaboyedoff et al. 2008)

remove the typical features associated with landsliding, in particular in case of ancient or dormant landslide. For these reasons, the landslide detection in strongly human–modified area needs to be supported by direct field investigations.

5.6.2 Shallow Landslides

The inventory provides information only on landslides that already occurred. To assess the potential extension of shallow instabilities and detect critical areas without any entry in the inventory, the SINMAP model (Pack et al. 1998) was used. This GIS-based approach allows a rapid and objective slope stability evaluation through a large territory based on relatively coarse information. SINMAP is limited to the detection of "starting zones" of shallow landslides (Fig. 5.9). It does not compute any runout. However, it is possible to couple with the propagation part of the models FLOW-R (Horton et al. 2008).

5.6.2.1 Methodology

The SINMAP (Stability INdex MAPping) methodology is based on the infinite slope stability model (Montgomery and Dietrich 1994) coupled with a steady state hydrological model, where the computed depth of saturated soil must be sufficient to sustain a lateral discharge proportional to the specific catchment area (Pack et al. 1998). The topographic wetness index used in SINMAP is a simplified version of the classical hydrological model as TOPMODEL (Beven and Kirby 1979). The main difference is that SINMAP does not account for a decreasing of the hydraulic conductivity with depth but assumes a uniform conductivity of the soil mantel (Pack et al. 1998). The factor of safety formulation became the following:

$$FS = \frac{C' + \cos^2\theta[1 - wr]\tan\phi}{\sin\theta\cos\theta} \tag{5.5}$$

Where C′ = dimensionless cohesion coefficient (integrating both soil and root cohesion and independent of soil thickness), θ = slope angle, ϕ = friction angle, r = water to soil density ratio and w = wetness index.

In Eq. 5.5 the wetness index defined for a given specific catchment area is represented by the ratio between the steady state recharge R [m/hr] and the soil transmissivity [m^2/hr], and is always equal or less than 1:

$$w = Min\left(\frac{Ra}{T\sin\theta}, 1\right) \tag{5.6}$$

Where a = Specific catchment area and θ = slope angle

SINMAP allows entering variables uncertainties through the specification of lower and upper bounds for hydrological and geotechnical parameters adopting uniform distribution. These introduce a probabilistic approach in the calculation of the factor of safety that allows proposing different possible scenarios. The derived dimensionless susceptibility index (SI) is given by Pack et al. (1998):

$$SI = \frac{C' + \cos\theta\left[1 - \min\left(\frac{Ra}{T\sin\theta}, 1\right)r\right]\tan\phi}{\sin\theta} \tag{5.7}$$

Where C′ = dimensionless cohesion coefficient, θ = slope angle, ϕ = friction angle, r = water to soil density ratio, a = specific catchment area, R/T = ratio corresponding to the steady state recharge relative to the effective rainfall quantity and the soil transmissivity.

The worst scenario is defined when tan ϕ, C parameters are close of the lower bound and R/T ratio close the upper bound (complete saturation). Areas under this worst case scenario, where FS is greater than 1, could be defined as unconditionally stable (SI > 1). Inversely, the best scenario is defined when the values for the parameters tan ϕ, C are close to the upper bounds and for the ratio R/T are minimal. Areas under this best case scenario, where FS is lower than 1, could be defined as unconditionally unstable (SI = 0). In between, different intermediate classes can be defined. Following Pack et al. (1998), six susceptibility classes are defined (Fig. 5.10).

5.6.2.2 Application of the SINMAP Model

The input data set for SINMAP consists of the DEM, the cohesion and the friction angle for the mechanical proprieties of the soil and the ratio R/T describing the hydrological conditions. The geomechanical parameter introduced in the model mainly emanate from the literature (Morrisey et al. 2001; Lan 2004; Salciarini et al. 2006) and from punctual in situ analyses (CPT and VAN test). The R (Recharge) parameter is more difficult to calculate. Hence, in our study, it was assumed to be the effective precipitation for 24 h rainfall with a return period of

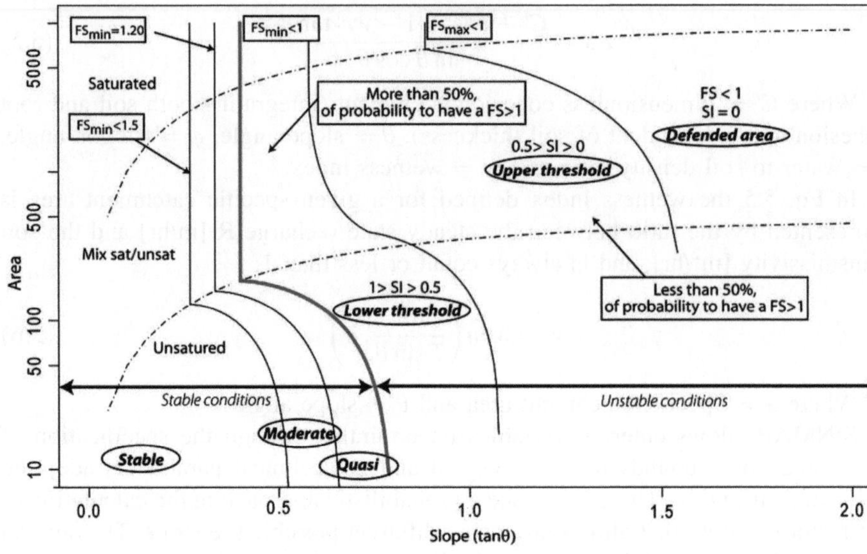

Fig. 5.10 Representation of the stability index in an Area-Slope space defining the different stability classes (modified after Pack et al. 1998)

100 years. The T (Transmissivity) parameter was derived from the hydraulic conductivity (minimal and maximal) of the different lithologies. These values have been chosen in order to give a maximal extension of the potential unstable area for rare event situations (Fig. 5.11).

The final stability map was compiled following an iterative approach (Fig. 5.11). In the first step of the calculation, the applied geomechanical and hydrological parameters were deduced based on the available literature and adapted to the geological and land use conditions. A first susceptibility map was created and compared to the location of the well-know shallow instability. In a second step, the geomechanical parameters were adjusted for each lithology until the stability fit the know landslide areas. During this adjustment process, all the parameters remained in the range of values suggested in the literature.

5.6.3 Debris Flows

5.6.3.1 Introduction

Physical modelling of debris flows in the framework of regional mapping is difficult because of their complexity and the variability of controlling factors. GIS-based approaches associating an automatic detection of the source areas to a simple estimation of the debris flow propagation provide a substantial basis for a preliminary

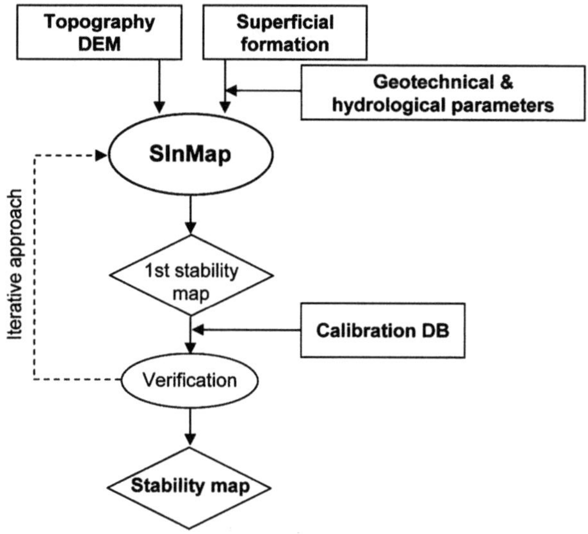

Fig. 5.11 Flow chart describing the input data and the iterative procedure used for construction of the final susceptibility map

susceptibility assessment at regional scale. Figure 5.12 illustrates the principles of the methods used in both case studies, the Vaud County and the Bagnes Valley.

5.6.3.2 Source Areas Identification

According to Rickenmann and Zimmermann (1993) and Takahashi (1981), three criteria in a critical combination are relevant for a debris flow initiation: sediment availability, water input and slope gradient. As not all the lithologies produce the same amount of sediments, a detailed study of the area was conducted by means of a geological or lithological map. The upslope contributing area can account for water input. The slope gradient is a determining factor in triggering of debris flows (Takahashi 1981). Most debris flows occur from terrain with a slope higher than 15° (Rickenmann and Zimmermann 1993; Takahashi 1981). Some initiation thresholds of other factors can be expressed as a relation with the slope angle, as for the contributive area. Such a relationship was first defined by Heinimann et al. (1998), and a second one was assessed on the basis of the 1987 observations of debris flows made by Rickenmann and Zimmermann (1993) after the extreme rainstorms of 1987. The 1987 events in Switzerland could be considered as extraordinary, so it may be advisable to distinguish the obtained limits as an approximation for different probabilities of occurrence (Horton et al. 2008). Thus, two new limiting curves were established: the first one for rare events, based on the Heinimann et al. (1998) limit, and the second one for extreme events, based on Rickenmann and Zimmermann (1993) observations (Fig. 5.13). Both curves are bounded by the theoretical 15° limit gradient. Every point above a curve is considered as critical. The new limit for extreme events is given by Eq. 5.8:

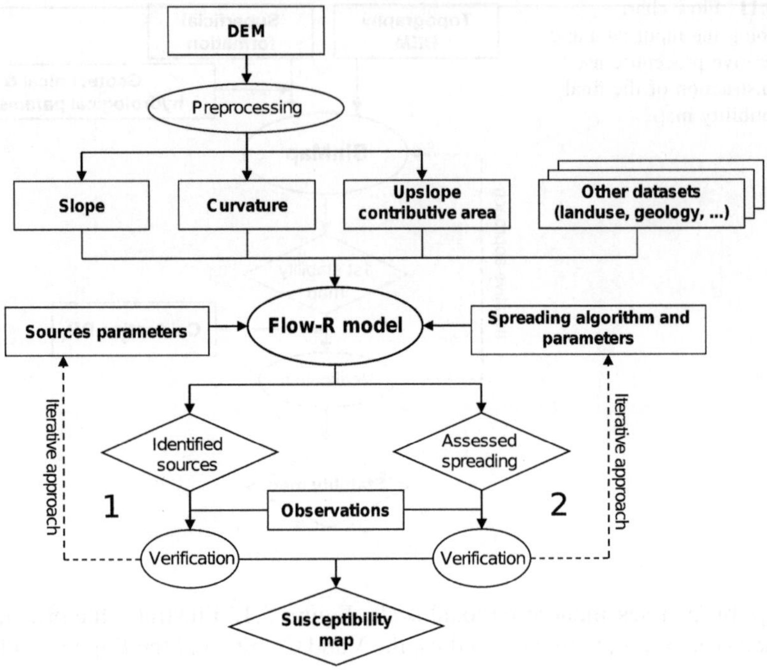

Fig. 5.12 Flow chart describing the input data and the iterative procedure used for construction of the final susceptibility map

$$\begin{cases} \tan \beta_{\mathrm{lim}} = 0.31 \cdot S_{\mathrm{UA}}^{-0.15} & \text{if } S_{\mathrm{UA}} < 2.5\mathrm{km}^2 \\ \tan \beta_{\mathrm{lim}} = 0.26 & \text{if } S_{\mathrm{UA}} \geq 2.5\mathrm{km}^2 \end{cases} \tag{5.8}$$

where $\tan\beta_{\mathrm{lim}}$ = slope gradient, S_{UA} = surface of the upslope contributing area.

Another potential morphological characteristic is the curvature, as debris flows are found where the slope is concave (Delmonaco et al. 2003; Wieczorek et al. 1997). To allow an identification of gullies, the plan curvature, which is perpendicular to the steepest slope, was considered. The contributing area, the slope and the curvature are morphological data processed on the DEM.

Other data can be added to improve the source area accuracy, like a geological map or a landuse map. Those layers help identifying the sources previously selected that are not accurate due to another criterion that is not morphological.

5.6.3.3 Spreading Area Assessment

In FLOW-R, several spreading algorithms and parameters can be selected for debris flows to characterize the flow direction. Holmgren's algorithm was used in both applications because it is a good compromise between highly dispersive and overly channelised flows.

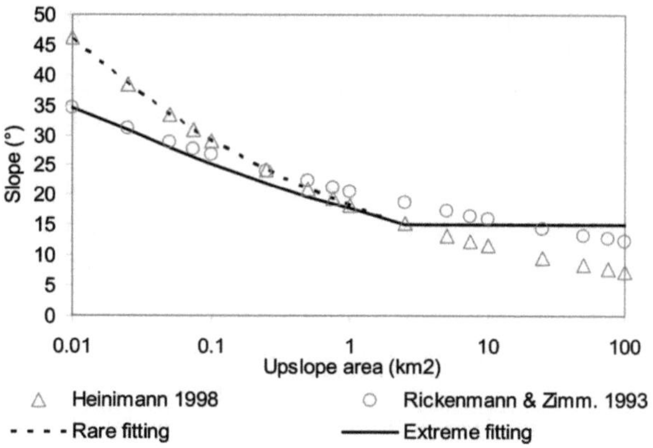

Fig. 5.13 Built gradient thresholds with regard to the upslope area for rare and extreme events. After Heinimann et al. (1998), Rickenmann and Zimmermann (1993), Horton et al. (2008)

For the energetic part of the propagation both the two parameters friction model and the constant loss function can be used for debris flow spreading.

5.6.4 Rockfall

A major issue in mapping rockfall hazard at regional scale is the identification of potential rockfall sources. Nowadays, high resolution topographic data (aerial/airborne LiDAR) can account for realistic landscape details even at a regional scale. For the preliminary assessment of rockfall runout areas, the integration of empirical or process-based methods in GIS environment have shown very promising results (Van Dijke and Van Westen 1990).

5.6.4.1 Identification of Potential Rockfall Source Areas

Potential rockfall initiation areas are identified by analysing the slope angle distribution (SAD) of the topography. A main factor required for the detachment of boulders is a slope greater than the frictional angle of the rock mass, hence, a steep slope (Heim 1932). According to Strahler (1950), the slope angle frequency distribution of a morphological unit (MU) of the relief varies randomly around its mean slope angle. The SAD expresses a range of slope angle values that are characteristic for a given morphology and rock type. The SAD can therefore be decomposed into several Gaussian slope angle frequency distributions that are characteristic of a specific morphological unit (GDMU) (Fig. 5.14a). In an Alpine topography for example, at least four morphological units (MUs) can be encountered:

a. Low slope angles units corresponding to the plains formed by fluvio-glacial deposits.
b. Mid/gentle slope angles units featuring the lower part of the hillslope, called here 'footslopes', and characterized by alluvial fans (debris flow) and landslides deposits.
c. Steep slope angles units corresponding to the valley flanks covered with till deposits or consisted of rock outcrops lightly covered with vegetation
d. Very steep slope angle units representing the cliff faces, outcropping areas and bare rock surfaces.

The sum of those Gaussian distribution must reproduce the SAD. The standard error can be minimized by using best-fitting methods, such as HISTOFIT (cf. Chapter Tools). Initial parameters can be defined according to the shape of the SAD, where the unsteadiness in the distribution tends to reveal a MU.

5.6.4.2 GDMU Interpretation

The slope angle distribution of each Gaussian curve in the SAD analysis can be seen as typical of the topography under consideration. Their mode can be considered as an average apparent slope angle of stability of their corresponding MU. Therefore, the SAD analysis and its decomposition in GDMU can be interpreted as follows (Fig. 5.14b):

1. A threshold angle is set at the intersection (noted A) between the two steepest MUs: the GDMUs "Steep slopes" and "Cliffs" (Rouiller et al. 1998). Above this slope angle, the cliffs MU become dominant over the steep slopes MU and can be potentially considered as rockfall sources. This is done independently of the local lithology and the land cover and includes therefore rocky slope surface lightly covered with vegetation.
2. There are some cases where the GDMU cliffs are missing within the SAD analysis. The highest GDMU is thus referred to the rocky steep slopes MU.
3. In very rugged landscape, such as the Alpine topography, the SAD analysis can reveal two GDMU cliffs. In this case, the lower value GDMU cliff is used to assess the threshold slope angle (see point 1).
4. When geo-thematic information (e.g. cliffs map, rocky outcrop map) is available, the SAD analysis can provide a second threshold angle taken at the mode of the GDMU steep slopes. Hence, cliffs zones and rocky outcropping areas lying above this threshold slope angle can be assumed as being above the average, often close to 35°, that is therefore more prone to be considered as potential rockfall source areas because these slopes will tend to readjust toward the average.

The SAD analysis provides finally two criteria to identify potential rockfall sources according to the morphology of the area under study. More details about this approach can be found in Loye et al. (2009).

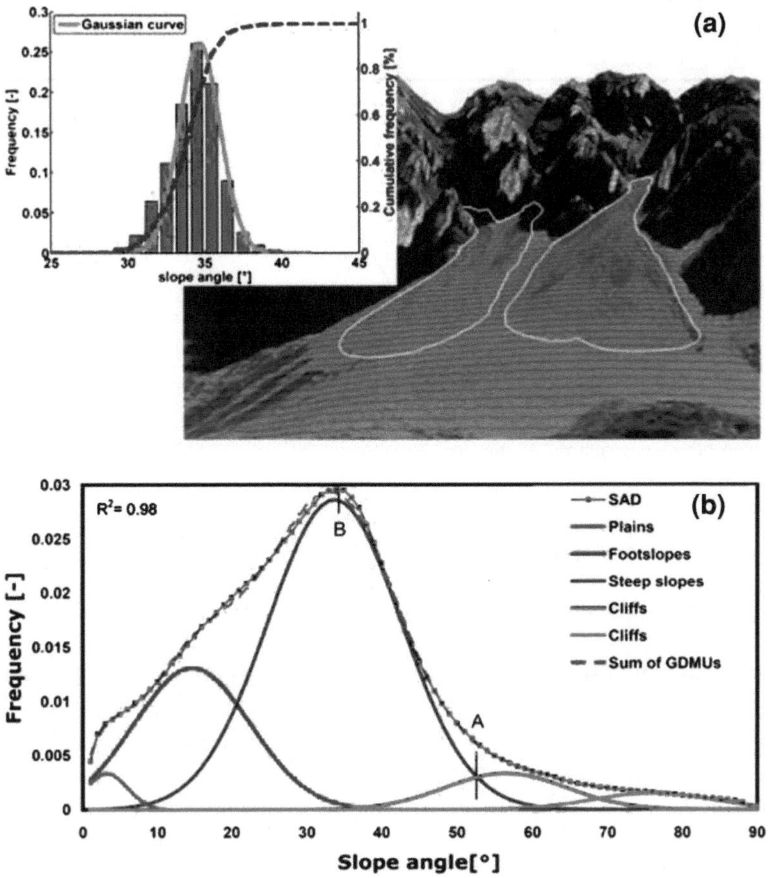

Fig. 5.14 **a** Example of the slope angle distribution of an undisturbed scree slope (delineated in white); The SAD shows that the slope angle values follow a normal distribution around an average close to 35° (orthophoto, swisstopo© 2005 SIT). **b** Example of the SAD of an alpine Valley featuring GDMU decomposition; **a** indicates the threshold angle above which the slope belongs dominantly to the cliffs and are therefore considered as potential rockfall source area; **b** indicates the mode of GDMU steep slopes. The two cliffs units represent two families of bare rock cliff faces lithologically distinct (Modified after Loye et al. 2009)

5.6.4.3 Assessment of the Maximum Runout Length

The maximum rockfall runout zones are estimated by a simple approach inspired from the shadow angle (Fahrböschung) method (Heim 1932) and generalized in 3D under the form of a cone. This model considers that an individual falling rock slides and rolls down the slope with a certain average friction angle ϕ_p (Scheidegger 1973; Evans and Hungr 1993). The angle of aperture ODF of the cone $90° - \phi_p$ then determines the runout and is estimated as follows:

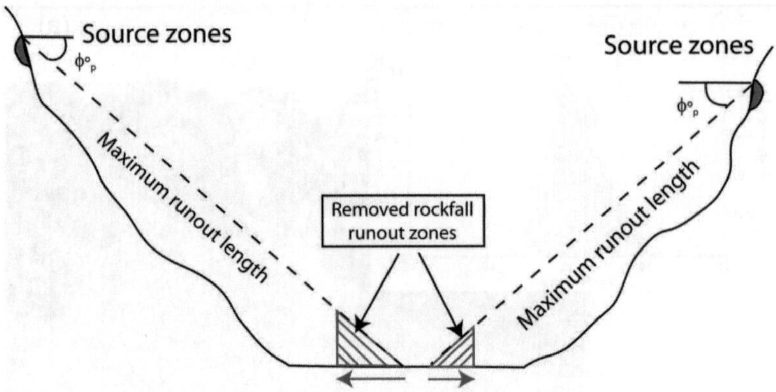

Fig. 5.15 Reduction of the length of the Fahrböschung line according to the rapid change of slope when boulders reach the bottom of the valley

$$\phi_p = \arctan\left(\frac{H_b}{X_b}\right) \tag{5.9}$$

where H_b is the height difference between the source area and the furthest runout reach and X_b the maximum runout length defined horizontally. The angle of aperture is defined empirically based on different aspects, such as information contained in the literature (Toppe 1987; Jaboyedoff and Labiouse 2003), comparison with known rockfall events taken from aerial photos and historical records or fieldwork. The results of several studies give an angle ϕ_p ranging between 28.5° and 35° (Crosta et al. 2001). For the lateral dispersion of the rockfall runout area, experience has shown that rockfall trajectories can be restricted to 15° from one side to the other of the greater slope gradient (Crosta and Agliardi 2003). As this approach doesn't require specific input parameters, such as coefficient of friction and bounding velocity, but is based exclusively on the topography (DEM), the cone angle method implemented in a 3D GIS environment is very convenient to be applied for large scale runout assessment. The runout area is then given as the maximum propagation zones that a boulder can reach. This first estimation can be further corrected for specific topography. For instance, source zones located high in steep mountain cliffs that overhang alluvial plains tend to model a maximum runout length that goes too far compared to field observations. Indeed, boulders reaching flat zones tend to greatly reduce their energy when the talus slope gets flat. Correction for valley-bottom can be then added to the previous consideration for particular topographical configuration (Fig. 5.15).

5.6.5 Snow Avalanches

To assess the susceptibility to snow avalanche, a two steps method was used: (1) Detection of the sources areas, (2) assessment of propagation zones. These avalanche susceptibility maps aim to provide a first overview of existing or potential danger without any information on the intensity or the probability of occurrence of the phenomenon.

5.6.5.1 Source Areas Identification

For the detection of the sources areas, three criteria are used: (1) the slope, (2) the elevation, (3) the landuse, (4) source surface area. The slope is the most important factor to define avalanches source areas. In fact, 90% of the avalanches take place in slopes between 30° and 50° (Salm 1983; McClung and Schaerer 1993; Lied and Kristensen 2003). Depending of the region, there is a lower elevation that provide a lower limit to the occurrence of source areas. This limit is located at 1'100 meters above sea-level for the Swiss Alps (Gruber and Bartelt 2007). A landuse digital cover is available for the whole Swiss territory. For snow avalanches, we are interested in forest areas which influence their triggering. Depending on the goal of the study, source areas in the forested zones may or may not be ignored. The DEM resolution is another important limiting factor that defines the minimum area that can be detected for the avalanche sources. If the DEM has a high resolution, than this minimum area may be too small to be significant; a filter must then be applied to ignore these areas. In our case study in the Bagnes Valley, a morphological filter (opening) was applied.

5.6.5.2 Propagation Zones Assessment

Two models were used to estimate the avalanches propagation areas. One is based on a alpha–beta method (RAS, Sect. 5.5.2.3) and the other one on a Perla multiple-flows model (FLOW-R, Sect. 5.5.2.4).

RAS (Alpha–Beta Methodology)

The propagation is defined by a cone angle of propagation, alpha, using the software RAS. Alpha is automatically calculated for each source cell of the DEM using the alpha–beta method of Lied and Bakkehoi (1980). To estimate the propagation area, the alpha–beta relationship (Eq. 5.10) must be calibrated using an inventory of events (using the maximum propagations of avalanches).

$$\alpha = m \times \beta + n \tag{5.10}$$

Several sets of empirical coefficients were tested and the differences in the final results were negligible. Finally the coefficients of Adjel (1996) for snow avalanches in the Haute-Tarrentaise (French Alps) were used because of the geographical proximity with our area and the large number of observations utilized (alpha = 0.82*beta + 2.82; N = 168). A minimum angle for alpha of 18° was used to avoid unrealistic long propagations.

Table 5.1 Susceptibility area classification methodology for snow avalanches

Susceptibility type	Propagation area		Known event
1. Probable	Flow-R and RAS	Or	Yes
2. Potential	Flow-R or RAS	And	No
3. Not susceptible	No one	and	No

FLOW-R

The details of FLOW-R, a multiple flow model with an energetic component, are described in Sect. 5.5.2.4. The main difference with the model RAS is that FLOW-R is much more sensitive to the topography. This has advantages, i.e. avoiding some overestimated lateral spreading, and some drawbacks, i.e. flat bottom valleys where it does not spread enough. Both models were used in the Bagnes Valley; both are very rough compared to the complexity of a snow avalanche and none of them allows the estimation of dynamic pressures, but they can be applied to large regions.

5.6.5.3 Avalanche Susceptibility Mapping

The susceptibility map of the Bagnes Valley was drawn combining the results of the two models: FLOW-R and RAS. The FLOW-R model tends to simulate the common avalanches (with a short return period) and the RAS model the extreme propagations. We chose to classify the susceptibility mapping in two categories, probable and potential. The Table 5.1 shows how we classify those two categories.

5.6.6 Flooding and Erosion

The goal of this chapter is to clarify the methods used to develop an indicative mapping of flood hazards on the territory of an Alpine valley. The study focuses on the main waterway and tributary streams crossing the valley and its purpose is to identify the areas subject to flooding. The proposed method is based on five distinct stages summarized in Fig. 5.16.

The fives stages allow to identify the potential sources of overflow and the model FLOW-R allows to simulate the propagation of those.

5.6.6.1 GIS Analysis

The analysis of geographical documents allows the characterization of the alluvial geomorphology of the waterways present in the studied area. GIS is extremely useful for support, which can improve the detection of possible source of

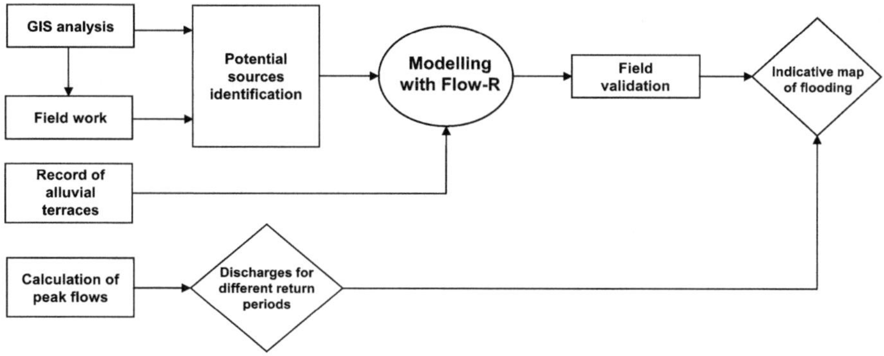

Fig. 5.16 Stages contributing to the development of the analysis

overflows. A high-resolution Digital Elevation Model allows the calculation of the flow direction, which helps to identify all the waterways that must be investigated next steps of the analysis. Using aerial photography observations, it is possible to identify eroded areas that may provide material which can dam the river. The geological maps are also an important indicator of potential erosion areas. A field survey is necessary to complete and validate the preliminary observations. Different factors could be extracted from the GIS analysis:

- alluvial plains and the different levels of alluvial terraces which correspond to different levels that can be reached by potential outburst;
- an assessment of bridge capacity is conducted.
- identify areas of bank erosion representing potentially mobilized material and areas of possible collapse;
- known elements, (i.e.: geological map, student works, and cadastre of events). The inventory of past events can improve the mapping and must be considered in the analysis. In fact, historical information is usually a good indicator for mapping hazard areas. It provides additional information on possible events and improves the evaluation of hazard.

5.6.6.2 Field Work

Field work is a necessary step to confirm or disprove the observations made during the GIS analysis. Moreover, it helps identifying pathologies of the hydrographical network, such as erosion banks, mobilizable material (trees with an eroded base, bank erosion, landslides along the waterway, important sediments deposits, etc.) or indices of past floods. These elements are used to calibrate the model FLOW-R described below.

5.6.6.3 FLOW-R Modelling

The model FLOW-R (Horton et al. 2008) is adapted to hydrological flows without material. It allows evaluating the propagation of flood. The location of sources of potential overflows areas is given according to two criteria:

- at the top of the channel for tributary streams and for the main river
- along the streams for simulation of dams.

Regarding the identification of the source areas of potential overflows, various criteria are chosen:

- the presence of a low bridge;
- the narrowing of the waterway;
- an obstacle to the flow;
- banks with a low level;
- the visual presence of past overflows;
- mobilized materials in abundance obstructing the waterway or threatening to obstruct it.

The algorithm used for spreading the water is the multiple flow direction algorithm, which makes possible to limit the lateral spreading by the topography only. Regarding the energy algorithm, the chosen approach is a linear decrease of the spreading energy.

5.6.6.4 Calculation of Maximum Discharge

The model Hydriff (OFEG 2003) of the Swiss Confederation is used to estimate the maximum discharge. It estimates the discharge based on the size of the watershed, the length of the hydrographical network and a growth factor, depending the considered return period and the catchment area deduced using standard GIS functions. This method is derived from multi-scale methods and it is applied for watersheds of medium size (1–500 km^2).

5.6.6.5 Record of Alluvial Traces

A geomorphlogical analysis of the alluvial terraces is necessary in order to assess the historic levels of the river is taken into account in hazard mapping as, infrastructures present in the major and minor stream bed can be potentially threatened by rising waters. A distinction is made between the three following levels: sedimentary deposits from the river, lower terraces (representing the first terrace level above the river) and the upper terraces (representing the second terrace level above the river). The data used for this analysis are aerial photos and geological maps. The result of the analysis is integrated in the susceptibility map.

5.6.6.6 Classes and Scenarios

To provide a focused analysis, two classes for the assessing hazard are defined. The first class is related to the propagation resulting from a simulation in the channel. The second class is related to the propagation simulated using a chosen source outside the channel, simulating a jam.

The methods described in this chapter are applied to four distinct geographical sections (Table 5.2), namely:

- tributary streams
- the threat of jams in tributary streams;
- the main river
- the threat of jams in the main river.

5.7 Results

5.7.1 Landslide Inventory Map

5.7.1.1 County of Vaud

In the county of Vaud, a former instability database was created in the 1990s by aerial photo analysis and field investigation (Noverraz 1995). This database contains 6455 landslides differentiated according to their depth (shallow, medium and deep landslides) and their activity (geomorphological evidences). The first step of the inventory updating was to check and redraw landslide perimeters based on the high-resolution DEM (HRDEM) and digital orthophotos. The second step was to identify and redraws instabilities indicated by the geological maps (Geological Atlas of Switzerland 1:25,000; www.swisstopo.ch) on the HRDEM. Finally, the HRDEM of the entire county was re-examined in order to identify new instabilities.

Finally the new database contains 8501 slope instabilities. The mapped instabilities cover around the 8% of the study area which is close to the 6% obtained for the whole Switzerland by Lateltin et al. (1997). 2718 landslides were identified as not correctly delimited on the previous inventory and they were redrawn based on the new HRDEM. For 608 landslides described in the former inventories, the HRDEM analysis could not provide any valuable delimitation (Fig. 5.17). This was usually the case for landslides in human-modified slopes or near urban centres. For these landslides, a distinctive specification in the attribute table has been introduced. The analysis of the available geological maps has made possible to identify 909 new instabilities. Most of these instabilities, identified during fieldwork, affect only small areas (<10,000 m^2) along river banks. During the

Table 5.2 Different scenarios analyzed and their characteristics for floods and erosion

Type of area	Tributary stream	Jam in a tributary stream	Main river	Jam in the main river
Type evaluation	Flow-R modelling and geodata analysis, Field work	Field work Flow-R to model the spread	Flow-R modelling and geodata analysis, Field work	Field work Flow-R to model the spread
Documents and support	Geographical data, geological map	Geographical material, geological map, existing hazard map	Geographical data, geological map	Geographical material, geological map, existing hazard map
Criteria	–	Bridges, shrinkage, accumulation of material in the stream, barriers to the flow, low banks	–	Bridges, shrinkage, accumulation of material in the stream, barriers to the flow, low banks
Source areas	Source of the river, in the drainage channel	Overflow over the top of the bank	Source of the river	Overflow over the top of the bank
Class	Probable	Potential	Probable	Potential

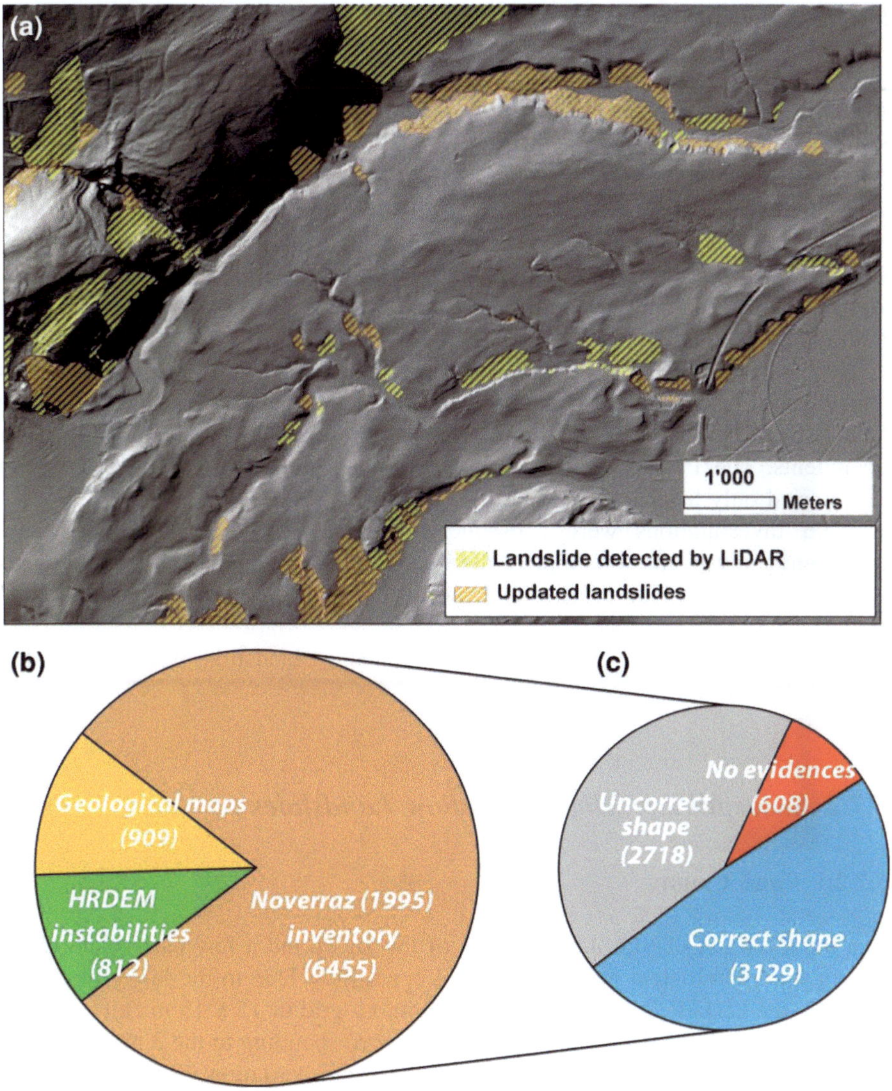

Fig. 5.17 Summary of the main results of the updating the landslide inventory map. **a** Origin of the different landslides listed in the database (hillshade, swisstopo© 2005 SIT). **b** Results of the revision of the Noverraz (1995) inventory

reanalysis of the entire Vaud County, 812 new instabilities were identified only using the HRDEM and the orthophotos. Most of these new detected instabilities were identified in the Plateau area (457 new instabilities) and in the Jura regions (183 new instabilities).

5.7.1.2 Valley of Bagnes

Contrary to the County of Vaud, no pre-existing landslide database was available for this region. The landslide inventory map of the Val de Bagnes was thus created based on the Geological Atlas of Switzerland 1:25,000 (www.swisstopo.ch), the HRDEM, the DEM25 and orthophotos. As the area is relatively small (300 km^2), some fieldwork was conducted to check this inventory.

Based on the geological maps, 13 shallow landslides and 132 medium landslides were inventoried, representing almost 40% of the total number of landslides. Among them, 42 cannot be confirmed by geomorphological evidences on DEM or field investigations.

By comparing HRDEM, DEM25 and orthophotos, 21 new shallow landslides and 98 new medium landslides were detected (34% of the total). 67% of the landslides detected on the geological maps were updated based on these other documents. The 15 deep-seated slope gravitational deformations (DSGSD) were inventoried only based on DEM analysis.

Field investigations were consequently performed in order to verify the document analysis and to complete the inventory. At the end, each landslide was classed as "proved" or "suspected", according to the criteria explained in the Sect. 5.6.1.1.

Finally, 15 DSGSD, 245 medium and 102 shallow landslides were identified in the Val de Bagnes.

5.7.2 *Susceptibility Map of Shallow Landslides*

5.7.2.1 Vaud County

The input dataset of SINMAP consists of the DEM and a few parameters quantifying the hydrological and geotechnical conditions. Due to the large area to be mapped, the LiDAR DEM was re-sampled into a grid of 15 x 15 m cell size. The study area was divided into three main zones corresponding to the 3 main tectonic subdivisions (Jura, Plateau and Alps). The lithology variation was introduced by the mean of the "Geotype maps" covering the entire Vaud area (Turberg et al. 2008). The Geotype map is a kind of lithological—genetic map in which the formations with similar rock or soil proprieties are merged together.

For the hydrological parameters, the effective precipitation for 24 h rainfall with a return period of 100 years was estimated for the different region based on the Hydrological Atlas of Switzerland (2006). The model calibration was performed based on pre-existing inventory maps (Noverraz 1995) and orthophoto interpretations. SINMAP model was originally developed to model translational landslides in a hilly topography. Its direct application to a more rugged topography like the Alps is not perfect. In fact, for steep mountain slope the SI will be very low even if the effective soil thickness is too thin to develop a landslide-type instability. To avoid this problem, all the bedrock outcrops were delimited with the 1:25,000 topographic vector map and the source area identified during the rock susceptibility map. In these zones the geomechanical parameter, the transmissivity and the recharge parameters were adapted to keep the SI high. The final raster map was cleaned and smoothed of small scale artefacts using a majority filter (ArcGIS®) based on four contiguous neighbourhoods.

According to the goal of the susceptibility mapping project, the results of the SINMAP analysis were reclassified in two different ways (Fig. 5.18):

A map containing a single susceptibility class for the pixels where the Stability index is lower than 1.

A map containing three susceptibility classes corresponding to a SI between 1 and 0.5, (lower threshold) between 0.5 and 0 (upper threshold) and equal to 0 (defended).

5.7.2.2 Bagnes Valley

As for the application to the Canton of Vaud, the input dataset for SINMAP consisted of a HRDEM degraded to a 10 m cell size grid below 2000 m a.s.l. and the DEM25 above 2000 m. a.s.l. However for this area, there was no numerical document which distinguishes the type of substratum such as the Geotypes of the Vaud County. The first step of the processing was then to identify the superficial formations. To perform this classification, the information of two documents was used: the land use occupation indicated by the Vector25 and the 1:500,000 vectorized geological maps. As shown in the Fig. 5.19, twelve classes of soil with distinct geotechnical parameters have been extracted: alluvium, breccia, bedrock, dense forest, flysch, high altitude soils, moraine, reworked soils, schist, scree deposits, sparse forest and sparse shrubs.

The geotechnical and hydrological parameters assigned to each class were taken from Morrisey et al. (2001), Lan (2004), Salciarini et al. (2006), Hydrological Atlas of Switzerland (2006). To calibrate the model, 67 shallow landslides that occurred during summer 2009 were inventoried in various types of superficial formations (Fig. 5.20).

The results of the SINMAP analysis were provided in one map containing six susceptibility classes corresponding to:

- Class 1: $0 < SI < 0.001$ Defended area
- Class 2: $0.001 < SI < 0.5$ Upper threshold
- Class 3: $0.5 < SI < 1$ Lower threshold
- Class 4: $1 < SI < 1.25$ Quasi-Stable
- Class 5: $1.25 < SI < 1.5$ Moderate Stable
- Class 6: $1.5 < SI$ Stable

Finally, 46% (almost 140 km^2) of the Bagnes Valley has a stability index lower than 1. The major part (52%) of the susceptibility areas ($SI < 1$) are in the class 3 "Upper threshold".

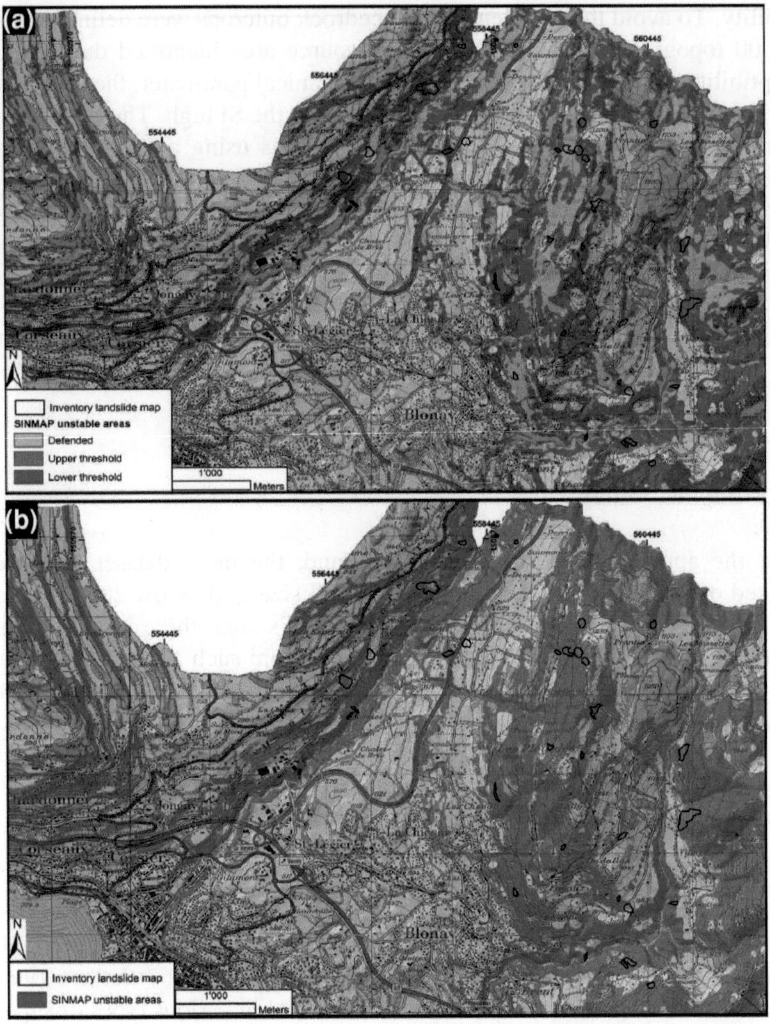

Fig. 5.18 Close up of the shallow landslide susceptibility map, showing the good agreement with the inventoried shallow landslide. **a** 3 classes map and **b** single class map (hillshade and topographic map, swisstopo© 2005 SIT)

5.7.3 Debris Flows

5.7.3.1 Vaud County

The DEM available for the Canton de Vaud territory is a laser DEM with a resolution of 1 meter. In order to reduce the processing time and to be more consistent with the phenomenon scale, it was degraded to a 10 m cell grid.

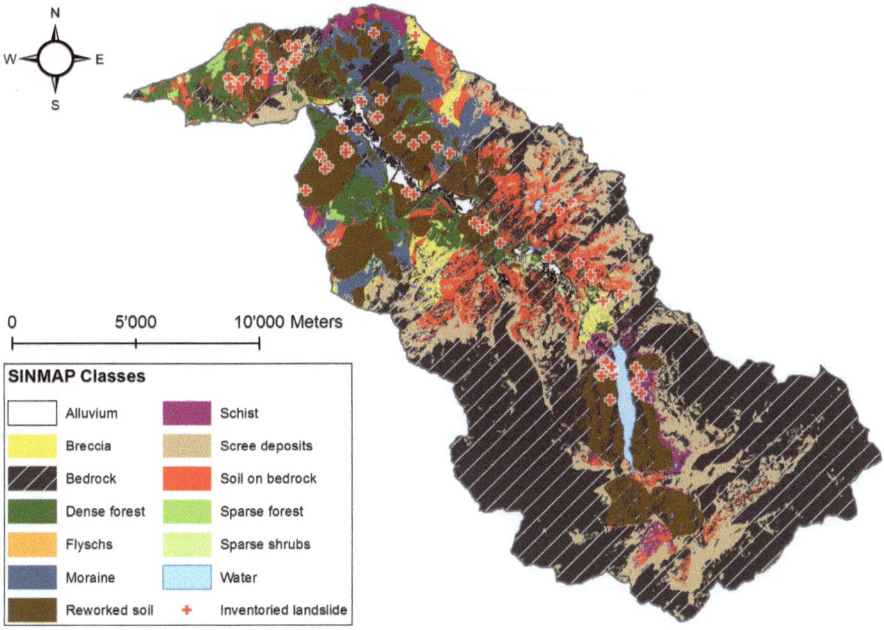

Fig. 5.19 Representation of the classified superficial formations issued from the Vector25 and the 1:500,000 geological atlas of Switzerland

Several types of processes were addressed in the study of this region: debris flows, deposits remobilization, mud flows and hyperconcentrated flows. Only the debris flows are presented hereafter.

Source Areas Identification

Although the curvature is often used to recognize the gullies, there is no admitted threshold. A limit had to be established on the basis of aerial photographs and the analysis of the 10 m DEM. For this study area, a curvature of $-2/100\ m^{-1}$ was found as optimal on the basis of the analysis of orthophotographs (Horton et al. 2008).

The minimum flow accumulation threshold chosen was of 1 ha, after calibration on observed debris flows. The extreme threshold was selected in accordance with the work objective which is to make an indicative map, supposed to cover the worst case scenario (Horton et al. 2008).

The lithology was taken into account by means of a “geotypes” map (Perret 2007; Turberg et al. 2008), which contains uniform and complete information about surface formations for the whole study area. The selected lithologies are debris flow prone rocks (marl, slate, siltstone) and slope deposits.

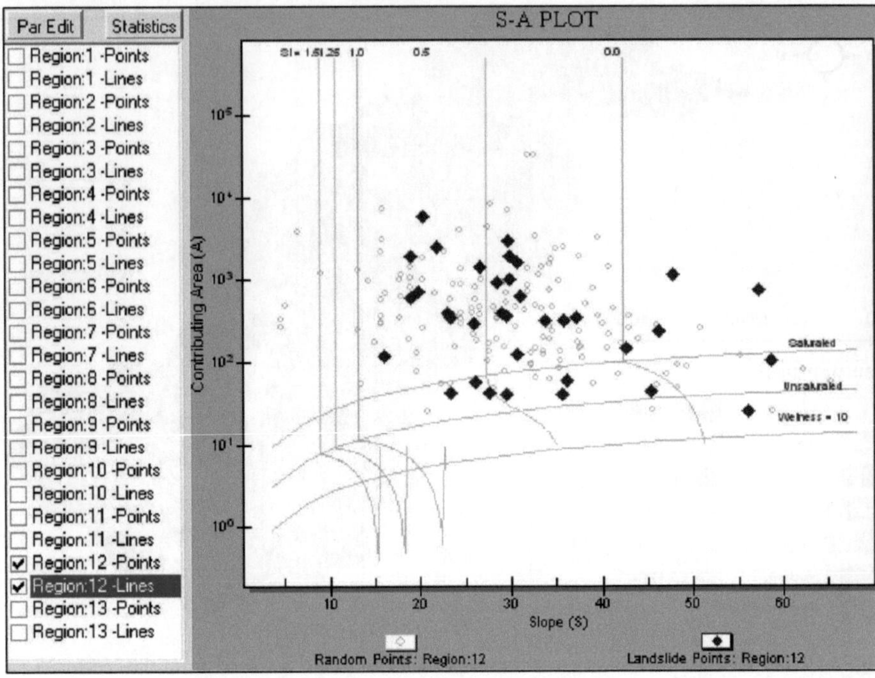

Fig. 5.20 S-A plot within SINMAP allowing the calibration of geotechnical and hydrological parameters for reworked soils according to shallow landslides inventoried

Landuse maps helped to detect certain inaccurate sources, located in developed areas or due to man-made infrastructures. Outcropping or suboutcropping rocks were also excluded from potential sources (Fig. 5.21).

Spreading Area Assessment

The Holmgren's algorithm was chosen because it best fits the events that can be observed on orthophotographs. Its exponent was set to 4, as proposed by Claessens et al. (2005) on the basis of field and laboratory measurements.

The probable maximum runout is characterized by an average slope gradient of 11° (Huggel et al. 2002). The chosen energy threshold was set to a maximum velocity of 15 m·s^{-1}. The observed maximum velocity among various debris flows events in Switzerland is 13 to 14 m·s^{-1} (Rickenmann and Zimmermann 1993).

5.7.3.2 Bagnes Valley

The main difference between this case study and the County of Vaud is that source areas were divided in two classes: proved and potential. In addition field investigation was carried out in order to evaluate the real debris flows susceptibility.

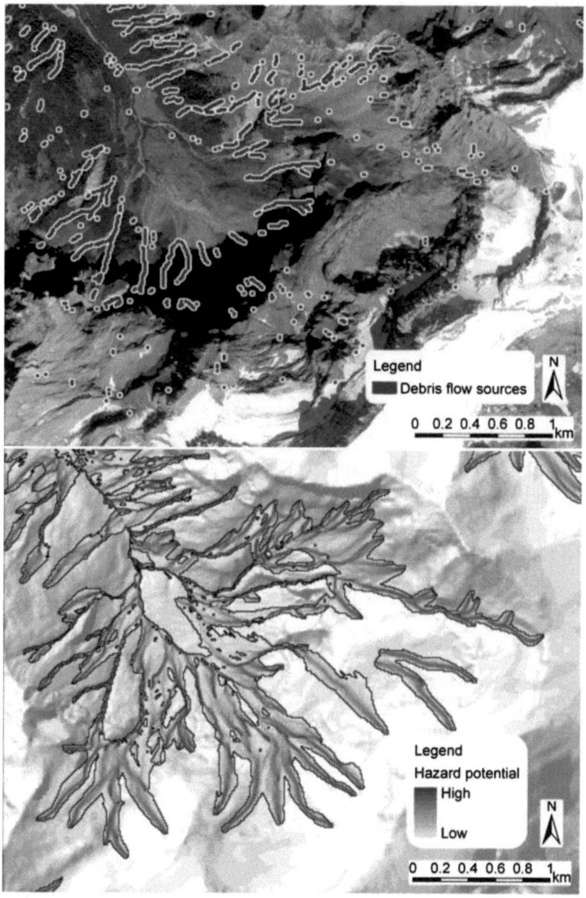

Fig. 5.21 Identified debris flows sources and corresponding spreading in the Diablerets region (hillshade, swisstopo© 2005 SIT)

The FLOW-R model was applied, and four test sites in the valley were chosen to calibrate the model parameters for the source areas detection and the propagation. The datasets used in this case are: a 10 m resolution DEM, the 25 m land-use map (Vector25), the topographic maps, the orthophotos and the local inventory of historical debris flows events.

Source Areas Detection

As describe in Chap. 6.3.2, at least the slope, the curvature (Delmonaco et al. 2003) and the water input (Rickenmann and Zimmermann 1993; Horton et al. 2008; Jaboyedoff et al. 2008) are needed to detect the debris flow sources. The criteria used are described in Table 5.3:

In the case of Val de Bagnes, the landuse and the geological information were also taken into account in order to suppress source areas detected on bedrock or

Table 5.3 Debris flows source detection criteria

Criteria	Slope	Curvature	Water input
Parameters	>15°	$-2/100\ m^{-1}$	1 ha

Table 5.4 Sources classification methodology

Source types	Past event	Phenomena evidences	Debris
Proved source area	Recorded	Recent evidences of debris flow activities	Debris stock sufficient for debris flow triggering
Potential source area	No	record	Fossil evidences of debris flow activities or suspected activities
Debris stock sufficient for debris flow triggering			
Incorrect detection	No	record	No activity evidences
Not enough debris stock for debris flow triggering			

Table 5.5 Spreading parameters

Criteria	Inertia	Flow direction algorithms	Runout distance calculation	
			μ	M/D
Parameters	Default mode	Holmgren exp. 6	0.09	30

man-made structures. The source areas classification is a crucial step in the creation of the debris flows susceptibility map. In this case, a distinction between proved, potential and incorrect detected source area has been made. The Table 5.4 presents the different parameters for the different kind of sources.

Propagation Area Assessment

Field observations and numerical data analysis (orthophotos, DEM, topographic maps) were used to calibrate the model parameters (Table 5.5). After several tests, the 2-parameters friction model (Perla et al. 1980) was chosen because it provides the best results comparing with the archived events.

The propagation area classification depends on the sources area classification. For example, if a source area is classified as "probable", the spreading area will be "probable" too. For the *probable* propagations, a distinction was made between the probabilities that are under 2% (*probable danger*) and higher than 2% (*strong probable danger*) (FLOW-R, Sect. 5.5.2.4).

In two watersheds, historical debris flow propagations are longer than expected within a normal parameterization of the model. This is due to the high amount of loose material that can be mobilized, their important size or hydrological characteristics. For these special cases, the method was adapted to obtain the known maximum runout distances. An average propagation slope angle of 5° with an energy threshold of 15 m/s (the other parameters are the same) were chosen.

Detection and Propagation: Example of the Merdenson

The Fig. 5.22 shows the result of the source detection and debris flow propagation simulation for one watershed of the Bagnes valley. This region contains active debris flows gullies and a main road is often affected by events (numbers 1 and 2 in Fig. 5.22). In this example, the sources detected by the model correspond well with the field observations, confirming a good correlation between the model's simulated propagation and field investigations (Figs. 5.22, 5.23), picture taken from the point 3, 4 and 5). Point 5 in Figs. 5.22 and 5.23 represents the debris flow triggering area and the black arrows shows the most probable propagation path.

5.7.4 Rockfall Susceptibility Map

5.7.4.1 Vaud Territory

The decomposition of the slope angle distribution in Gaussian populations was performed with the 1 m cell size DEM of the Vaud County (2800 km^2). The potential source zones were aggregated to a cell size of 25 m and the run-out model was run with a DEM of same cell size. The runout areas were computed with CONEFALL. The territory was divided into 5 five distinct zones: the Alpine part is composed of the Helvetic, Ultrahelvetic and Prealpines Nappes; the Jura Mountains and the molassic Plateau are the last two zones. The slope angle distribution decomposition in Gaussian distribution was performed with HISTOFIT. Results of the threshold slopes angle are summarized in Table 5.6.

Therefore, potential rockfall sources were defined by all slopes belonging to the units cliffs defined according to the slope angle histogram decomposition. Moreover, all rocky outcrops and cliffs available from the 1:25,000 topographic vector map where their slope angle is steeper than the mode of the unit steep slopes were added to the map of potential rockfall sources. Finally, a minimum size of 10 m^2 was set in order to consider a surface as rockfall source zones. This had the advantage of filtering out possible artefacts contained in the DEM without removing any essential information.

For the runout area assessment, CONEFALL was applied to each potential source defined previously. The angle of aperture $\phi_p = 33°$ was the best compromise with all the information collected through the available documents Fig. 5.24. Furthermore, an angle of 33° has the advantage of including a potential

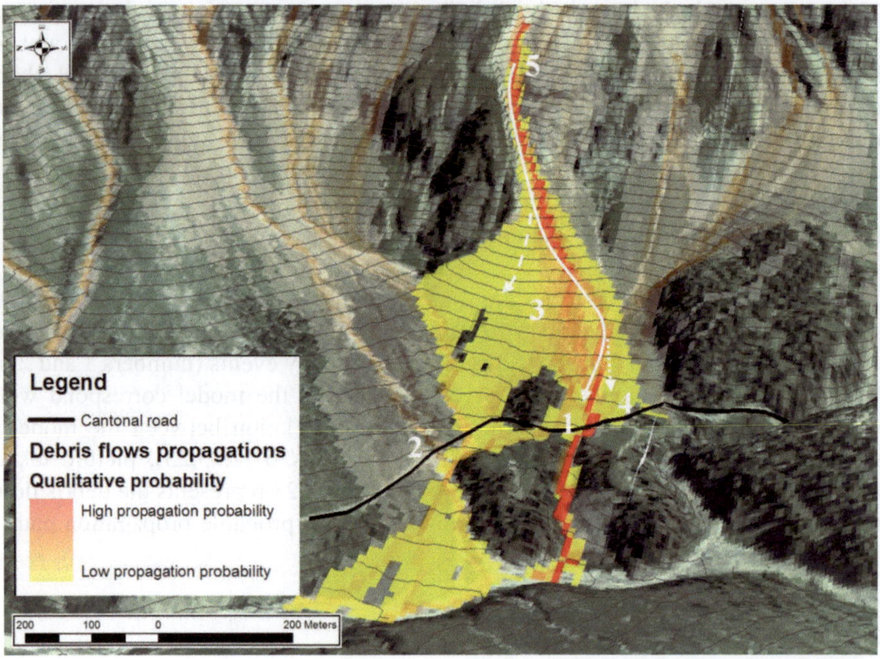

Fig. 5.22 Example of spreading assessment in the Mauvoisin region (orthophoto and isolines: ©swisstopo)

Fig. 5.23 Field investigation. Correspondences with the spreading assessment in Fig. 5.22

Table 5.6 Threshold slope angles above which rockfall source areas are potentially considered

Location	HMA	Threshold angles for	
		A. Minimum threshold angle for the unit cliffs	B. mode of the unit steep slopes
Alps	Helvetic	54°	36°
	Préalpes Médianes	53°	34°
	Ultrahelvetic	49°	33°
Plateau	Molassic Plateau	46°	30°
Jura	Jura Mountains	46°	32°

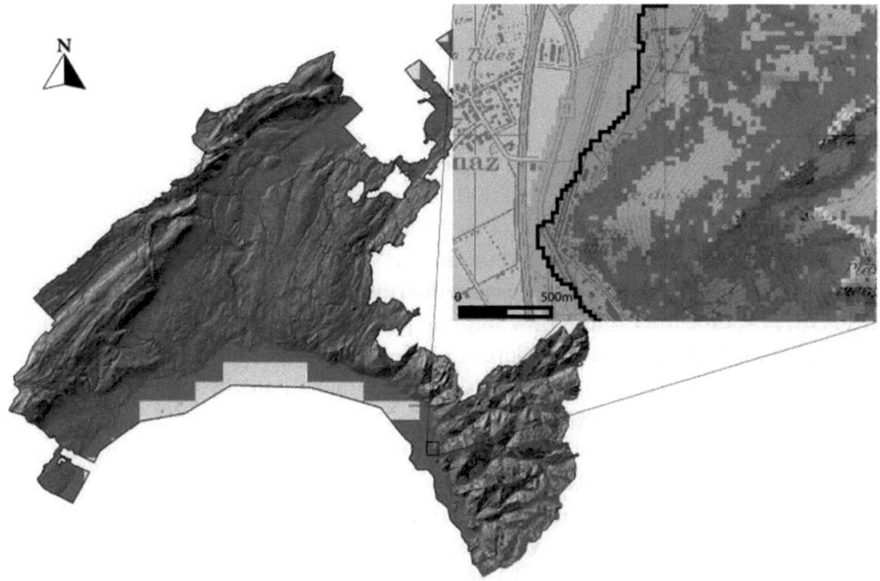

Fig. 5.24 (*Left*) Susceptibility rockfall hazard map for the canton of Vaud (Switzerland). Rockfall sources zones are drawn in red and the runout perimeters in brown. (*Right*) Correction for the Valley-bottom performed on the maximum runout area, limiting the very unlikely propagation of boulders far into the flat lands (light beige) (hillshade and topographic map, swisstopo© 2005 SIT)

remobilization of the screes located beneath the rockfall sources area, which usually have a mean slope angle of 35° (Loye et al. 2009).

The correction for the flat bottom valleys was performed when a surface bigger than 2.5 km^2 has a mean slope gradient lower than 11°. Based on orthophotos analyses of scree deposits along flat zones, maximum runout length of boulder reaching those plains was limited to 60 m for tributaries valley-bottom. This limit was expanded to 100 m in the alluvial plain of the Rhône Valley due to its higher falling velocities resulting from the greater size of its valley flanks (Fig. 5.24).

Table 5.7 Threshold slope angles above which rockfall source areas are potentially considered in the Val de Bagne

HMA	HRDEM		DEM25	
	A. Minimum threshold angle for the unit cliffs	B. mode of the unit steep slopes	A. Minimum threshold angle for the unit cliffs	B. mode of the unit steep slopes
Acid rocks	47°	36°	46°	35°
Basic rocks	Not present	48°	32°	
Conglomerate	52°	34°	45°	32°
Flysch	52°	36°	47°	35°
Limestone	52°	37°	46°	36°
Moraine	51°	30°	38°	30°
Marble and breccia	50°	35°	46°	33°
Schists	49°	35°	48°	32°

5.7.4.2 Bagnes Territory

To identify to potential rockfall source areas, the homogeneous morphometric areas HMA were extracted from the 1:500,000 vectorised geological Atlas of Switzerland. Then the Slope Angle Distribution for each HMA area was extracted on the HRDEM for altitudes lower than 2000 m and on the DEM25 for the entire study area.

The SAD decomposition in Gaussian distributions was performed with the tool HISTOFIT. As for the Canton of Vaud, the potential rockfall source areas resulted from a combination of two sources defined by minimum threshold angles (Table 5.7): Areas which have a slope angle higher than the threshold angle of slopes belonging to the population cliffs;

Areas of the Vector25 mapped as crops which have a slope angle higher than the mode of the population steep slopes.

The differences of threshold slope angles detected with the HRDEM and the DEM25 were expected. As shown in Loye et al. (2009), the bigger the resolution, the lower the threshold angle for the same cliff. Source areas less than 10 m^2 were filtered out to avoid artefacts due to potential local errors of the DEM.

Two propagation simulations were performed in order to estimate two susceptibility classes, one with CONEFALL and one with FLOW-R (Fig. 5.25). Both computations were performed on the DEM25 to ensure continuous and homogeneous results. According to previous detailed works on the valley of Bagnes and the experience of the county of Vaud, the propagation angle used was 33°. Even if FLOW-R is based on constant friction loss which relates this model to the shallow angle method, the algorithms simulates far fewer propagations than pure geometrical models as CONEFALL. Furthermore, the multiple flow direction algorithm (Holmgren 1994) used in FLOW-R calculates more channelized runouts.

The final results were presented in two classes: (1) Probable propagation areas, corresponding to runout assessed by FLOW-R with higher susceptibility, (2) potential propagation areas, corresponding to runout assessed only by CONEFALL with lower susceptibility.

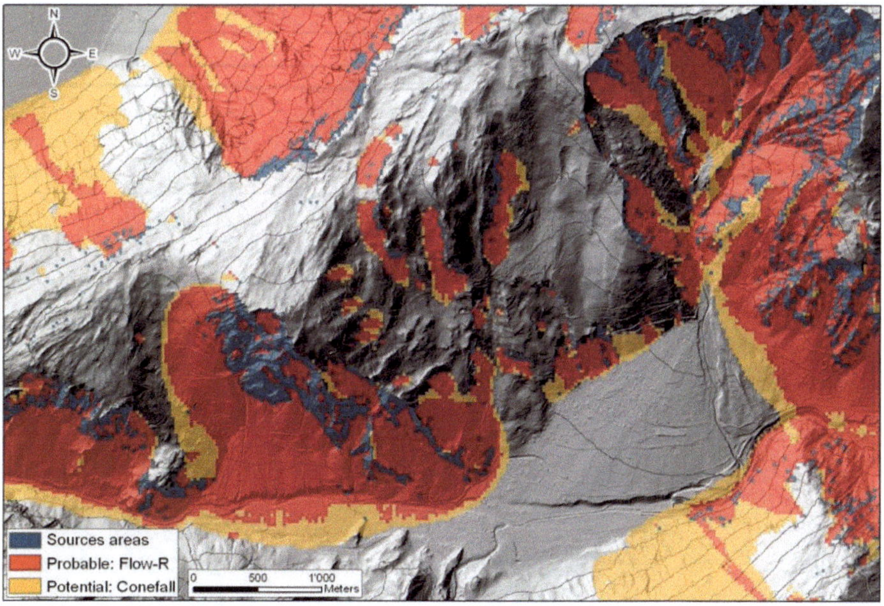

Fig. 5.25 Rockfall susceptibility map. The runout was separated in two classes: probable susceptibility, which belongs to FLOW-R processing, and potential susceptibility, which belongs to CONEFALL processing (hillshade and isolines: ©swisstopo)

5.7.5 Snow Avalanches

5.7.5.1 Bagnes Valley

The data upon which this case is based consist of a 2 m resolution DEM under 2000 m elevation and the DEM25 above 2000 m, the 1:25,000 landuse map (Vector25), the topographic maps and the inventory of historical avalanches propagations. The used resolution extracted from the 2 m DEM was 10 m for the entire study area in order with the phenomenon scale and to reduce the processing time (24 h of processing with this resolution). Moreover, we conducted a smoothing operation on the DEM to better simulate the inertia of this phenomenon (smoothing of the topographic irregularities).

Source Ares Identification

The Table 5.8 shows the parameters used for this analysis case for the four detection criteria:

Table 5.8 Avalanche source detection parameters

Detection criteria	Parameters
Slope	30° to 50°
Altitude	>1,000 meters high
Landuse	Outside of forest area
Minimum surface of avalanches triggering	Sources area >900 m^2

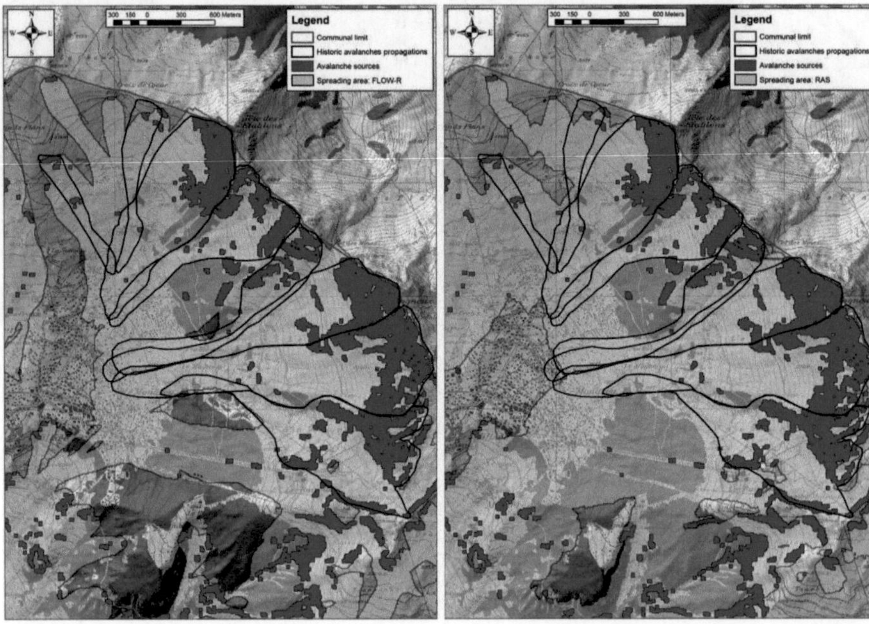

Fig. 5.26 Example of propagations. At *left*: Flow-R propagations. At *right*: RAS propagations. The dark lines are the maximum avalanche propagations recorded in the archives (hillshade and topographic map: ©swisstopo)

Spreading Area Assessment

To use the model RAS (Fig. 5.26, right), we first had to calibrate the relationship between α and β. After some tests, we decided to use the coefficients determined by Adjel (1996) in Haute-Tarentaise in the French Alps (Table 5.9). In the Bagnes valley itself, only 5 to 10 sectors have good inventories; this is not enough sufficient for making good calibrations.

For the model FLOW-R (Fig. 5.26, left), (as explained in Sect. 5.5.2.4), we must calibrate the inertia, energy and direction algorithms on past, well archived, propagation events. We selected four test areas homogeneously distributed in the study area and we chose to use the following propagation parameters (Table 5.10):

Table 5.9 Law alpha–beta used for the avalanche spreading assessments with the RAS model

Location	Number of event	Alpha–beta relationship	r	σ
Haute-Tarentaise	168	$\alpha = 0.82 * \beta + 2.82°$	0.81	2.6

Table 5.10 Spreading parameters used with the FLOW-R model

Flow direction algorithm	Runout distance calculation (Perla et al. 1980)	
	μ	M/D
Holmgren exp. 4	0.28	2,500

Detection and Spreading, Example of the Folorsi

To illustrate our choices and the ability of our method to create an avalanche susceptibility map, we now present the case of the Folorsi corridor (Fig. 5.27). This corridor is marked by annual avalanches cutting the forest roads (3 and 5). In this example we can see that there is a good correspondence with the detection of the triggering areas (1, 2 and 4), the spreading area and the event shown in the picture on the left. In addition we can observe that an avalanche with a large volume has the potential to flow beyond the point 5, down to the valley bottom.

5.7.6 Flooding

5.7.6.1 Bagnes Territory

The methodology presented in the Chap. 6.6 was applied to create the indicative map of flood hazard in Val de Bagnes (Valais, Switzerland). The Val de Bagnes has a particular hydrological context because of the presence of glaciers in the watershed and the strong human impact on rivers. For example, the Mauvoisin dam, with a retention capacity of over 210 million m^3, plays an important role in flood control for the valley. Many conducts redirecting water of various watersheds have an impact on discharges.

For this study, it was decided to take into account the 'natural' watersheds, which correspond to the boundaries of an extreme event in case of flooding. In-depth field work was conducted to identify potential overflow areas feeding the model FLOW-R. These include the various weakness of the hydrographical system such as shrinkage, bridges, debris that could create a jam or other obstructions to the watercourse (Fig. 5.28). The indicative map of danger does not take into account the intensity of the phenomenon.

The final map shows that the main danger comes from tributary streams (Fig. 5.29). Indeed, the watersheds composing the valley are large and present relatively strait outlets, resulting in many opportunities for overflows. Moreover,

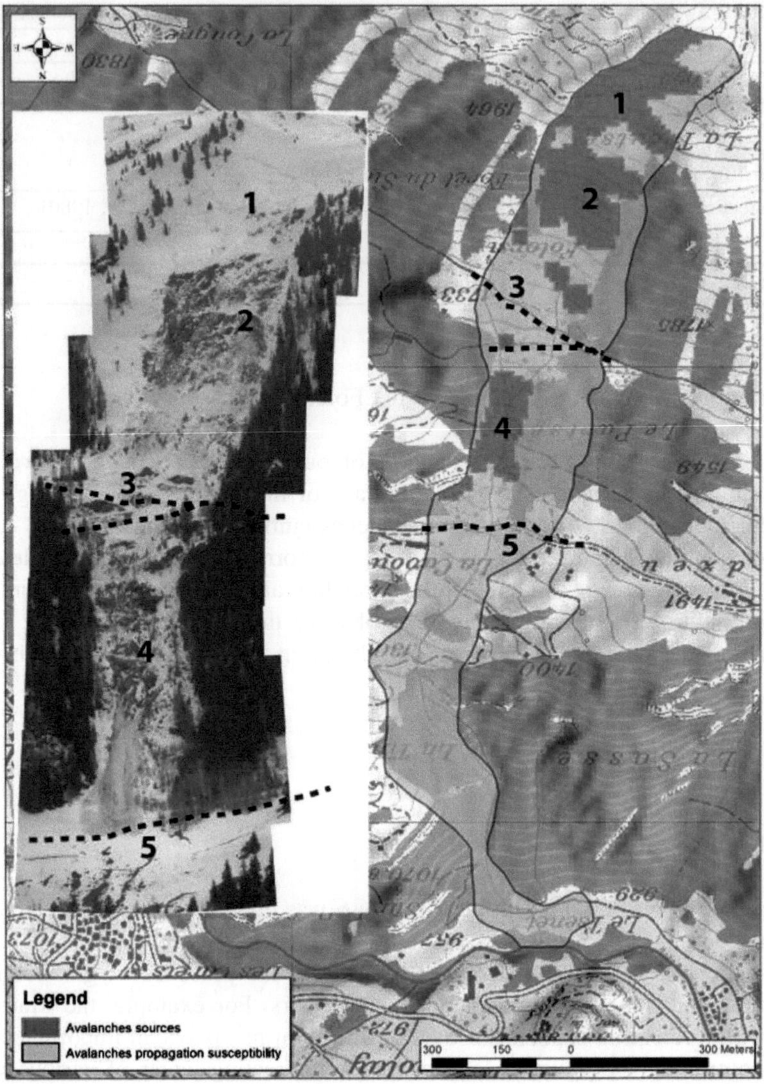

Fig. 5.27 Correspondence between the sources detected, the spreading area assess and the reality (Folorsi sector), the numbers indicate the same location on the picture and the map (hillshade and topographic map: ©swisstopo)

the human impact on rivers has led to a change in their natural course, creating an additional danger. Propagations performed with the software FLOW-R highlight those areas to be considered in a subsequent study of risk.

Fig. 5.28 An example of a potential source of overflow ant its corresponding modeled spreading (orthophoto: ©swisstopo)

5.8 Discussion

The above-described multi hazard susceptibility assessment is based on quite simple models, but they are cross-checked with field observations and inspections of other sources of information such as orthophotos, partial inventories, and feedback from local people in charge of natural hazards. This point makes this approach quite flexible. One of the major advantages is the fact the parameters used in the model are few, except for SINMAP, and are tangible. Yet this approach has its own limits and does not reflect the local controlling factors and specific conditions. The specific conditions have to be integrated in more detailed studies when the scale of study is more precise than 1:2,500; this scale corresponds for instance to the "danger maps" in Switzerland, contrasting with the present approach that correspond to the "indicative danger map" (Lateltin 1997; Loat and Petraschek 1997). Nevertheless, the proposed method demonstrates the efficiency of such a simplified approach, with results in good agreements with the observations. It must be noticed also that the original results of the models are usually kept, but a relative rating of the "plausibility" of these results is assigned, creating a susceptibility scale.

Because the models are mainly based on DEM, the DEM is the source of any issues that may arise. For instance, DEM generated topography is sometimes too detailed; source areas containing bridges or ditches do not permit to assess the potential area of propagation of debris-flows or floods, because the flow is channelize and thus to simulate flooding or debtis-flow propagation the bridges or ditches must be artificially broken, which must be prefomred manually.

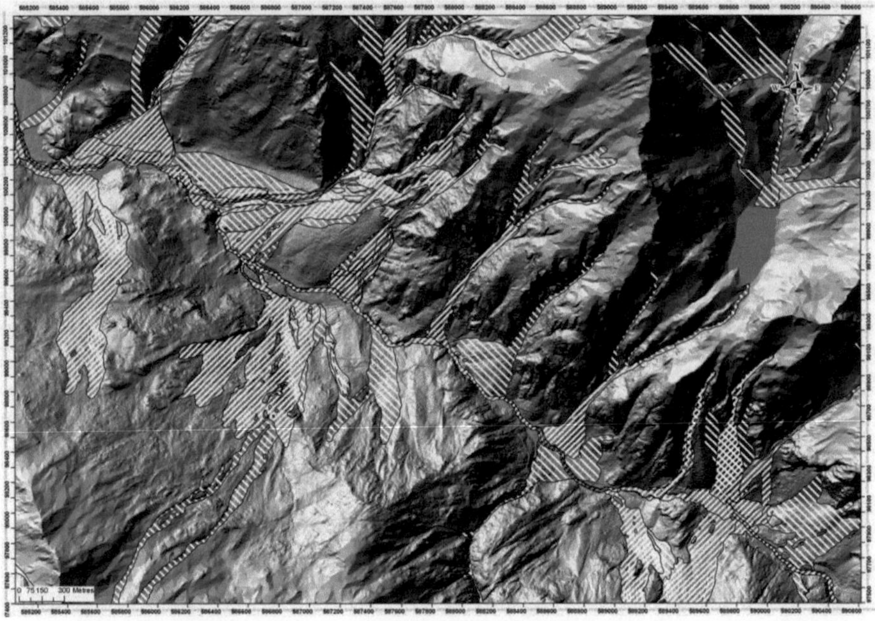

Fig. 5.29 Lots of tributary streams could be a danger for infrastructure. The final map illustrates the propagation of possible overflows (hillshade: ©swisstopo)

5.8.1 Landslide Inventory Maps and Shallow Landslides

Both examples coming from Vaud and Bagnes show different situations. In the first case, an existing inventory was available (DUTI 1985) and 1:25,000 geological maps were available in addition to orthophotos and vectorized topographic maps. In the second case, no inventory was available, only a partial cover of geological maps, but the local knowledge was used. This discrepancy shows the problem of data collection and treatment. The large area did not permit to make a detailed field check but the available inventory was the base of the investigations. The main issue was to link the former inventory based on relief interpretation at 1:25,000 (DUTI inventory + info from geological maps) to the Lidar-DEM hillshade with a resolution of 1 m. As shown by Ardizzone et al. (2007), this creates a problem of merging information without losing information. As the time allocated for this work was short it was not possible to create a completely new inventory; instead rules were used to obtain a scale of "plausibility". This problem occurred also in Bagnes to a lesser extent as little information existed from previous studies. However it was possible to validate the inventory according to local geologists' knowledge because of the relatively small size of the concerned area. According to the local experts, the landslide inventory map created for the Val de Bagnes was quite reliable. Nevertheless, this study has demonstrated the necessity to use a

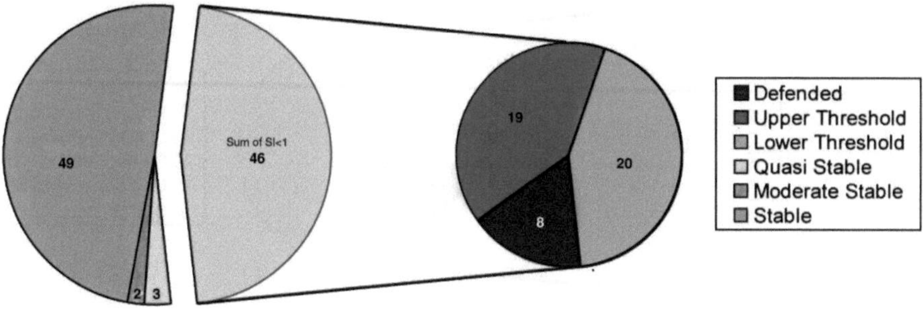

Fig. 5.30 Percentage of the total Val de Bagnes territory classed according to the different stability classes

HRDEM in order to identify landslides. Another issue linked to the data is that above 2000 m, mapping is based on the DEM25, because the Lidar-DEM does not exist. The use of orthophotos and aerial photo observations were the only way to check the results.

Shallow landslide modeling is one the most standard and used assessment method. Nevertheless it faces the usual problems: lack of information on soil and hydrological parameters and also limitations as the model only considers a constant soil thickness. Nonetheless, on average the results are valuable and reliable if they are connected with other information.

In the County of Vaud the comparison between the SINMAP results and the inventory landslide map shows that 78% of the inventoried landslides are contained in the predicted unstable zone. In general, 85% of inventoried landslides are found in a buffer of 50 m around the predicted unstable areas. The main differences between SINMAP results and the inventory map are found in the Jura region where the particular hydrological system (mainly karstic) makes the application of SINMAP model more difficult. Compared to the total surface of the study area, the SINMAP analysis shows that 18% (576 km^2) of the County of Vaud can be potentially affected by shallow landslides. The most susceptible region is the alpine region were the susceptible area increases to 55% of the overall surface. For the Val de Bagnes area 46% (almost 140 km^2) of the territory shows a stability index lower than 1. The main portion of the detected unstable areas have been classified in the susceptibility classes 2 and 3 respectively "Lower" and "Upper threshold" (Fig. 5.30). If only surface areas below 2000 m a.s.l. are considered for statistics, the area showing a stability index lower than 1 increases to 71% (Fig. 5.31). This is related to the presence of steep slopes glacier and rock-glacier surfaces above 2000 m a.s.l., where no shallow landslide can be initiated.

Some interesting points can be deduced form the relationship between shallow landslides, landslides and DSGD and other features:

- The DSGSD, which represent 13% of the Val de Bagnes territory (almost 40.5 of 300 km^2), are destabilizing entire slopes and promoting the formation of

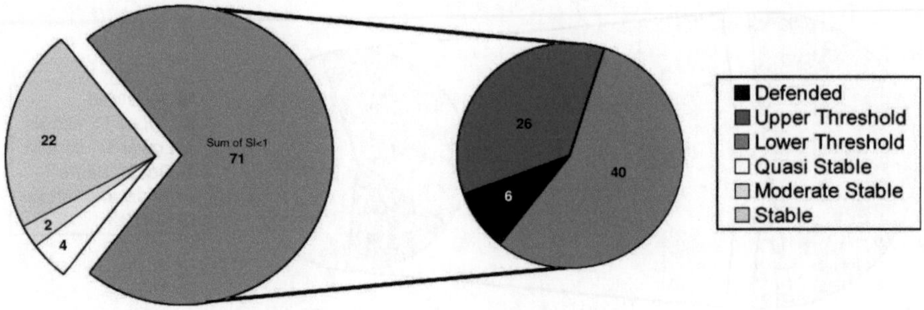

Fig. 5.31 Percentage of the surfaces above 2000 m a.s.l. in the Val de Bagnes classed according to SINMAP results

shallow and medium landslides. Indeed, 42% of the surfaces of shallow and medium landslides (almost 8.8 of 21.3 km^2) are linked to DSGSD.

- Main structural features (faults and thrusts) increase the development of slope instabilities. Indeed, considering a buffer of 200 m along ground features, it represents 19% of the total surface of the Val de Bagnes (58 km^2). Nevertheless, almost 34% of sliding surfaces are located at less than 200 m of a fault or thrust.
- 32% of the landslides are concentrated in moraine formations, which include up to 12% of the territory (30 km^2).

SINMAP or other similar models are very powerful methods to obtain a first overview of the most susceptible area for shallow landslides. However, the modeling results need to be critically interpreted. In particular three main points need to be considered for an objective interpretation:

The accuracy and the artifacts present in the DEM (HRDEM or DEM25) have a lot of influence on the SINMAP results especially in urbanized or forested areas;

Reliability of results is strongly related to the quality of the geomechanical parameters;

SINMAP results are useful and could be qualitatively or semi-quantitatively analyzed for small to intermediate scale (1:50,000–1:25,000) but they cannot be used at a greater scale than 1:25,000 without an important local parameter calibration.

5.8.2 Debris Flows

Again the difference in total area of both territories reveals contrasting results. For Vaud County the results cannot reflect local controlling factors and specific conditions, nevertheless good correlations exist between simulations and field observations performed on specific catchments where historical events occurred. Because of the large area (2,800 km^2) it is impossible to take into account the volume and the type of material of the various debris sources, requiring average data for large areas.

More generally, the propagation area modelling is strongly dependent on the quality of the DEM. In case of a wrong representation of the real topography, the propagation area will contain nonsense results. For example, a stream flowing under a high bridge will act as a dam; if the DEM is too detailed the channel can not be overtopped which can occur if the debris-flow erodes or if it is dammed promoting an outburst. These effects are not simulated by the model.

For the Bagnes Valley more than 50% of the sources detected by the model are confirmed by field observations, while 35% of the sources are classified "potential" (i.e. without evidence of past event) and 14% are classified as incorrect. This last "false" class is mainly due to the fact that the outcropping bedrock information was not of high quality. In this case, some actual bedrock areas were included in the source areas where no debris-flow triggering is possible as there is not any loose material (soil, moraine, debris, etc.).

The chosen methods show realistic results and allow a first fast assessment of debris-flow susceptibility over a large region, despite a limited knowledge of the local controlling factors. Again the results are contrasted:

1. For the Vaud County, both observations and the model are in good agreement, using a unique source of information, i.e., the DEM and regional geologic information.
2. The possibility to perform a qualitative classification of the source areas permits to obtain a good first overview of the debris-flow impacts in the communal territory of Bagnes Valley, which is directly useful for local authorities to prioritize remediation actions.
3. The propagation assessment based on probabilistic and basic energy calculations results in debris-flow susceptibility close to observed events.

This model has been applied with success also in Italy and France (Blahut et al. 2010; Lari et al. in review; Kappes et al. 2011), using a similar approach.

5.8.3 Rockfall

The detection of source areas using the slope angle histogram decomposition by normal distributions is efficient, but it can be improved by additional data such as geological maps, vectorized topographic maps and orthophotos. The morphological units that are identified by decomposing the slope classes in cliff faces and steep slopes are significant for identifying the morphology (Strahler 1950; Loye et al. 2009). When using a Lidar-DEM the detection is very detailed even in the zone covered with vegetation.

The angle ϕ_p of the cone aperture was set to 33° for both studies performed here. The results show to be in good conformity with an orthophotos analysis and field work. Yet when using CONEFALL at regional scale, a valley-bottom correction is required to consider local areas where such an angle of propagation would be too large. In addition, the unlimited lateral aperture of the cone does not seem to overestimate the lateral propagation of the rockfall runout. Using the Flow-R model, the runout length is taken into account. The multiple flow direction algorithm (Holmgren 1994) used for propagation in Flow-R takes into account channelized topography. The use of both models permits to create a classification of susceptibility with the result from Flow-R classified as probable and those from CONEFALL as potential.

Such a combined approach provides a fast and cost-effective way of identifying rockfall prone areas, without taking into account the structural setting and mechanical parameters in detail. This approach can be based on a DEM of poorer resolution as well, as demonstrated by Loye et al. (2009).

5.8.4 Snow Avalanches

Mapping snow avalanches is a tedious task especially at the regional scale, because in mountainous areas more or less all slopes from 30° to 60° are susceptible to snow avalanches, albeit sometimes with a very low frequency. Hazard assessments in the case of snow avalanches is mainly based on inventories and fine tuning modelling in specific sites (PPR 2011). This means that implicitly the frequency is taken into account in the choice of the sites, because snow avalanches are possible everywhere but in most cases only exceptionally.

Although the methodology used in this study to create an avalanche susceptibility map shows good correspondence between the archived events and the simulated one, some limitations should be noted. First there is a problem with the quantity of avalanche sources detected by the model. Indeed, the detection parameters are only topographic (slope, altitude) and for a steep alpine region this means that a lot of sources are detected. To solve this problem, first a reduction of the quantity of the sources detected according to the source sizes was tried. Then the spreading area was limited using the forest areas. Those tests were not conclusive because in both cases, there are some avalanche sources or propagation areas observed in reality that were ignored by the models. Further investigations still need to be conducted to decrease the number of potential sources in such steep valleys.

5.8.5 Floods in Bagnes Territory

In general, the application of the methods presented here has allowed covering the entire territory in a homogeneous and coherent way. Some remarks can be made about some specific points. The computation of discharge for important return periods must be applied to all locations where streams flow under bridges. An estimation of bridge capacity would give major information for the development of a hazard map; this requires improving mapping procedures if implemented for regional susceptibility mapping. The human impact on hydrological network complicates the representation of reality. In the case of the susceptibility map of Val de Bagnes, the most critical situations were always considered. The analysis was particularly focused on the tributary streams of the Dranse Bagnes, where hazard is higher rather than on the main stream. In fact, it has never been adapted to raise the possibility of overflow.

In any case, the Laser-DEM geomorphic interpretation is an important input, but it still needs to be used in a more rigorous approach.

5.9 Conclusion

The possibility to treat large amount of data is mainly due to availability of powerful computers and GIS technology for hazard assessment at regional scale by mainly using DEM. In addition, more readily available digital documents such as topographic maps with attributes such as landuse, geological maps, orthophotos, etc. greatly improves the simple model used for such purposes. Models must be simple for two reasons: (1) it is still difficult to use complex models over a large area if the DEM is a HRDEM having a resolution higher that 1 m; (2) complex models often require several unknown parameters. In this sense, the usefulness of SINMAP is borderline when it is applied to large inhomogeneous areas.

Depending on the time available the proposed simple modelling can be improved, especially based on comparisons with field observations or inspecting other data. Yet this interaction between the different types of information is of primary importance to create reliable susceptibility maps. The susceptibility classification depends on the ability to mixed models with different inputs that are in accordance with different confidence levels of the input data.

The greatest problem is linked to updating former inventories. The information produced by Lidar-DEM is often far more detailed than what we can expected from field observations, especially in forested areas. As a result, ad hoc solutions must be found in order to maintain past information and to improve the final product.

Nowadays, the availability of Lidar-DEM is the major input that permits to obtain a rapid overview of potential unstable areas or to perform more detailed studies as well. It provides input for geomorphic studies as well as for modelling.

5.10 Summary/Futuristic Vision

As already indicated, the future is linked to the power of computers and to the availability of new types of mapping documents. We expect that Lidar-DEM will become available more or less everywhere around the world, and a standard procedure for hazard mapping in the near future. At present, the method of regional hazard assessment is evolving very fast, because new techniques such as Interferometric Synthetic Aperture Radar (InSAR) are completely revolutionising the monitoring of mass movements, flood observations etc. Yet InSAR shows also a lot of limitations due to vegetation cover, slope orientations or mass movement directions. This example underlines the necessity for hazard assessment methods to always be validated by another one, which can be simply field investigations.

Thus the greatest current progress in hazard mapping arises from the Lidar technique, which gives high quality images of the ground even under dense vegetation. Lidar resolution and information content (full wave forms and intensities) are constantly improving, opening new fields of investigation. It seems that the geomorphology and soil and rock mechanics have not yet taken full advantage of the information that can be extracted from the high resolution Lidar-DEM, i.e., to characterize the mechanical properties of the ground or curvature of scars, which contain important information on failure mechanisms.

In addition, the modelling of natural hazards such as landslides, debris-flows, rockfall, snow avalanches, and floods has to take advantage of the fine resolution offered by new HRDEM, but as the data are huge and efforts must be made to find simpler models with few parameters that are suitable for modelling hazard on large areas.

Acknowledgments We are grateful to the authority of the Canton de Vaud for their permission to publish these results and for the helpful discussion we had with D. Moratel and P. Fouvy (SFFN), N. Christinet (DSE), C. Gerber (Cantonal geologist), D. Giorgis (OIT). We thank also the Bagnes municipality and his geologist Dr. F.J. Baillifard for allowing us to publish maps and results and for the useful discussions that make it possible to validate the models. We are also grateful to our colleagues K. Sudmeier for improving the English and to M. Charrière for the creation of the reference list. We are grateful to the National Swiss Foundation for Science project Number 200021-118105 and from the EU project Safeland.

References

Adjel G (1996) Méthodes statistiques pour la détermination de la distance d'arrêt des avalanches. Ph.D. thesis, Université Joseph-Fourier, Grenoble, France

Agliardi F, Crosta GB (2003) High resolution three-dimensional numerical modeling of rockfalls. Int J Rock Mech Min Sci 40:455–471

Aksoy H, Ercanoglu M (2006) Determination of the rockfall source in an urban settlement area by using a rule-based fuzzy evaluation. Nat Hazards Earth Syst Sci 6:941–954

Aleotti P, Chowdhury R (1999) Landslide hazard assessment: summary review and new perspectives. Bull Eng Geol Environ 58:21–44

Ancey C, Bain V, Bardou E, Borrel G, Burnet R, Jarry F, Kölbl O, Meunier M (2006) Dynamique des avalanches. Presses polytechniques et universitaires romandes (Lausanne, Suisse) and Cemagref (Antony, France), p 338

Ardizzone F, Cardinali M, Galli F, Guzzetti F, Reichenbach P (2007) Identification and mapping of recent rainfall-induced landslides using elevation data collected by airborne lidar. Nat Hazards Earth Syst Sci 7:637–650

Bai B, Wang J, Pozdnoukhov A, Kanevski MF (2009) Validation of logistic regression models for landslide susceptibility maps, IEEE computer society. In: Proceedings of the 2009 WRI world congress on computer science and information engineering, vol 2. Washington, DC, pp 355–358

Baillifard F, Jaboyedoff M, Sartori M (2003) Rockfall hazard mapping along a mountainous road in Switzerland using a GIS-based parameter rating approach. Nat Hazards Earth Syst Sci 3:435–442

Baillifard F, Jaboyedoff M, Rouiller JD, Couture R, Locat J, Locat P, Robichaud G, Hamel G (2004) Towards a GIS-based hazard assessment along the Quebec city Promontory, Quebec, Canada. In: Lacerda WA, Ehrlich M, Fontoura AB, Sayao A (eds) Landslides evaluation and stabilization. Balkema, Rotterdam, pp 207–213

Baum RL, Savage WZ, Godt JW (2002) TRIGRS—a fortran program for transient rainfall infiltration and grid-based regional slope stability. Open-file report 02-424, U.S. geological survey, p 61

Beven KJ, Kirkby MJ (1979) A physically based variable contributing area model of basin hydrology. Hydrol Sci Bull 24:43–69

Beven KJ, Lamb R, Quinn P, Romanowicz R, Freer J (1995) Topmodel. In: Singh VP (ed) Computer models of watershed hydrology. Water Resource Publications, Highlands Ranch, pp 627–668

Blahut J, Horton P, Sterlacchini S, Jaboyedoff M (2010) Debris flow hazard modeling on medium scale: Valtellina di Tirano, Italy. Nat Hazards Earth Syst Sci 10:2379–2390

Burrough P, McDonnell RA (1998) Principles of geographical information systems. Oxford University Press, Oxford, p 330

Carrara A, Guzetti F (1995) Geographical information systems in assessing natural hazards. Kluwer Academic Publisher, Dordrecht, p 353

Carrara A, Catalano E, Sorriso-Valvo M, Reali C, Osso I (1978) Digital terrain analysis for land evaluation. Geologia Applicata e Idrogeologia 13:69–127

Carrara A, Cardinali M, Detti R, Guzzetti F, Pasqui V, Reichenbach P (1991) GIS techniques and statistical models in evaluating landslide hazard. Earth Surf Proc Land 16:427–445

Cascini L (2008) Applicability of landslide susceptibility and hazard zoning at different scales. Eng Geol 102:164–177

Chacón J, Irigaray C, Fernández T, El Hamdouni R (2006) Engineering geology maps: landslides and geographical information systems. Bull Eng Geol Environ 65:341–411

Chigira M, Duan F, Yagi H, Furuya T (2004) Using an airborne laser scanner for the identification of shallow landslides and susceptibility assessment in an area of ignimbrite overlain by permeable pyroclastics. Landslides 1:203–209

Chung CJ, Fabbri AG (2008) Predicting landslide for risk analysis–spatial models tested by cross-validation technique. Geomorphology 94:438–452

Chung CJ, Fabbri AG, Van Westen CJ (1995) Multivariate regression analysis for landslide hazard zonation. In: Geographical information systems in assessing natural hazards. In: Carrara A, Guzetti F (ed) Kluwer Academic Publisher, Dordrecht, pp 107–133

Claessens L, Heuvelink GBM, Schoorl JM, Veldkamp A (2005) DEM resolution effects on shallow landslide hazard and soil redistribution modelling. Earth Surf Proc Land 30:461–477

Consuegra D, Joerin F, Vitalini F (1995) Flood delineation and impact assessment in agricultural land using GIS technology. In: Carrara A, Guzetti F (eds) Geographical information systems in assessing natural hazards. Kluwer Academic Publisher, Dordrecht, pp 107–133

Crosta GB, Agliardi F (2002) How to obtain alert velocity threshold for large rockslides. Phys Chem Earth 27:1557–1565

Crosta GB, Agliardi F (2003) A methodology for physically based rockfall hazard assessment. Nat Hazards Earth Syst Sci 3:407–422

Crosta GB, Frattini P, Sterlacchini S (2001) Valutazione e gestione del rischio da frana. Regione Lambardia, Milano, p 169

Cruden DM, Thomson S (1987) Exercises in terrain analysis. The Pica Pica Press, The University of Alberta Press, Edmonton, Alberta, p 185

De Moel H, Van Alphen J, Aerts JCJH (2009) Flood maps in Europe–methods, availability and use. Nat Hazards Earth Syst Sci 9:289–301
Delmonaco G, Leoni G, Margottini C, Puglisi C, Spizzichino D (2003) Large scale debris-flow hazard assessment: a geotechnical approach and GIS modelling. Nat Hazards Earth Syst Sci 3:443–455
Desmet PJJ, Govers G (1996) Comparison of routing algorithms for digital elevation models and their implications for predicting ephemeral gullies. Geog Inf Syst 10:311–331
Directive 2007/60/EC of the European parliament and of the council of the 23 Oct 2007 on the assessment and management of flood risks, p 8
Domínguez-Cuesta M, Montserrat JS, Colubi A, González-Rodríguez G (2009) Modelling shallow landslide susceptibility: a new approach in logistic regression by using favourability assessment. Int J Earth Sci. doi: 10.1007/s00531-008-0414-0
Dorren LKA, Seijmonsbergen AC (2003) Comparison of three GIS-based models for predicting rockfall runout zones at a regional scale. Geomorphology 56:49–64
DUTI (1985) Détection et Utilisation des terrains instables—projet d'Ecole DUTI. Rapport final. Ecole Polytechnique Fédérale de Lausanne, Lausanne, Suisse, p 229
Einstein HH (1988) Special lecture: landslide risk assessment procedure. In: Bonnard C (ed) Proceedings of the 5th international symposium on landslides, vol 2. Balkema, Lausanne, pp 1075–1090, 10–15 July 1988
Endreny TA, Wood EF (2003) Maximizing spatial congruence of observed and DEM-delineated overland flow networks. Int J Geog Inf Sci 17:699–713
Ercanoglu M, Gokceoglu C (2002) Assessment of landslide susceptibility for a landslide-prone area (north Yenice, NW Turkey) by fuzzy approach. Env Geol 41:720–730
Ercanoglu M, Gokceoglu C (2004) Use of fuzzy relations to produce landslide susceptibility map of a landslide-prone area (west black sea region, Turkey). Eng Geol 75:229–250
Erskine R, Green T, Ramirez J, MacDonald L (2006) Comparison of grid-based algorithms for computing upslope contributing area. Water Res Res, p 42
Evans S, Hungr O (1993) The assessment of rockfall hazard at the base of talus slopes. Can Geotech J 30:620–636
Fairfield J, Leymarie P (1991) Drainage networks from grid digital elevation models. Water Resour Res 27:709–717
Fell R, Corominas J, Bonnard C, Cascini L, Leroi E, Savage WZ (2008a) Guidelines for landslide susceptibility, hazard and risk zoning for land-use planning. Eng Geol 102:85–98
Fell R, Corominas J, Bonnard C, Cascini L, Leroi E, Savage WZ (2008b) Guidelines for landslide susceptibility, hazard and risk zoning for land-use planning—commentary. Eng Geol 102:99–111
Frattini P, Crosta G, Carrara A, Agliardi F (2008) Assessment of rockfall susceptibility by integrating statistical and physicallybased approaches. Geomorphology 94:419–437
Freeman TG (1991) Calculating catchment area with divergent flow based on a regular grid. Comput Geosci 17:413–422
Galli M, Ardizzone F, Cardinali M, Guzzetti F, Reichenbach P (2008) Comparing landslide inventory maps. Geomorphology 94:268–289
Gamma, P. (2000). dfwalk-Ein Murgang-Simulationsprogramm zur Gefahrenzonierung. Inaugural dissertation, Geographisches Institut der Universität Bern
Gilard O, Gendreau N (1998) Inondabilité: une méthode de prevention raisonnable du risque d'inondation pour une gestion mieux intégrée des bassins versants. Revue des Sci de l'Eau 3:429–444
Glenn NF, Streutker DR, Chadwick DJ, Thackray GD, Dorsch SJ (2006) Analysis of LIDAR-derived topographic information for characterizing and differentiating landslide morphology and activity. Geomorphology 73:131–148
Gokceoglu C, Sonmez H, Ercanoglu M (2000) Discontinuity controlled probabilistic slope failure risk maps of the Altindag (settlement) region in Turkey. Eng Geol 55:277–296
Gruber U, Bartelt P (2007) Snow avalanche hazard and modelling of large areas using shallow water numerical methods and GIS. Environ Model Softw 22:1472–1481

Günther A (2003) SLOPEMAP: programs for automated mapping of geometrical and kinematical properties of hard rock hill slopes. Comput Geosci 29:865–875
Günther A, Carstensen A, Pohl W (2004) Automated sliding susceptibility mapping of rock slopes. Nat Hazards Earth Syst Sci 4:95–102
Gupta RP, Saha AK, Arora MK, Kumar A (1999) Landslide hazard zonation in part of the Bhagirathi valley, Garhwal mimalyas, using integrated remote sensing–GIS. Himalayan Geol 20:71–85
Guzzetti F, Carrara A, Cardinali M, Reichenbach P (1999) Landslide hazard evaluation: a review of current techniques and their application in a multi-scale study, central Italy. Geomorphology 31:181–216
Guzzetti F, Cardinali M, Reichenbach P, Carrara A (2000) Comparing landslide maps: a case study in the upper Tiber River Basin, Central Italy. Environ Manage 25:247–363
Guzzetti F, Crosta G, Detti R, Agliardi F (2002) STONE: a computer program for the three-dimensional simulation of rock-falls. Comput Geosci 28:1079–1093
Guzzetti F, Reichenbach P, Wieczorek GF (2003) Rockfall hazard and risk assessment in the Yosemite Valley, California, USA. Nat Hazards Earth Syst Sci 3:491–503
Haugerud RA, Harding DJ, Johnson SY, Harless JL, Weaver CS (2003) High-resolution LIDAR topography of the Puget Lowland, Washington–a bonanza for earth science. GSA Today 13:4–10
Heim A (1932) Bergsturz und Menschenleben. Fretz and Wasmuth Verlag, Zurich, p 218
Heinimann HR, Hollenstein K, Kienholz H, Krummenacher B, Mani P (1998) Methoden zur analyse und Bewertung von Naturgefahren, Bundesamt für Umwelt, Wald und Landschaft (BUWAL), Bern
Holmgren P (1994) Multiple flow direction algorithms for runoff modeling in grid based elevation models: an empirical evaluation. Hydrol Process 8:327–334
Horton P, Jaboyedoff M, Bardou E (2008) Debris flow susceptibility mapping at a regional scale. In: Locat J, Perret D, Turmel D, Demers D, Leroueil S (ed) Proceedings of the 4th canadian conference on Geohazards. From causes to management. Presse de l'Université Laval, Québec, p 594
Huggel C, Kääb A, Haeberli W, Teysseire P, Paul F (2002) Remote sensing based assessment of hazards from glacier lake outbursts: a case study in the Swiss Alps. Can Geotech J 39:316–330
Huggel C, Kääb A, Haeberli W, Krummenacher B (2003) Regional-scale GIS-models for assessment of hazards from glacier lake outbursts: evaluation and application in the Swiss Alps. Nat Hazards Earth Syst Sci 3:647–662
Hydrological Atlas of Switzerland (2006) Universität Bern and OFEV, Switzerland
Jaboyedoff M (2003) CONEFALL 1.0: a program to estimate propagation zones of rockfall based on cone method. Quanterra, www.quanterra.ch
Jaboyedoff M, Derron MH (2005) Integrated risk assessment process for landslides. In: Hungr O, Fell R, Couture RR, Eberhardt E (eds) Landslide risk management. On CD included in the book, A.A. Balkema, Leiden
Jaboyedoff M, Labiouse V (2003) Preliminary assessment of rockfall hazard based on GIS data. ISRM 2003–technology roadmap for rock mechanics, symposium series—South African Institute of Mining and Metallurgy, vol 1. pp 575–578
Jaboyedoff M, Baillifard F, Philippossian F, Rouiller JD (2004) Assessing the fracture occurrence using the "weighted fracturing density": a step towards estimating rock instability hazard. Nat Hazards Earth Syst Sci 4:83–93
Jaboyedoff M, Labiouse V (2011) Technical note: preliminary estimation of rockfall runout zones. Nat Hazards Earth Syst Sci 11:819−828
Jaboyedoff M, Pedrazzini A, Horton P, Loye A, Surace I (2008) Preliminary slope mass movements susceptibility mapping using LIDAR DEM. In: Proceedings of the 61th Canadian geotechnical conference and 9th joint CGS/IAH-CNC Grounwater conference. Edmonton, Canada, pp 419–426

Jaboyedoff M, Michoud C, Mazotti B, Choffet M, Dubois J, Breguet A, Métraux V, Derron MH, Horton P, Loye A, Pedrazzini A (2010) Cartes Indicatives de dangers pour le Val de Bagnes et de Vollèges. Carnet méthodologique, p 89

Jones JL (2004) Mapping a flood… before it happens. Fact sheet 2004–3060, U.S. geological survey, p 2

Kappes M, Malet JP, Remaître A, Horton P, Jaboyedoff M (2011) Assessment of debris flow susceptibility at medium-scale in the Ubaye valley, France. Nat Hazards Earth Syst Sci 11:627–641

Kasai M, Ikeda M, Asahina T, Fujisawa K (2009) LiDAR-derived DEM evaluation of deep-seated landslides in a steep and rocky region of Japan. Geomorphology 113:57–69

Lan HX (2004) Landslide hazard spatial analysis and prediction using GIS inthe xiaojiang watershed, Yunnan, China. Eng Geol 76:109–128

Lan H, Martin CD, Lim CH (2007) RockFall analyst: a GIS extension for three-dimensional and spatially distributed rockfall hazard modeling. Comput Geosci 33:262–279

Lateltin O (1997) Prise en compte des dangers dus aux mouvements de terrain dans le cadre des activités de l'aménagement du territoire. Recommandations, Office fédéral de l'Environnement, des forêts et du paysage

Lari S, Frattini P, Crosta GB, Jaboyedoff M, Horton P (in review) Rockfall and debris flow societal and economic risk assessment at the regional scale. Rendiconti Lincei Scienze Fisiche e Naturali

Lied K (1977) Rockfall problems in Norway. Instituto sperimental modelli e stutture Publication 90:51–53

Lied K, Bakkehøi S (1980) Empirical calculations of snow avalanche run-out distances based on topographic parameters. J Glaciol 26:165–177

Lied K, Kristensen K (2003) Snøskred—Håndbook om snøskred. Vett and Viten AS, Nesbru, Norway, p 200

Loat R, Pertrascheck A (1997) Prise en compte des dangers dus aux crues dans le cadre des activités de l'aménagement du territoire. Recommandations, dangers naturels. Office fédéral de l'économie des eaux, Office fédéral de l'aménagement du territoire, Office fédéral de l'environnement, des forêts et du paysage

Loye A, Jaboyedoff M, Pedrazzini A (2009) Identification of potential rockfall source areas at a regional scale using a DEM-based geomorphometric analysis. Nat Hazards Earth Syst Sci 9:1643–1653

Malamud BD, Turcotte DL, Guzzetti F, Reichenbach P (2004) Landslide inventories and their statistical properties. Earth Surf Proc Land 29:687–711

Marco JB (1994) Flood risk mapping. In: Rossi G, Harmancioğlu N, Yevjevich V (ed) Proceedings of the NATO advanced study institute on coping with floods, vol 257. NATO applied sciences series, pp 353–374

McClung DM, Schaerer PA (1993) The avalanche handbook. The Mountainers, Seattle, p 271

McKean J, Roering J (2004) Objective landslide detection and surface morphology mapping using high-resolution airborne laser altimetry. Geomorphology 57:331–351

Merz B, Thieken AH, Gocht M (2007) Flood risk mapping at the local scale: concepts and challenges. In: Begum S, Stive MJF, Hall JW (eds) Advanced in natural and technological hazards research, vol 25. Springer Publication, New York, pp 231–251

Metzger R (2003) Modélisation des inondations par approches déterministe et stochastique avec prise en compte des incertitudes topographiques pour la gestion des risques liés aux crues. Ph.D Thesis, Ecole Polytechnique de Lausanne, Lausanne, Suisse, p 167

Montgomery DR, Dietrich WE (1994) A physically-based model for the topographic control on shallow landsliding. Water Resour Res 30:1153–1171

Morrissey MM, Wieczorek GF, Morgan BA (2001) A comparative analysis of hazard models for predicting debris flows in Madison county, Virginia. Open file report 01-0067. Geological Survey, US

Noverraz F (1995) Carte des instabilités de terrain du Canton de Vaud. Rapport conclusif et explicatif des travaux de levé de cartes. Ecole Polytechnique Fédérale de Lausanne, p 33

OFEG (2003) Evaluation des crues dans les bassins versants de Suisse. Guide pratique. Rapport de l'OFEG, Série Eaux, 4. Berne, Suisse, p 114

Pack R, Tarboton DG, Goodwill CN (1998) The SINMAP approach to terrain stability mapping. 8th congress of international association of engineering geology, Vancouver

Perla RI, Cheng TT, McClung DM (1980) A two-parameter model of snow avalanche motion. J Glaciol 26:197–207

Perret J (2007) Géotypes, une relecture, Tracés. Société des éditions des associations techniques universitaires

PPR (1999) Plans de prévention des risques naturels—Risques d'inondation. Guide méthodologique. La documentation Française, Paris, France, p 123

PPR (2011) Plans de prévention des risques d'avalanches. Guide méthodologique. http://archives.prim.net

Pradhan B, Lee S (2010) Regional landslide susceptibility analysis using back-propagation neural network model at Cameron Highland, Malaysia. Landslide 7:13–30

Quinn P, Beven K, Chevallier P, Planchon O (1991) The prediction of hillslope flow paths for distributed hydrological modelling using digital terrain models. Hydrol Process 5:59–79

Rickenmann D, Zimmermann M (1993) The 1987 debris flows in Switzerland: documentation and analysis. Geomorphology 8:175–189

Roering JJ, Kirchner JW, Dietrich WE (2005) Characterizing structural and lithologic controls on deep-seated landsliding: Implications for topographic relief and landscape evolution in the Oregon coast range, USA. Geol Soc Am Bull 117:654–668

Rouiller JD, Jaboyedoff M, Marro C, Philippossian F, Mamin M (1998) Pentes instables dans le Pennique valaisan. Matterock: une méthodologie d'auscultation des falaises et de détection des éboulements majeurs potentiels. Rapport final du PNR31. VDF Hochschulverlag AG, ETH Zürich, Switzerland, p 238

Salciarini D, Godt JW, Savage WZ, Conversini P, Baum RL, Michael JA (2006) Modeling regional initiation of rainfall-induced shallow landslide the eastern Umbria region of central Italy. Landslide 3:181–194

Salm B (1983) Guide pratique sur les avalanches. Club Alpin Suisse, Suisse, p 148

Sartori M, Gouffon Y, Marthaler M (2006) Harmonisation et définition des unités lithostratigraphiques briançonnaises dans les nappes penniques du Valais. Eclogae Geologicae Helvetiae 99:363–407

Scheidegger AE (1973) On the prediction of the reach and velocity of catastrophic landslides. Rock Mech 5:231–236

Schulz WH (2004) Landslides mapped using LIDAR imagery, Seattle, Washington. Open-file report 04-1396. Geological survey, US, p 11

Schulz WH (2007) Landslide susceptibility revealed by LIDAR imagery and historical records, Seattle, Washington. Eng Geol 89:67–87

Shan J, Toth K (2008) Topographic laser ranging and scanning–principles and processing. CRC Press, LLC, p 590

Soeters R, Van Westen CJ (1996) Slope instability recognition, analysis, and zonation. In: Turner AK, Schuster RL (ed) Landslides—investigation and mitigation—special report 247, Transportation research board, National research council. National Academy Press, Washington, pp 129–177

Strahler AN (1950) Equilibrium theory of erosional slopes approached by frequency distribution analysis. Am J Sci 248(673–696):800–814

Swisstopo (2004) MNT25—Le modèle numérique du terrain de la Suisse. Factsheet, Office fédéral de la topographie, p 15

Swisstopo (2005) MNT-MO—Les géodonnées de la Suisse proposées par l'Office fédéral de la topographie pour une utilisation à caractère professionnel. Factsheet, Office fédéral de la topographie, p 3

Takahashi T (1981) Estimation of potential debris flows and their hazardous zones: Soft countermeasures for a disaster. Nat Disaster Sci 3:57–89

Tarboton DG (1997) A new method for the determination of flow directions and upslope areas in grid digital elevation models. Water Resour Res 33:309–319

Thélin P, Gouffin Y, Allimann M (1994) Caractéristiques et métamorphisme des phyllosilicates dans la partie occidentale de la “super” nappe du Grand St-Bernard (Val d’Aoste et Valais). Bulletin Géologique, vol 327. Lausanne

Toppe R (1987) Terrain models—a tool for natural hazard mapping. In: Salm B, Gubler H (eds) Avalanche formation, movement and effects, vol 162. IAHS Publication, Wallingford, pp 629–638

Trümpy R (1980) Geology of Switzerland—a guide book, part A: an outline of the geology of Switzerland. Wepf & Co, Basel, p 104

Turberg P, Parriaux A, Kalbermatten M, Golay F, Lance JM (2008) The geotype concept to develop GIS oriented analysis in engineering geology application. Swiss Geoscience Meeting conference, Lugano

Van Alphen J, Martini F, Loat r, Slopm R, Passchier R (2009) Flood risk mapping in Europe, experiences and best practices. J Flood Risk Manage 2:285–292

Van Den Eeckhaut M, Poesen J, Verstraeten G, Vanacker V, Nyssen J, Moeyersons J, Van Beek LPH, Vandekerckhove L (2007) Use of LIDAR-derived images for mapping old landslides under forest. Earth Surf Proc Land 32:754–769

Van Dijke JJ, van Westen CJ (1990) Rockfall hazard, a geomorphological application of neighbourhood analysis with ILWIS. ITC J 1:40–44

Van Westen CJ, Van Asch TWJ, Soeters R (2006) Landslide hazard and risk zonation–why is it so difficult? Bull Eng Geol Environ 65:167–184

Varnes DJ (1984) Landslide hazard zonation: a review of principles and practice. United Nations Educational, Scientific and Cultural Organization, Paris, p 63

Wieczorek GF, Mandrone G, DeCola L (1997) The influence of hillslope shape on debris-flow initiation. In: ASCE (ed) First international conference water resources engineering division, San Francisco, CA, pp 21–31

Wyllie DC, Mah CW (2004) Rock slope engineering–civil and mining, 4th edn. Spon Press, New York, p 431

Zeng-Wang X (2001) GIS and ANN model for landslide susceptibility mapping. J Geog Sci 11:374–381

Chapter 6
Application of GIS and RS for Mapping Landslides at the Watershed Level

Jiann-Yeou Rau, Kang-Tsung Chang, Chi-Chung Lau, Liang-Chien Chen, Yi-Chen Shao and Jin-King Liu

Abstract Assessment and inventory of natural hazards such as landslides are essential for effective watershed management and sustainable development. In Taiwan, a typhoon (tropical cyclone) or earthquake event can trigger hundreds to thousands of shallow landslides in mountainous areas with steep slopes and rapid streams. Therefore, how to improve the efficiency and accuracy of landslide mapping by means of GIS (geographic information system) and remote sensing techniques is an important research issue. This study proposes a novel, semiautomatic method for mapping and editing landslides at a watershed level. Data sources include airborne laser scanner (ALS) data and color/near infrared ortho-imagery: the ALS data provide topographic features such as elevation, slope, surface roughness, and object height, and the ortho-imagery furnishes the

J.-Y. Rau (✉)
National Cheng Kung University, Tainan, Taiwan
e-mail: jyrau@mail.ncku.edu.tw

K.-T. Chang
Kainan University, Taoyuan, Taiwan
e-mail: ktchang@ntu.edu.tw

C.-C. Lau
Industrial Technology Research Institute, Hsin-Chu, Taiwan
e-mail: cclau@itri.org.tw

L.-C. Chen
National Central University, Taoyuan, Taiwan
e-mail: lcchen@csrsr.ncu.edu.tw

Y.-C. Shao
China University of Science and Technology, Taipei, Taiwan
e-mail: ycshao@gmail.com

J.-K. Liu
LIDAR Technology Co., Ltd., Hsin-Chu, Taiwan
e-mail: jkliu@lidar.com.tw

B. Pradhan and M. Buchroithner (eds.), *Terrigenous Mass Movements*,
DOI: 10.1007/978-3-642-25495-6_6,

radiometric characteristic of land cover such as greenness index or NDVI (normalized difference vegetation index) for identifying bare grounds. Based on the derived topographic and radiometric parameters, the method first uses a global, automatic algorithm to interpret and delineate landslides. Then it uses a local region growing algorithm and a 3D eraser to edit and compile landslide maps. To explore the causes of mass movement, these landslide maps can also be registered with other geospatial data in a GIS for data visualization and analysis. Experimental results indicate that the method is highly efficient and accurate compared with results of human interpretation from the stereo pairs of aerial photographs. Because Taiwan experiences an average of four or five typhoons every year, this new, semiautomatic method is expected to provide a useful tool for watershed management.

Keywords Landslides mapping · Watershed · Remote sensing

6.1 Introduction

Landslides can result in serious property damage and human casualties in the hills and steep mountainous regions. It is also known that landslides, especially shallow landslides, can be easily triggered by rainfall and earthquake (Caine 1980; Crozier 1999; Hong and Adler 2007; Chang et al. 2007; Chang et al. 2008; Chiang and Chang 2009). In the meantime, landslides can deliver large amounts of sediment into reservoirs and decrease their water reservation capacity (Dadson et al. 2004; Mikos et al. 2006). Thus the assessment and inventory of landslides are essential for effective watershed management and sustainable development.

In disaster mitigation, preparation, and risk assessment, a landslide susceptibility map is a must in order to prevent human casualty and property damage. Normally, the risk map is predicted or modeled by means of GIS overlay analysis between existing landslide inventory map and related topographic and geologic factors, such as geology, soil type, slope aspect, land cover, vegetation, land surface temperature, under-ground water level, elevation, surface roughness, semi-variance, and fractal dimension (Dhakal et al. 2000; Komac 2004; Glenn et al. 2006; Weirich and Blesius 2007). Different methods or algorithms for landslide susceptibility map generation have been published including several GIS based methods (van Westen 1994; Wang et al. 2005).

The island of Taiwan is located in a relatively new mountain-building belt. The natural rock bodies are highly fractured with numerous joints and cleavages. The majority of the area is covered with Oligocene, Miocene argillite, and shale, which are soft and fragile, especially when saturated with moisture. In Taiwan a typhoon (tropical cyclone) or earthquake event can trigger hundreds to thousands of shallow landslides in mountainous areas with steep slopes and rapid streams (Lin and Jeng 2000; Cheng et al. 2005; Lin et al. 2006).

The Chi–Chi Earthquake (also known as the 921 Earthquake), measured 7.3 on the Richter scale, occurred on September 21st, 1999 in central Taiwan. The Chi–Chi Earthquake triggered more than 20,000 landslides on the island of Taiwan. The co-seismic weakening of the substrate material increased the danger of landslides after a typhoon (Dadson et al. 2004; Cheng et al. 2005; Lin et al. 2006). Five years after the Chi–Chi Earthquake, on Aug. 23–25, 2004, typhoon AERE brought an accumulated rainfall of 1,607 mm within the Shihmen Reservoir watershed and triggered 222 landslides covering 393 hectares (Rau et al. 2007). Due to the steep terrain in the watershed, most landslides were unreachable and on-site surveying was not possible during a short period of time. This kind of condition has called for the use of remote sensing technology for the monitoring of landslides, a critical task for the purpose of emergency response and decision making. And, how to improve the efficiency and accuracy of landslide monitoring and mapping by means of GIS (Geographic Information Systems) and remote sensing techniques has become an important research issue (Raju and Saibaba 1999; Hervas et al. 2003; Tarchi et al. 2003; Rau et al. 2007).

6.2 Problem Identification/Conceptual Background

To improve the efficiency and accuracy of landslide mapping by means of GIS and remote sensing techniques, this study proposes a novel, semiautomatic method for mapping and editing landslides at a watershed level. Data sources include airborne laser scanning (ALS) data and color/near infrared ortho-imagery. The ALS data provide topographic features such as elevation, slope, surface roughness and above terrain object model whereas the ortho-imagery furnishes the radiometric characteristics of the land cover such as greenness index or NDVI (normalized difference vegetation index) for the identification of bare grounds. Based on the derived topographic and radiometric parameters, the proposed scheme uses a global and automatic algorithm to detect landslides. Then, it employs a local region growing algorithm and a 3D eraser for the editing and compilation of landslides to complete the mapping process.

6.3 Review of Literature

Remote sensing data is normally used in three phases of landslide related studies: (1) detection and identification, (2) monitoring, and (3) spatial analysis and hazard prediction (Metternicht et al. 2005). For detection and identification of landslides, many kinds of remote sensing data have been discussed in the literature including terrestrial-based photography or laser scanning (TLS), air-based stereo- or ortho-imagery and aerial laser scanning (ALS), and satellite-based optical or SAR imagery. Among them, digital stereoscopic aerial photograph is the most frequently used data because not only landslide diagnostic features (e.g., disruptive vegetation cover, and scarps) but also landslide qualitative characteristics (e.g., number, distribution, and type) can be derived from the data (Metternicht et al. 2005). The recognition procedure using stereoscopic aerial photograph is normally performed by human visual interpretation and manual digitization (Donati and Turrini 2002; He et al. 2003). Although such procedure can achieve a landslide map with high accuracy and reliability, it is time consuming. Thus, to reduce manpower in photo interpretation, researchers have turned to the ortho-rectified aerial imagery or multi-spectral satellite imagery in conjunction with the use of image filtering, differencing, band ratio techniques and NDVI calculation (Lin et al. 2002; van Westen and Getahun 2003; Herva's et al. 2003).

The detail of geomorphologic features extracted depends on the image scale or spatial resolution. For slope instability study the best image scale would be 1:15,000, because the disrupted vegetation cover, scarp, rotated blocks, etc. could be identified clearly (Mantovani et al. 1996). An image scale of 1:25,000 is considered as the smallest scale to interpret slope instability phenomena from aerial photographs (Singhroy 2002).

Airborne Laser Scanning (ALS) or Light Detection And Ranging (LiDAR) techniques is an optical remote sensing technique popular in digital terrain modeling (Kruas and Pfeifer 1998; Vosselman 2000; Sithole and Vosselman 2004), city modeling (Haala and Brenner 1999; Schwalbe et al. 2005; Hofmann 2004), archaeology (Willis et al. 2009), geology (Webster et al. 2006), geomorphology (McKean and Roering 2004), seismology (Cunningham et al. 2006; Chen et al. 2006), remote sensing (Bork and Su 2007; Chen et al. 2009), and atmospheric physics (Pappalardo et al. 2004; Zhao et al. 2008). The principle of distance measurement using LiDAR is based on time-of-flight from transmission of a laser pulse to the detection of its returning signal. During the traveling of laser pulse, it may encounter many kinds of objects and return with a certain magnitude of laser signal. Normally, it can provide a maximum of four returns (i.e., the coordinates of four ground objects). Thus, an enormous three-dimensional point cloud can be obtained including terrain, vegetation canopy, building roof-top, road, etc. A 5–20 cm of elevation accuracy has been reported by using the integrated GPS/INS (inertia navigation system) approach (Ussyshkin and Smith 2006). Normally, the LiDAR data are scanned with high frequency (33–100 kHz), high density (1–2 points/m^2), and small footprint (5–30 cm) capable of describing local surface roughness and identifying individual morphological domains within the landslide complex (McKean and Roering 2004; Optech ALTM 2009).

The integration of airborne digital imagery with ALS data is useful in many applications, such as building modeling (Chen and Teo et al. 2006), coastal mapping (Elaksher 2008), vegetation detection (Bork and Su 2007) and disaster assessment (Li et al. 2008). So far as we know, this study is the first case that fusions both data for landslide mapping.

6.4 Tools/Materials

This study developed indigenous software to demonstrate the feasibility of automated mapping of landslides. The Visual Studio.NET framework was used as the programming platform and Managed C++ with OpenGL library were integrated to realize the GUI interface. The product of the software is a shapefile that contains 3D landslide polygons and some topographic features, which can be exported to a GIS for further spatial analysis.

6.5 Methodology

This study adopted the fusion of airborne LiDAR and color ortho-imagery for landslide detection after considering such factors as the operational range, accuracy, resolution, efficiency, and cost. LiDAR data can be used to derive high resolution geometric features of the ground surface (digital surface model or DSM) and terrain (digital elevation model or DEM) from high density point clouds. On the other hand, the color ortho-imagery can provide spectral information of ground objects. These two types of data allowed this study to develop a global and local supervised classification algorithm using the object height model (OHM), slope, surface roughness and greenness/NDVI indices. An interactive manual 3D Eraser/Painter was also developed for editing landslides and minimizing commission/omission errors.

This study first classified the LiDAR data semi-automatically as surface and terrain points by TerraScan© and interpolated the point data into 1 m DSM and DEM, respectively. Then it followed the flowchart in Fig. 6.1 to map landslides. At the start, the color ortho-imagery and LiDAR DSM and DEM were imported into the designed system. Four topographical (slope, roughness and OHM) and spectral (greenness) indices were generated from the input data. The detection procedure contained the global and local approaches. Global landslide detection uses a supervised classification method. Since the omission and commission errors were unavoidable using the global approach, primarily due to diverse geologic and topographic environments in which landslides occurred, local landslide detection and the editing scheme was required to increase the accuracy of the produced landslide map. For error analysis, the user accuracy, producer accuracy, average accuracy, and overall accuracy were calculated from a

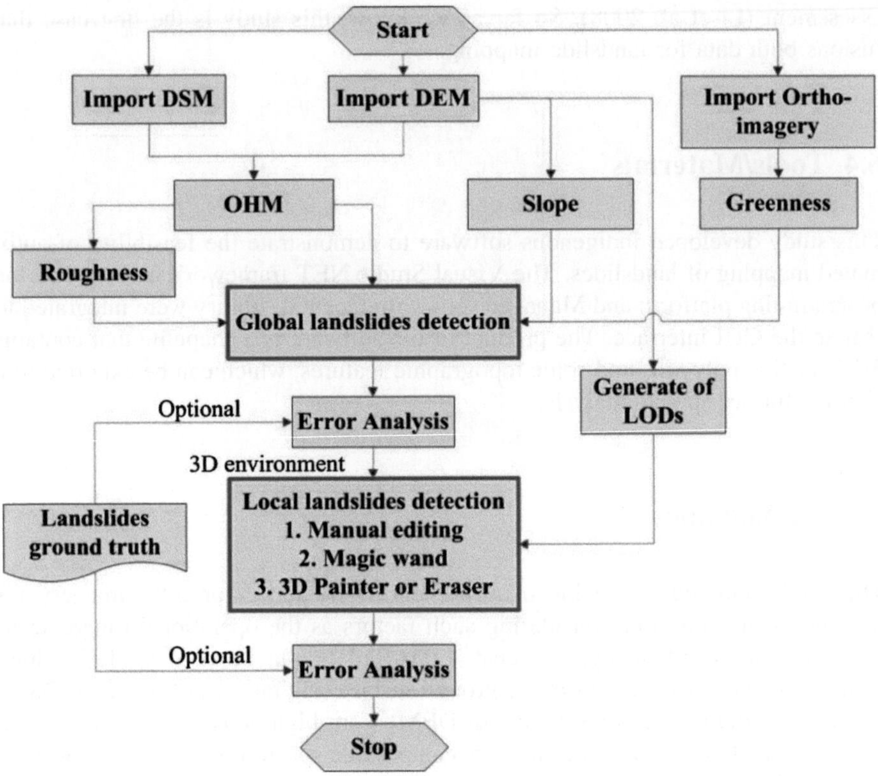

Fig. 6.1 Flowchart for landslide detection and mapping

Table 6.1 Confusion matrix and formulas for error analysis

T = A+B + C+D		Producer (Detection Results)	
		Landslides	Non-Landslides
User (Ground truth)	Landslides	A	C
	Non-Landslides	B	D
	Detection of landslides	Detection of non-landslides	
Producer accuracy	PA1 = 100 * A/(A + C)	PA2 = 100 * D/(B + D)	
Omission error	OE1 = 100 − PA1	OE2 = 100 − PA2	
User accuracy	UA1 = 100 * A/(A + B)	UA2 = 100 * D/(C + D)	
Commission error	CE1 = 100– UA1	CE2 = 100 − UA2	
Overall accuracy	OA = 100 * (A + D)/Total		
Average accuracy	AA = (PA1 + PA2)/2		

confusion matrix (Kohavi and Provost 1998), provided that independent ground truth data were available. Table 6.1 shows the matrix and related formulas for error analysis. A more detailed description of the procedure is presented in the following sections.

6.5.1 Pre-Processing

This session describes the algorithms for calculating the object height model, roughness, slope, and greenness.

6.5.1.1 Object Height Model

The OHM describes the heights of above ground objects in raster format. Objects with height close to zero may represent new landslides with bare soil. This study derived OHM by subtracting DEM from DSM, as shown in Eq. 6.1.

$$\text{OHM} = \text{DSM} - \text{DEM} \tag{6.1}$$

6.5.1.2 Roughness

Most new landslides are bare soil areas, where the surface is smoother than forested areas. A surface roughness index can therefore be used to detect new landslides. To account for the high terrain variation in mountainous areas, this study used object heights (h) rather than surface heights for calculating surface roughness. In Eq. 6.2, the surface roughness index γ is the standard deviation of h within a local window, $\bar{h}$ is the mean height of all pixels within the local window, and n is the number of pixels $\bar{\mathrm{h}}$.

$$\gamma = \sqrt[2]{\frac{\sum (\mathrm{h} - \bar{\mathrm{h}})^2}{\mathrm{n} - 1}} \tag{6.2}$$

6.5.1.3 Slope

A variety of methods are available for terrain slope estimation. Many of them use a fixed window size to estimate the terrain slope. Since a small mask may contain man-made artificial objects in a high resolution DEM, a flexible window size denoting multiple scale of terrain feature is deemed more suitable for the characterization of landslides. This study adopted a method proposed by Parker (1997), who used derivatives of the Gaussian function to convolute with the DEM in the x and y direction and combines them together to estimate the slope. Equation 6.3 illustrates a one-dimensional Gaussian function in which the sigma (σ) value represents the scale of the terrain feature. Equation 6.4 is the result of the derivative of the Gaussian function in the x direction. The terrain slope can thus be calculated using Eq. 6.5. In the case of raster DEM, an odd-number Gaussian kernel is used with a suitable scale factor (σ). For example, Fig. 6.2 illustrates a

Fig. 6.2 An example of *x*-directional, Gaussian kernel with a window size of 5 × 5 and a scale factor (σ) of 0.8

0.137	0.715	0.000	-0.715	-0.137
0.137	0.715	0.000	-0.715	-0.137
0.137	0.715	0.000	-0.715	-0.137
0.137	0.715	0.000	-0.715	-0.137
0.137	0.715	0.000	-0.715	-0.137

5 × 5 Gaussian kernel in the *x*-direction with a scale factor of 0.8. A larger scale factor with a larger window size tends to produce smoother output.

$$f(x) = e^{-(\frac{x^2}{2\sigma^2})} \tag{6.3}$$

$$f_x = f'_x(x) = (-\frac{x}{\sigma^2})e^{-(\frac{x^2}{2\sigma^2})} \tag{6.4}$$

$$slope = \tan^{-1}\left(\sqrt{f_x^2 + f_y^2}\right) \tag{6.5}$$

6.5.1.4 Greenness

A number of vegetation indices, such as NDVI (Normalized Difference Vegetation Index) (Jackson et al. 1983), EVI (Enhanced Vegetation Index) (Liu and Huete 1995) and LAI (Leaf Area Index) (Chen and Black 1992) have been used in remote sensing for analyzing vegetation cover on the ground. Among them NDVI is the standard method for comparing relative biomass and vegetation greenness in remote sensing images. Equation 6.6 shows the calculation of NDVI. A higher NDVI indicates a higher level of healthy vegetation cover. Greenness, as shown in Eq. 6.7, is similar to NDVI except that the green band substitutes for the near-infrared band. Both NDVI and greenness values range from −1 to +1. Since the aerial images for this study might not be radiometrically calibrated, the estimated NDVI/greenness index should be considered as a relative, instead of absolute, index.

$$\mathrm{NDVI} = \frac{\mathrm{NIR}-Red}{\mathrm{NIR}+Red} \tag{6.6}$$

$$Greeness = \frac{\mathrm{Green} - Red}{\mathrm{Green} + Red} \tag{6.7}$$

6.5.2 Global Landslide Detection

Assuming that landslide sites are characterized by low surface height variation, bare soil without vegetation or buildings, and steep terrain, this study used the four topographic and spectral indices discussed above to first detect landslides within the whole target area. Four threshold values (T_1–T_4) were defined a priori. Since a universal or empirical threshold definition for those four indices was impossible or impractical, we adopted a supervised mechanism to estimate these indices from training areas that were recognized and digitized manually. The mean and standard deviation values of each index within the training areas were calculated and its threshold values were set to be the mean ±3 standard deviations (Eqs. 6.8–6.11).

$$T_1 = \overline{OHM} + 3 \times \sigma_{OHM} \tag{6.8}$$

$$T_2 = \overline{Roughness} + 3 \times \sigma_{Roughness} \tag{6.9}$$

$$T_3 = \overline{Slope} - 3 \times \sigma_{Slope} \tag{6.10}$$

$$T_4 = \overline{Greenness} + 3 \times \sigma_{Greenness} \tag{6.11}$$

To classify a pixel as landslide or non-landslide, this study used the query expression, *(OHM* $< T_1$*) & (Roughness* $< T_2$*) & (Slope* $> T_3$*) & (Greenness* $< T_4$*)*. A pixel was classified as landslide, if the expression was true. Since global landslide detection was pixel-based, isolated and small groups of landslide pixels were unavoidable. Thus, a morphological filtering (e.g., opening and closing) was applied to remove the pepper and salt effect. In the meantime, small landslides were eliminated by assigning the minimum mapping unit. Finally, the detected landslide pixels were converted into vector-based polygons and stored in ERSI's shapefile format.

6.5.3 Local Landslide Detection

For compiling landslide maps through interactive manipulation by an inexperienced operator, this study designed three kinds of tools: (1) landslide polygon editing, (2) magic wand, and (3) 3D painter or eraser.

6.5.3.1 Landslide Polygon Editing

The landslide polygon editing tools allow the operator to:

1. Delete a landslide polygon.
2. Select a landslide polygon vertex, move its position, erase it, or add one more vertex close to it.
3. Add a new landslide polygon.

6.5.3.2 Magic Wand

The use of the magic wand tool involves three major steps. First, the operator selects a seed point for region growing by visual inspection. This operation is relatively easy given a 3D browsing environment and the draping of the ground surface texture with

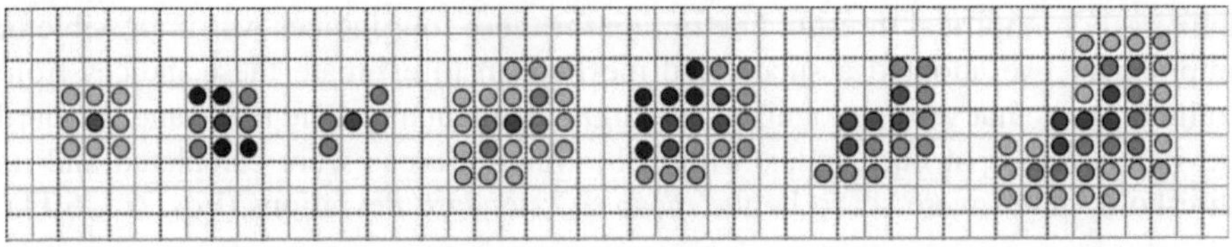

Fig. 6.3 Example of region growing

the color ortho-imagery. Second, the tool determines the similarity between the seed point with its neighborhood based on the initial threshold values of OHM, slope, roughness, and greenness from a 5 × 5 window. In case any pixels have similar geometric and spectral features as the seed point, they are considered to be landslide pixels, such as red pixels in Fig. 6.3. The last step is to search for additional landslide pixels. During the growing process, the threshold values are updated adaptively. With the updated threshold values, candidate pixels (grey in Fig. 6.3) are checked again to evaluate their similarity to landslide pixels even though they have been classified as non-landslide (blue in Fig. 6.3) in the previous iteration. Region growing continues until all candidate pixels are non-landslide.

6.5.3.3 3D Painter or Eraser

This tool allows the operator to edit 3D objects on the 2D screen. Figure 6.4 illustrates how the tool works. On the viewer's screen, 3D objects are projected onto the screen in 2D. Conventionally this is known as back-projection or bottom-up projection using the photogrammetric collinear equation. To edit 3D objects on the 2D screen, the viewer's exterior orientation parameters and DEM are indispensable. When the operator clicks one point on the screen, its corresponding 3D position (denoted as E, N, and H in Fig. 6.4) on the terrain can be calculated using the ray-tracing or top-down algorithm. Accordingly, a landslide polygon and its location can be projected onto a horizontal plane. The landslide polygon is then converted into raster format and the editing on it is thus the same as 2D image editing. The size of the pen can also be adjusted manually according to the curvature of landslide boundary. Figure 6.5 demonstrates the effects of editing (i.e., erasing and expanding landslide), using the tool with two different-sized painters.

6.6 Results

This section consists of two parts. The first part evaluates landslide detection and editing in a case study. The second part performs an error analysis for a study area with four scenes of 1:5,000 base maps.

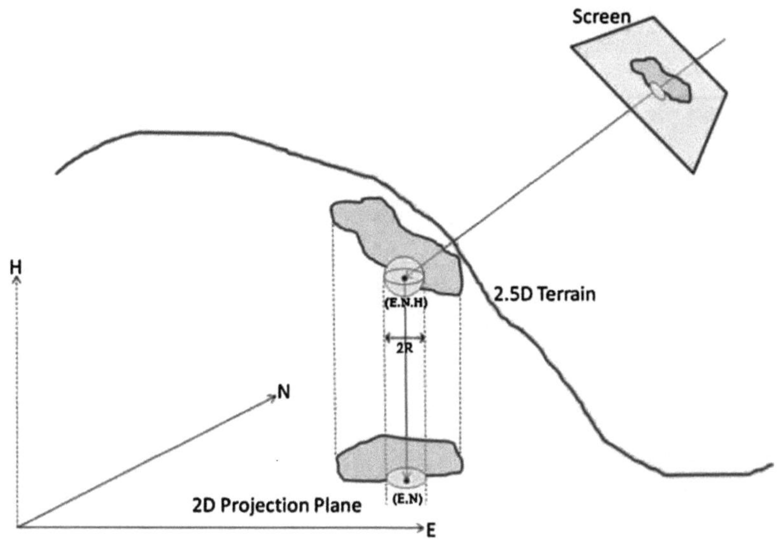

Fig. 6.4 The working of 3D painter

6.6.1 Landslide Detection and Editing

In this case study, the airborne ortho-imagery was taken by an RMK-TOP15 analogue camera on Sep., 2004 after typhoon AERE with an image scale of 1:14,000. The film was scanned as digital images and ortho-rectified with a ground sampling distance (GSD) of 1 m. The airborne laser scanning data acquired by Leica ALS50 on Dec. 16th, 2005 had a point density of 2.5 points per m^2. The 15-month time difference between two data sources was due to data availability.

6.6.1.1 Global Landslides Detection

Figure 6.6 illustrates the training areas selected from bare ground regions that are suspected to be landslides for global landslides detection. The thresholds determined by Eqs. 6.8–6.10 were 5.0, 5.37, 20.0, and −0.25 for T_1 to T_4, respectively. Figure 6.7 shows the global landslides detection results, and Figs. 6.8 and 6.9 plot the detection results on OHM and the greenness index, respectively. The figures show the omission error of one new landslide and the commission errors of three old landslides covered with vegetation. The greenness value was −0.15 and −0.35 for the omitted and commissioned landslide, respectively. Since the greenness threshold value was −0.25, the results were interpreted correctly.

Other than the above errors, the global approach performed well with high reliability and accuracy. In Fig. 6.7, a new landslide located on the left-hand side was accurately detected without being mixed with the road close to it. Also, most

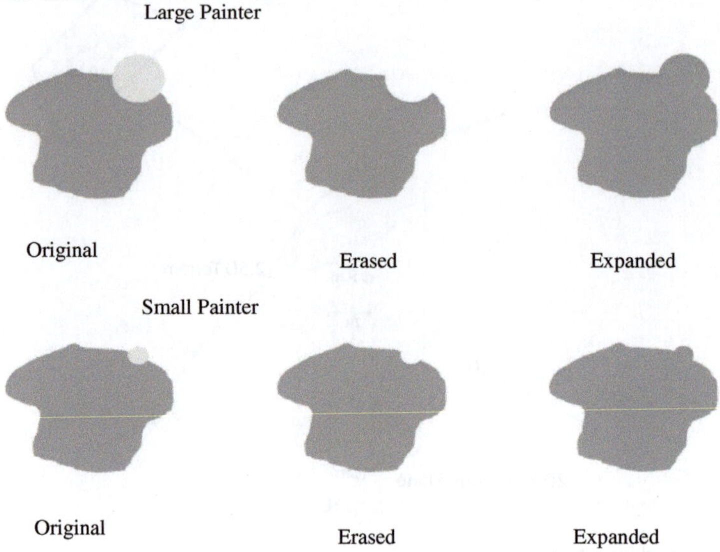

Fig. 6.5 Landslide editing using two different-sized painters

Fig. 6.6 Training areas

landslides along the river were correctly detected without being overlooked or mixed with the riverbed.

For further reliability evaluation, Fig. 6.10 plots globally detected landslides on a terrestrial photography, taken along the river on Nov. 5th 2006 (two years after the aerial photo), and a 3D visual simulation. The largest landslide located on the right hand of Fig. 6.10 shows a good correspondence between the detection result and the on-site surveying. A large landslide on the left hand of Fig. 6.10 was not included in the detection result because it occurred after the acquisition date of aerial photograph.

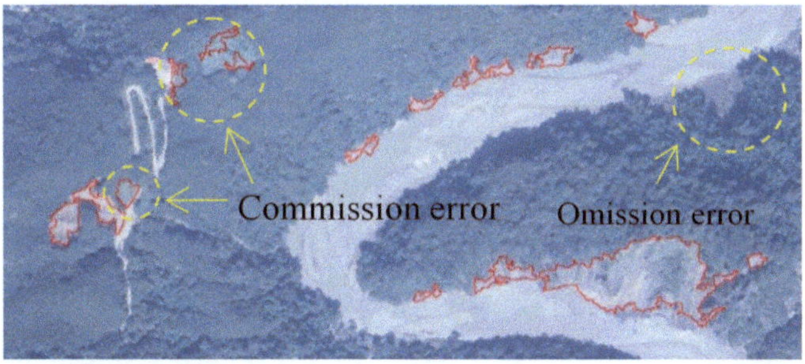

Fig. 6.7 Global detection results

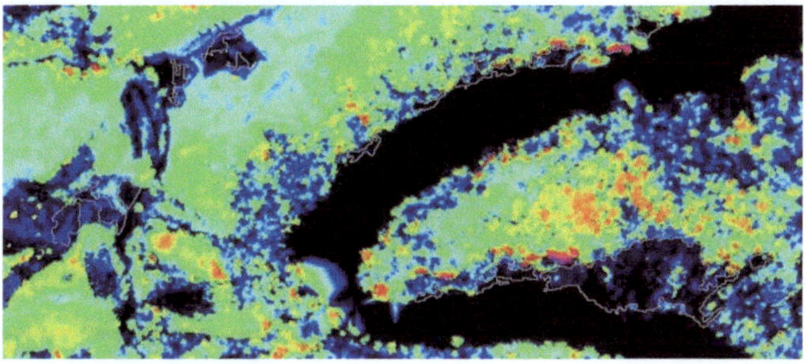

Fig. 6.8 Detection results superimposed on OHM

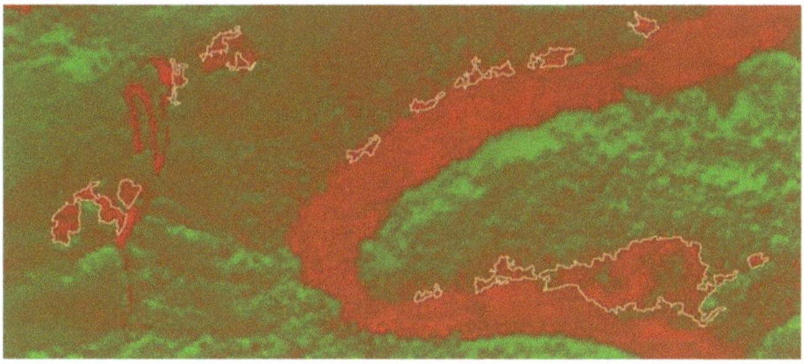

Fig. 6.9 Detection results superimposed on the greenness index

6.6.1.2 Local Landslides Detection

Figure 6.11 illustrates a series of 3D painting operations. Figure 6.11a displays globally detected results. Figures 6.11 b, c, d and e show the erasing and addition operations and the change of the painter size by the operator. Figure 6.11f is the final result after 3D painting. Notice that the road cross the region in Fig. 6.11 was not classified as landslide during the global detection process.

Fig. 6.10 Comparison between global detection results and on-site surveying

Figures 6.12a and b display the use of the magic wand in a sequence. Notice that the operator needs only to double-click where a landslide is located on the screen. The detected landslide will be displayed in cyan for verification. If the result is incorrect, the operator may choose to cancel the current result and select another seed point until a reasonable detection is achieved. Figure 6.12c denotes the region (landslide) growing results after two steps of the magic wand operation.

6.6.2 Error Analysis

For error analysis, this study used four scenses of ortho-imagery and LiDAR data. Each scene had an area of 6.75 km^2 about the same size as a 1:5,000 base map. Table 6.2 shows the basic information for the four scenes. Notice that the acquisition dates for ALS data are the same for those four scenes but are different for aerial imagery.

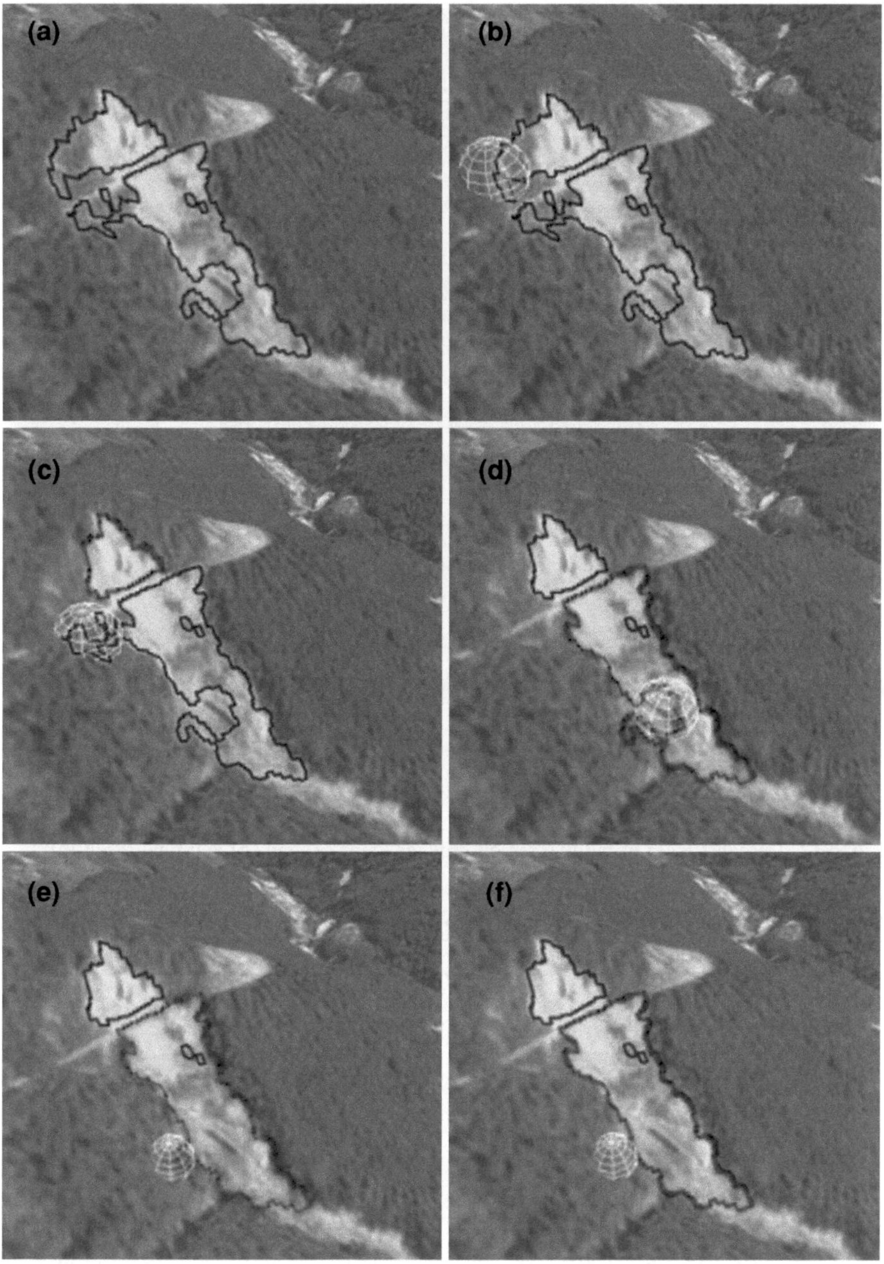

Fig. 6.11 Demonstration of 3D painting. **a** original, **b**, **c** and **e** erasing, **d** adding, and **f** final result

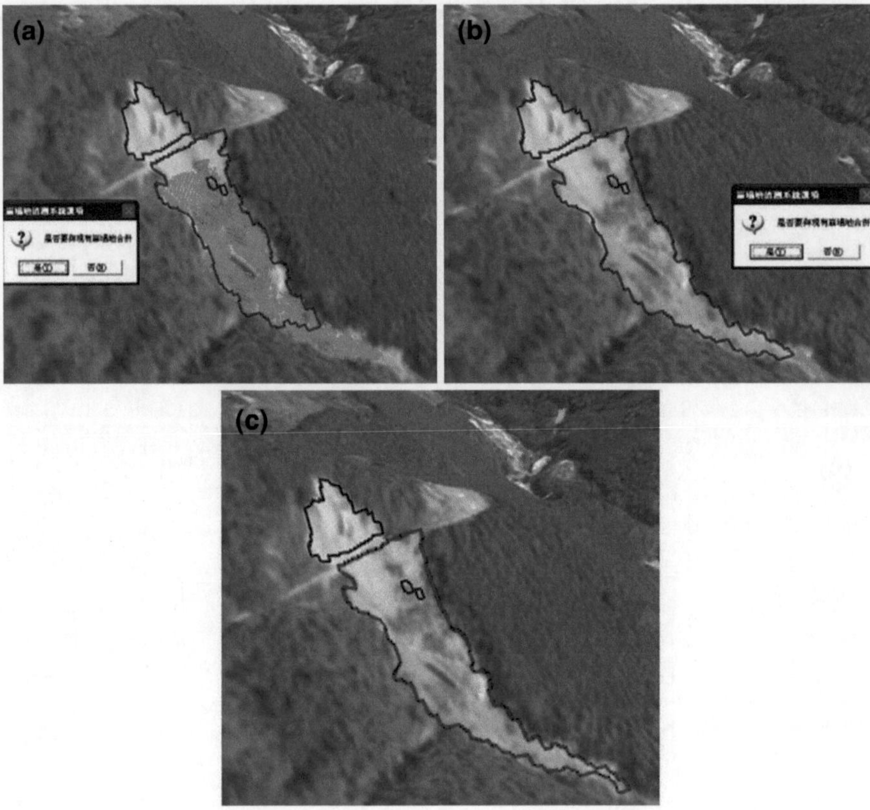

Fig. 6.12 Demonstration of the two region growing steps (**a**) and (**b**) and final result (**c**)

Table 6.2 Aerial imagery and ALS data for error analysis

	Scene 1 & 2	Scene 3 & 4
Aerial camera type	RMK-TOP 15	RMK-TOP 15
Acquisition date of aerial imagery	Sep. 2004	Jan. 26, 2007
ALS type	Leica ALS50	Leica ALS50
Acquisition date of ALS data	Dec. 16th, 2005	Dec. 16th, 2005
Average point cloud density	2.5 pts/m^2	2.5 pts/m^2

Table 6.3 tabulates the results of error analysis based on the confusion matrix. In the table, the producer accuracy, user accuracy, omission error and commission error belong to the landslide class only whereas the overall accuracy and average accuracy include both landslide and non-landslide. The threshold values estimated from the training areas are also listed for reference. The commision error is high in some cases primarily because Eqs. 6.8–6.10 use three standard deviations.

Figure 6.13 and 6.14 demonstrate the global landslide detection results for scene no. 1–2 and scene no. 3–4, respectively. The figures also show landslides

Table 6.3 Threshold values for global landslide detection and error analysis results

	Scene 1	Scene 2	Scene 3	Scene 4
Scene No.	96221094	96221095	95202082	95202083
T_1 (OHM)	5.0	5.0	4.71	7.0
T_2 (Roughness)	5.37	7.94	4.21	10.92
T_3 (Slope)	10.0	8.0	3.0	8.0
T_4 (Greenness)	−0.25	−0.3	−0.26	−0.46
Producer accuracy (PA1)	78.12%	95.25%	94.32%	76.53%
Omission error (OE1)	21.88%	4.75%	5.68%	23.47%
User accuracy (UA1)	61.33%	49.46%	86.9%	93.76%
Commissioned error (CE1)	38.67%	50.54%	13.1%	6.24%
Overall accuracy (OA)	99.41%	98.80%	99.2%	97.49%
Average accuracy (AA)	88.85%	97.04%	96.86%	88.02%

Fig. 6.13 Global landslides detection results (*yellow polygons*) for scene 1 (*left*) & 2 (*right*), overlaid with ground truth (*red polygons*)

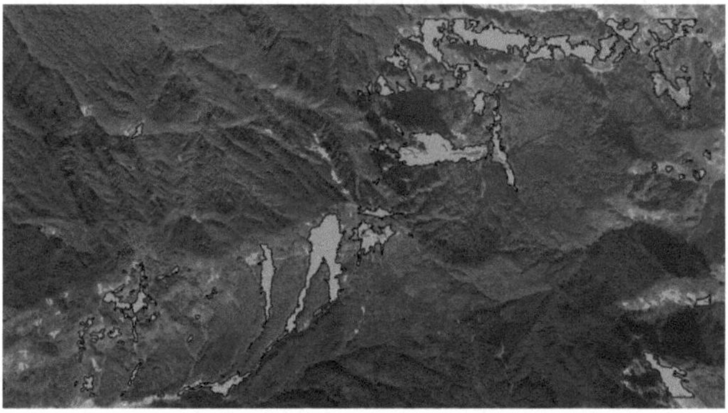

Fig. 6.14 Global landslides detection results (*yellow polygons*) for scene 3 (*left*) & 4 (*right*), overlaid with ground truth (*red polygons*)

Table 6.4 Reasons for failures in landslide detection

	Reasons
Greenness	Radiometry inconsistence due to deterioration of aerial films, weather condition, shadow, different types and densities of vegetation, and different soil type
	Image distortion after ortho-rectification in high terrain relief areas
Roughness	Commission errors in farming areas without vegetation or with low height plant varieties
	Omission errors for landslides with rough surface
	Errors in OHM
OHM	Classification errors of DSM and DEM, especially in densely forested areas
Slope	DEM errors
	Mud-flows on flat terrain

from manual interpretation of aerial photograph using the proposed local tools. In Table 6.3, average accuracy rates are all above 88% and overall accuracy rates above 97%. Depending on the quality of the ortho-imagery, DSM and DEM, producer accuracy can be as high as 95%.

6.7 Discussion

Table 6.4 summarizes reasons that might have affected global landslide detection. Additionally, inconsistency of data acquisition dates between the ortho-imagery and LiDAR data might have caused changes in the land cover information. Shadow introduced by trees and mountainous relief could also cause problems in missing landslides or delineating landslide boundaries.

Several LiDAR classification methods have been published for the categorization of raw data into surface and terrain (Vosselman 2000; Axelsson 2000; Vosselman and Maas 2001; Sithole 2001; Pfeifer et al. 2001). However, in dense forested areas LiDAR data may not reflect from the terrain; thus, the derived OHM may not be able to describe the above ground objects precisely. On the other hand, if an improper LiDAR classification algorithm, which can cause large height variations in bare ground areas, is chosen, it will introduce omission errors.

The surface geometric and spectral information will be occluded under dense cloud, causing problems in obtaining accurate DEMs, DSMs, and ortho-imagery. Sometimes, the ortho-imagery is mosaicked from multiple aerial images taken at different locations and viewing directions. This will cause radiometric inconsistencies and errors in calculating the greenness or NDVI index. In the meantime, because the ortho-imagery is made by aerial film camera, its radiometric response can easily deteriorate by the atmospheric effect.

Due to different soil types or soil moisture contents, some omitted landslides look darker compared with other new landslides. This can affect the calculated greenness index, thus making it difficult to distinguish between bare ground and vegetation.

Mountainous areas with high terrain relief can cause geometric distortion during the process of ortho-rectification. This will affect the spectral information and introduce errors in both global landslide detection and the visual inspection part of local landslide detection.

6.8 Conclusions

This paper has presented the global and local approaches for landslides mapping using the airborne laser scanning data and color ortho-imagery as the data sources. In addition to the spectral information that is conventionally used in human interpretation, this study has used topographic features derived from LiDAR DSM and DEM for landslide detection. These topographic features of slope, surface roughness, and object height play an important role during automatic, global landslide detection. Experimental results have demonstrated that they are more significant than the greenness index derived from color ortho-imagery.

This study has also developed tools for landslide mapping and editing at the local level. Those tools operate in a 3D interactive environment, which provides terrain relief and spectral information concurrently. They are therefore intuitive and user friendly to an inexperience operator. We believe that the landslide mapping procedure proposed in this study is cost effective and practical compared with other fully- or semi-automatic approaches.

Many hybrid systems are currently available on the market that can integrate airborne laser scanning with medium format digital camera (Skaloud et al. 2005; Optech DSS 2009; Optech AIC 2009). The data acquisition time is consistent and the digital imagery has better image quality than film camera. Some of these systems also provide near infrared imagery that is useful for landslide detection. We believe that these hybrid systems can play an important role in future landslide mapping.

Acknowledgments This project was supported by the Forestry Bureau of the Council of Agriculture, Taiwan. The authors thank their support for funding and providing data for research.

References

Axelsson P (2000) DEM generation from laser scanner data using adaptive TIN models. Int Arch Photogrammetry Remote Sens Spatial Inf Sci XXXIII(Pt. B4/1):110–117

Bork EW, Su JG (2007) Integrating LIDAR data and multispectral imagery for enhanced classification of rangeland vegetation: a meta analysis. Remote Sens Environ 111(1):11–24

Caine N (1980) The rainfall intensity-duration control of shallow landslides and debris flows. Geografiska Annaler 62A:23–27

Chen JM, Black TA (1992) Defining leaf area index for non-flat leaves. Agric Meteorol 57:1–12

Chen LC, Teo TA, Hsieh CH, Rau JY (2006a) Reconstruction of building models with curvilinear boundaries from laser scanner and aerial imagery. Lect Notes Comput Sci LNCS 4319:24–33

Chen RF, Chang KJ et al (2006b) Topographical changes revealed by high-resolution airborne LiDAR data: the 1999 Tsaoling landslide induced by the Chi–Chi earthquake. Eng Geol 88(3–4): 160–172

Chen Y, Su W et al (2009) Hierarchical object oriented classification using very high resolution imagery and LIDAR data over urban areas. Adv Space Res 43(7):1101–1110

Cheng JD, Huang YC et al (2005) Hydrometeorological and landuse attributes of debris flows and debris floods during typhoon Toraji, July 29–30, 2001 in central Taiwan. J Hydrol 306(1–4): 161–173

Chiang SH, Chang KT (2009) Application of radar data to modeling rainfall-induced landslides. Geomorphology 103:299–309

Chang KT, Chiang S, Hsu M (2007) Modeling typhoon- and earthquake-induced landslides in a mountainous watershed using logistic regression. Geomorphology 89:335–347

Chang KT, Chiang S, Feng L (2008) Analysing the relationship between typhoon triggered landslides and critical rainfall conditions. Earth Surf Proc Land 33:1261–1271

Crozier MJ (1999) Prediction of rainfall-triggered landslides: a test of the antecedent water status model. Earth Surf Proc Land 24:825–833

Cunningham D, Grebby S et al (2006) Application of airborne LiDAR to mapping seismogenic faults in forested mountainous terrain, southeastern Alps, Slovenia. Geophys Res Lett 33(20):5

Dadson SJ, Hovius N, Chen H et al (2004) Earthquake-triggered increase in sediment delivery from an active mountain belt. Geology 32:733–736

Dhakal AS, Amada T, Aniya M (2000) Landslide hazard mapping and its evaluation using GIS: an investigation of sampling schemes for a grid-cell based quantitative method. Photogrammetric Eng Remote Sens 66(8):981–989

Donati L, Turrini MC (2002) An objective method to rank the importance of the factors predisposing to landslides with the GIS methodology: application to an area of the Apennines (Valnerina: Perugia, Italy). Eng Geol 63:277–289

Elaksher AF (2008) Fusion of hyperspectral images and lidar-based dems for coastal mapping. Opt Lasers Eng 46(7):493–498

Glenn NF, Streutker DR, Chadwick DJ et al (2006) Analysis of LiDAR-derived topographic information for characterizing and differentiating landslide morphology and activity. Geomorphology 73(1–2):131–148

Haala N, Brenner C (1999) Virtual city models from laser altimeter and 2D map data. Photogrammetric Eng Remote Sens 65(7):787–795

He Y, Xie H, Cui P et al (2003) GIS-based hazard mapping and zonation of debris flows in Xiaojiang Basin, southwestern China. Environ Geol 45:285–293

Herva′s J, Barredo JI et al (2003) Monitoring landslides from optical remotely sensed imagery: the case history of Tessina landslide, Italy. Geomorphology 54(1–2):63–75

Hofmann AD (2004) Analysis of TIN-structure parameter spaces in airborne laser scanner data for 3-D building model generation. Int Arch Photogrammetry Remote Sens Spatial Inf Sci 35(B3):302–307

Hong Y, Adler RF (2007) Towards an early-warning system for global landslides triggered by rainfall and earthquake. Int J Remote Sens 28:3713–3719

Jackson RD, Slater PN, Pinter PJ (1983) Discrimination of growth and water stress in wheat by various vegetation indices through clear and turbid atmospheres. Remote Sens Environ 15:187–208

Kohavi R, Provost F (1998) The case against accuracy estimation for comparing classifiers. In: Proceedings of the 15th international conference on machine learning, Morgan Kaufmann, San Francisco, CA

Komac M (2004) Statistical landslide prediction map as a basis for a risk map Rhodes. WIT Press, Greece

Kraus K, Pfeifer N (1998) Determination of terrain models in wooded areas with airborne laser scanner data. ISPRS J Photogrammetry Remote Sens 53:193–203

Li M, Cheng L et al (2008) Post-earthquake assessment of building damage degree using LiDAR data and imagery. Sci China Ser E: Technol Sci 51(2):133–143

Liu HQ, Huete AR (1995) A feedback based modification of the NDV I to minimize canopy background and atmospheric noise. IEEE Trans Geosci Remote Sens 33:457–465

Lin ML, Jeng FS (2000) Characteristics of hazards induced by extremely heavy rainfall in Central Taiwan–Typhoon Herb. Eng Geol 58(2):191–207

Lin PS, Lin JY, Hung HC, Yang MD (2002) Assessing debris flow hazard in a watershed in Taiwan. Eng Geol 66:295–313

Lin CW, Liu SH et al (2006) Impacts of the Chi–Chi earthquake on subsequent rainfall-induced landslides in central Taiwan. Eng Geol 86(2–3):87–101

Mantovani F, Soeters R, van Westen C (1996) Remote sensing techniques for landslide studies and hazard zonation in Europe. Geomorphology 15:213–225

McKean J, Roering J (2004) Objective landslide detection and surface morphology mapping using high-resolution airborne laser altimetry. Geomorphology 57:331–351

Metternicht G, Hurni L, Gogu R (2005) Remote sensing of landslides: an analysis of the potential contribution to geo-spatial systems for hazard assessment in mountainous environments. Remote Sens Environ 98:284–303

Mikos M, Fazarinc R et al (2006) Sediment production and delivery from recent large landslides and earthquake-induced rock falls in the Upper Soca River Valley, Slovenia. Eng Geol 86(2–3): 198–210

Optech ALTM (2009) ALTM 3011EA system specifications. Available: http://www.optech.ca/pdf/Brochures/ALTM3100EAwspecsfnl.pdf (2009.05)

Optech DSS (2009) DSS 322 digital camera. Available: http://www.optech.ca/pdf/ALTMApplanixPC.pdf (2009.05)

Optech AIC (2009) Rollei AIC digital camera. Available: http://www.optech.ca/pdf/ALTMRolleiPC.pdf (2009.05)

Parker JR (1997) Algorithms for image processing and computer vision. Wiley, New York

Pappalardo G, Amodeo A et al (2004) Aerosol lidar intercomparison in the framework of the EARLINET project 3. Raman lidar algorithm for aerosol extinction, backscatter, and lidar ratio. Appl Opt 43(28):5370–5385

Pfeifer N, Stadler P, Briese C (2001) Derivation of digital terrain models in the SCOP__ environment. In: Proceedings of the OEEPE workshop on airborne laserscanning and interferometric SAR for detailed digital elevation models, Stockholm, Sweden, OEEPE Publication No. 40, unpaginated CD–ROM, 01–03 March

Rau JY, Chen LC, Liu JK, Wu TH (2007) Dynamics monitoring and disaster assessment for watershed management using time-series satellite images. IEEE Trans Geosci Remote Sens 45:1641–1649

Raju P, Saibaba J (1999) Landslide hazard zonation mapping using remote sensing and geographic information system techniques-a case study of Pithoragarh area. U.P. Int Geosci Remote Sens Symp (IGARSS) 1:577–579

Skaloud J, Vallet J et al (2005) HELIMAP: Rapid large scale mapping using handheld LiDAR/CCD/GPS/INS sensors on helicopters, Long Beach. Institute of Navigation, United States

Schwalbe E, Maas HG and Seidel F (2005) 3D building model generation from airborne laser scanner data using 2D GIS data and orthogonal point cloud projections. Int Arch Photogrammetry Remote Sens Spatial Inf Sci 36(3/W19):209–214

Sithole G (2001) Filtering of laser altimetry data using a slope adaptive filter. Int Arch Photogrammetry Remote Sens Spatial Inf Sci 36(Pt. 3/W4):203–210

Sithole G, Vosselman G (2004) Experimental comparison of filter algorithms for bare-earth extraction from airborne laser scanning point clouds. ISPRS J Photogrammetry Remote Sens 59:85–101

Singhroy V (2002) Landslide hazards: CEOS, the use of earth observing satellites for hazard support: assessments and scenarios. Final report of the CEOS Disaster Management Support Group, NOAA, p 98

Tarchi D, Casagli N et al (2003) Landslide monitoring by using ground-based SAR interferometry: an example of application to the Tessina landslide in Italy. Eng Geol 68(1–2): 15–30
Ussyshkin RV, Smith B (2006) Performance analysis of ALTM 3100EA: instrument specifications and accuracy of lidar data. Revue Francaise de Photogrammetrie et de Teledetection 182:66–72
van Westen CJ (1994) GIS in landslide hazard zonation: a review, with examples from the Andes of Colombia. In: Price MF, Heywood DI (eds) Mountain environments and geographic information systems. Taylor and Francis, London, pp 135–165
van Westen C, Getahun F (2003) Analyzing the evolution of the Tessina landslide using aerial photographs and digital elevation models. Geomorphology 54:77–89
Vosselman G (2000) Slope based filtering of laser altimetry data. Int Arch Photogrammetry Remote Sens Spatial Inf Sci XXXIII(Pt. B3):935–942
Vosselman G and Maas H (2001) Adjustment and filtering of raw laser altimetry data. In: Proceedings of the OEEPE workshop on airborne laserscanning and interferometric SAR for detailed digital elevation models, Stockholm, Sweden, OEEPE Publication No. 40, unpaginated CD-ROM, 01–03 March
Wang H, Lin G, Xu W, Wang G (2005) GIS-based landslide hazard assessment: an overview. Rogress Phys Geogr 29:548–567
Webster TL, Brendan MJ et al (2006) The application of lidar-derived digital elevation model analysis to geological mapping: an example from the Fundy Basin, Nova Scotia, Canada. Can J Remote Sens 32(3):173–193
Weirich F, Blesius L (2007) Comparison of satellite and air photo based landslide susceptibility maps. Geomorphology 87(4):352–364
Willis A, Sui Y et al (2009) Design and implementation of an inexpensive LIDAR scanning system with applications in archaeology. In: Proceedings of SPIE—The international society for optical engineering 7239
Zhao PT, Zhang YC et al (2008) Capability of Raman lidar for monitoring the variation of atmospheric CO_2 profile. Chin Phys B 17(1):335–342

Chapter 7
Ensemble-Based Landslide Susceptibility Maps in Jinbu Area, Korea

Saro Lee and Hyun-Joo Oh

Abstract Ensemble techniques were developed, applied and validated for analysis of landslide susceptibility in Jinbu area, Korea using the geographic information system (GIS). Landslide occurrence areas were detected in the study by interpreting aerial photographs and field survey data. Landslide locations were randomly selected in a 70/30 ratio for training and validation of the models, respectively. Topography, geology, soil and forest databases were also constructed. Maps relevant to landslide occurrence were assembled in a spatial database. By using the constructed spatial database, 17 landslide-related factors were extracted. The relationships between the detected landslide locations and the factors were identified and quantified by frequency ratio, weight of evidence, logistic regression and artificial neural network models and their ensemble models. The relationships were used as factor ratings in the overlay analysis to create landslide susceptibility indexes and maps. Then, the four landslide susceptibility maps were used as new input factors and integrated using the frequency ratio, weight of evidence, logistic regression and artificial neural network models as ensemble methods to make better susceptibility maps. All of the susceptibility maps were validated by comparison with known landslide locations that were not used directly in the analysis. As the result, the ensemble-based landslide susceptibility map that used the new landslide-related input factor maps showed better accuracy (87.11% in frequency ratio, 83.14% in weight of evidence, 87.79% in logistic regression and 84.54% in artificial neural network) than the individual

S. Lee (✉)
Geoscience Information Center,
Korea Institute of Geoscience and Mineral Resources (KIGAM),
92 Gwahang-no, Yuseong-gu, Daejeon 305-350, Korea
e-mail: leesaro@kigam.re.kr

H.-J. Oh
Mineral Resources Research Department,
Korea Institute of Geoscience and Mineral Resources (KIGAM),
92 Gwahang-no, Yuseong-gu, Daejeon 305-350, Korea

B. Pradhan and M. Buchroithner (eds.), *Terrigenous Mass Movements*,
DOI: 10.1007/978-3-642-25495-6_7,

landslide susceptibility maps (84.94% in frequency ratio, 82.82% in weight of evidence, 87.72% in logistic regression and 81.44% in artificial neural network). All accuracy assessments showed overall satisfactory agreement of more than 80%. The ensemble model was found to be more effective in terms of prediction accuracy than the individual model.

Keywords Landslide · Susceptibility · Ensemble · GIS · Korea

7.1 Introduction

Landslides are a major hazard, often causing property damage and economic losses and creating high maintenance costs. Landslides are triggered by different factors, either natural or related to human activities. Among natural factors, rainfall is certainly one of the most frequent causes of shallow landslide occurrence of the flow type in granular soil. Thus, it is necessary to assess landslide susceptibility to support forecasting of the phenomena. Landslide susceptibility is defined as areas likely to have slope failures in the future. It is estimated by correlating some of the principal factors that have contributed to past landslides (Brabb 1984; Guzzetti et al. 2005). In mathematical form, it is defined by correlating landslide density with different combinations of the factors (Clerici et al. 2002; Guzzetti et al. 2005).

Many methods have been proposed to assess landslide susceptibility, with increasing use of geographic information systems (GIS) using different models. These examples, many of these studies have applied probabilistic models such as frequency ratio, weight of evidence, etc. (Audisio et al. 2009; Dahal et al. 2008; Lee and Min 2001; Lee and Pradhan 2006; Lee et al. 2004a; Mousavi et al. 2009; Oh et al. 2009; Ozdemir 2009; Pirasteh et al. 2009; Regmi et al. 2010; Vahidnia et al. 2009; Yalcin 2008; Yilmaz 2009b, c, 2010). One of the statistical models available, the logistic regression model, has also been applied to landslide susceptibility mapping (Akgun and Bulut 2009; Bai et al. 2010, 2011; Chauhan et al. 2010; Das et al. 2010; Dominguez-Cuesta et al. 2010; Dong et al. 2009; Falaschi et al. 2009; Lee 2005, 2007a; Legorreta Paulin and Bursik 2009; Mathew et al. 2009; Nandi and Shakoor 2010; Oh and Lee 2010; Pradhan and Lee 2010a, b). More sophisticated assessments involved fuzzy logic, artificial neural network, AHP, Dempster-Shapfer theory of evidence, Monte Carlo methods also have been applied (Akgun and Türk 2010; Chen et al. 2009a, b; He and Fu 2009; Kanungo et al. 2008, 2009; Kawabata and Bandibas 2009; Lee 2007b; Lee and Evangelista 2006; Lee et al. 2006, 2004b; Liu et al. 2009; Melchiorre et al. 2008; Miles and Keefer 2009; Muthu et al. 2008; Park 2011; Poudyal et al. 2010; Pradhan and Lee 2010c, 2009, 2007; Prabu and Ramakrishnan 2009; Pradhan et al. 2010; Shafri et al. 2010; Tangestani 2009; Wang et al. 2009; Yilmaz 2009a). For the same study area,

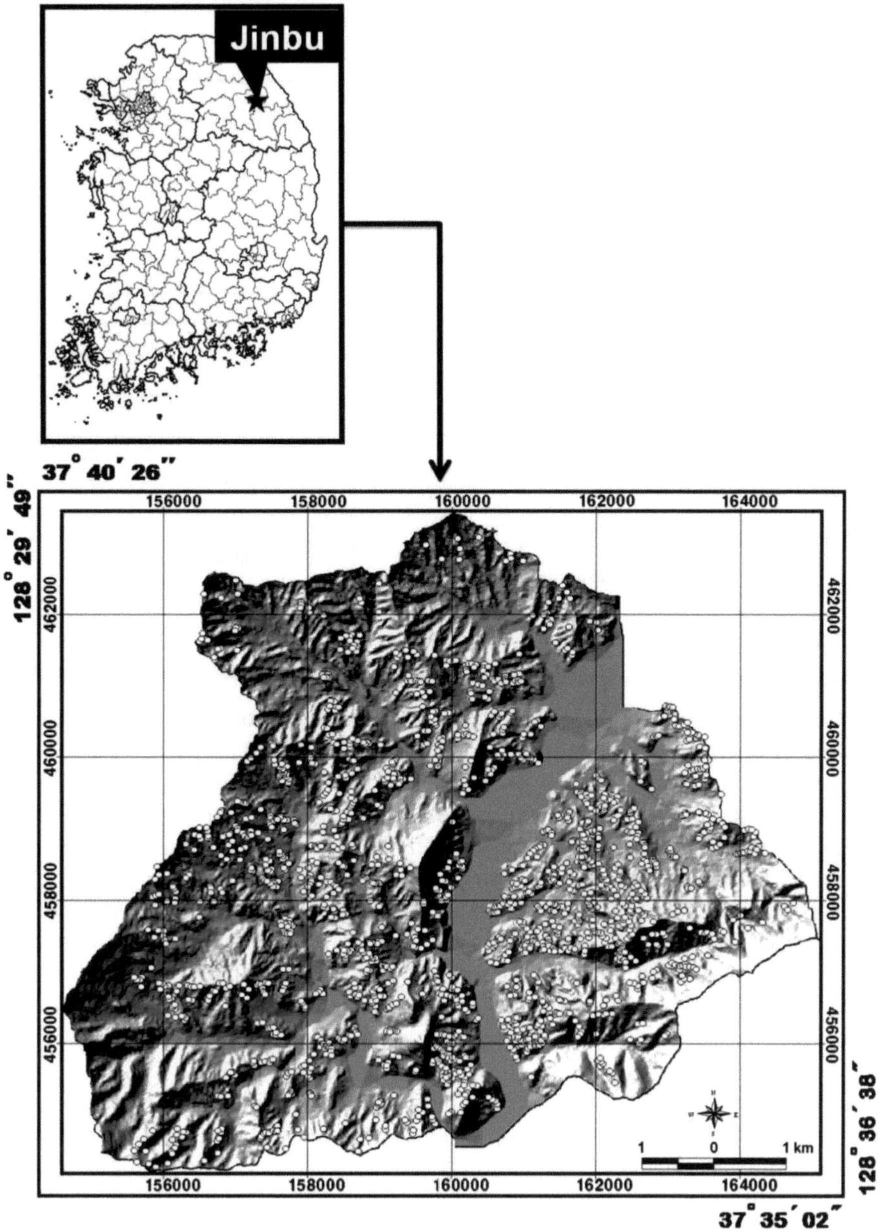

Fig. 7.1 Hillshade and landslide location map of the study area

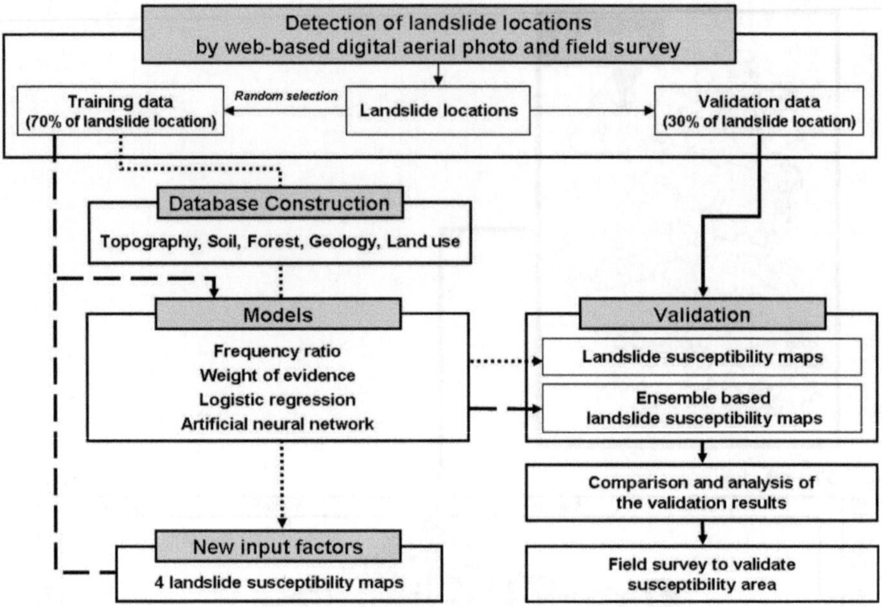

Fig. 7.2 Flow chart of this study

the frequency ratio, weight of evidence, logistic regression and, artificial neural network models were applied and compared (Lee et al. 2011).

Any attempt to quantitatively evaluate landslide susceptibility needs validation (Guzzetti et al. 2006). Validation uses the same geographic data, with independent landslides not used to construct the model. Most previous research presents an individual susceptibility model, discusses factors, and validates the susceptibility maps as an assessment of the model's performance. Although various models and their application to landslide susceptibility have been published, the ensemble study involving combination of susceptibility maps from frequency ratio, weight of evidence, logistic regression and artificial neural network models has not been used for landslide susceptibility mapping purposes previously to improve prediction accuracy.

In this paper, we provide an ensemble-based methodology in order to obtain more accurate and reliable estimates than can be obtained from using an individual model. More specifically, the main difference of the study is to integrate individual landslide susceptibility maps made using frequency ratio, weight of evidence, logistic regression and artificial neural network models, and so to make a better landslide susceptibility map in Jinbu area, Korea (Fig. 7.1).

The study flow is (Fig. 7.2):

1. The landslide locations were detected from web-based digital aerial photographs, and the locations were checked in the field.

2. The landslide data were randomly divided into training data (70% of landslide locations) and validation data (30% of landslide locations).
3. Topography, soil, forest, geology and land-use datasets were compiled in a spatial database. Then the slope, aspect, curvature, topographic wetness index (TWI), stream power index (SPI), soil topography, soil drainage, soil material, soil thickness, soil texture, timber type, timber age, timber diameter, timber density, distance from lineament, geology and land use were extracted from the spatial database as potential contributing factors.
4. Using frequency ratio, weight of evidence, logistic regression and artificial neural network models and the landslide locations selected for training, the relationships between landslide and each factor were calculated quantitatively. Then, using the relationships, four landslide susceptibility maps were made.
5. The four landslide susceptibility maps were considered as four new input factors. Ensemble-based landslide susceptibility maps were again made using the frequency ratio, weight of evidence, logistic regression and artificial neural network models.
6. Each original landslide susceptibility map and ensemble-based landslide susceptibility map were validated using the landslide locations that were not used for training. The individual and ensemble-based methods were compared in terms of their prediction accuracy.

7.2 Study Area

Recently, increased rainfall and more frequent hurricanes associated with global warming have increased the incidence of landslides. In Korea, due to the impact of typhoons YANNI in 1998, RUSA in 2002 and MAEMI in 2003, landslides and forest damage have sharply increased. In 2006, Jinbu area was severely affected by landslides following the typhoons EWINIAR, BILIS and KAEMI, all with strong economic and social impact. The study area of around 60 km^2 lies between the 37° 40′ 26″ N and 37° 35′ 02″ N, and 128° 29′ 49″ E and 128° 35′ 36″ E, and is composed of two basins (Fig. 7.1). It is covered by a topographic map of 1:5,000 scale.

During the Korea rainy season from 14 June to 29 July 2006, very intense precipitation episodes also occurred, causing many shallow landslides of the flow type in granular soil on slopes. Jinbu, Pyeongchan district is the rainiest area in Korea, its total precipitation is more 750 mm than the annual ones of 340 mm in rainy season. Especially heavy rainfall over 29 h on 15–16 July 2006 totaled 429 mm in the study area. This caused a lot of landslides, collapse of embankments and flooding of farmland because of water level increases at the confluence of rivers. The property damage in the study area amounted to about 449 billion USD (http:\\www.waterjournal.co.kr).

Table 7.1 Geological section including Jinbu (Geological society of Korea 1962)

Era	Period	System	Symbol	Description	
Cenozoic	Quaternary	Quaternary	Qr	Recent River Deposits	Greyish green fine sandstone, white or grey coarse sandstone, grey sandy shale and shale
Mesozoic	Jurassic	–	Jigr	Imgye Granite	–
	Triassic	Pyeongan	Intrusion	Nok-Am Formation	Grey, dark grey, milky white lime stone intercalated with thin layers of grey shale and sandstone
			TRn2		
			- - -		
Paleozoic	Ordovician	Joseon	Unconformity	Chongson Limestone Formation	Biotite granite(porphyritic and shistose in pants)
			Oj		
			- - -		
Pre Cambrian		–	Unconformity	Biotite Gneiss	Banded biotite gneiss, mignatitic gneiss, limestone, amphibolite
			PCEbgn		

The geology of the area includes the Imgye Granite, which is an extensive intrusion of granitoids by the Daebo Orogeny and is distributed over most of the study area. The Granite intruded into the Chongson Limestone and Nok-Am Formation (Table 7.1). These are located on the Pre-Cambrian biotite gneiss with discontinuity. The Daebo Orogeny continued from Early Jurassic to Early Cretaceous as established by the dating of the Daebo Granite in South Korea and the Tanch'on Granite in North Korea. The Daebo Orogeny was the most severe in intensity, so that all the previous formations were intensely deformed and some were mildly metamorphosed. In this text, for simplicity, the Daebo Granite is used to represent the Jurassic Granite. The Daebo Granite occupies more than 16% of the land area, and is dispersed all over the country. The granite batholiths are diagonally aligned in a direction of southwest to northeast, originating in the Yangtze Paraplatform of southern China, and then proceeding across the Korean Peninsula (Geological Society of Korea 1962; Lee 1988).

To detect landslide locations, the digital aerial photographs serviced from Internet portal site Daum (www.daum.net) were collected with a ground resolution element of 0.5 m. The web-based aerial photos of all over Korea available on the website. For the checking, the photographs were compares with analog aerial photographs provided from NGII and field survey. The photographs from portal website were taken on 27, May 2008 using UltraCam-X sensor by Samah Aerial Survey Co., Ltd (www.samah.com) after landslide occurred during raining season, 2006. The analog photos were taken on 4, Arial 2005 before landslide occurrence. The total number of landslide was 1,801 in the study area.

7.3 Ensemble Method for Integration

The main idea of ensemble modeling is to build an effective method for improving prediction accuracy by integrating multiple outputs from a set of models (Rokach 2005). More specifically, it means a better-integrated method, with more accuracy and reliable estimates for decisions than can be obtained from using a single model such as frequency ratio, weight of evidence, logistic regression, artificial neural network, etc. The analysis part of the method could include various models, which could be used to make and find new information many times. In other words, a set of models in the 1st analysis part could be run N times using N^{th}–1 output or N^{th} new input data in order to improve the quality of information (Fig. 7.3).

In this study, the four models' frequency ratio, weight of evidence, logistic regression and artificial neural network were used in the 1st analysis part. The landslide susceptibility map from each model in the 1st analysis part was considered as a new input factor for the 2nd analysis part and ensemble-based landslide susceptibility maps were made from the models in the 2nd analysis part. In other words, the relationships between landslide locations and the four new input factors were identified and quantified again by frequency ratio, weight of evidence,

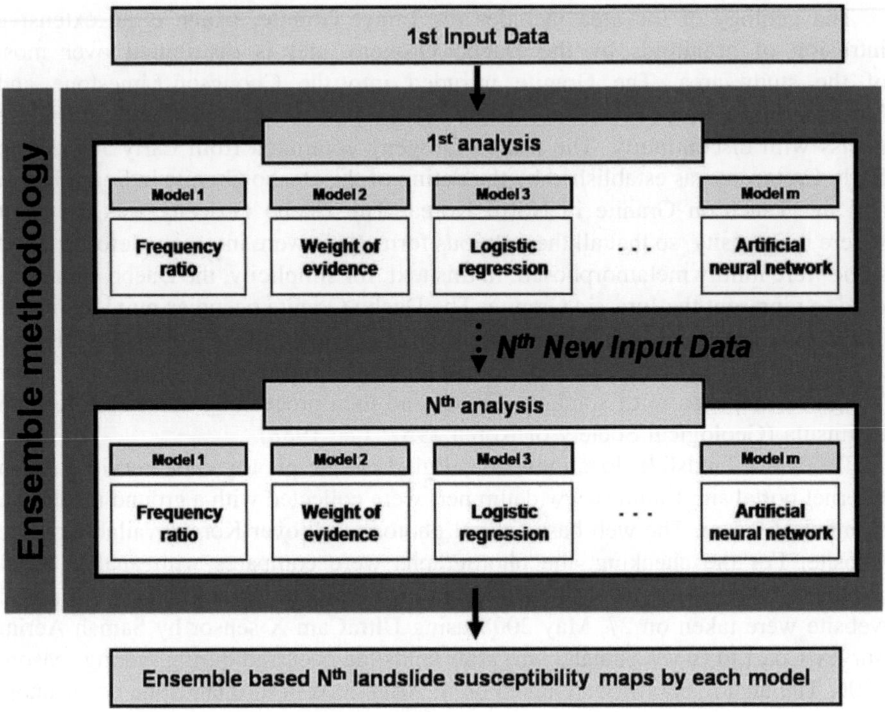

Fig. 7.3 Flow chart of ensemble methodology

logistic regression and artificial neural network models in the 2nd analysis part to make better susceptibility maps.

7.4 Spatial Database

To make landslide susceptibility map, a spatial database was considered as related factors that influence landslides such as topography, soil, forest, geology and land use. Landslide occurrence areas were detected in the study area, by interpreting the digital aerial photograph and field survey data. In the study area, rainfall-triggered debris flows and shallow soil slide are the most abundant. Maps relevant to landslide occurrence were constructed in a vector format spatial database using the ArcGIS software package. These included topographic maps (1:5000 scale), soil maps (1:25000 scale), forest maps (1:25000 scale), geological map (1:50000 scale), and land use map (1:5000 scale) (Table 7.2). A Digital Elevation Model (DEM) was constructed using contour and survey base points that had an elevation value read from a topographic map were extracted. The DEM has a 10 m

Table 7.2 Data layer related to landslide of study area

Category	Factors	Data Type	Scale
Hazard map	Landslide	Point	–
Topographic map	Slope Aspect Curvature TWI SPI Distance from lineament	GRID	1:5,000
Soil map	Topography Soil drainage Soil material Soil thickness Soil texture	Polygon	1:25,000
Forest map	Timber type Timber age Timber diameter Timber density	Polygon	1:25,000
Geological map	Geology	Polygon	1:50,000
Land use map	Land use	Polygon	1: 5,000

resolution and was used to calculate the slope, aspect, curvature, TWI and SPI. Lineament structure was detected from hillshade map by a structural geologist. Soil database includes soil topography, drainage, material, thickness and texture. Soil texture is classified as code including name of area, texture, and slope of topography (Table 7.3). Forest database include timber type, age, diameter and density. The geology was extracted from the geological database.

The causes of landslide events are directly or indirectly related to slope, aspect, curvature, topographic wetness index (TWI), stream power index (SPI), lineament, soil, forest, geology and land use. As the entire area receives nearly the same amount of precipitation, the rainfall intensity effect on the landslide events was ignored. The topography has a vital role in the spatial variation of hydrological conditions such as soil moisture, groundwater flow and slope stability. Topographic indices have therefore been used to describe spatial soil moisture patterns (Moore et al. 1991). The SPI measures the erosion power of the stream, and is also considered as a factor contributing towards stability within the study area. Another topographic factor within the run-off model is the TWI (Beven and Kirkby 1979).

Both the calculated and extracted factors were converted to form a 10×10 m grid (ArcGIS GRID type), and then were converted to ASCII data for use with the logistic regression and artificial neural network program. The dimensions of the study area grid were 923 rows by 1,053 columns, and so the total number of cells was 579,884. Landslides occurred in 1,801 cells.

Table 7.3 Description of soil texture

Code	Area name	Texture	% Slope
Water	Water	Water	–
Gbd	Gakha	Pebblely silty clay loam	15–30
MuC	Mui	Stony loam	7–15
MuD	Mui	Stony loam	15–30
MuE	Mui	Gravelly loam	30–60
MtD	Mitan	Gravelly loam	15–30
MtE	Mitan	Sandy loam	30–60
Bo	Bonryang	Loamy soils	0–2
Sod	Songjeong	Sandy loam	15–30
SoE	Subuk	Sandy loam	2–7
SpC	Subuk	Stony loam	7–15
SiC	Sinbul	Loam with rock dome	7–15
SlC	Sinbul	Stony loam	7–15
SiD	Sinbul	Loam with rock dome	15–30
SlD	Sinbul	Loam with rock dome	15–30
SlE	Sinbul	Rocky loam	30–60
OdE	Odae	Rocky loam	30–60
OdF	Odae	Pebblely silty clay loam	60–100
UgC	Ungyo	Pebblely silty clay loam	7–15
UgE	Ungyo	Sandy loam	30–60
WjE	Woljeong	Sandy loam	30–60
WjF	Woljeong	Sandy loam	60–100
ImB	Imok	Sandy loam	2–7
ImC	Imok	Loamy soils	7–15
Jd	Jungdong	Rocky silty clay loam	0–2
JsE	Jangseong	Rocky silty clay loam	30–60
JsF	Jangseong	Loamy soils	60–100
ChD	Chahang	Loamy soils	15–30
ChE	Chahang	Stony silty clay loam	30–60
CsE	Cheongshim	Stony silty clay loam	30–60
CsF	Cheongshim	Pebblely clay loam soil	60–100
PaC	Pyeongan	Gravelly loam	7–15
HgB	Hogye	Gravelly loam	2–7
HgC	Hogye	Gravelly loam	7–15
Hr	Hwangnyong	Gravelly sandy loam	0–2
RL	Rock outcrop	Rock outcrop	–
RC	Flood passed land	Flood passed land	–
Gt	Gangseo	Fine sandy loam	0–2
Gz	Gangseo	Sandy loam	0–2
GpB	Gopyeong	Silty clay loam	2–7
Ng	Namgye	Sandy loam	0–2
NkB	Noegok	Sandy loam	2–7
Dq	Doekcheon	Fine sandy loam	2–7

Table 7.3 (continued)

Code	Area name	Texture	% Slope
DEB	Dogye	Gravelly loam	2–7
DEC	Dogye	Gravelly loam	7–15
DED	Dogye	Gravelly loam	15–30
SIB	Sangye	Loamy soils	2–7
SIC	Sangye	Loamy soils	7–15
PuB	Pungcheon	Gravelly loam	2–7
HjN	Hwadong	Silty clay loam	2–7

7.5 Landslide Susceptibility Mapping Using Various Models

Using the probability model, the spatial relationship between landslide occurrence location and each related factor was derived. The rating of each factor type or range was assigned as the relationship between landslide location and each factor, this being the ratio of landslide-free to event-evident cells as shown in A. 1. The landslide susceptibility index (LSI_{FR}), Eq. 7.1, is calculated by a summation of each factor ratio value (Lee and Min 2001):

$$LSI_{FR} = FR_1 + FR_2 + FR_3 + \ldots + FR_n \tag{7.1}$$

Where FR_n is frequency ratio of each factor type or range (Fig. 7.4a).

The relation analysis is the ratio of the area where landslides occurred to the total area, so a value of 1 means an average value. If the value is greater than 1, there is a high correlation, and lower than 1 means a lower correlation. If the probability is high, there is a greater susceptibility to landslides; a lower value indicates a lesser susceptibility.

The following formulation of the Bayesian probability model, known as the weights-of-evidence model, was applied to landslide susceptibility analysis as synthesized from Bonham-Carter et al. (1989), The weights of evidence analysis result in a set of probabilistically derived values reflecting the spatial association between a factor map and landslide location. To generate binary predictor patterns for each factor, the spatial database was classified into a binary map by calculating W^+ and W^- from Eqs. 7.2 and 7.3, which show favorable and unfavorable areas.

$$W^+ = \log_e \frac{P(B|D)}{P(B|\bar{D})} \tag{7.2}$$

$$W^- = \log_e \frac{P(\bar{B}|D)}{P(\bar{B}|\bar{D})} \tag{7.3}$$

where P is probability, B is presence of factor, $\bar{B}$ is absence of factor, D is presence of landslide occurrence and $\bar{D}$ is absence of landslide occurrence. W^+ and W^- are the weights of evidence when a factor is present and absent, respectively (Bonham-Carter 1994). The binary predictor patterns were assigned weights (A. 1)

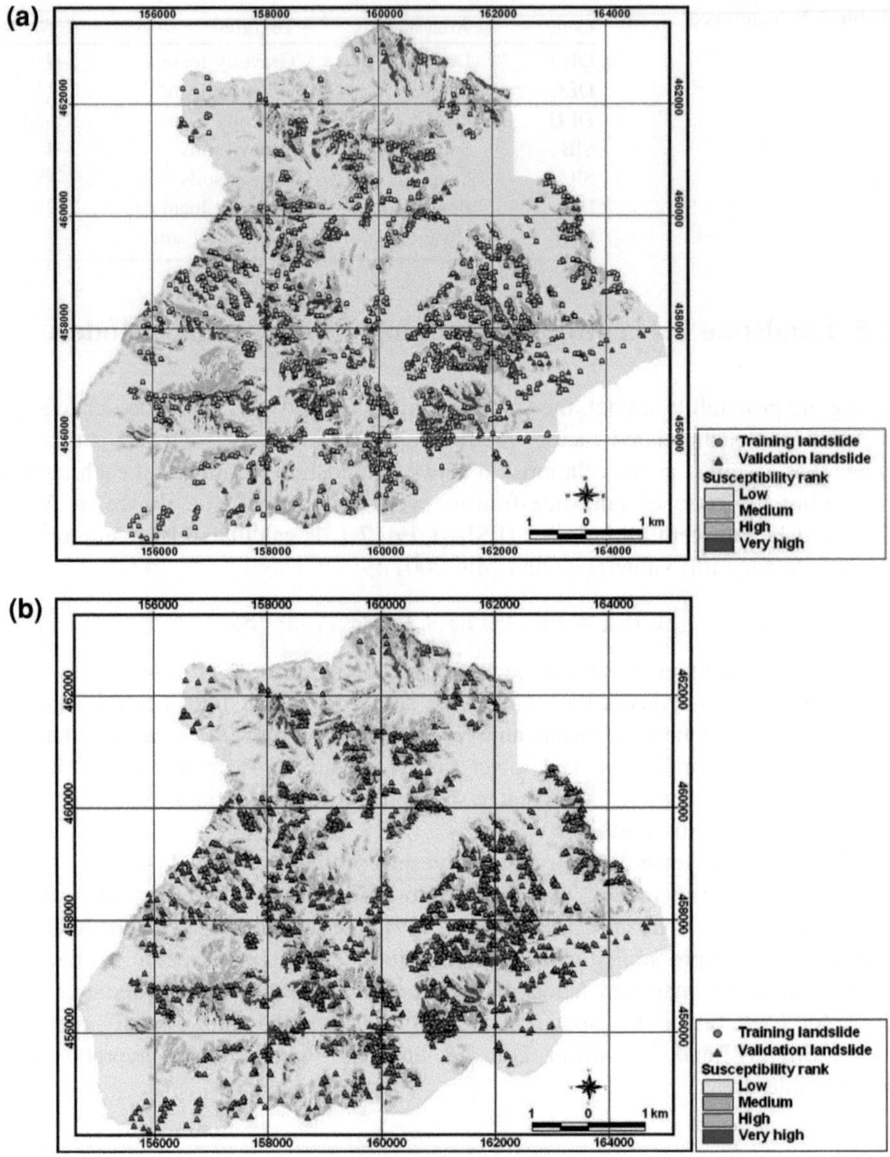

Fig. 7.4 Landslide susceptibility map based on individual models. **a** Frequency ratio. **b** Weight of evidence. **c** Logistic regression. **d** Artificial neural network

with maximum studentized value of contrast, C/S(C), and were calculated according to Eq. 7.4.

$$LSI_{WOE} = WOE_1 + WOE_2 + WOE_3 + \ldots + WOE_n \tag{7.4}$$

(c)
(d)

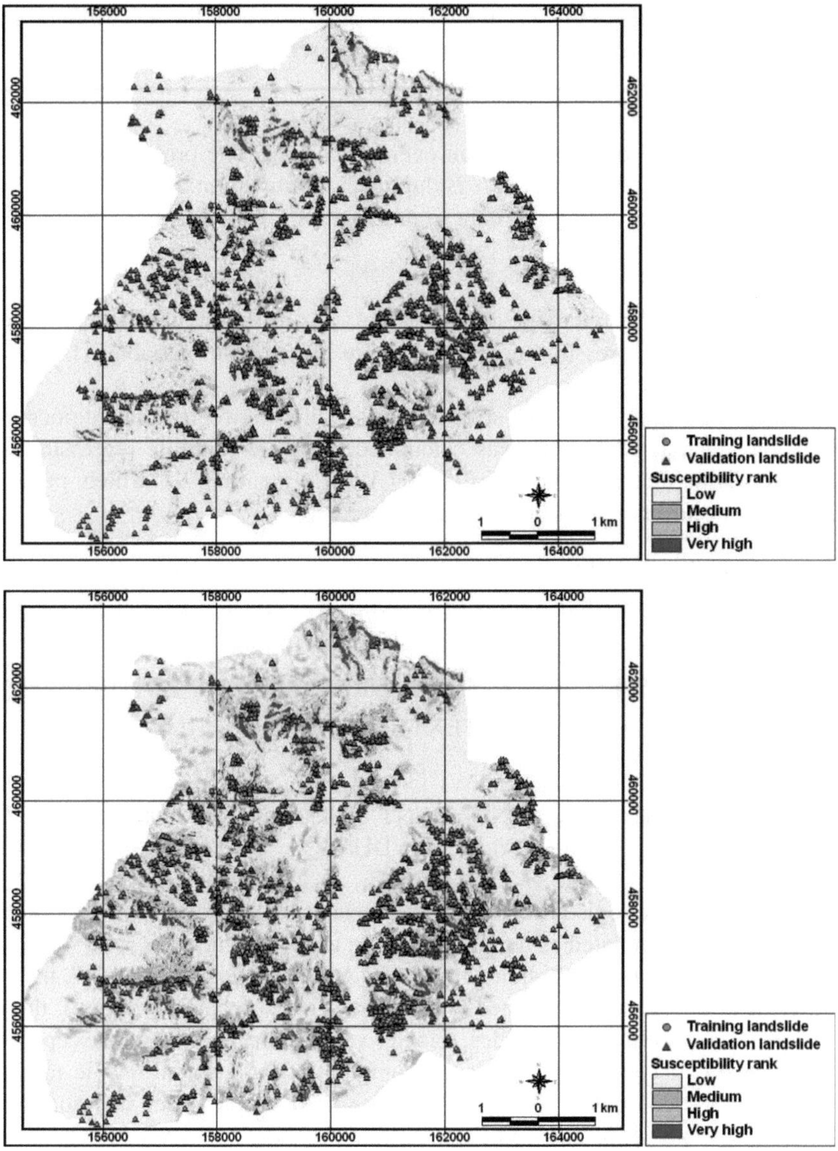

Fig. 7.4 (continued)

Where WOE_n is W^+ and W^- of the binary pattern for a range of each factor values or factor class at Max. C/S(C) (Fig. 7.4b).

Logistic multiple regression allows one to form a multivariate regression relation between a dependent and several independent variables. In the present situation, the dependent variable is binary, representing the presence or absence of landslides. Quantitatively, the relationship between the occurrence and its dependency on several variables can be expressed as:

$$p = 1/(1 + e^{-z}) \text{ or } p = e^{z}/(1 + e^{z}) \tag{7.5}$$

Where p is the probability of an event occurring. The probability varies from 0 to 1 on an S-shaped curve and z is the linear combination. It follows that logistic regression involves fitting the data to an equation of the form:

$$z = b_0 + b_1x_1 + b_2x_2 + \ldots + b_nx_n \tag{7.6}$$

Where b_0 is the intercept of the model, b_i are the slope coefficients of the logistic multiple regression model, and x_i are independent variables (Dai and Lee 2002).

Using this approach, the spatial relationship between landslide-occurrence location and related factors was calculated. Logistic multiple regression coefficients of the factors were calculated (A. 1). Equation 7.7 which predict the landslide-occurrence possibility, were created.

$$\begin{aligned} z = {} & (0.023 \times \mathrm{SLOPE}) + (-0.030 \times \mathrm{TWI}) + (-0.005 \times \mathrm{SPI}) \\ & + (-0.005 \times \mathrm{LINEAMENT}) + \mathrm{ASPECT_a} + \mathrm{CURVATURE_a} \\ & + \mathrm{GEOLOGY_a} + \mathrm{LANDUSE_a} + T\_\mathrm{DIAMETER_a} + \mathrm{T\_TYPE_a} \\ & + \mathrm{T\_DENSITY_a} + \mathrm{T\_AGE_a} + \mathrm{S\_DRAINAGE_a} + \mathrm{S\_MATERIAL_a} \\ & + \mathrm{S\_THICKNESS_a} + \mathrm{S\_TEXTURE_a} - 57.710 \end{aligned} \tag{7.7}$$

where SLOPE is slope value; TWI is topographic wetness index value; SPI is stream power index value; LINEAMENT is distance from lineament; $\mathrm{ASPECT_a}$, $\mathrm{CURVATURE_a}$, $\mathrm{GEOLOGY_a}$, $\mathrm{LANDUSE_a}$, $\mathrm{T\_DIAMETER_a}$, $\mathrm{T\_TYPE_a}$, $\mathrm{T\_DENSITY_a}$, $\mathrm{T\_AGE_a}$, $\mathrm{S\_DRAINAGE_a}$, $\mathrm{S\_MATERIAL_a}$, $\mathrm{S\_THICKNESS_a}$, $\mathrm{S\_TEXTURE_a}$ are logistic regression coefficient values listed in A. 1; $\mathrm{S\_TOPOGRAPHY_a}$ was eliminated as redundant factor; z is a parameter.

Using the logistic regression coefficient (A. 1) and Eqs. 7.5 and 7.7, the probability of landslide was computed (Fig. 7.4c). If the probability is high, there is a greater susceptibility to landslides; a lower value indicates a lesser susceptibility.

An artificial neural network is a "computational mechanism able to acquire, represent, and compute a mapping from one multivariate space of information to another, given a set of data representing that mapping" (Garrett 1994). The back-propagation training algorithm is the most frequently used neural network method and was the method used in this study. Areas where landslides have not occurred and where the slope is 0 were classified as "areas not prone to landslide" and areas where landslides were known to have occurred were assigned to the "areas prone to landslide" training set. The algorithm then was applied to calculate the weights between the input and the hidden layers, and between the hidden (Zhou 1999) and the output layers, by modifying the number of hidden nodes and the learning rate. The Weights that represent the contribution or importance of each factor were determined. A program developed by Hines (1997) was used for weight calculation, and for the interpretation of the weight we used a newly developed program using MATLAB software.

In this study, a 17 × 34 × 1 structure was selected for the networks, with input data normalized in the range 0.1–0.9. The nominal and interval class group data were converted to continuous values ranging between 0.1 and 0.9. The learning rate was set to 0.01, and the initial weights were randomly selected. The weights calculated from 10 test cases were compared to determine whether the variation in the final weights was dependent on the selection of the initial weights. The number of epochs was set to 5,000, and the root mean square error value used for the stopping criterion was set to 0.01. All the iterations met the 0.01 root mean square error goal in less than 5,000 epochs. After training, the weights were determined, and are shown in A. 2. For easy interpretation, the average values were calculated and divided by the weight of the minimum value. For example, the soil texture had the minimum value of 1.000, and the aspect had the maximum value of 1.503. Finally, the weights were applied to the entire study area. That is, the calculated weights from each dataset were applied to all datasets (Fig. 7.4d).

7.6 Ensemble-Based Landslide Susceptibility Mapping

The ensemble methodology is same to integrate the landslide-related factors and make landslide susceptibility map. The input factors were changed from slope, aspect, curvature, TWI, SPI, soil topography, soil drainage, soil material, soil thickness, soil texture, timber type, timber age, timber diameter, timber density, distance from lineament, geology and land use to the four landslide susceptibility maps from frequency ratio, weight of evidence, logistic regression and artificial neural network models. The ensemble-based susceptibility index values were classified in the same manner as for the landslide susceptibility maps generated by each model as described in Sect. 7.5.

In the case of the ensemble by frequency ratio, the landslide susceptibility maps from each model were compared with landslide location data using the frequency ratio model again. The frequency ratios for each landslide susceptibility index (Table 7.4) were summed to calculate $ELSI_{FR}$ (Ensemble-based Landslide Susceptibility Index), as shown in Eq. 7.8, and $ELSI_{FR}$ was mapped (Fig. 7.5a).

$$ELSI_{FR} = EFR_{LSM\_FR} + EFR_{LSM\_WOE} + EFR_{LSM\_LR} + EFR_{LSM\_ANN} \quad (7.8)$$

Where EFR is frequency ratio for a class range of each landslide susceptibility maps (Fig. 7.4a).

In the case of the ensemble by weight of evidence, the landslide susceptibility maps from each model were compared with landslide using the weight of evidence model again. The weight values for range of each landslide susceptibility index

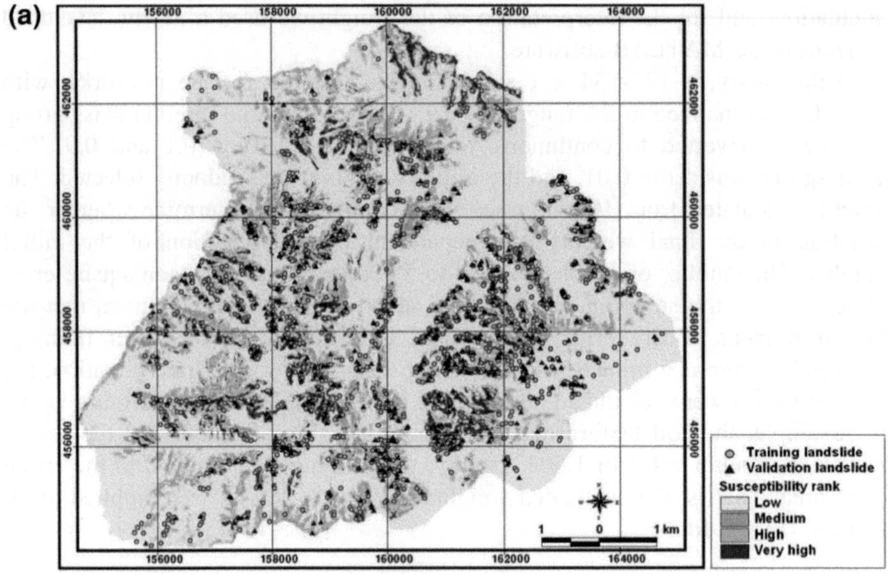

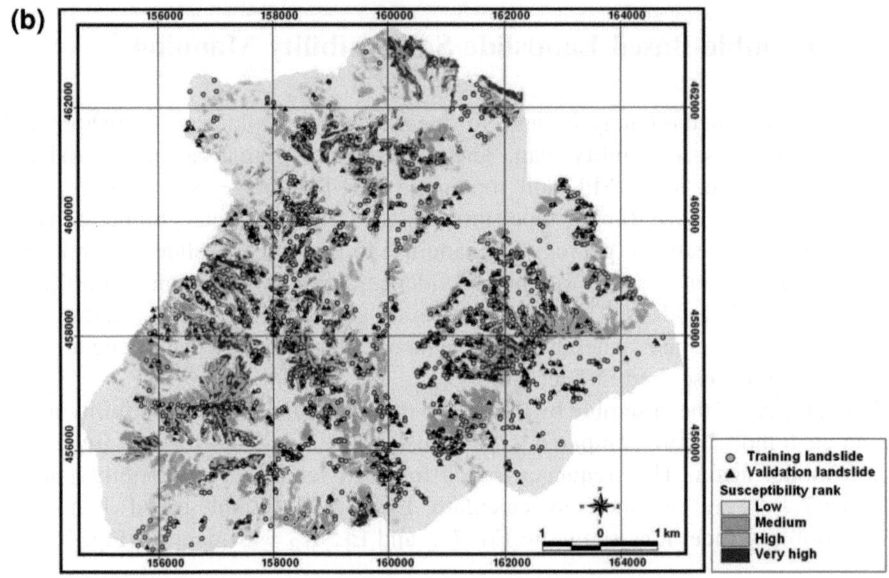

Fig. 7.5 Ensemble based landslide susceptibility map. **a** Frequency ratio. **b** Weight of evidence. **c** Logistic regression. **d** Artificial neural network

(Table 7.4) were summed to calculate $ELSI_{WOE}$ (Ensemble based Susceptibility Index), as shown in Eq. 7.9, and $ELSI_{WOE}$ was mapped (Fig. 7.5b).

$$ELSI_{WOE} = EWOE_{LSM\_FR} + EWOE_{LSM\_WOE} + EWOE_{LSM\_LR} + EWOE_{LSM\_ANN}. \quad (7.9)$$

(c)

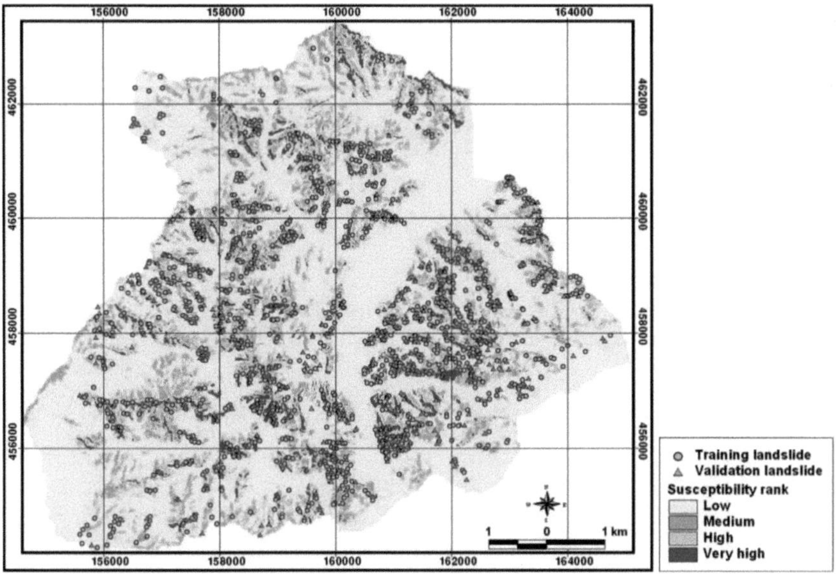

(d)

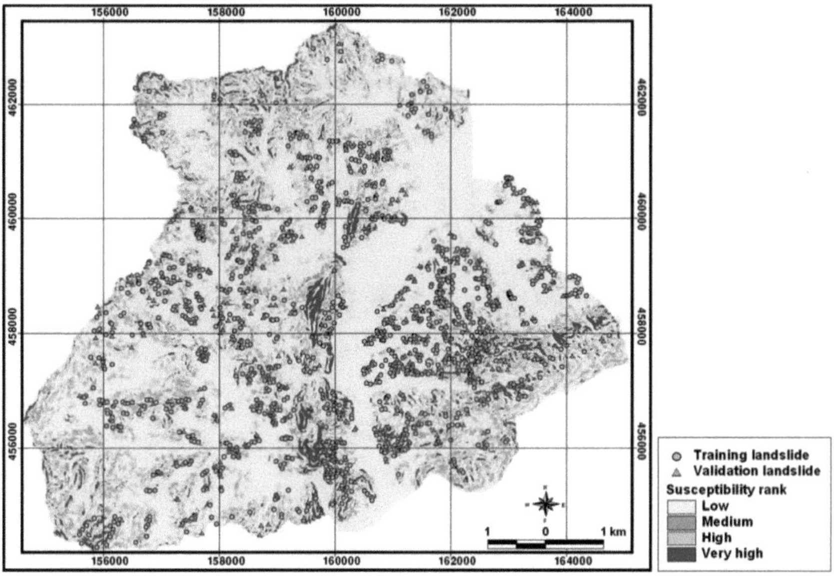

Fig. 7.5 (continued)

Where EWOE = W^+ and W^- weights at Max. C/S(C) of each landslide susceptibility maps (Fig. 7.4b).

In the case of the ensemble by logistic regression, the spatial relationship between landslide-occurrence location and new input factors (landslide

Table 7.4 Spatial relationships between landslide and landslide susceptibility maps from each model

Models	Landslide susceptibility index	No. of landslide[a]	% of landslide	No. of pixels in domain[b]	% of pixels in domain	FR	W+	W-	C	C/S(C)	LRC
LSM_FR	2.76–8.59	0	0.00	59856	10.01	0.00	NaN	0.11	NaN	NaN	2.881
	8.60–11.60	7	0.56	59915	10.02	0.06	−2.89	0.10	−2.99	−7.90	
	11.61–14.66	37	2.93	59890	10.02	0.29	−1.23	0.08	1.30	−7.81	
	14.67–16.64	46	3.65	59833	10.01	0.36	−1.01	0.07	1.08	−7.17	
	16.65–18.13	106	8.41	59836	10.01	0.84	−0.17	0.02	0.19	−1.89	
	18.14–19.58	149	11.82	59773	10.00	1.18	0.17	−0.02	0.19	2.15	
	19.59–20.66	120	9.52	59959	10.03	0.95	−0.05	0.01	0.06	−0.61	
	20.67–21.49	164	13.01	59875	10.01	1.30	0.26	−0.03	0.30	3.52	
	21.50–22.26	256	20.30	59641	9.98	2.04	0.71	−0.12	0.83	11.89	
	22.27–24.92	376	29.82	59306	9.92	3.01	1.10	−0.25	1.35	21.92	
LSM_WOE	−8.25–6.24	0	0.00	61124	10.22	0.00	NaN	0.11	NaN	NaN	0.236
	−6.22–5.12	6	0.48	60048	10.04	0.05	−3.05	0.10	−3.15	−7.70	
	−5.11–3.94	43	3.41	59861	10.01	0.34	−1.08	0.07	−1.15	−7.40	
	−3.93–2.58	110	8.72	60496	10.12	0.86	−0.15	0.02	−0.16	−1.64	
	−2.57–1.07	104	8.25	61744	10.33	0.80	−0.22	0.02	−0.25	−2.42	
	−1.06–0.21	101	8.01	59315	9.92	0.81	−0.21	0.02	−0.24	2.26	
	0.22–1.29	188	14.91	59423	9.94	1.50	0.41	−0.06	0.46	5.85	
	1.30–2.42	195	15.46	60215	10.07	1.54	0.43	−0.06	0.49	6.30	
	2.43–3.43	260	20.62	58053	9.71	2.12	0.75	−0.13	0.88	12.66	
	3.44–5.95	254	20.14	57605	9.63	2.09	0.74	−0.12	0.86	12.26	
LSM_LR	0.000000–0.000000	0	0.00	84859	14.19	0.00	NaN	0.15	NaN	NaN	4.643
	0.000001–0.000265	13	1.03	57177	9.56	0.11	−2.23	0.09	−2.32	−8.31	
	0.000266–0.000596	16	1.27	57089	9.55	0.13	−2.02	0.09	−2.11	−8.37	
	0.000597–0.001034	51	4.04	57014	9.54	0.42	−0.86	0.06	−0.92	−6.41	

(continued)

Table 7.4 (continued)

Models	Landslide susceptibility index	No. of landslide[a]	% of landslide	No. of pixels in domain[b]	% of pixels in domain	FR	W+	W-	C	C/S(C)	LRC
	0.001035–0.001617	62	4.92	57055	9.54	0.52	−0.66	0.05	−0.71	−5.47	
	0.001619–0.002397	103	8.17	57022	9.54	0.86	−0.15	0.02	−0.17	−1.65	
	0.002399–0.003422	121	9.60	56955	9.53	1.01	0.01	0.00	0.01	0.08	
	0.003425–0.004830	192	15.23	56947	9.52	1.60	0.47	−0.07	0.53	6.81	
	0.004835–0.007103	270	21.41	56925	9.52	2.25	0.81	−0.14	0.95	13.85	
	0.007110–0.038865	433	34.34	56841	9.51	3.61	1.28	−0.32	1.60	27.05	
LSM_ANN	0.0494–0.4260	1	0.08	59802	10.00	0.01	−4.84	0.10	−4.94	−4.94	–
	0.4261–0.6616	19	1.51	59817	10.00	0.15	−1.89	0.09	−1.98	−8.58	
	0.6617–0.7638	49	3.89	59860	10.01	0.39	−0.95	0.07	−1.01	−6.95	
	0.7639–0.8140	89	7.06	59831	10.01	0.71	−0.35	0.03	−0.38	−3.47	
	0.8141–0.8498	106	8.41	59880	10.02	0.84	−0.18	0.02	−0.19	−1.90	
	0.8499–0.8784	166	13.16	59930	10.02	1.31	0.27	−0.04	0.31	3.70	
	0.8785–0.9021	175	13.88	59963	10.03	1.38	0.32	−0.04	0.37	4.52	
	0.9022–0.9220	233	18.48	59764	10.00	1.85	0.61	−0.10	0.71	9.83	
	0.9221–0.9390	206	16.34	59566	9.96	1.64	0.49	−0.07	0.57	7.45	
	0.9391–0.9821	217	17.21	59471	9.95	1.73	0.55	−0.08	0.63	8.47	

[a] Number of landslide cells : 1,261
[b] Number of total cells : 579,884
FR: Frequency Ratio; LRC: Logistic Regression coefficient
LSM_FR : Landslide Susceptibility Map based on Frequency Ratio
LSM_WOE : Landslide Susceptibility Map based on Weight of Evidence
LSM_LR : Landslide Susceptibility Map based on Logistic Regression
LSM_ANN : Landslide Susceptibility Map based on Artificial Neural Network

Table 7.5 Weight of each factors calculated by neural network

Run Factors	1	2	3	4	5	6	7	8	9	10	Mean	S.D.	N. W.*
LSM_FR	0.2980	0.2604	0.3236	0.3765	0.3343	0.3578	0.2540	0.2691	0.3161	0.3356	0.3125	0.0415	1.9192
LSM_WOE	0.2673	0.2277	0.2003	0.2386	0.2039	0.2133	0.2941	0.2749	0.2251	0.2222	0.2367	0.0317	1.4537
LSM_LR	0.1378	0.1530	0.2032	0.1499	0.2234	0.1778	0.1127	0.1084	0.1834	0.1789	0.1629	0.0374	1.0000
LSM_ANN	0.2969	0.3589	0.2729	0.2351	0.2384	0.2510	0.3392	0.3475	0.2754	0.2634	0.2879	0.0458	1.7677

N.W.: Normalized Weight
S.D.: Standard deviation
* Normalized weight with respect to weight of evidence

susceptibility maps) was calculated. Thereafter, logistic multiple regression coefficients of the factors were calculated (Table 7.4). After interpretation, Eq. 7.10 which
predict the landslide-occurrence possibility, were created and using the Eqs. 7.5 and 7.10, the ensemble landslide susceptibility map was made (Fig. 7.5c).

$$Z = (2.881 \times LSM\_FR) + (0.236 \times LSM\_WOE) + (4.643 \times LSM\_LR) - 10316 \quad (7.10)$$

Where LSM_FR is landslide susceptibility map by frequency ratio model; LSM_WOE is landslide susceptibility map using weight of evidence model; LSM_LR is landslide susceptibility map using logistic regression model; LSM_ANN was eliminated as redundant factor; Z is a parameter.

In the case of the ensemble by artificial neural network, the back-propagation algorithm was applied and a $4 \times 8 \times 1$ structure was selected. The other parameters such as normalization of input data, learning initial weights, number of training, number of epochs and RMS were set to same to Sect. 7.5. After training, the weights were determined, and are shown in Table 7.5. The ensemble based landslide susceptibility map by weight of evidence had the minimum value of 1.000, and the one by frequency ratio had the maximum value of 1.9192. Finally, the ensemble landslide susceptibility map was made (Fig. 7.5d) using the weights applied to the entire study area.

The landslide susceptibility maps were quantitatively made using the frequency ratio, weight of evidence, logistic regression and artificial neural network models (Figs. 7.4a–d) and their ensemble models (Figs. 7.5a–d). The susceptibility index values were classified into four classes based on area for visual and easy interpretation; very high 5%, high 10%, medium 15% and low reminding 70%.

7.7 Validation

For validation of the landslide susceptibility calculation methods, two basic assumptions are needed. One is that landslide occurrences are related to spatial information such as topography, geology, soil, forest and land use, and the second is that future occurrences will be precipitated by a specific impact factor such as rainfall. In this study, these assumptions are satisfied because landslides are related to spatial information factors by analyzing the relationship between landslide and each factor, and are initiated by a single event of heavy rainfall in the study area.

The results of success rate validation were derived by comparing the landslide occurrence location which is not used for training the model and the susceptibility maps which created using the frequency ratio, weight of evidence, logistic regression, artificial neural network models and their ensemble models, and appear as a line graph in Figs. 7.6 and 7.7. The success rate illustrates how well the estimators perform. To obtain the relative ranks for each prediction pattern, the

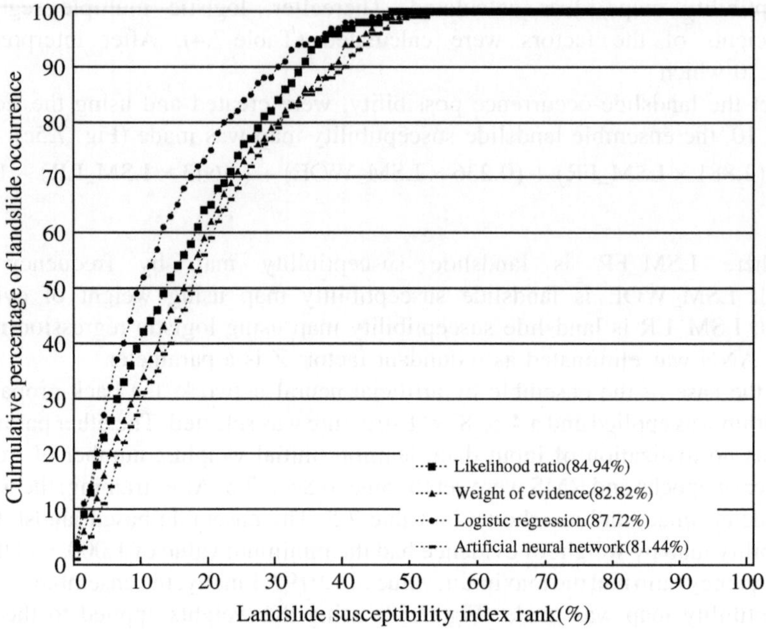

Fig. 7.6 Validation results of landslide susceptibility map using individual models

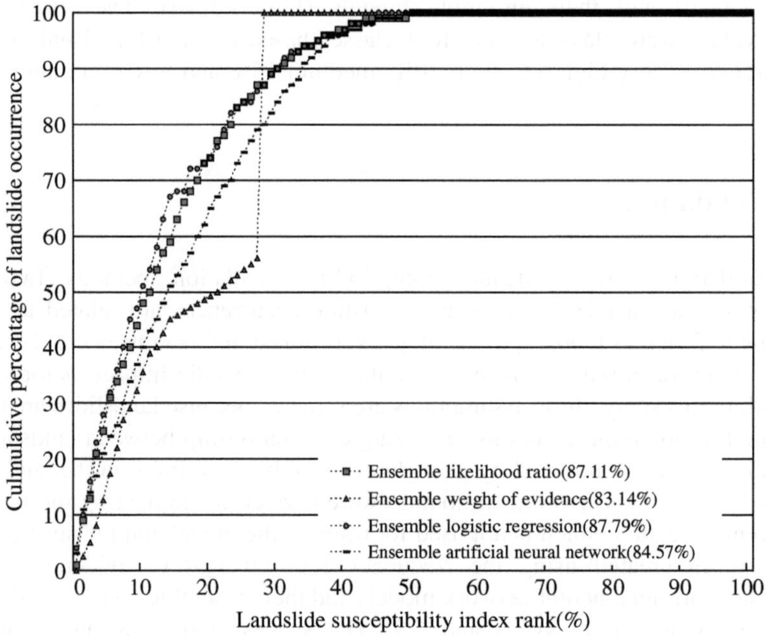

Fig. 7.7 Validation results of landslide susceptibility map using integrated models

Table 7.6 The result of validation of each landslide susceptibility map

Model	Accuracy LSM (%)	Difference of accuracy (ELSM – LSM) ELSM (%)	
FR	84.94	87.11	2.17%
WOE	82.82	83.14	0.32%
LR	87.72	87.79	0.07%
ANN	81.44	84.57	3.13%

ELSM: Ensemble based Landslide Susceptibility Map

calculated index values of all cells in the study area were sorted in descending order and divided into 100 classes with a 1% interval. To quantitatively compare the results, the areas under the curve (AUC) were recalculated for when the total area is represented by 1, which means perfect prediction accuracy.

As the result, in the case of ensemble by frequency ratio, weight of evidence, logistic regression and artificial neural network models, the area ratio was 0.8711, 0.8314, 0.8779 and 0.8457 and that the prediction accuracy is 87.11, 83.14, 87.79 and 84.57%, respectively. As references, the results of landslide susceptibility map using 17 factors and frequency ration, logistic regression and artificial neural network models are 84.94, 82.82, 87.72 and 81.44% respectively, after application the AUC method (Table 7.6).

7.8 Discussion and Conclusions

In this study, an ensemble-based predictive method was developed and applied to landslide susceptibility mapping. Factors associated with the landslide were assembled in a spatial database and the landslide susceptibility maps were made using the frequency ratio, weight of evidence, logistic regression, and artificial neural network models. The maps were validated using landslide location data that had not been used for training the model for the prediction. This study derived the following conclusions:

The location of landslides using aerial photographs, which have 50 cm spatial resolution, was checked in the field and found to be very accurate.

Using the frequency ratio, weight of evidence, logistic regression, and artificial neural network models, landslide susceptibility maps and ensemble-based landslide susceptibility maps were made. For the validation of the maps, landslide locations that were not used in the training were used. In the case of individual models, the logistic regression model showed 87.72% accuracy, which is best, and the artificial neural network model showed 81.44%, which is worst. In the case of ensemble methods, the logistic regression model showed 87.79% accuracy, which is best, and the weight of evidence model showed 83.14%, which is worst.

Comparing cases of individual models and ensemble models, the ensemble models showed the better result. The difference of accuracies is 2.17% in frequency ratio, 0.32% in weight of evidence, 0.07% in logistic regression, and

3.13% in artificial neural network models. Therefore, the ensemble model could be concerned as better than the individual model.

Comparing ensemble models, the logistic regression model showed the best result (87.79% accuracy) compared with frequency ratio (87.11%), weight of evidence (83.14%), and artificial neural network (84.57%) models.

This study showed that integrating the output of multiple models improves the predictive accuracy, as the difference of accuracy (ELSM minus LSM) showed positive values. Therefore, the ensemble method can enhance the quality of landslide susceptibility mapping from a set of models because of mutually complementary relations between result maps and different predictive models. Indeed, it can be compared and estimated that the predictive preference of each model. The logistic regression model showed the best prediction accuracy of 87.72% and 87.79% of individual and ensemble-based landslide susceptibility maps, respectively. The difference of accuracy is 0.07%, which is the lowest value among the four models. The frequency ratio model showed the second highest prediction accuracy of 84.94 and 87.11% for individual- and ensemble-landslide susceptibility maps, respectively. Indeed, the difference in their accuracy is 2.17%, indicating the second-highest gain of accuracy. As the result, the frequency ratio model showed sensitivity to the input data, and the ensemble method was effective in improving the prediction accuracy. Weight of evidence was found to be not sensitive to the input data, with a low accuracy difference of 0.32%, while it also generally showed lower prediction accuracy than frequency ratio and logistic regression. Artificial neural network modeling also showed low prediction accuracy, similar to weight of evidence. However, the difference in accuracy of the artificial neural network model showed the highest value of 3.13%. This suggests the model is sensitive to input data.

Because the ensemble model was applied using the same procedure for landslide susceptibility index mapping as was used in the individual models, the prediction accuracy can be improved using no more data. However, more case studies are required for checking the ensemble model, because the ensemble model is not widely used. For effective performance prediction, the ensemble model can be tested in other application areas such as mineral potential mapping, land subsidence hazard mapping, etc.

Using the landslide susceptibility map, we can predict future landslide locations where the landslide has not previously occurred. Therefore, based on the map, it can be prepared for the landslide hazard, and plan land development and make land-use policies efficiently. In addition, if the same analysis is performed in other areas having similar topographic and geological conditions, time and cost can be saved in predicting the landslide efficiently.

Acknowledgments This research was supported by the Basic Research Project of the Korea Institute of Geoscience and Mineral Resources (KIGAM) funded by the Ministry of Knowledge and Economy of Korea.

References

Akgun A, Bulut F (2009) GIS-based landslide susceptibility for Arsin-Yomra (Trabzon, North Turkey) region. Env Geol 9:339–404

Akgun A, Türk N (2010) Landslide susceptibility mapping for Ayvalik (Western Turkey) and its vicinity by multicriteria decision analysis. Environ Earth Sci 61:595–611

Audisio C, Nigrelli G, Lollino G (2009) A GIS tool for historical instability processes data entry: an approach to hazard management in two Italian Alpine river basins. Comput Geosci 35:1735–1747

Bai S, Lü G, Wang J, Zhou P, Ding L (2011) GIS-based rare events logistic regression for landslide-susceptibility mapping of Lianyungang, China. Environ Earth Sci 62:139–149

Bai S, Wang J, Lu GN, Zhou PG, Hou SS, Xu SN (2010) GIS-based logistic regression for landslide susceptibility mapping of the Zhongxian segment in the Three Gorges area, China. Geomorphology 115:23–31

Beven KJ, Kirkby MJ (1979) Physically based, variable contributing area model of basin hydrology. Hydrol Sci Bull des Sci Hydrologiques 24:43–69

Bonham-Carter GF (1994) Geographic information systems for geoscientists, modeling with GIS. Pergamon Press, Oxford

Bonham-Carter GF, Agterberg FP, Wright DF (1989) Weights of evidence modeling: a new approach to mapping mineral potential. In: Agterberg FP, Bonham-Carter GF (eds) Statistical applications in the Earth sciences. Geological Survey of Canada, Ottawa, pp 171–183

Brabb EE (1984) Innovative approaches to landslide hazard and risk mapping. In: Proceedings, fourth international symposium on landslides, vol 1. Canadian Geotechnical Society, Toronto, pp 307–324

Chauhan S, Sharma M, Arora MK (2010) Landslide susceptibility zonation of the Chamoli region, Garhwal Himalayas, using logistic regression model. Landslides 7:411–423

Chen CH, Ke CC, Wang CL (2009a) A back-propagation network for the assessment of susceptibility to rock slope failure in the eastern portion of the Southern Cross-Island Highway in Taiwan. Env Geol 57:723–733

Chen YR, Chen JW, Hsieh SC, Ni PN (2009b) The application of remote sensing technology to the interpretation of land use for rainfall-induced landslides based on genetic algorithms and artificial neural networks. IEEE J Sel Top Appl Earth Observations Remote Sens 2:87–95

Clerici A, Perego S, Tellini C, Vescovi P (2002) A procedure for landslide susceptibility zonation by the conditional analysis method. Geomorphology 48:349–364

Dahal RK, Hasegawa S, Nonomura A, Yamanaka M, Takuro M, Nishino K (2008) GIS-based weights-of-evidence modelling of rainfall-induced landslides in small catchments for landslide susceptibility mapping. Env Geol 54:311–324

Dai FC, Lee CF (2002) Landslide characteristics and slope instability modeling using GIS, Lantau Island, Hong Kong. Geomorphology 42:213–228

Das I, Sahoo S, van Westen C, Stein A, Hack R (2010) Landslide susceptibility assessment using logistic regression and its comparison with a rock mass classification system, along a road section in the northern Himalayas (India). Geomorphology 114:627–637

Dominguez-Cuesta MJ, Jimenez-Sanchez M, Colubi A, Gonzalez-Rodriguez G (2010) Modelling shallow landslide susceptibility: a new approach in logistic regression by using favourability assessment. Int J Earth Sci 99:661–674

Dong JJ, Lee CT, Tung YH, Liu CN, Lin KP, Lee JF (2009) The role of the sediment budget in understanding debris flow susceptibility. Earth Surf Process Landform 34:1612–1624

Falaschi F, Giacomelli F, Federici PR, Puccinelli A, D'Amato Avanzi G, Pochini A, Ribolini A (2009) Logistic regression versus artificial neural networks: Landslide susceptibility evaluation in a sample area of the Serchio River valley, Italy. Nat Hazards 50:551–569

Garrett J (1994) Where and why artificial neural networks are applicable in civil engineering. J Comput Civil Eng 8:129–130

Geological Society of Korea (1962) Changdong-Hajinburi geological map sheet

Guzzetti F, Reichenbach P, Ardizzone F, Cardinali M, Galli M (2006) Estimating the quality of landslide susceptibility models. Geomorphology 81:166–184
Guzzetti F, Reichenbach P, Cardinali M, Galli M, Ardizzone F (2005) Probabilistic landslide hazard assessment at the basin scale. Geomorphology 72:272–299
He YR, Fu WJ (2009) Application of fuzzy support vector machine to landslide risk assessment. J Nat Disasters 18:107–112
Hines JW (1997) Fuzzy and neural approaches in engineering. Wiley, New York
Kanungo DP, Arora MK, Gupta RP, Sarkar S (2008) Landslide risk assessment using concepts of danger pixels and fuzzy set theory in Darjeeling Himalayas. Landslides 5:407–416
Kanungo DP, Arora MK, Sarkar S, Gupta RP (2009) A fuzzy set based approach for integration of thematic maps for landslide susceptibility zonation. Georisk 3:30–43
Kawabata D, Bandibas J (2009) Landslide susceptibility mapping using geological data, a DEM from ASTER images and an Artificial Neural Network (ANN). Geomorphology 113:97–109
Lee DS (1988) Geology of Korea (second ed.), Kyohak-Sa, Seoul
Lee S (2005) Application of logistic regression model and its validation for landslide susceptibility mapping using GIS and remote sensing data. Int J Remote Sens 26:1477–1491
Lee S (2007a) Comparison of landslide susceptibility maps generated through multiple logistic regression for three test areas in Korea. Earth Surf Proc Land 32:2133–2148
Lee S (2007b) Application and verification of fuzzy algebraic operators to landslide susceptibility mapping. Env Geol 52:615–623
Lee S, Choi J, Woo I (2004a) The effect of spatial resolution on the accuracy of landslide susceptibility mapping: A case study in Boun, Korea. Geosci J 8:51–60
Lee S, Evangelista DG (2006) Earthquake-induced landslide-susceptibility mapping using an artificial neural network. Nat Hazards Earth Syst Sci 6:687–695
Lee S, Min K (2001) Statistical analysis of landslide susceptibility at Youngin, Korea. Env Geol 40:1095–1113
Lee S, Pradhan B (2006) Probabilistic landslide hazards and risk mapping on Penang Island, Malaysia. J Earth Syst Sci 115:661–672
Lee S, Ryu JH, Lee MJ, Won JS (2006) The application of artificial neural networks to landslide susceptibility mapping at Janghung, Korea. Math Geol 38:199–220
Lee S, Ryu JH, Won JS, Park HJ (2004b) Determination and application of the weights for landslide susceptibility mapping using an artificial neural network. Eng Geol 71:289–302
Lee S, Song KY, Oh HJ, Choi J (2011) Detection of landslide using web-based aerial photographs and landslide susceptibility mapping using geospatial analysis. Int J Remote Sensing, Accepted 7 Jun 2011
Legorreta Paulin G, Bursik M (2009) Logisnet: A tool for multimethod, multiple soil layers slope stability analysis. Comput Geosci 35:1007–1016
Liu CN, Dong JJ, Peng YF, Huang HF (2009) Effects of strong ground motion on the susceptibility of gully type debris flows. Eng Geol 104:241–253
Mathew J, Jha VK, Rawat GS (2009) Landslide susceptibility zonation mapping and its validation in part of Garhwal Lesser Himalaya, India, using binary logistic regression analysis and receiver operating characteristic curve method. Landslides 6:17–26
Melchiorre C, Matteucci M, Azzoni A, Zanchi A (2008) Artificial neural networks and cluster analysis in landslide susceptibility zonation. Geomorphology 94:379–400
Miles SB, Keefer DK (2009) Evaluation of CAMEL–comprehensive areal model of earthquake-induced landslides. Eng Geol 104:1–15
Moore ID, Grayson RB, Ladson AR (1991) Digital terrain modelling: a review of hydrological, geomorphological, and biological applications. Hydrol Process 5:3–30
Mousavi SR, Pirasteh S, Shattri M, Amani A (2009) Landslides and Active Faults Using Remote Sensing and GIS Techniques in Central Alborz Mountains, Iran. Disaster Adv 2(3):24–29
Muthu K, Petrou M, Tarantino C, Blonda P (2008) Landslide possibility mapping using fuzzy approaches. IEEE Trans Geosci Remote Sensing 46:1253–1265
Nandi A, Shakoor A (2010) A GIS-based landslide susceptibility evaluation using bivariate and multivariate statistical analyses. Eng Geol 110:11–20

Oh HJ, Lee S, Chotikasathien W, Kim CH, Kwon JH (2009) Predictive landslide susceptibility mapping using spatial information in the Pechabun area of Thailand. Env Geol 57:641–651

Oh HJ, Lee S (2010) Cross-validation of logistic regression model for landslide susceptibility mapping at Geneoung areas, Korea. Disaster Adv 3(2):44–55

Ozdemir A (2009) Landslide susceptibility mapping of vicinity of Yaka Landslide (Gelendost, Turkey) using conditional probability approach in GIS. Env Geol 57:1675–1686

Park NW (2011) Application of Dempster-Shafer theory of evidence to GIS-based landslide susceptibility analysis. Environ Earth Sci 62:367–376

Pirasteh S, Pradhan B, Mahmoodzadeh A (2009) Stability mapping and landslide recognition in zagros mountain South West Iran: a case study. Disaster Adv 2(1):47–53

Poudyal CP, Chang C, Oh HJ, Lee S (2010) Landslide susceptibility maps comparing frequency ratio and artificial neural networks: a case study from the Nepal Himalaya. Environ Earth Sci 61:1049–1064

Prabu S, Ramakrishnan SS (2009) Combined use of socio economic analysis, remote sensing and GIS data for landslide hazard mapping using ANN. J Indian Soc Remote Sensing 37:409–421

Pradhan B, Lee S (2007) Utilization of optical remote sensing data and GIS tools for regional landslide hazard analysis by using an artificial neural network model at Selangor, Malaysia. Earth Sci Front 14:143–151

Pradhan B, Lee S (2009) Landslide risk analysis using artificial neural network model focussing on different training sites. Int J Phys Sci 4:1–15

Pradhan B, Lee S, Buchroithner MF (2010) A GIS-based back-propagation neural network model and its cross-application and validation for landslide susceptibility analyses. Comput Environ Urban Syst 34:216–235

Pradhan B, Lee S (2010a) Delineation of landslide hazard areas on Penang Island, Malaysia, by using frequency ratio, logistic regression, and artificial neural network models. Environ Earth Sci 60:1037–1054

Pradhan B, Lee S (2010b) Landslide susceptibility assessment and factor effect analysis: backpropagation artificial neural networks and their comparison with frequency ratio and bivariate logistic regression modelling. Environ Model Softw 25:747–759

Pradhan B, Lee S (2010c) Regional landslide susceptibility analysis using back-propagation neural network model at Cameron Highland, Malaysia. Landslides 7:13–30

Regmi NR, Giardino JR, Vitek JD (2010) Modeling susceptibility to landslides using the weight of evidence approach: Western Colorado, USA. Geomorphology 115:172–187

Rokach L (2005) Ensemble methods for classifiers. In: Maimon O, Rokach L (eds) The data mining and knowledge discovery handbook. Springer, Berlin, pp 957–958

Shafri HZM, Zahidi IM, Bakar SA (2010) Development of landslide susceptibility map utilizing remote sensing and Geographic Information Systems (GIS). Disaster Prev Manage 19:59–69

Tangestani MH (2009) A comparative study of Dempster-Shafer and fuzzy models for landslide susceptibility mapping using a GIS: an experience from Zagros Mountains, SW Iran. J Asian Earth Sci 35:66–73

Vahidnia MH, Alesheikh AA, Alimohammadi A, Hosseinali F (2009) Landslide hazard zonation using quantitative methods in GIS. Int J Civil Eng 7:176–189

Wang WD, Xie CM, Du XG (2009) Landslides susceptibility mapping based on geographical information system, GuiZhou, south-west China. Env Geol 58:33–43

Yalcin A (2008) GIS-based landslide susceptibility mapping using analytical hierarchy process and bivariate statistics in Ardesen (Turkey): Comparisons of results and confirmations. Catena 72:1–12

Yilmaz I (2009a) A case study from Koyulhisar (Sivas-Turkey) for landslide susceptibility mapping by artificial neural networks. Bull Eng Geol Environ 68:297–306

Yilmaz I (2009b) GIS based statistical and physical approaches to landslide susceptibility mapping (Sebinkarahisar, Turkey). Bull Eng Geol Environ 68:459–471

Yilmaz I (2009c) Landslide susceptibility mapping using frequency ratio, logistic regression, artificial neural networks and their comparison: a case study from Kat landslides (Tokat-Turkey). Comput Geosci 35:1125–1138

Yilmaz I (2010) Comparison of landslide susceptibility mapping methodologies for Koyulhisar, Turkey: conditional probability, logistic regression, artificial neural networks, and support vector machine. Environ Earth Sci 68:821–836
Zhou W (1999) Verification of the nonparametric characteristics of backpropagation neural networks for image classification. IEEE T Geosci Remote 37:771–779

Chapter 8
Geoinformatics and Mass Movements: A Study on Li-shan Landslide, Taiwan

Keh-Jian Shou

Abstract In this study, spatial risk of a large landslide area was analyzed by two approaches, implemented by coupling geography information system (GIS), limit equilibrium analysis, and Monte Carlo analysis. With the GIS, the three-dimensional surface topography, underground geomaterial distribution and ground water level can be processed for further analysis. Then the safety of slopes can be evaluated by limit equilibrium analysis. The limit equilibrium analyses are different for those two GIS based approaches. One is for a sliding body with a well-defined sliding plane and the other is for an infinite slope sliding on an assumed sliding plane. Taking spatial uncertainties into consideration, mechanical properties of geomaterial were considered as random variables instead of single values. And, ground water profiles for stability analysis are also randomly adopted. Through stability analysis with Monte Carlo sampling, a distribution of safety factor can be obtained to determine the probability of failure. Those two GIS based approaches were applied to the Li-shan landslide in central Taiwan.

Keywords Li-shan landslide · Hazard estimation · Spatial uncertainty · GIS · Probabilistic model · Monte Carlo analysis

8.1 Introduction

Geographic Information System (GIS) possesses strong capability in processing and analysing spatial data; therefore, it is useful and popular for the assessment of natural disasters (Hungr et al. 1989; Miller 1995; Haneberg 2000). It has been integrated

K.-J. Shou (✉)
Department of Civil Engineering, National Chung-Hsing University,
250 Kuo-Kuang Road, 402 Taichung, Taiwan
e-mail: kjshou@dragon.nchu.edu.tw

B. Pradhan and M. Buchroithner (eds.), *Terrigenous Mass Movements*,
DOI: 10.1007/978-3-642-25495-6_8,

with other analytical models (Hammond et al. 1992; Guzzetti et al. 1999; Reid et al. 2000; Van Westen 2004), such as GIS-probabilistic infinite slope model (Pack et al. 1998; Zhou et al. 2003; Haneberg 2004; Qiu et al. 2005) and GIS-infinite slope probabilistic seismic landslide model (Jibson et al. 1999; Khazai and Sistar 2000; Haneberg 2006), etc. And a GIS based probabilistic analysis approach was developed in a preliminary study (Shou and Chen 2005). In this probabilistic analysis approach, mechanical properties of geomaterial are considered as random variables, and Monte Carlo sampling is coupled with GIS and limit equilibrium analysis.

In this study, based on the prototype model, two advanced probabilistic analysis approaches were implemented that can evaluate the hazard of landslides with consideration of spatial uncertainties. One is for a sliding body with well defined sliding plane and the other is for a sliding area with assumed sliding plane. Those GIS based approaches were applied to spatially evaluate the hazard of Li-shan landslide in which there are well-defined major sliding bodies and poorly defined minor sliding bodies.

8.1.1 Li-Shan Landslide

In mid April 1990, after a heavy rain, the sites in Li-shan village near the highway 7A (mileage 73 km + 150 m) and the highway 8 (mileage 82 km) began to subside, as their foundations are located on one of the major sliding blocks in Li-shan landslide area. Since Li-shan is an important village located at the midway of the east–west cross-island highway (the highway 8) in central Taiwan (see Fig. 8.1), it is essential to keep the highway open and the town secure for living. The government had executed the first phase emergency treatment followed by the second remediation treatment since July 1990. A drainage system including surface ditches, drainage wells and two drainage galleries was constructed and completed in early 2003.

Geology and Instability

Geologically, Li-shan area is located in colluvial formations originally from the Miocene Lushan slate formation (see Fig. 8.1). Because of tectonic activities as well as the high precipitation, the surfacial slate formations in this area are highly weathered; it is strongly supported by the occurrence of slaty cleavages, foliation shears, and interlayers of silty residual soil (Huang 2002; Shou 2002). Topographically, as situated at the western rim of Hsueh-shan ridge, Li-shan area dips toward the northwest and down to Tachia river.

The landslide area is about 76 hectares in size, and can be divided into three regions, i.e. the west, northeast, and southeast regions. And there are 14 major sliding bodies in those three regions (see Fig. 8.2). Except the southeast region, most of the unstable sliding bodies possess sliding planes about 9–26 m below surface (Energy and Resources Laboratories, ITRI 1993; Shou 2002; Shou and Su

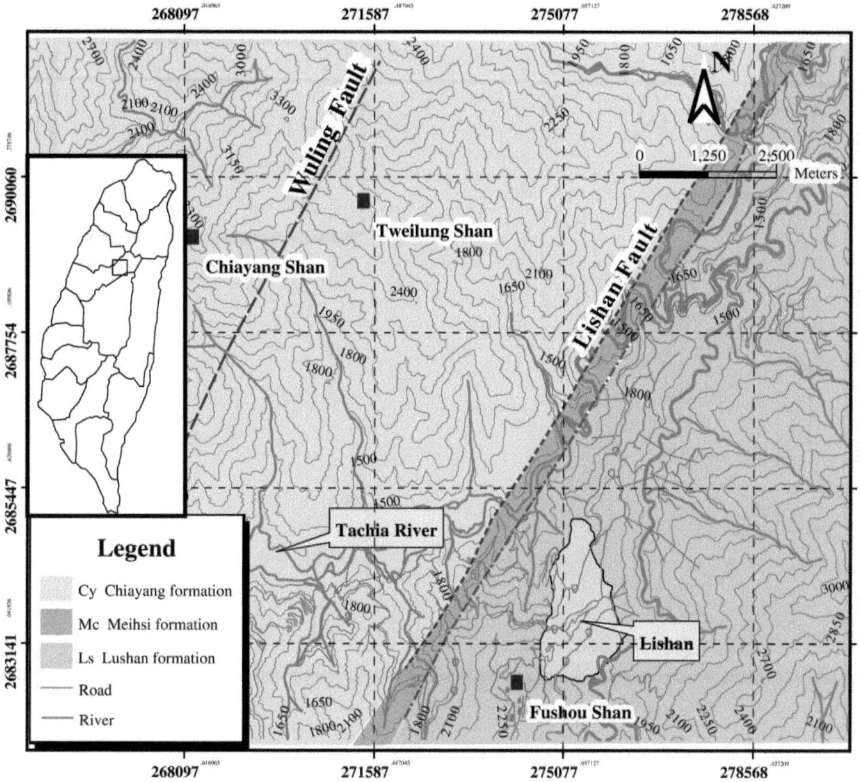

Fig. 8.1 Geologic map showing regional geology near Li-shan area

2002). However, according to the core logs, there is an old sliding body within the southeast region (sliding body B5), and its sliding plane is more than 40–60 m below surface.

The results of back analysis show that those slopes are stable for dry condition as the safety factor is 1.21–1.35; however, they become critical with high ground water level as the safety factor drops to 0.99–1.15 (Shou and Chen 2005). It strongly suggests that there is more than one activity in this area, because the precipitation is quite high in this area.

Beside the major sliding bodies, minor shallow sliding bodies might be exist as the geomaterial is highly weathered and the precipitation is very high in this area. Therefore, shallow sliding bodies without clear boundaries could appear and cause problems more frequently than the major ones.

Remediation Work

As heavy rain and poor drainage are the major factors triggering the Li-shan landslide, ground water control is essential for slope stabilization in this area.

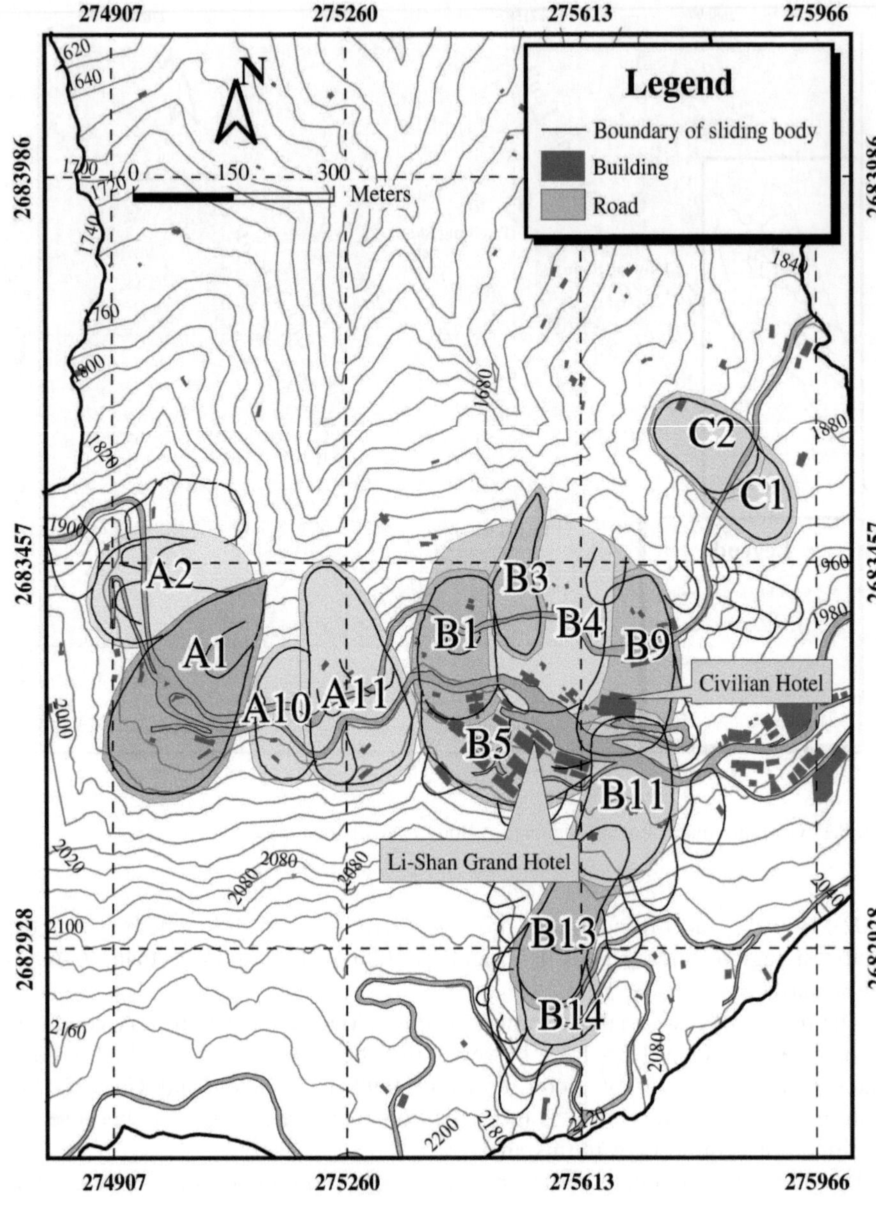

Fig. 8.2 The 14 major sliding bodies for detailed analyses

A drainage system, comprised of surface and subsurface subsystems, was designed as a remediation treatment (Energy and Resources Laboratories, ITRI 1993).

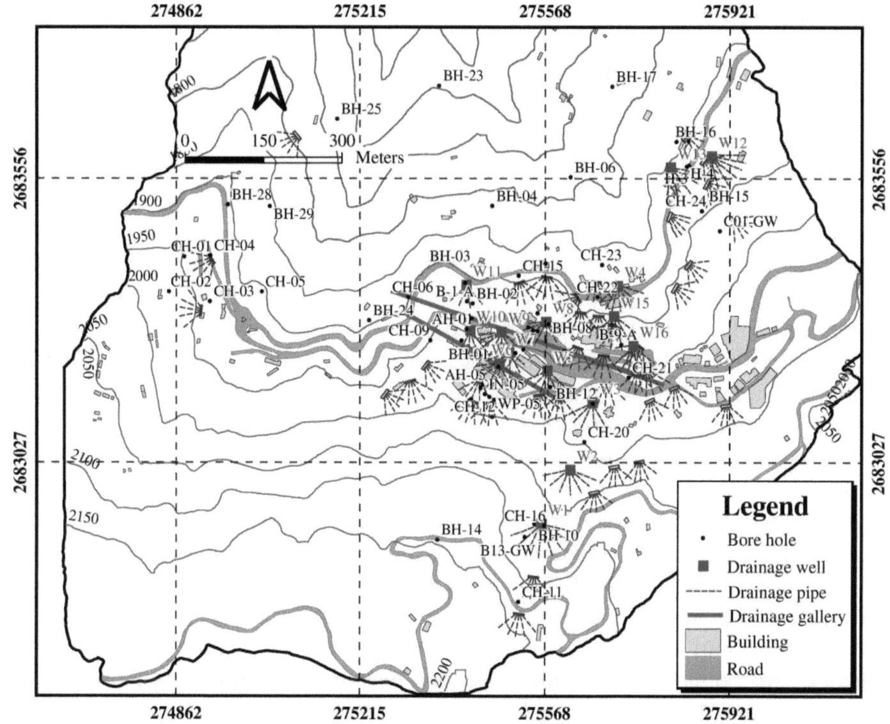

Fig. 8.3 The drainage system designed for remediation treatment

For the surface drainage subsystem, existing ditches were integrated as a system to divert the surface water to non-problem area, and to prevent excessive water infiltration near tension cracks. To more efficiently control the ground water level, a subsurface drainage sub system is also applied. It comprises three major ingredients: (1) 15 horizontal drainage sites, 7–9 pipes (30–60 m in length) in each site, (2) 13 drainage wells, located mainly in the heads of slopes, and (3) 2 drainage galleries, excavated below the sliding planes (see Fig. 8.3).

From the results of ground water level monitoring, the ground water level has been successfully reduced about 10–20 m after the drainage galleries in operation. By this improvement, the stability of slopes is reasonably increased as expected (Shou and Chen 2005). Besides, this landslide area survived during the 1999 Chi–Chi earthquake (ML = 7.3) as well as the typhoons for the past few years. During the Chi–Chi earthquake, with horizontal acceleration around 0.15–0.20 g, the ground water level rose no more than 1 meter. It fairly reveals the effectiveness of the remediation treatment.

8.2 Methodology

8.2.1 Basis and Improvements

The new approaches in this study are based on a GIS-based probabilistic analysis model developed in a preliminary study (Shou and Chen 2005). In this probabilistic analysis approach, mechanical properties of geomaterial are considered as random variables instead of single values, and Monte Carlo sampling on those mechanical properties is coupled with limit equilibrium analysis. Within the GIS, a Kriging process (Cressie 1988) is applied to determine the three-dimensional elevation model as well as the slope profiles for stability analysis. Through stability analysis with Monte Carlo sampling, a distribution of safety factor can be obtained to determine the probability of failure for a sliding body. However, this model is for hazard analysis of sliding bodies in a landslide area only.

To modify this prototype model, changes were made including more failure mechanism options, various methods to obtain three-dimensional elevation model, etc. Two new GIS-based probabilistic analysis approaches were implemented that can evaluate the hazard of landslides. One is for a sliding body with well-defined sliding plane and the other is for a sliding area with assumed sliding plane. Besides, two boundary programs were also developed to more efficiently handle the calculations. Those GIS based approaches were applied to evaluate the hazard of Li-shan landslide with comparisons.

8.2.2 Three-Dimensional Elevation Model

For the stability analyses, it is necessary to have the three-dimensional elevation model, i.e., the surface topography, underground geomaterial distribution and ground water level. And it is obtained by the Kriging estimation on the drilling and monitoring data from 61 holes in the landslide area. The geomaterial is simply characterized to three types, i.e., colluvium, regolith and bed rock.

The ground water level is measured in the bore holes as well as the drainage wells. As it is strongly influenced by the remediation construction, four benchmark points are adopted: before drainage well construction (1997/2), after drainage well construction (1999/3), after typhoon Toraji (2000/9) and after drainage gallery construction (2002/5). And the specific monthly average ground water level was taken to represent the ground water level of the specific benchmark point.

Based on the above mentioned data for geomaterial distribution and ground water level, the GIS software ArcGIS (ESRI 2005) is applied to obtain the three-dimensional elevation model (see Fig. 8.4). As one of the central applications in

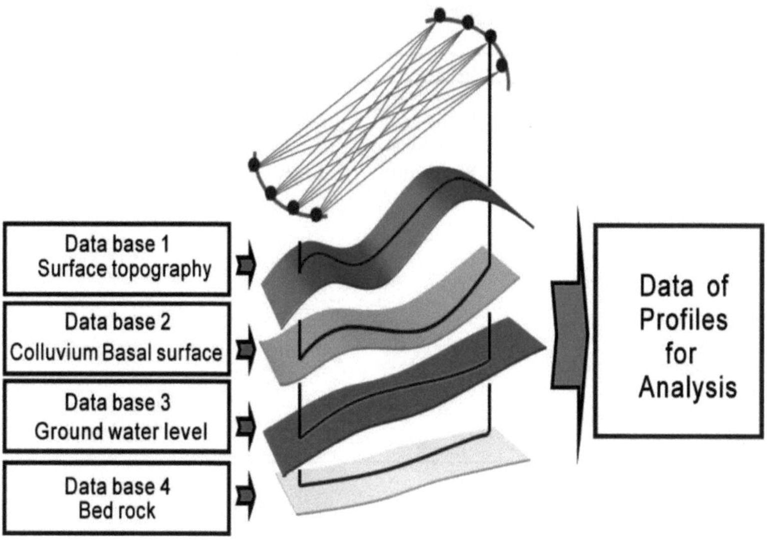

Fig. 8.4 Slope profiles are obtained from the databases in GIS

ArcGIS, ArcMap is good for all map-based tasks including cartography, map analysis, and editing. And the spatial analysis function of ArcMap is applied to build up the surfaces based on bore hole data.

In this study, nine Kriging options were tested to determine the best one. Three different spatial analysis methods, i.e., simple Kriging, ordinary Kriging, and general Kriging, are applied with spherical, Guassian, and exponential semi-variogram evaluation. To obtain an unbiased estimation, the cross validation, in which the minimum root mean square error (RMSE) of the estimated and actual values can be obtained to determine the best searching method and searching radius. RMSE can be defined as

$$\mathrm{RMSE} = \sqrt{\sum_{i=1}^{n} \frac{(Z(x_i) - y_i)^2}{n}} \tag{8.1}$$

in which n is number of sample, Z(xi) is the estimated value, yi is the actual value.

By this unbiased estimation method, the surfaces of lower boundary of colluvium, lower boundary of regolith, and ground water level can be properly estimated as shown in Tables 8.1, 8.2 and 8.3. As underground geology and ground water conditions are different in three regions, i.e., west, southeast, and northeast regions, Kriging estimation was performed separately for those regions.

Table 8.1 The best estimation mode for lower boundary of colluvium

Mode	Area							
OKG	⊛	(neighbors, at least)	(5, 2)					
		Searching Radius (m)	300	<u>350</u>	400	500	600	700
		RMSE (m)	12.59	12.3	12.61	13.14	13.07	13.06

1.OKG: Ordanary Kriging with Gaussian semi-variogram evaluation
2. ⊛: searching in eight areas
3. Neighbors:number of related data considered
4. At least: minimum number of data considered
5. The best searching radius is highlighted

Table 8.2 The best estimation mode for lower boundary of regolith

Mode	Aree							
OKS	⊗	(neighbors, at least)	(4, 2)					
		Searching Radius (m)	500	600	700	800	<u>900</u>	1000
		RMSE (m)	30.38	30.34	30.25	30.24	30.19	30.2

1.OKS: Ordanary Kriging with Spherical semi-variogram evaluation
2. ⊗: searching in four areas
3. Neighbors: number of related data considered
4. At least: minimum number of data considered
5. The best searching radius is highlighted

8.2.3 Stability Analysis

According to the core log records, the failure mechanisms of sliding bodies can be characterized to two types, i.e., deep sited and shallow sited (see Fig. 8.5). For the deep sited sliding, the analysis is focused on the major sliding bodies of which boundary is well-defined and sliding plane is located at the boundary of regolith and bedrock. For shallow sited slidings, the sliding plane is considered to be at the

Table 8.3 The best estimation mode for ground water level in Feb., 1997

Mode	Area							
OKG		(neighbors, at least)	(5, 2)					
	○	Searching Radius(m)	200	300	400	500	<u>600</u>	700
		RMSE(m)	24.87	25.24	24.53	23.71	23.67	23.67

1. OKG: Ordanary Kriging with Gaussian semi-variogram evaluation
2. ○: searching in one area
3. Neighbors: number of related data considered
4. At least: minimum number of data considered
5. The best searching radius is highlighted

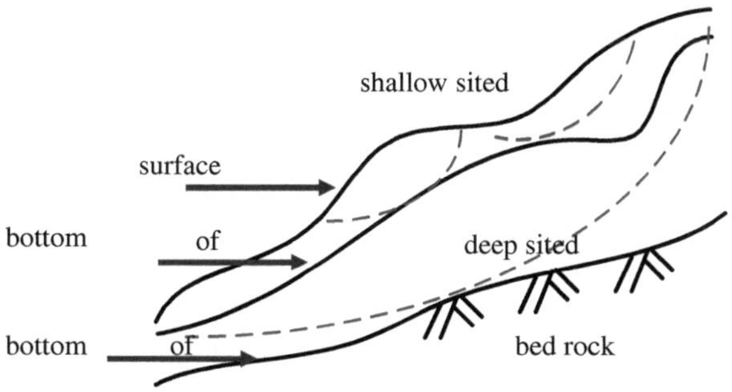

Fig. 8.5 The failure modes of landslides in Li-shan

lower boundary of colluvium and. stability analysis is executed to get the safety factors of the grids in whole landslide area.

To assess the risk of deep sited sliding bodies, the limit equilibrium analysis program PC-STABL (Bandini and Salgado 1999) is applied. The profiles for analysis are determined by GIS program ArcMap. For each sliding body, a segment of upper boundary and a segment of lower boundary are adopted; then, equally spaced 10 points can be found on those two segments. 100 testing profiles can be determined by connecting those boundary points and data from the three-dimensional elevation model got by GIS. And the 10 most dangerous profiles are used for further risk analysis.

For shallow sited slidings, the infinite slope concept (Hammond et al. 1992) is adopted, in which the safety factor can be formulated as the ratio of stabilizing and

Table 8.4 The mechanical properties of geomaterials

		Cohesion (kN/m^2)	Friction angle (°)
Colluvium	Range	0 ~ 22.0	20 ~ 30
	Mean	8.0	32.68
	Standard deviation	8.0	5.05
Regolith	Range	3.0 ~ 38.0	15.67 ~ 38
	Mean	16.9	26.81
	Standard deviation	10.2	6.35

destabilizing forces. The stability analysis is performed for each of the predefined grids comprising the landslide area. The size of 10 meter by 10 meter is chosen for the grids, and safety factor is calculated for each grid. However, the three-dimensional elevation model for each grid is also obtained by GIS.

8.2.4 Probabilistic Risk Evaluation

To consider the uncertainty of mechanical properties of geomaterial, input parameters are considered as random variables to be determined by Monte Carlo sampling process. In this study, according to data from literature review and laboratory tests, the probability density functions of cohesion and friction angle are considered as normal distributions. The mean value and standard deviation of mechanical parameters of colluvium and regolith are shown in Table 8.4. Besides, based on literature review and monitoring data (NCHU 2000; Su and Chen 2002), the ground water level is considered as exponential distribution with variance set to be 1.9 meters.

Considering the uncertainty of those mechanical parameters, Monte Carlo process randomly samples those parameters from their probability distributions and a safety factor can be calculated based on the chosen data set. By repeating the above sampling and safety factor calculation, for 1000 times in this study, we can obtain a series of safety factors. The probability of failure Pf can also be defined as

$$P_f = P(F_s \leq 1.0) \qquad (8.2)$$

where P(Fs $\leq$ 1.0) denotes the probability of safety factor less than one. And P(Fs $\leq$ 1.0) can be determined by the ratio of the area under the distribution curve for safety factor less than one divided by the total area under the distribution curve.

The hazard evaluation problem is now deduced to determine the distribution of factor of safety, which depends upon a number of random variables. Although it is not always true in nature, input parameters are considered as independent variables for simplicity and based on sensitivity analysis (Chen 2003). For slope engineering, probability of failure is generally considered as a simple index for risk evaluation. Although, the cost of failure should be implemented with the probability of failure to obtain the risk, which is in theory defined as probability of loss.

To make the calculations more efficient, two interface programs, DeepLandslide for deep sited landslide and ShallowLandslide for shallow sited landslide, were written in Visual Basic 6.0. And the random sampling function of Monte Carlo analysis is also embedded in those programs.

8.3 Results and Discussion

For the analysis of deep sited slidings, 14 major sliding bodies, i.e., sliding bodies A1, A2, A10 and A11 for the west region, sliding bodies B1, B3, B4, B5, B9, B11, B13, and B14 for the southeast region, and sliding bodies C1 and C2 for the northeast region, are adopted (see Fig. 8.2). The hazards of those sliding bodies are evaluated for different phases. Then the hazard of a region can be considered as the mean hazard of the major sliding bodies inside.

Considering the groundwater surface before and after the construction of drainage wells, it shows the hazard changes from 59.42% to 99.97% for the west region, from 34.26% to 27.94% for the southeast region, and from 15.77% to 14.93% for the northeast region (see Fig. 8.6). The unreasonable results for the west region show the inaccuracy caused by poor-quality and insufficient drilling data in this region.

Another evidence for the performance of remediation could be the redistribution of hazard in landslide area. The hazards of sliding bodies for different phases are ranked and illustrated in Table 8.5 and Figs. 8.7 and 8.8. In the pre-remediation phase, the critical sliding bodies with smaller ranking numbers concentrate in the center area (Fig. 8.7). Because of the remediation components are mainly in the center area of landslide area, hazard dissipates from center area in the post-remediation phase showed by larger ranking numbers of sliding bodies (Fig. 8.8). The hazard of sliding bodies can also be regionally ranked to suggest priority for further treatment.

For the shallow sited slidings in the post-remediation phase, the distribution of safety factor for cohesion and friction angle set to be 16.9 kN/m^2 and 26.81° is shown in Fig. 8.9. And the distribution of probability of failure is shown in Fig. 8.10. It shows the similar trend as the analysis of deep sited slidings, i.e., hazard dissipates from center area in the post-remediation phase (Fig. 8.11).

The results of risk analysis of shallow sited slidings are summarized in Table 8.6. It shows that there is about 25.56% of the landslide area with safety factor less than 1.0 and about 41.42% of the landslide area with safety factor less than 1.2. For the probability of failure, there is about 51.97% of the landslide area with probability of failure greater than 10%.

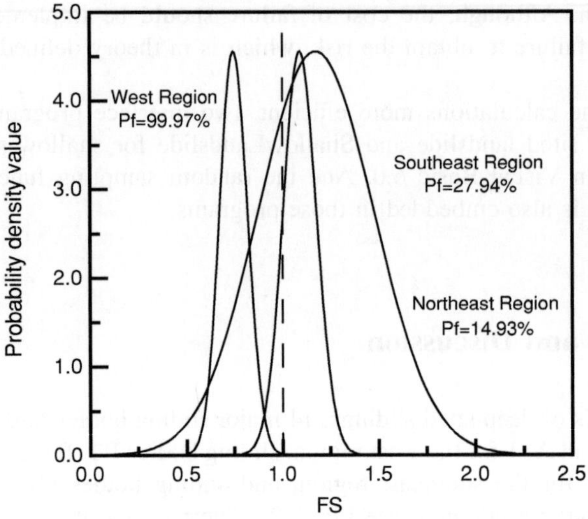

Fig. 8.6 Regional probability distributions of safety factor in post-remediation phase

Table 8.5 Regional rank of hazard for sliding bodies before and after remediation

Regional Rank	Pre-remediation		Post-remediation	
	Sliding body	Mean value of Safety factor	Sliding body	Mean value of Safety factor
1	A11	0.979	A1	0.824
2	A10	1.022	A2	0.862
3	A1	1.203	A10	0.898
4	A2	1.227	A11	1.073
1	B3	0.797	B3	0.837
2	B4	0.912	B4	1.020
3	B5	1.054	B5	1.143
4	B9	1.070	B1	1.273
5	B1	1.163	B9	1.419
6	B13	1.260	B13	1.464
7	B11	1.941	B11	1.949
8	B14	2.018	B14	1.974
1	C2	1.093	C1	1.140
2	C1	1.113	C2	1.159

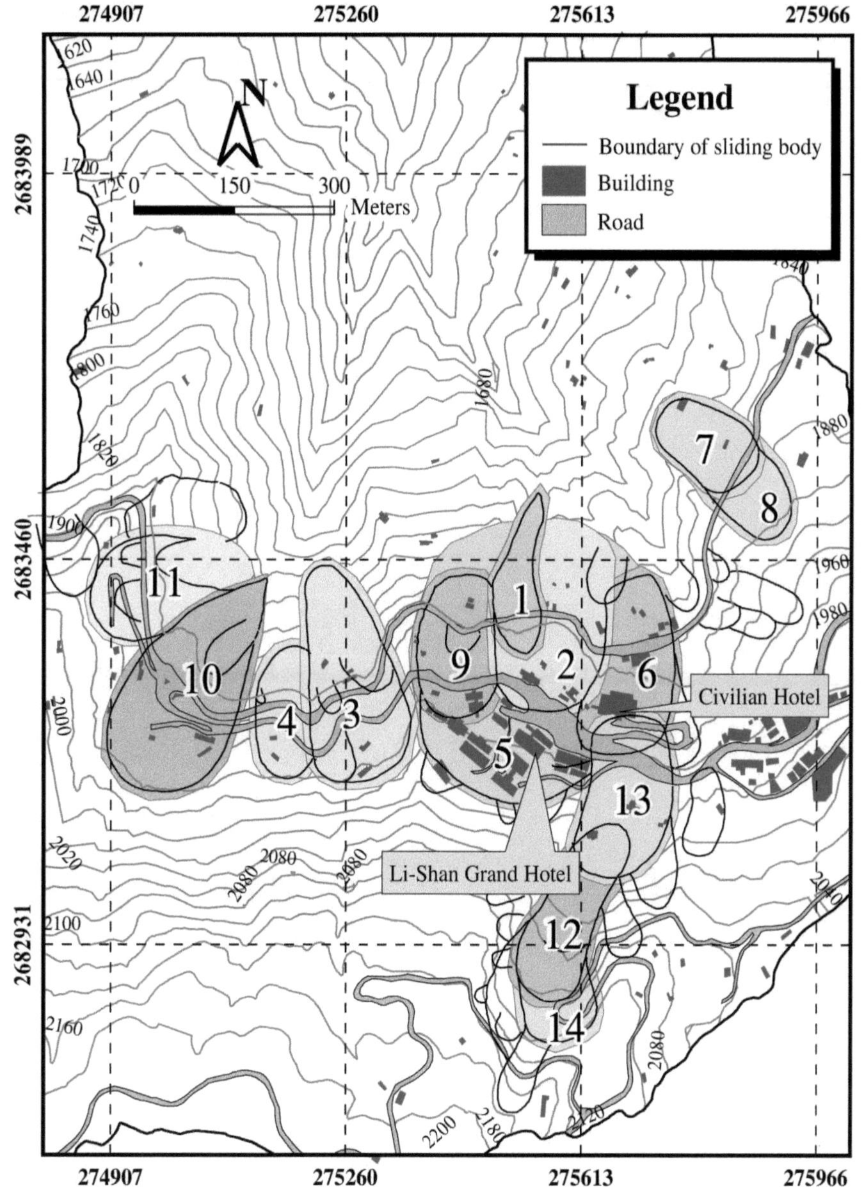

Fig. 8.7 Rank of risk for major sliding bodies in pre-remediation phase. The top 6 critical sliding bodies are concentrated in center area

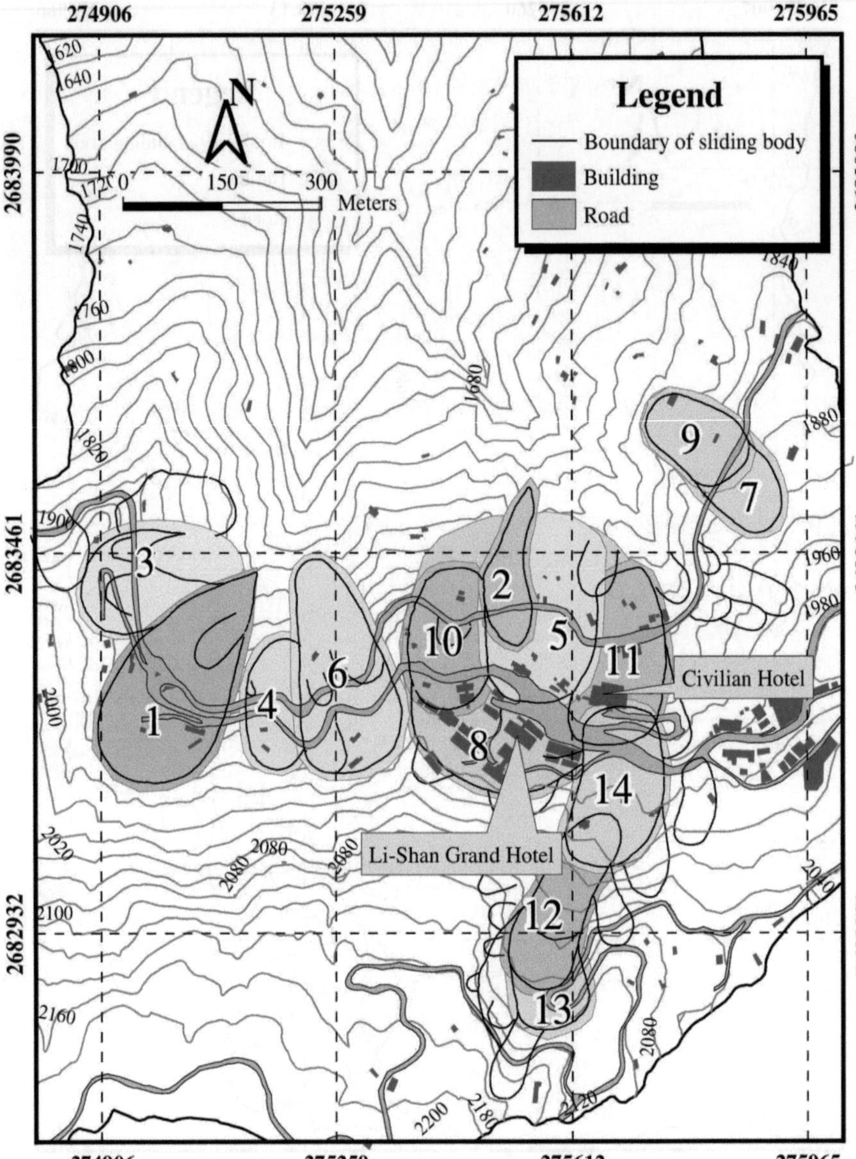

Fig. 8.8 Rank of risk for major sliding bodies in post-remediation phase. Risk has been dissipated from center area

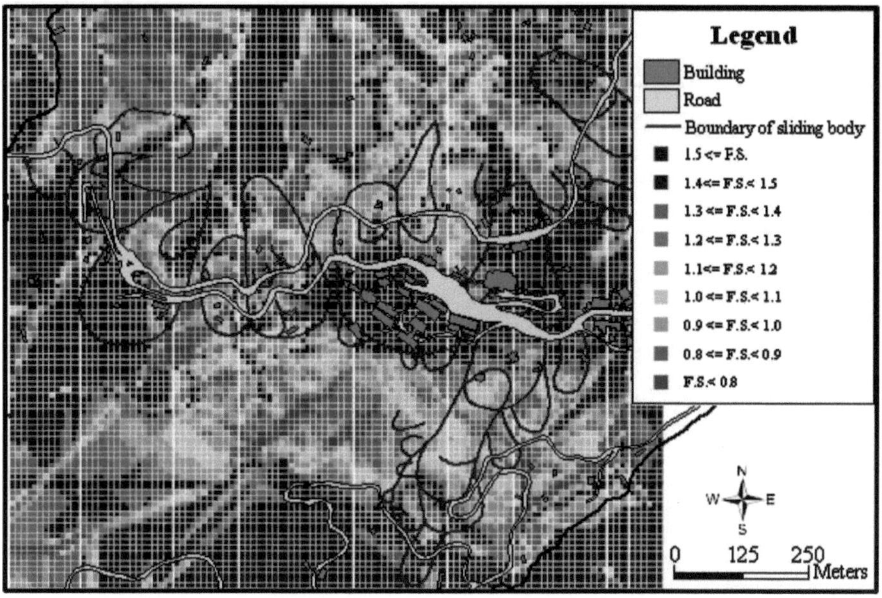

Fig. 8.9 The distribution of safety factor for shallow slidings (with 10 m*10 m grids, c = 16.9 kN/m^2; $\phi = 26.81^o$)

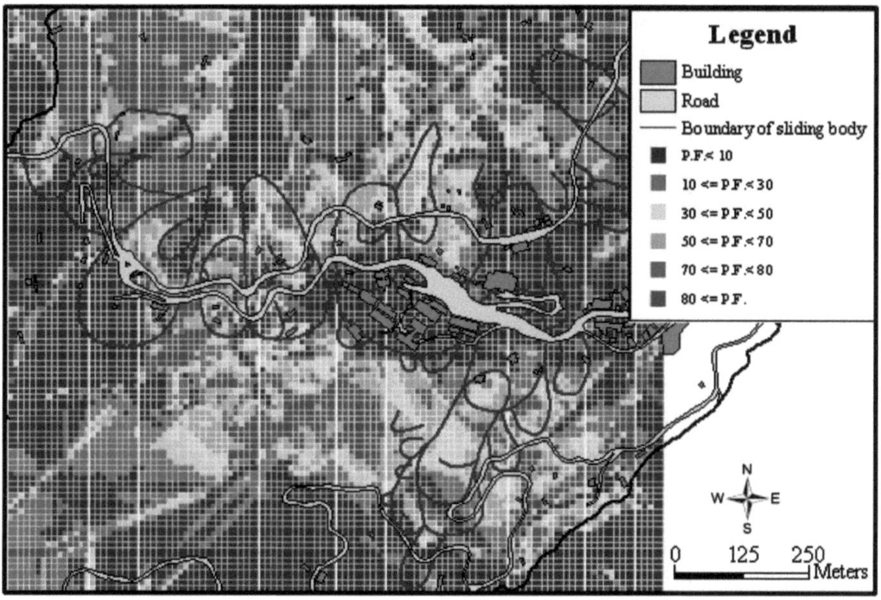

Fig. 8.10 The distribution of probability of failure for shallow slidings (with 10 m*10 m grids)

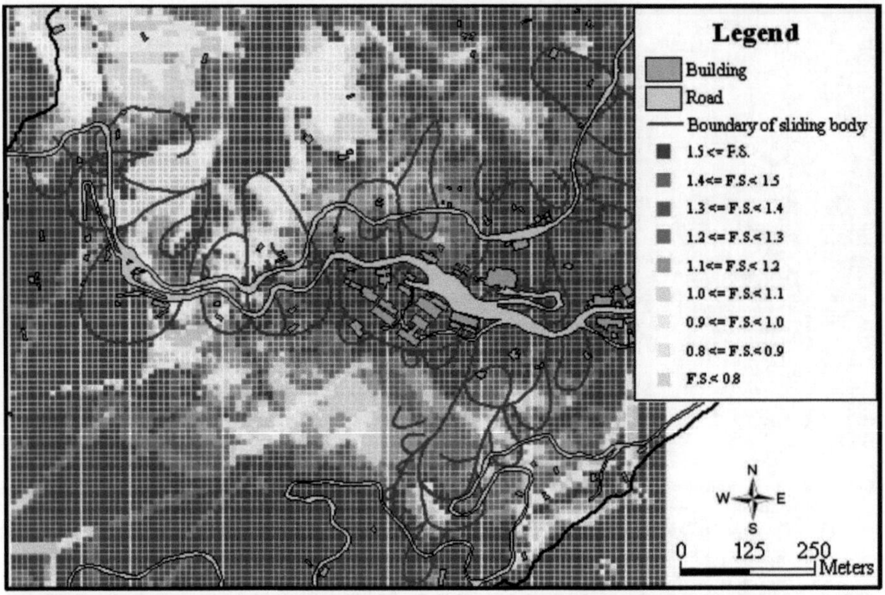

Fig. 8.11 The distribution of safety factor for shallow slidings (with 10 m*10 m grids, c = 10.0 kN/m^2; = 33.0^o)

Table 8.6 Risk analysis results of shallow sited slidings (c = 16.9 kN/m^2; ϕ = 26.81^o)

Distribution of Safety Factor (F.S.)					
F.S.	1.5< F.S	1.4≦F.S.<1.5	1.3≦F.S.<1.4	1.2≦F.S.<1.3	1.1≦F.S.<1.2
% in area	39.02	6.36	6.31	6.90	7.60
accumulative %	100.00	60.98	54.62	48.31	41.42
F.S.	1.0≦F.S.<1.1	0.9≦F.S.<1.0	0.8≦F.S.<0.9	0.7≦F.S.<0.8	F.S.≦0.7
% in area	8.25	9.00	8.83	4.23	3.50
accumulative %	33.82	25.56	16.57	7.74	3.50
Distribution of Probability of Failure (P.F.)					
P.F. (%)	P.F.< 5	5≦P.F.<10	10≦P.F.<20	20≦P.F.<30	30≦P.F.<40
% in area	40.49	7.54	9.78	7.11	4.59
Accumulative (%)	100.00	59.51	51.97	42.19	35.08
P.F.(%)	40≦P.F.<50	50≦P.F.<60	60≦P.F.<70	70≦P.F.<80	80≦P.F.
% in area	5.57	4.99	4.99	5.19	9.75
Accumulative (%)	30.49	24.92	19.93	14.94	9.75

8.4 Conclusions

In this study, coupling GIS, limiting equilibrium analysis, Monte Carlo analysis, and interface programs written in Visual Basic, probabilistic hazard analysis methods were implemented to analyze the hazard of deep sited and shallow sited landslides. And those methods were applied for the analysis of Li-shan landslide.

Through a series of analyses, the hazards of deep sited sliding bodies as well as shallow sited slidings in Li-shan landslide are estimated. It shows the influence of groundwater level is significant, which reveals the importance of ground water control and monitoring. The results also show the stability of center area is significantly improved as the ground water levels are lowered by drainage treatment. Comparatively, it is more dangerous for the west area.

As drillings in the west and northeast areas are not sufficient, the three-dimensional elevation model cannot be accurately obtained by the GIS. More drillings in those areas are necessary in the future. Besides, more investigation and analysis are necessary to make clear the history of sliding activities in this landslide area.

Acknowledgments This study was made possible through the support of the National Science Council and the Soil and Water Conservation Bureau, Taiwan, R.O.C. The authors are also indebted to our colleague Prof. M. Su and Mr. Y. Chen for their very kind support.

References

Bandini P, Salgado R (1999) STABLE 6 for DOS 6.0 User's Manual. Purdue University, US

China Engineering Consultants Inc (1996) Basic design and subsidiary investigation for Li-shan landslide, final report to the Soil and Water Conservation Bureau, Taiwan

Chen YL (2003) Study on risk analysis of slopes considering spatial variability-a case study of Li-shan. Master thesis, NCHU, Taichung, Taiwan

Cressie N (1988) Spatial prediction and ordinary kriging. Math Geol 20(4):405–421

Energy and Resources Laboratories, ITRI (1993) Planning on investigation and remediation for li-shan landslide, final report to the Soil and Water Conservation Bureau. Taiwan

ESRI (2005) ArcGIS 9.0—the complete geographic information system

Guzzetti F, Carrara S, Cardinali M, Reichenbach P (1999) Landslides hazard evaluation: a review of current techniques and their application in multi-scale study, Central Italy. Geomorphology 31:181–216

Hammond C, Hall D, Miller S, Swetik P (1992) Level I Stability Analysis (LISA). Documentation for version 2.0: USDA, Intermountain Research Station

Haneberg WC (2000) Deterministic and probabilistic approaches to geologic hazard assessment. Environ Eng Geosci 6(3):209–226

Haneberg WC (2004) A rational probabilistic method for spatially distributed landslide hazard assessment. Environ Eng Geosci 10(1):27–43

Haneberg WC (2006) Effects of digital elevation model errors on spatially distributed seismic slope stability calculations: an example from seattle, Washington. Environ Eng Geosci 12(3):247–260

Huang CS (2002) Personal discussion on the incomplete geological map of li-shan, 1:50000. Central Geological Survey, MOEA, Taiwan

Hungr O et al (1989) Evaluation of a three-dimensional method of slope stability analysis. Can Geotech J 26:679–686

Jibson RW, Harp EL, Michael JA (1999) A modeling procedure to produce probabilistic seismic landslide hazard maps, abstracts with programs. Geol Soc Am, 1999 Annu Meet 31(7):195

Khazai B, Sistar N (2000) Assessment of seismic slope stability using GIS modeling. Geogr Inf Sci 5(2):121–128

Miller DM (1995) Coupling GIS with physical models to aassess deep-seated landslide hazards. Environ Eng Geosci 1(4):263–276

NCHU (2000) Establishment and application of databases for li-shan landslide, monitoring data section. Report to the Soil and Water Conservation Bureau, Taiwan

Pack RT, Tarboton DG, Goodwin CN (1998) The SINMAP approach to terrain stability mapping. 8th congress of the IAEG, Vancouver, British Columbia

Qiu C, Esaki T, Xie M, Mitani Y, Wang C (2005) A GIS system development for evaluating 3D slope stability. Proceedings of the 25th annual ESRI user conference

Reid ME, Christian SB, Brien DL (2000) Gravitational stability of three-dimensional startovolcano edifices. J Geophys Res 105:6043–6056

Shou KJ (2002) Geology in li-shan area. Workshop on performance evaluation of li-shan landslide remediation, Taichung, Taiwan, pp 121–142 (in Chinese)

Shou KJ, Su MB (2002) On the failure mechanism and remediation of the li-shan landslide in Taiwan. International conference mountain environment and development, Chengdu

Su MB, Chen DT (2002) Monitoring system and databases. Workshop on performance evaluation of li-shan landslide remediation, Taichung, Taiwan, pp 121–142 (in Chinese)

Shou KJ, Chen YL (2005) Spatial risk analysis of li-shan landslide in Taiwan. Eng Geol 80:199–213

van Westen CJ (2004) Geo-information tools for landslide risk assessment: an overview of recent developments. In: Lacerda WA, Ehrlich M, Fontoura AB, Sayo A (eds) Landslides evaluation and stabilization. A.A. Balkema, Rotterdam, pp 39–56

Zhou G, Esaki T, Mitani Y, Xie M, Mori J (2003) Spatial probabilistic modeling of slope failure using an integrated GIS Monte Carlo simulation approach. Eng Geol 68:373–386

Chapter 9
Landslide Inventory, Hazard and Risk Assessment in India

Cees J. van Westen, Pankaj Jaiswal, Saibal Ghosh, Tapas R. Martha and Sekhar L. Kuriakose

Abstract The recent census in India revealed that India is now housing 17% of the world's population, and India is on the way to become the most populated country. Landslides are an increasing concern in India due to the rapid population expansion in hilly and mountainous terrain. Landslides affect vast areas within India, in particular in the Himalayan chain in the North and Eastern part of the country and the Western Ghats in the Southwest. The Geological Survey of India (GSI) has been designated as the nodal agency for landslides by the Indian government, and they are responsible for landslide inventory, susceptibility and hazard assessment. Until recently their landslide susceptibility assessment was based on a heuristic approach using fixed weights or ranking of geofactors, based on guidelines of the Bureau of Indian Standards (BIS). However, this method is disputed as it doesn't provide accurate results. This paper gives an overview of recent research on how the existing methods for landslide inventory, susceptibility and hazard assessment in India could be improved, and how these could be used in (semi)quantitative risk assessment. Due to the unavailability of airphotos in large parts of India, satellite remote sensing data has become the standard data input for

C. J. van Westen (✉)
Faculty of Geo-Information Science and Earth Observation (ITC), University of Twente, Enschede, Netherlands
e-mail: westen@itc.nl

P. Jaiswal · S. Ghosh
Geological Survey of India, Kolkata, India

T. R. Martha
National Remote Sensing Center, Indian Space Research Organization, Hyderabad, India

S. L. Kuriakose
HVRA Cell, Department of Revenue and Disaster Management, Government of Kerala, Trivandrum, India
e-mail: sekhar.lk@gmail.com

B. Pradhan and M. Buchroithner (eds.), *Terrigenous Mass Movements*,
DOI: 10.1007/978-3-642-25495-6_9, © Springer-Verlag Berlin Heidelberg 2012

landslide inventory mapping. The National Remote Sensing Center (NRSC) has developed an approach using semi-automatic image analysis algorithms that combine spectral, shape, texture, morphometric and contextual information derived from high resolution satellite data and DTMs for the preparation of new as well as historical landslide inventories. Also the use of existing information in the form of maintenance records, and other information to generate event-based landslide inventories is presented. Event-based landslide inventories are used to estimate the relation between temporal probability, landslide density and landslide size distribution. Landslide susceptibility methods can be subdivided in heuristic, statistical and deterministic methods. Examples are given on the use of these methods for different scales of analysis. For medium scales a method is presented to analyze the spatial association between landslides and causal factors, including those related to structural geology, to select the most appropriate spatial factors for different landslide types, and combine them using the multivariate methods. For transportation corridors a method is presented for quantitative hazard and risk assessment based on a landslide database. Deterministic methods using several dynamic slope-hydrology and slope stability models have been applied to evaluate the relation between land use changes and slope stability in a steep watershed. The paper ends with an overview how the susceptibility maps can be combined with the landslide databases to convert them into hazard maps which are subsequently used in (semi) quantitative risk assessment at different scales of analysis, and how the results can be used in risk reduction planning.

Keywords Landslide inventory · India · Historical landslide data · Object oriented image classification · Event-based landslide maps · Statistical analysis · Physical modeling · Hazard assessment · Risk assessment

9.1 Introduction

Landslides are an increasing concern in India due to the ongoing expansion of the population into hilly terrain (Kuriakose et al. 2009a). Vast areas within India, in particular the entire Himalayan chain in the North and Eastern part of the country and the Western Ghats in the Southwest are affected by landslides.

For instance, the densely populated state of Kerala in India is prone to debris flows due to its geomorphic setting as 40% of it lies in the most prominent orographic feature of peninsular India, The Western Ghats. The west facing scarps that runs the entire extent of the Western Ghats mountain chain in Kerala experience several types of landslides, of which shallow landslides and consequent debris flows are the most common (Kuriakose et al. 2009b). Many of the roads and railroads that run through the Himalayas and the Western Ghats (Jaiswal and van Westen 2009; Sreekumar 2009) are highly prone to landslides. Over the period of 1975–1995, it is estimated that landslides killed about 100 people and rendered about 600 families homeless along the Western Ghats (Thakur 1996) whilst floods and landslides together have caused an estimated damage of 12 billion Euros in Kerala in 2007 alone (Murali Kumar 2007).

The Geological Survey of India (GSI) has been designated as the nodal agency for landslide studies by the Government of India. The GSI was established in 1851 though it was only in 1945 that the engineering geology and ground water division of GSI was created which had landslide investigation as one of its main responsibilities. Since its inception GSI has investigated many disastrous landslides, such as the Nainital landslide in 1880, the blockade of Birahi Ganga by a massive rockfall at Gohna in Garhwal in 1893, the Malpa landslide in 1998 and the Uttarkashi landslide in the 1990s', just to name a few.

Though there are several site specific studies, spatial landslide hazard and risk assessment in India has always been hampered by the lack of data. Along the international border regions of the country including the entire Himalayan range, the use of topographical maps and aerial photos has been restricted for national security reasons. Thus, for long the generation of landslide inventory maps faced significant lacuna. In lieu of aerial photos for large parts of the country, satellite remote sensing data has become the standard input for landslide mapping. The National Remote Sensing Center (NRSC) under the Indian Space Research Organisation (ISRO) has been very active in the use of satellite data for landslide inventory mapping and susceptibility assessment. High resolution stereo imagery from satellites such as Cartosat 1 has proven to be very useful for landslide studies (Vinod Kumar et al. 2006).

In 1998, the Bureau of Indian Standards (BIS) formulated guidelines for landslide susceptibility zonation on macro scale (1:50,000) for the whole country (BIS 1998). These guidelines propose an indirect approach to landslide susceptibility mapping based on a generalized heuristic system of fixed weighting or ranking of geofactors without directly or indirectly considering the landslide inventory data (Anbalagan 1992; Anbalagan and Singh 1996; Sarkar et al. 2008). Its' direct applicability in all the landslide prone regions of the country is a matter of dispute amongst landslide researchers owing to the wide variability of the geo-climatic conditions that prevails across the country.

Due to the difficulty in obtaining base maps and multi-temporal landslide inventory maps, the development of new approaches for landslide hazard and risk assessment, including the use of spatially distributed physically-based landslide initiation models, run-out models and statistical models, have always been impeded in India. In this paper we present the results of the application of some of these new techniques for landslide inventory, susceptibility mapping and, hazard and risk assessment as applied to four test sites in different regions of India.

9.2 Study Areas

Figure 9.1 shows the locations of the four test sites. Two of these are located in the Western Ghats and two in the Himalayas. The test site in Kerala (A in Fig. 9.1) is

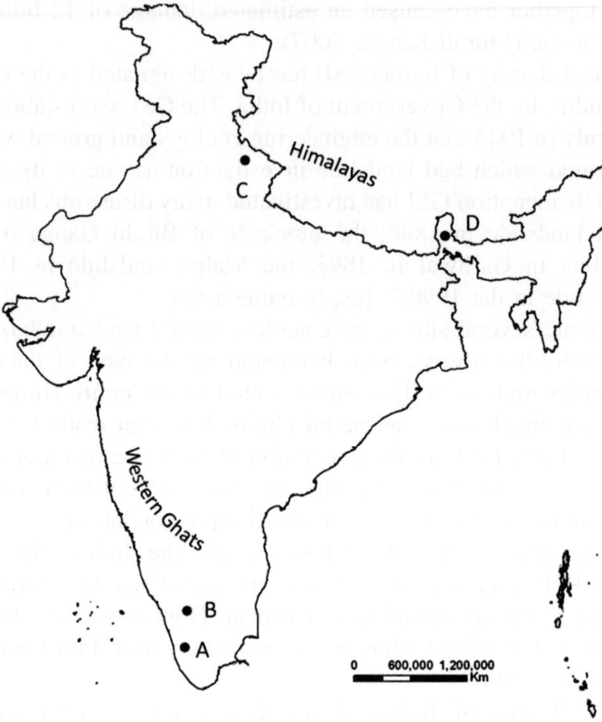

Fig. 9.1 Location of the test sites in India. **a** Upper Meenachil River basin. **b** Nilgiri. **c** Okhimath. **d** Kurseong

the upper Meenachil River basin upstream of Erattupetta Town. The region is administratively part of Kottayam and Idukki districts (Fig. 9.2). The area experiences numerous shallow landslides leading to debris flows almost every year during the monsoon season. Two separate study areas (Fig. 9.2) namely Aruvikkal catchment and Peringalam landslide in the region were selected to test a physically based dynamic shallow landslide model and a run-out model, respectively. As per measurements at Pullikanam Tea Estate's Upper Division, which is thc closest long term rainfall recording station, the region experienced an average annual rainfall of 5315 mm during the period from 1952 to 1999. Underlain by Precambrian charnockites the region is predominantly covered with shallow sandy soils over a thin layer of sparolite interleaved by lithomargic clay (Kuriakose et al. 2009c). Anthropogenic land disturbances in the area started in the late 1880s (Victor 1962). The predominant land use of the region is rubber plantations, covering an area 3.6 km^2. Rubber has an average crop life of 20 years after which the trees have to be felled, thus exposing the land to the high intensity rainfall until a new set of saplings are planted and they achieve significant canopy cover. Both for cassava and rubber planting, slopes are terraced often ignoring ephemeral streams thereby

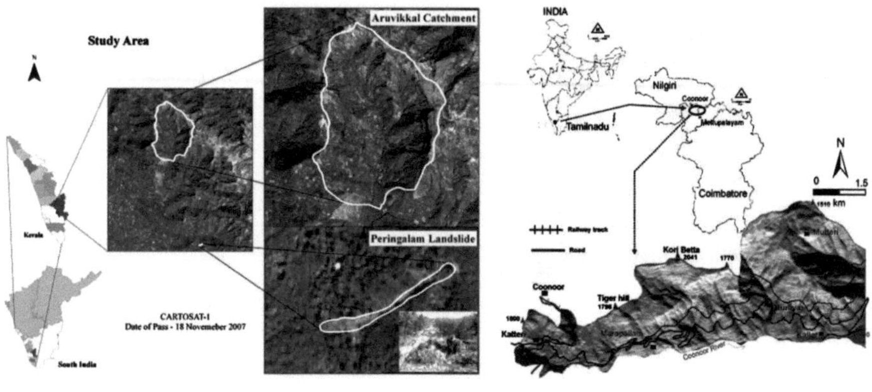

Fig. 9.2 Study areas in the Western Ghats. *Left*: Aruvikkal catchment and Peringalam Landslide. *Right*: Nilgiri area

obstructing natural drainage channels that act as conduits for the discharge of excessive surface flow during high intensity rainfall (Thampi et al. 1998).

The second study area (B in Fig. 9.1) is located in Nilgiri (Tamil Nadu) and includes a 19 km long historic railway track, which is declared as a world heritage site by UNESCO and a 26 km long national highway connecting Mettupalayam and Coonoor in the state of Tamil Nadu in southern India (Fig. 9.2). The railway was constructed in the late nineteenth century and became operational in 1899. Both transportation lines run parallel to each other on the southern slopes of the Nilgiri plateau. The routes are cut through soil and laterite, underlain by charnockite and garnetiferrous quartzo-felspatic gneisses belonging to the Charnockite Group of Archaean age (Seshagiri and Badrinarayanan 1982). The regional strike of the foliation is ranging from ENE–WSW to E–W direction with moderate to steep dips. The sub-tropical climate and intense physical and chemical weathering have resulted in a thick yellowish to reddish brown soil. The regolith thickness varies from less than one meter to 20 m, as observed in the cut slopes along the road and railroad. The study area forms a part of the Nilgiri plateau with steep slopes to the south and gentle slopes to the north and near ridge tops. The area has an elevation difference of 1641 m with lowest areas near Kallar farm (400 m) and highest at Kori Betta ridge (2041 m). Most part of the transportation corridor is either under reserved forest or tea plantation and settlements are very few and sporadic. Landslides are abundant in the area and occur mostly in cut slopes of the transportation routes. These are mostly shallow translational debris slides and flows and are invariably triggered by rainfall (Jaiswal and van Westen 2009).

The third study area (C in Fig. 9.1) is a part of Mandakini river valley in the High Himalayas, located around the town of Okhimath in the Rudraprayag district of Uttarakhand state, India (Fig. 9.3). Okhimath is situated at an average elevation of 1300 m at the confluence of the Mandakini and Madhyamaheshwar rivers. The Mandakini River is a tributary of the Ganges River. The region has a highly variable land cover and terrain with elevation ranging from 718 to 4510 m.

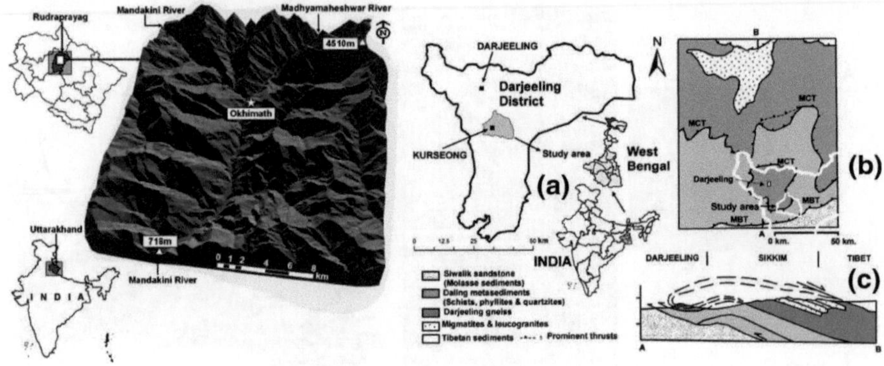

Fig. 9.3 Study areas in the Himalayas. *Left* Okhimath. *Right* Kurseong

The northeastern part of the area is perennially snow covered. This area is dominated by low altitude oak forest. The terrain is dominantly steep and rugged with few flat fluvial terrace areas along the Mandakini River. Since the rugged topography is controlled by the geological structure, there are many fault-related south facing escarpments in this area. The northeastern and western parts of the area are very difficult to access.

The fourth study area (D in Fig. 9.1) is located in Darjeeling district, West Bengal, within the eastern part of the Himalayas (Fig. 9.3). The eastern Himalayas represent a complicated geological and tectonic environment, in which rocks are juxtaposed along certain E–W trending Tertiary regional thrusts. Along the foothills to the south, coarse to very coarse-grained clastics (conglomerate-sandstone–siltstone) of the so called Siwalik Group of Tertiary age are exposed and are bordered by a frontal thrust (Himalayan Foothill Thrust—HFT or Himalayan Frontal Thrust HFT). To the north these are thrusted over by sandstone-shale (±coal) sequence of the Gondwanas (Mesozoic) along the Main Boundary Thrust (MBT). Further to the north, low grade meta-psammo-pelitic lithoassemblages of the Precambrian Daling Group are thrusted over the Younger Gondwana/Siwalik sediments. And further north in the Middle to Higher Himalaya, granite gneisses and high-grade meta-sediments belonging to the Central Crystalline Gneissic Complex (CCGC) are thrusted over the low-grade metamorphics of the Daling Group along the Main Central Thrust (MCT). The overall relief difference in the studied area varies from 250 m to as high as 2650 m. The general trend of the mountain ranges is E–W. A number of NE–SW and NW–SE trending ridges and spurs are carved out of this trend and form high mountain ranges. The average rainfall in Darjeeling Himalaya to the west of the Tista River fluctuates between 2000 mm and 4000 mm. Landslides are perennial problems in Darjeeling Himalayas during monsoon (June to October) resulting in frequent closures of important communication corridors, destruction of limitedly-available agricultural land, house and loss of human life. The earliest recorded landslide event in Darjeeling Himalayas, India dates back on 24th September 1899, which was triggered by a 1065 mm precipitation (in 3 days) and resulted in devastation in Darjeeling town and its surrounding areas with loss of 72 lives. Since then, this part of the Himalayas has experienced such comparable devastating landslide events in at least 10 times till 2003, although the intensity and distribution may differ considerably for individual extreme rainfall events.

9.3 Landslide Inventory Mapping

Landslides are generally isolated natural processes, which individually may not be of very large in size but can occur with a high frequency in an area (van Westen et al. 2006). Landslide inventories can be prepared through various methods (Guzzetti et al. 2000; Hansen 1984; van Westen et al. 2008; Wieczorek 1984) such as historical archive studies, interviews, detailed geomorphologic fieldwork, and mapping from remote sensing data and topographic maps. Each of the methods indicated above has its drawbacks. Due to the lack of sufficient historical information on landslides, stereoscopic interpretation of aerial photographs or satellite images from the past is often used as the main source for obtaining a multi-temporal landslide inventory (Rib and Liang 1978). Event-based inventories are prepared just after a prominent triggering event which depicts all slope failures caused due to that particular triggering event (Carro et al. 2003; Guzzetti et al. 2004; Harp and Jibson 1996).

A landslide inventory contains the location, classification, volume, run-out distance, date of occurrence and other characteristics of landslides in an area (Fell et al. 2008). Techniques used to prepare landslide inventory maps depend on the quality and accessibility of desired information, the scope and the extent of the study area, the scales of base maps and the resources available to carry out the work. In this case study an approach is presented to obtain a complete multi-temporal landslide inventory by interpreting historical records coupled with participatory mapping. There are different techniques for landslide inventory mapping, three of which are illustrated in the following sections: using historical data, image interpretation and automatic classification.

9.3.1 Collecting Historical Landslide Information

In the Nilgiri area (B in Fig. 9.1) it was possible to generate a very detailed landslide inventory based on historical data. The data sources used to obtain landslide information can be grouped into three main categories:

- *Railroad maintenance records* such as the railroad maintenance registers (locally called 'railway slip register') and a summary table of landslides along the railroad. The data were present in an analog (paper) form recorded in a register or table and maintained by the Southern Railway office at Coonoor. The railway

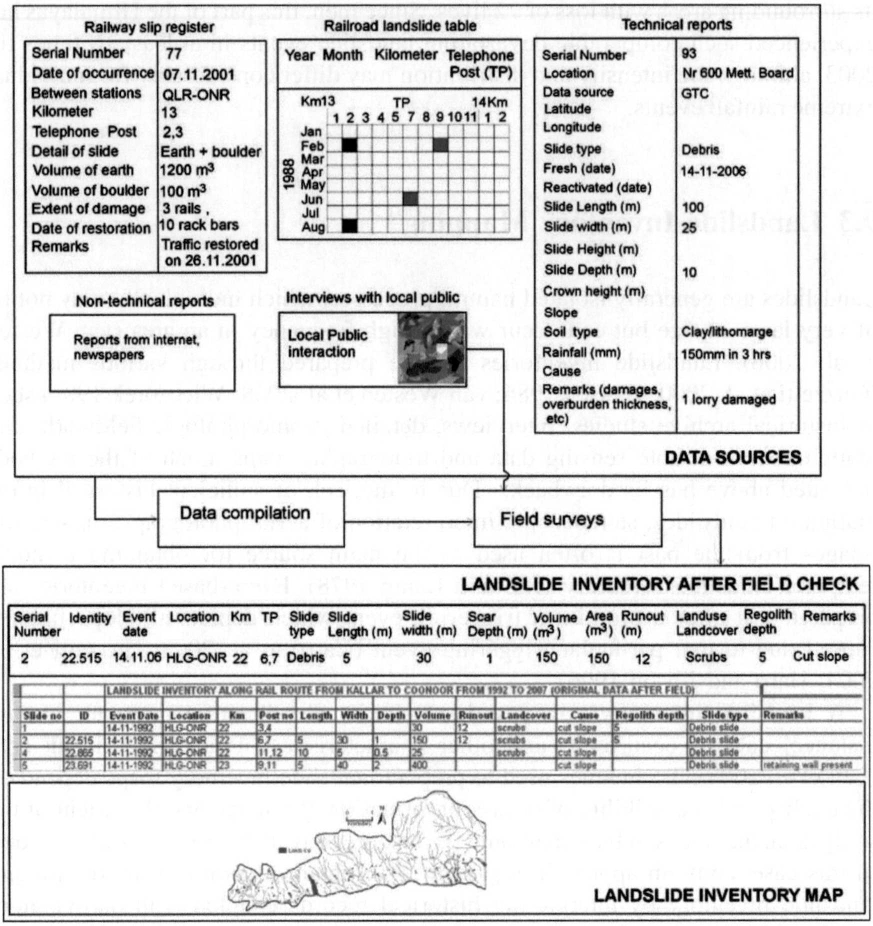

Fig. 9.4 Different types of data sources and methodology used to prepare landslide inventory

slips register is updated soon after the occurrence of a landslide triggering event and is used for tendering contracts for railroad clearance. It contains data on the spatial distribution of landslide debris on the railroad for the period since 1992. The other form of historical records was a summary table of landslides along the railroad, which provided the spatial distribution of debris on the railroad in different months and sectors from 1987 to 1991. Landslides prior to 1987 were also recorded in the form of a landslide table but for the study area older records were not available in the railway office. The data format and an example of the type of data available in the records are shown in Fig. 9.4. The records also provide additional information on damages and the date of restoration of the railroad for traffic.

- *Technical reports* such as published and unpublished technical documents of landslide investigations. Publications on landslides in the Nilgiri area

(e.g., Seshagiri et al. 1982; Ramasamy et al. 2003) provided detailed information including the spatial distribution of the landslides during the major events in 1978 and 1979. In the unpublished technical reports the oldest information on landslides dates back to 1824. Most of the reports contain information on detailed geotechnical investigation of the major landslides that have occurred in the Nilgiri area. Besides, they also contain inventory of landslides that have affected the road in the period since 1987. Before 1987, no systematic record of landslides along the road is available with the technical offices. The data format and type of information extracted from the technical reports along with an example is shown in Fig. 9.4.
- *Non-technical sources* such as internet (NDM 2009), newspapers, and interviews with local people. They provided additional information on the date of occurrence and damages caused by landslides. Interviews with local people have helped in estimating the indirect economic loss resulting from road blockages.

The landslide characteristics was compiled in standardized forms, and rearranged based on the location description. This data formed the basis for field mapping where all the landslide dates related to one specific location was listed in a tabular form. All landslide sites reported in the historic archives were visited and an attempt was made to identify the landslide scars. Some of the landslide scars and run-out areas were not clearly discernable due to the removal of debris and remedial works. During the field work, local residents were interviewed and questions were asked pertaining to their livelihood, and any information regarding landslides and damages. For example, during the field mapping, an old worker from the railway office has helped in locating older landslides based on the clearance work that he had carried out. After identifying the exact location of a landslide it was then mapped on a 1:10,000 scale topographic map and its initiation (source) and run-out area were separately marked. The morphological parameters were plotted after measuring them in the field. Additional data such as type of landslide, run-out distance, present land use and land cover, probable cause, regolith thickness, and damage details were also added to the inventory. The mapped landslides were digitized as polygons or points and entered in a geo-database of ArcGIS. Separate layers were prepared for the landslides associated with cut slopes and natural slopes. The smaller landslides were digitized as points in a separate layer. A unique identification number (ID) was assigned to each landslide (polygon or point), which provided a link between the spatial and non-spatial attributes.

In total 1040 landslides were compiled from the historical records and field work within a 22 km^2 area. The inventory was nearly complete for the period 1987–2007. Landslides were triggered on 116 different dates. From the total of 1040 landslides, 643 landslides (62%) were obtained from the railway slips register, 259 (24%) from the landslide summary table along the railroad, 132 from technical reports and six from the other sources. Through field mapping it was possible to identify 67% of the compiled landslides. Some of the smaller landslides were not identifiable in the field due to possible reactivations which have obliterated the earlier morphology. The volume of these small landslides was therefore taken directly from the original source data. Since they were small and located along the road or the railroad, it was presumed that most of the released material from these landslides was accumulated on the road and the railroad. Therefore, the measured volume from the maintenance records was considered a good representation of the size of these landslides. Landslides were classified as debris slide and debris flowslide following the classification proposed by Cruden and Varnes (1996). Landslides initiating as slide and then converting to flow under saturated condition were grouped under 'debris flowslide'. About 97% of the landslides were debris slides. Most of these were shallow translational with a depth of the slip plane less than 5 m. Only three landslides were found to have a depth of more than 5 m. The landslides are further regrouped into cut slope and natural slope failures based on the location of their source area. Most of the landslides (96%) were recorded in cut slopes. Smaller landslides in the cut slopes were found to have a short run-out as the road and the railroad acts as a platform for the accumulation of the debris. In terms of the volume of material displaced, most of these landslides (91%) lie within the range of 2–500 m^3.

The landslide distribution map is shown in Fig. 9.5. At some places (e.g., Fig. 9.5b) the railroad has cut slopes on both sides. The cut slopes on the valley side act as a barrier and prevent landslide debris moving downslope. All the debris falling from the slopes accumulates on the railroad. The annual distribution of recorded landslides in the past 21 years is shown in Fig. 9.5e. Landslides occur annually in the area (except in 1995) with an average rate of 20 landslides per year. At some locations the same slope is affected by landslides in different periods (e.g., Fig. 9.5c, d). On November 14, 2006 about 205 landslides occurred mostly in the eastern part of the area. This occurrence corresponds to high intensity rainfall (150 mm rainfall in 3 h). In terms of the monthly distribution of recorded landslides, November is the severest month (Fig. 9.5f) containing 58% of the landslides. This month also receives the highest rainfall each year due to the retreating monsoon.

Interpretation of historical data pertaining to the Nilgiri area was carried out for generating a complete multi-temporal landslide inventory. An inventory remains incomplete when location references of landslides are not available, and in most cases features left by landslides cannot be recognized in the field or through the interpretation of aerial photographs, as they are often obscured by erosion, vegetation and human interferences. In literature there is no unique measure of completeness of an inventory, but if an inventory is prepared soon after the triggering event then there is a greater possibility of recording all landslides in an area. For the study area a similar inventory was available in the form of a railroad maintenance archive. It is updated soon after the occurrence of a landslide triggering event and is used for tendering contracts for railroad clearance. Completeness of an inventory can also be tested by studying the magnitude-frequency relation of landslides. In most cases, the structure of the magnitude-frequency relationships were found to have a power law distribution over two orders of magnitude (landslide area) with a flattening of the curve at lower magnitudes, termed as 'rollover'. This rollover or less frequent occurrence of smaller landslides

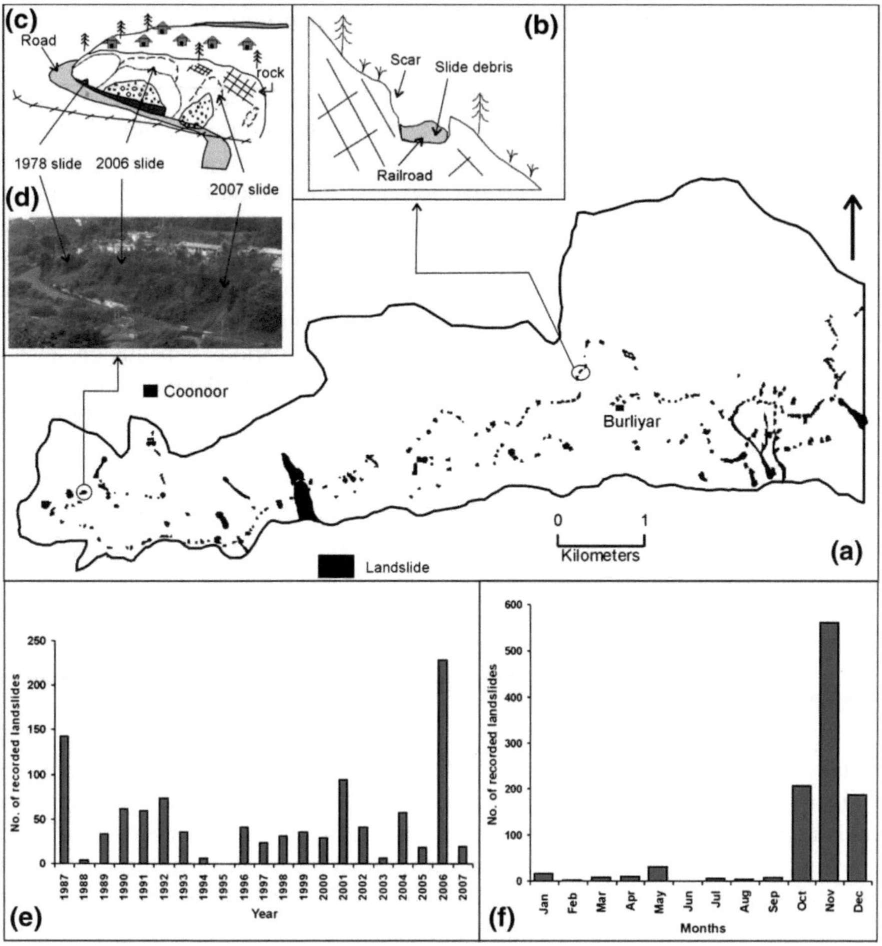

Fig. 9.5 **a** Landslide inventory map. **b** Sketch of the railway having cuts lopes on both sides. **c** Sketch showing landslides in different years in one slope. **d** Field photograph of landslides in cut slope at Katteri. **e** Distribution of landslides over the past 21 years. **f** Annual distribution of landslides

is debated as a real effect reflecting slope stability processes (Guthrie and Evans 2004) or due to the incompleteness of the inventory (Malamud et al. 2004). For the study area magnitude-frequency analysis was also performed and the structure showed a power law distribution with β equals -1.6259 for all landslides with volume ranging from two to more than 104 m^3. For the same dataset, the probability distribution for landslide area also showed power law structure with β equals -1.6764. The power function obtained from this study did not show any rollover effect, implying that the inventory did not suffer from the under-sampling of small landslides and thus is practically complete (Catani et al. 2005; Malamud et al. 2004).

9.3.2 Mapping Landslides from Multi-Temporal Images

In order to be able to express the spatial and temporal probability of landslides in regional hazard maps, it is important to know the location characteristics of the landslides and their behavior through time. One of the techniques used to generate such maps following a combined heuristic/statistical method is to generate multi-temporal landslide maps, linking the landslides to triggering events and use the temporal probability of the event in the hazard assessment combined with the spatial probability resulting from the statistical analysis. The success of generating event-based multi-temporal landslide inventories through stereoscopic interpretation of images/aerial photographs from different time periods depends on several factors. First of all the terrain characteristics, combined with the types and volume of landslides determine the period over which landslides that have occurred will still be recognizable in imagery of a later date. For instance, landslide will quickly become obliterated by vegetation in tropical environments, requiring many images with short time intervals in order to be able to map them. Secondly, the relation between triggering events, such as rainfall or earthquakes, and the landslides caused by these, can only be mapped if imagery is taken shortly after the triggering event. Thirdly the image interpretation skills of the geoscientists that make the image interpretation are important. Although well-trained and experienced geoscientists can make a good interpretation of recent landslides, they will have more difficulty in characterizing old and dormant ones (Carrara 1993). Errors in estimating the dimensions of landslides do exist due to inaccurate base maps (Malamud et al. 2004). Through time, multiple small slope failures may merge into larger landslides causing problems in the analysis of the frequency–magnitude statistics of landslides. Furthermore, for many of the mapped landslides, the exact date of occurrence remains unknown, thus making it difficult to relate such landslides with triggering events, especially as different landslide types have different relations with the meteorological triggers. In some situation, the post-event maps were even not prepared or no detailed information on such event-based landslides is available. The lack of such temporal information (exact date of occurrence) in landslide inventories, and incompleteness of past landslide information adds to serious bottlenecks in the determination of temporal probability and therefore pose difficulties in quantitative hazard assessment (van Westen et al. 2006).

The landslides were mapped using various data sources of 1968–2007 in a highly landslide-prone area around Kurseong in Darjeeling Himalaya, India (D in Fig. 9.1). The first step of generation of a multi-temporal landslide inventory was the collection of all available data on past landslide occurrences, such as, spatial data from high resolution satellite images, topographic sheets, old landslide maps and reports of field investigations and old aerial photographs. For this study, the oldest available data was of 1968. In that year, a major rainfall event occurred in Darjeeling Himalaya between 2 and 5 October with an amount of 1100 mm in 3 days, which caused a large number of landslides in different parts of Darjeeling

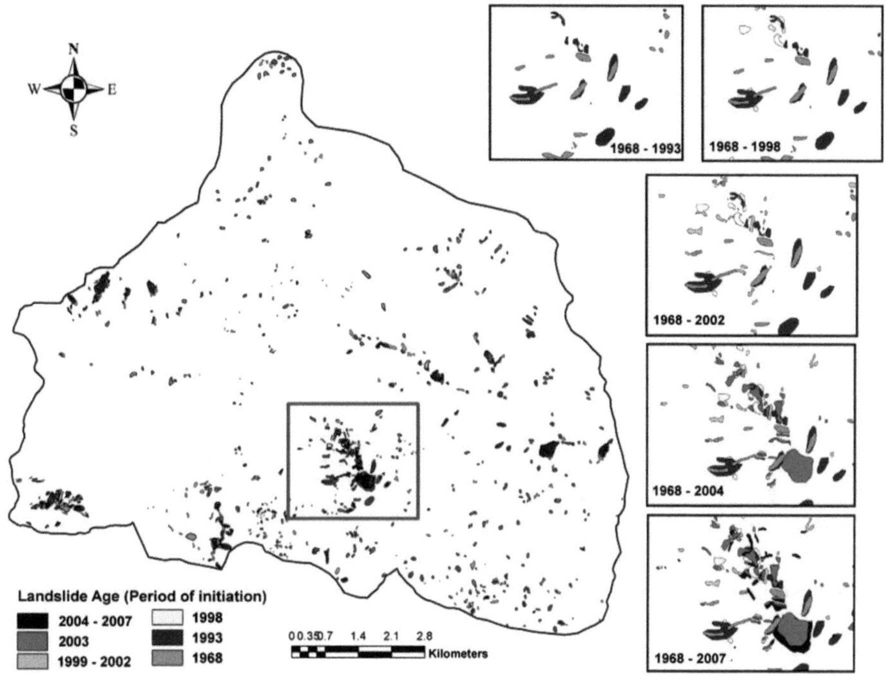

Fig. 9.6 Multi-temporal landslide inventory map of Kurseong area, Darjeeling Himalaya, India

Himalayas (Basu and De 2003). During 1969–70, just after the event of 1968, the Survey of India (SOI) updated their topographic survey and prepared new 1:25,000 topographic maps. In these topographic maps, the locations of prominent and active landslides from 1968 were included. The next available data source is a field-based landslide inventory map of 1993 prepared by the Geological Survey of India (GSI) just after a landslide event that happened between 1 and 3 July, 1993 (Sengupta 1995). Unfortunately the field map of 1993 only covered the south-eastern part of the study area (56 km^2). The third data source represents another event-based landslide inventory map prepared by the Geological Survey through field investigations just after a prominent landslide event occurring between 6 and 8 July, 1998. Also this landslide inventory map covers only a part (central portion $\sim$20 km^2) of the entire study area, along a major communication corridor (NH-55) and around Kurseong town (Bhattacharya et al. 1998). Apart from these maps, high resolution satellite images namely, IRS 1D-PAN merged LISS III image (5.8 m resolution) of 2002, IRS P6-LISS IV MX image (5.8 m resolution) of 2004 and Cartosat 1 stereo pair (2.5 m resolution) of 2006 were also available. These were used together with a digital elevation model to produce stereoscopic images that were interpreted visually to map the landslide incidences from each set of imagery. The last data source used was a detailed fieldwork carried out in 2007,

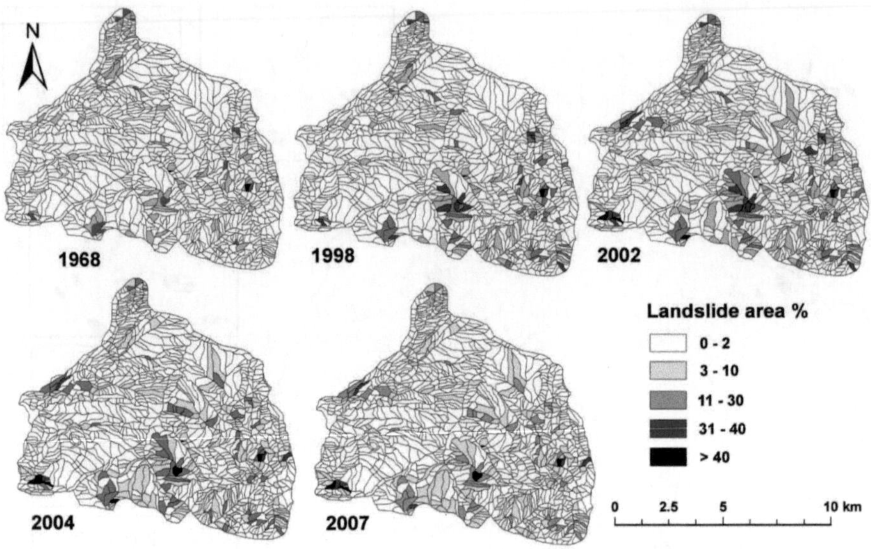

Fig. 9.7 Spatio-temporal variation of landslide area percentage from 1968 to 2007

which allowed distinguishing the landslides that happened as a result of a rainfall event in 2007. Apart from the 1968, 1993, 1998, 2003 and 2007 event landslides, none of the data sources contained information on the exact dates of the landslide events, and thus, for some inventory maps; we could only attach a time period and not an exact date or year of the event.

All the spatial data were projected to UTM; WGS 84; 45N and the landslides were digitized as polygons. The landslides mapped were further updated and verified through extensive field observations in 2007–2008. During the field investigation in 2007 a more recent, post-event (July and September 2007) landslide map was prepared and added to the inventory database. The resulting map is shown in Fig. 9.6.

Despite constraints in the source database and the resultant inventories, we could compare in a GIS, the spatial locations of landslides of different time periods to know the frequency and pattern of new and reactivated landslides. Apart from the above, an analysis of landslide area percentage within different topographic slope units using inventories of different periods can also be observed to study the spatio-temporal changes in landslide distribution (Fig. 9.7). This inventory database can also suitably be used for the quantitative hazard analysis through identification of the recurrence interval of triggering events through multivariate modeling (Ghosh et al. 2009a) and by quantitative prediction of the probability of occurrences of landslides of different dimensions.

9.3.3 Automatic Landslide Mapping from Satellite Images

Aerial photo interpretation and field investigation are the traditional techniques for landslide mapping. Stereo-images are not only useful for the derivation of height information but also for landslide inventory mapping as it provides a 3-dimensional visualization opportunity. With the availability of high quality data products from new generation high resolution satellites and advancement in image processing techniques, satellite images are being increasingly used for landslide mapping. The main advantages of landslide mapping from remote sensing images are the synopticity, repetivity and sensor agility. The multi temporal images can be used to prepare a landslide activity map.

Based on the landslide diagnostic criteria, an expert prepares landslide inventory map from remote sensing images by visual interpretation. These diagnostic criteria can be effectively incorporated in a pixel or object based classification technique to automatically detect landslides. This helps in getting the result faster and the methodology is repeatable. Other remote sensing approaches of landslide inventory mapping include shaded relief images produced from Light Detection and Ranging (LiDAR) DEM and Synthetic Aperture Radar (SAR) interferometry. Detection of landslides include recognition and classification (Mantovani et al. 1996). One of the common methods in automated landslide mapping is change detection. In this method, time series data of an area are analysed to detect any change in the state of a matter e.g., landslide. Seven different categories of change detection techniques, their application and selection of the most suitable method have been summarized by Lu et al. (2004). Although difficult, spectrally similar matters such as bare rock and soil were successfully differentiated from landslides using the Maximum Likelihood Classification (MLC) method by Nichol and Wong (2005) using medium resolution SPOT data. Landslide crowns and trails as small as 7–10 m width were detected in the SPOT change images, and 70% classification accuracy achieved compared with an existing landslide inventory in the area. This method holds well if there is a clear spectral heterogeneity between landslide and its' surroundings e.g., landslides occurring in woodland. Merging of multispectral data with better resolution panchromatic data using image fusion techniques enhances the interpretability of the image and enables mapping of small landslides. Nichol and Wong (2005) applied different image fusion techniques to IKONOS data and found that PAN sharpening technique provides the best result and the fused image is useful for mapping very small landslides.

To overcome the limitations of pixel based methods, other researchers have attempted to use a combination of both satellite imagery and digital elevation models (DEMs) (e.g., Giles and Franklin 1998; McDermid and Franklin 1994). McDermid and Franklin (1994) noted that, in many cases, per pixel reflectance patterns are unrelated to geomorphic processes, and that classification schemes based on these data would fail. They suggested that a combination of geomorphometric criteria as well as spectral data would yield better results in identifying mass movement features. The automated landslide inventory mapping by image segmentation technique using high resolution satellite data and DEMs will not only save time but also add a new dimension in further refinement of hazard assessment techniques by rapidly mapping the inventories after any triggering event on routine basis. Image segmentation is a preliminary step in object-oriented image classification that divides the image pixels into homogeneous, contiguous objects based on scale, colour, shape, compactness and smoothness. Object-oriented classification was introduced in the 1970s (De Kok et al. 1999). The initial application was limited by hardware, software, poor resolution of images and interpretation theories (Flanders et al. 2003). Since the mid-1990s, with an increase in hardware capability and availability of high spatial resolution images, the demand for object-oriented analysis has also increased (De Kok et al. 1999). Object based classification is superior to per-pixel classification (Blaschke and Strobl 2001). Segmentation of pixels into objects provides opportunity to apply geographical and morphological concepts to subsequent image processing such as classification. Thus it provides an information rich environment to work within any application domain. Object-oriented classification is potentially of great value in the detection of landslide scars because it allows spatial characteristics, such as the length to width ratio, to be included as class discriminators. So far the image segmentation approach has been restricted to landslides larger than 1 ha due to limitations in spatial resolution (Barlow et al. 2003). But with the availability high resolution satellite data (Table 9.1) there is a scope to identify even smaller landslides (100 m^2). The recognition and classification of individual process types (Cruden and Varnes 1996) using an automated approach has been less successful (Barlow et al. 2003). However, this limitation can be overcome by the use of curvature from DTM in the object based classification. Initially, spectral information with a suitable scale parameter was used to segment an image which produced image objects adequate to define the landslide boundary (Fig. 9.8). NDVI was used to separate landslides and its false positives from others. Subsequently, using morphometric parameters such as slope and flow direction, shape parameters such as compactness and asymmetry, and spectral parameters such as layer mean and ratios, the false positives were sequentially eliminated (Fig. 9.8a) and the remaining are the landslides. Based on the adjacency conditions such as high relative border to rocky land or weathered zone, a landslide was classified as a rock slide or debris slide, respectively. After landslides were classified based on material, the landslide objects were resegmented using the curvature data (Fig. 9.8c). A positive value for curvature indicates that landslide surface was convex upwards and negative value indicates that it was concave upwards. These criteria were used to classify landslides on the basis of the failure mechanism i.e. rotational and translational. Thus, using the object based classification method landslides were recognized and classified from remote sensing data and DEM.

The initial results indicate that a combination of high resolution satellite data and DEM is useful to prepare a landslide inventory map automatically. Object based detection is more promising in comparison to pixel based detection of landslides. This method is very helpful for rapid damage assessment and decision making process.

Table 9.1 Landslide magnitude class for debris slide and debris slide/flow

Magnitude class	Landslide type	Occurrence probability	Criteria used to define magnitude	
			Characteristic features	Damages and human perception
I	Slide	0.722	Shallow translational slides associated with cut/fill slopes; volume <100 m^3; scar depth <1 m; occur instantly; short run-out <10 m; depth of accumulated mass ~1 m; very high occurrence probability.	Minor or no damage to infrastructure (e.g., road or railroad or house); one can escape unhurt; controlled using simple retaining structures (e.g., retaining wall)
II	Slide and flow	0.248	Shallow translational slides associated with cut or natural slopes; volume of 102–103 m^3; scar depth <2 m; occur instantly as a single slide or more from a single cutslope; run-out <50 m; depth of accumulated mass ~1.5 m; high occurrence probability.	No damage to tarmaced road but can damage rail and non-RCC structures, one can escape; injuries minor; society live with it and accept risk; controlled using specially designed retaining structures (e.g., retaining wall with soil nailing).
III	Slide and flow	0.026	Shallow or deep translational slides associated with natural slopes; volume of 103–104 m^3; scar depth <8 m; confined or unconfined high velocity flow; run-out (~200 m) up to gentle slopes and deposit as fan of depth ~5–8 m; low occurrence probability.	No major damage to tarmaced road but complete damage to rail and buildings; uprooting of trees along path of flow; scouring of channel; escape possible; injuries major or even death in some cases; society live with it and tolerate risk; controlled by specially designed retaining structures of high cost.
IV	Slide and flow	0.002	Deep translational slides associated with natural slopes; volume of 104–105 m^3; scar depth >8 m; confined or unconfined high velocity flow; run-out up to gentle slopes and deposit as fan of depth >5 m; very low occurrence probability.	Complete damage to rail, road and RCC structures; uprooting of trees along path of flow; deep scouring of channel; no reaction time and difficult to escape; intolerable risk; area declared unsafe; controlled by specially designed structures of very high cost.
V	Slide and flow	<0.002	Deep translational slides associated with natural slopes; volume of >105 m^3; scar depth >20 m; confined or unconfined high velocity flow; run-out up to gentle slopes and deposit as fan of depth >10 m; minor slides may precede before the event; very rare event.	Catastrophic event; total damage of infrastructure and properties of all types; complete uprooting of trees along path of flow; deep scouring of channel leaving hollow topography behind; no reaction time and difficult to escape; devastation and death; complete resettlement to other places; no further settlements allowed; controlled by specially designed structures of extremely high cost.

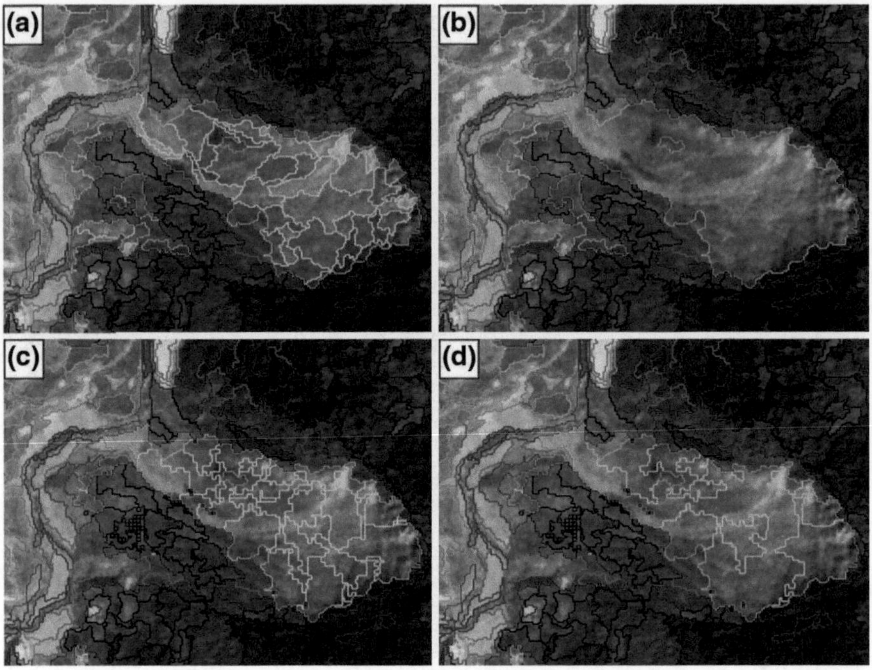

Fig. 9.8 Results of image segmentation and object based classification for landslides using IRS-P6 LISS-IV Mx images and Cartosat-1 derived DTM. **a** Multiresolution segmentation with scale parameter as 10. *Yellow outline* shows landslides. **b** Landslide objec landslide objects classified based on material type and merged to give a single outline (*yellow colour*) to each landslide. **c** Resegmentation of landslide objects using curvature layer. **d** Classification of landslides using curvature value to rotational (*cyan outline*) and translational (*yellow outline*)

9.4 Landslide Hazard Assessment Approaches

Once landslide inventories have been made using one of the techniques presented above they can be used in different ways for landslide hazard assessment. In this section several methods are presented: direct hazard calculation based on complete landslide records, statistical analysis and dynamic modeling. An overview of the approaches and their applicability at different scales is given by Guidelines of the Technical Committee on landslides and Engineered Slopes, JTC-1 (Fell et al. 2008).

9.4.1 Direct Hazard Estimation Along Transportation Routes

In this section we present the options that can be used to quantify landslide hazard if a complete inventory is available. Here we used the earlier described complete multi-temporal landslide inventory prepared for the study area in Nilgiri hills (B in Fig. 9.1). Landslide hazard studies along a transportation line (road or railroad) focus on the landslides that may directly affect the infrastructure. The method requires assessment of two essential parameters:

1. The probability that a landslide affecting the infrastructure are of a given magnitude, and
2. The total number of landslides per kilometer affecting the infrastructure in a given return period.

In literature no established classification for landslide magnitude is available (Guzzetti et al. 2002). Some researchers have used landslide area or volume as a proxy for magnitude, for certain landslide types such as slides or flows (Guzzetti et al. 2005). In this study, a different landslide magnitude class is proposed. The landslides in the study area were grouped into five magnitude classes ranging from I (less severe) to V (catastrophic). The classification is essentially based on landslide type and volume, but it also addresses other characteristics such as the location of the source, damage potential and human perception about the risk related to landslides. The classification is semi-quantitative and derived on the basis of the historical information obtained during the inventory mapping (Table 9.1). The probability that a landslide affecting the infrastructure is of a given magnitude class was estimated using magnitude-frequency relationships. The probability was obtained from the probability density, for which a scatter plot was generated with landslide volume (in m^3) on the *x-axis* and probability density on the *y-axis*. The probability density function of landslide volume was found to have a good correlation with a power law distribution of type:

$$p(V_L) = k(V_L)^{-\beta} \tag{1}$$

where k is a constant and β is the power-law scaling exponent. Annual probability of landslide occurrence can be estimated from the observation of the frequency of past landslide events. The method generally used for estimating probability is by determining the annual exceedance probability (*AEP*), which is the estimated probability that an event will be exceeded in any year (Fell et al. 2005). The Poisson and Binomial distribution models are the two most commonly used models for such analysis (e.g., Coe et al. 2004; Guzzetti et al. 2005). The estimation of landslide risk, particularly indirect risk resulting from the blockage of transportation line, requires estimation of the number of landslides reaching the infrastructure per annum. The above model provides estimate of probability of experiencing one or more landslides and not the specific number of landslides. The number of landslides is required to calculate the blockage period based on clearance time needed per cubic meter of debris. The relation between the annual probability of occurrence of landslides (or return period) and the number of landslides of a magnitude class per kilometer can be established using a Gumbel extreme value distribution (Gumbel 1958).

For this study the annual probability was estimated from the total number of landslides per section of a transportation line in one year. The total number of landslides in a year per section of the transportation lines was selected from the inventory covering 21 years from 1987 to 2007. The yearly values were ranked from low to high, such that lowest rank 1 was assigned to the lowest data value and the highest rank 21 to the highest data value. At each section of the road and the railroad the expected number of landslides in 1, 3, 5, 15, 25 and 50 years return period were estimated. The Gumbel's distribution allows estimating the probability for return periods of 100 years or even more depending on the total length of the data. But, as a general rule, frequency analysis should be limited to expected events within a period twice the record length. In this study we estimated probability only up to 50 years return period, which is slightly more than twice the record length available for the study.

After obtaining all the relevant information such as the probability and magnitude of the landslide reaching the transportation line, the total number of landslides affecting the infrastructure in a given return period and its annual probability, the quantitative landslide hazard assessment was conducted for different sections of the transportation line. The landslide hazard, expressed as the number of landslides of a given magnitude per kilometer of the transportation line in a given return period, was calculated by multiplying the total number of landslides with the probability that the landslides are of a given of magnitude. The hazard estimation can be performed for a number of scenarios using different combination of landslide magnitude class and return period. For this study 24 scenarios were generated using four magnitude classes and six return periods.

Along the railroad Gumbel's analysis was carried out for segments of one kilometer thereby producing 19 Gumbel's plots for the total 19 km railroad. During the period from 1987 to 2007, the entire railroad was affected by 898 landslides of which the lowest was recorded in the vicinity of the kilometer stone number (ksn hereafter) 26 (14 landslides) and the highest around ksn-12 (101 landslides). During the same period the road was affected by 124 landslides with an average of 4.76 landslides per kilometer. The frequency of landslides in a year for different return intervals along the railroad line is shown in Fig. 9.9. A five kilometer stretch (from ksn-9 to ksn-13) is relatively more prone to be hit by landslides, as is the 10 km section (from ksn-390 to ksn-399) along the road.

Landslide hazard was estimated for 24 scenarios using the combination of four magnitude classes (class I–IV) and six return periods (1, 3, 5, 15, 25 and 50 years). An example of the scenario with 50 years return period along the railroad is given in Table 9.2. The results indicate that on an average once in 50 years (annual probability of 0.02) the entire railroad will be affected by 164, 56, 6, and 0.5 landslides of magnitude class I, II, III and IV, and the road by 5, 2, 0.2, 0.02 landslides of these classes.

The study showed that a direct landslide hazard assessment can be carried out if a complete landslide inventory is available. Hazard estimation in terms of number of landslides per kilometer and per year is possible only if the rate of occurrence of

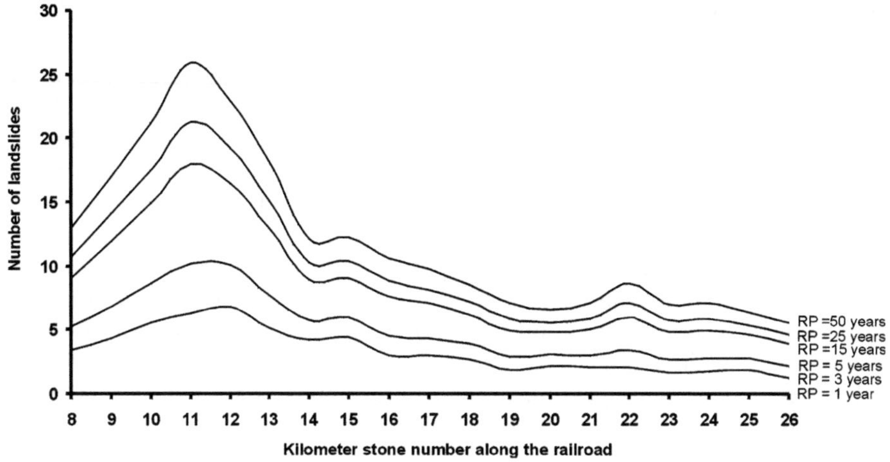

Fig. 9.9 Frequency of landslides in a year for different return periods along the railroad

Table 9.2 Landslide hazard along the railroad in 50 years return period

Kilometer stone number	Total number of landslides of different magnitude class that can occur in a year			
km	I	II	III	IV
8	9.31	3.20	0.34	0.03
9	12.25	4.21	0.44	0.03
10	15.26	5.24	0.55	0.04
11	18.68	6.42	0.67	0.05
12	16.56	5.69	0.60	0.05
13	13.15	4.52	0.47	0.04
14	8.79	3.02	0.32	0.02
15	8.82	3.03	0.32	0.02
16	7.67	2.64	0.28	0.02
17	7.03	2.42	0.25	0.02
18	6.14	2.11	0.22	0.02
19	5.09	1.75	0.18	0.01
20	4.74	1.63	0.17	0.01
21	5.10	1.75	0.18	0.01
22	6.22	2.14	0.22	0.02
23	5.03	1.73	0.18	0.01
24	5.10	1.75	0.18	0.01
25	4.62	1.59	0.17	0.01
26	4.01	1.38	0.14	0.01
Total	163.56	56.18	5.89	0.45

landslides is known and for such analysis continuous records of landslide incidences over a period of time is required. Any gap in the record may result in the over or underestimation of the probability. The Gumbel's distribution used in this study is very appropriate in modeling extreme events such as incidences of large number of landslides. It helps in establishing a relation between the return period and number of landslides. The inclusion of the proposed magnitude class in the hazard assessment will help in analyzing the phenomena both in terms of risk to life and property. Ideally it should be quantified based on absolute values of landslide velocity, its intensity, its peak discharge, etc. But such parameters are very site specific and vary with local conditions such as channel geometry, terrain roughness, and land use, etc. and thus difficult to obtain and integrate in the hazard map. Due to this limitation and the complexity of landslide phenomena, the proposed classification is considered the most pragmatic solution. For the risk analysis, hazard assessment based on the number of landslides expected to hit per kilometer in a given return period is a workable solution. The assessment of direct and indirect risk becomes possible once the total volume of the expected landslide material is known.

9.4.2 Rock Failure Modeling Using Gis

Many of the landslides in the Himalayas are rock slides, or are in weathering soil with a clear structural control. Structure represents nature and extent of discontinuity in the rock mass. For rock slides, since failure propagates along a near-planar surface (planar) or triggered along the intersection of two planes (wedge), presence/absence of any planar discontinuity, its nature, extent, orientation and frequency of occurrence in relation to topography are crucial deciding geofactors. The principle attributes considered in the rock structure are (i) blockiness or degree of rock dissection, (ii) geometric shape, orientation and form of the blocks and fragments and (iii) degree of looseness of the potentially unfavourable rock mass (Varnes and IAEG Commission on Landslides and other Mass-Movements 1984). Apart from the effect of predominance in the concentration of joints, faults and shears, locally, the geometric or kinematic interrelationships between the attitudes of bedding/foliation/joint planes and topography could be pivotal in deciding the mode of movement of rock slides (Briggs 1974; Günther et al. 2004; Hocking 1976; Hoek and Bray 1981; Meentemeyer and Moody 2000).

To determine the unfavourable *discontinuity-topography/structure domains,* we mapped different topographic segments after establishing the geometric interrelationships of the orientations between topography and prominent discontinuity surface following the classification proposed by Meentemeyer and Moody (2000). For the calculation of the angular interrelationship, raster maps of topographic slope (S: 0–90°), topographic aspect (A: 0–360°), discontinuity dip (θ: 0–90°) and discontinuity dip direction (α: 00–360°) were used. Slope and Aspect maps were directly derived from the 10 m $\times$ 10 m DEM. Dip and dip direction raster maps of prominent discontinuities were generated through interpolation (Inverse Distance Weighted) of discrete dip/dip direction values of foliation planes, measured at different point locations. After this, *structure domains* (Fig. 9.10) were derived by

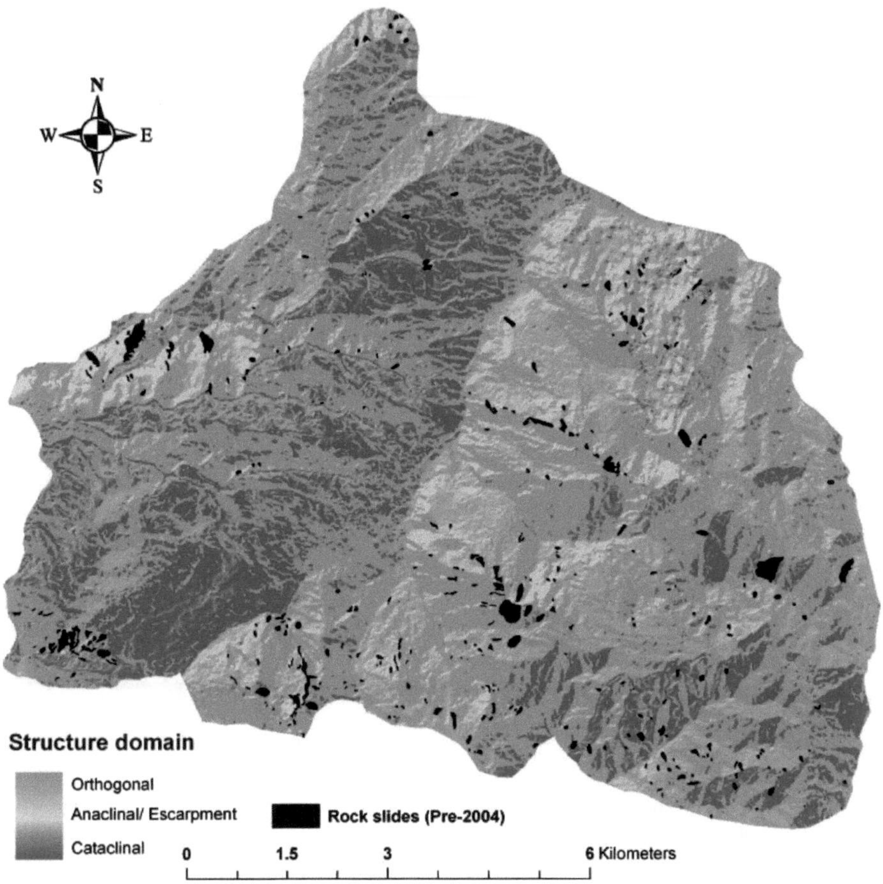

Fig. 9.10 Map showing different structure domains of Kurseong (D in Fig. 9.1)

combining the above four raster maps and calculating their angular relationships. This map can be used as a predictor maps in susceptibility assessment.

Different rock slide failure mechanisms can also be successfully identified spatially by studying the kinematic interrelationships between the attitudes of distributed bedding/foliation/joint planes and topography through spatial correlation (Günther et al. 2004), which follows the criterion suggested by Hoek and Bray (1981).

The RSS-GIS extension of ArcView® 3.X is one of the very few software capable of automated spatial analysis with distributed rock structure data in a GIS platform (Günther 2005). This package was applied in the Kurseong study area (D in Fig. 9.11) to delineate the wedge and planar failure modes. The best approach would be to parameterize all these structure domains per failure mechanism spatially and utilize them for the detailed quantitative rock slope susceptibility analysis.

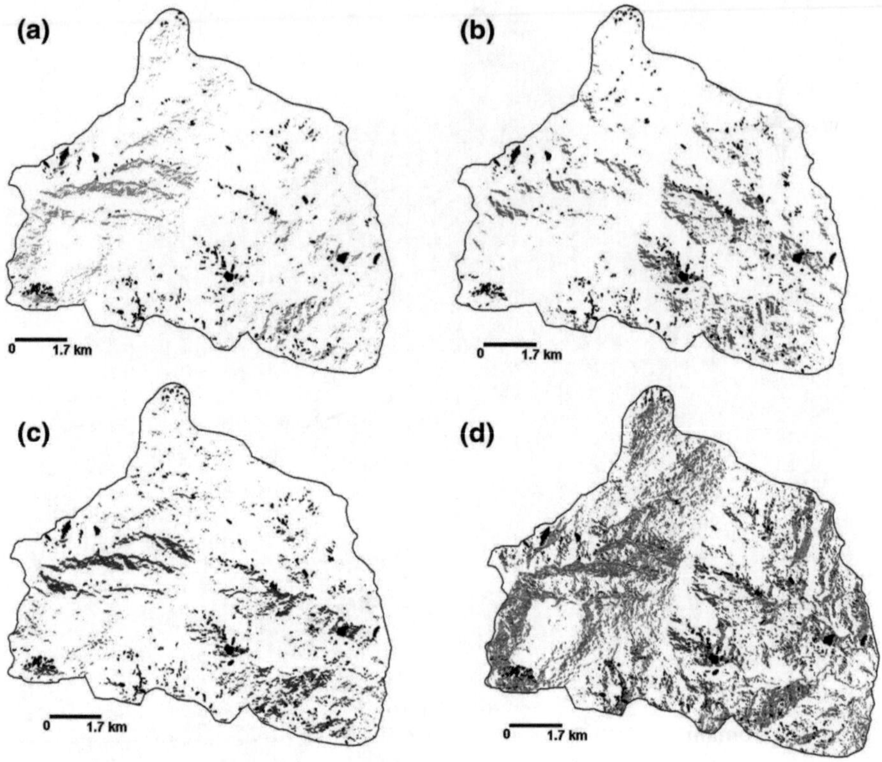

Fig. 9.11 Preliminary results of the application of RSS-GIS in Kurseong area, Darjeeling district, India. **a** Planar failure modes caused by prominent foliation orientation. **b** Planar failure modes caused by one prominent joint plane. **c** Wedge failure modes caused by foliation and joint plane and **d** Combination of all failure modes (Black polygons are rock slides of 1968–2007)

9.4.3 Multivariate Statistical Analysis

Landslide susceptibility can quantitatively be determined by a number of statistical and mathematical methods in a GIS using spatial relations of landslides and their relevant causal predictors. Amongst these, multivariate quantitative methods such as discriminant analysis (Carrara et al. 1991) and logistic regression (Mark and Ellen 1995) have proven to lead to better prediction results than other statistical methods, although the interpretation of the contribution of each causal parameter is less straightforward. Several multivariate susceptibility methods are available however there exists considerable difference between the scope of these methods and the usability of produced susceptibility maps (Aleotti and Chowdhury 1999; Brabb 1984; Chung and Fabbri 2008; Guzzetti et al. 1999; Soeters and van Westen 1996; van Westen et al. 1997). Thus, to ensure the quality of any quantitative prediction cross validation is necessary, which is best done by separating the

Table 9.3 Contingency table (rock slides—discriminant analysis, 76.5% original grouped cases classified)

Observed		Non-landslide slope units	Landslide slope units	Total
Count	Non-landslide slope units	667	188	855
	Landslide slope units	77	194	271
%	Non-landslide slope units	78.0	22.0	100
	Landslide slope units	28.4	71.6	100

landslide population into two independent groups—one for model calibration/training and the other for validation (Chung and Fabbri 1999, 2008; Chung 2003).

Multivariate statistical analysis of landslide susceptibility was carried out for the Kurseong area, where both logistic regression (LR) and discriminant analysis (DA) were applied. For this study, 342 shallow translational rock slides of pre-2004 period were used as the training data and 183 rock slides of 2004–2007 periods were used for cross-validation. Multivariate model calibration with such landslide data was carried out through establishing the multivariate spatial signature of different causal parameters/geofactors as relevant independent/explanatory/predictor variables in each terrain mapping unit against the presence or absence of calibrating landslides as grouping or dependent variables. In the study area, 1126 slope units were used as terrain mapping unit, which were semi-automatically derived through sub-division of the terrain into different small hydrological subunits. For this subdivision, digital topographic information from a 10 m × 10 m digital elevation model (DEM) was used along with ridge lines and drainages. This type of terrain sub-division has been successfully implemented by a number of authors in for predicting new landslides (Cardinali et al. 2002; Carrara 1999; Galli et al. 2008; Guzzetti et al. 2006; Van Den Eeckhaut et al. 2009).

Density of landslides in each of the slope units (1126) was calculated using the pre-2003 landslide inventory. The slope units (855) with landslide density >2% was considered as landslide prone and the rest (271) were grouped as landslide free. Nine relevant causal geofactor maps comprising different morphometric and geo-environmental parameters were used as explanatory variables. These include slope material, structure, geomorphology, land use, fracture density and aspect as categorical variables and minimum, maximum, range, mean and standard deviation of elevation, slope, rainfall, wetness index, curvature as continuous variables. The categorical explanatory variables were converted to their continuous equivalents by calculating the cumulative area percentage of each geofactor class for each mapping unit (slope unit).

The application of the DA and LR methods for rock slides resulted in an overall model classification of slope units (both landslide-bearing and non-landslide-bearing slope units) of 76.5% (Table 9.3) and 81.2% (Table 9.4) respectively with a receiver operator characteristic (ROC) area of 0.82 and 0.84 respectively. The resultant maps are shown in Fig. 9.12 and the ROC curves in Fig. 9.13. Both susceptibility maps shows moderate to high rate of *goodness-of-model-fit* and *prediction*, though, the performance of LR is comparatively better than the DA

Table 9.4 Contingency table (rock slides—logistic regression, 81.2% original grouped cases classified)

Observed		Non-landslide slope units	Landslide slope units	Total
Count	Non-landslide slope units	800	55	855
	Landslide slope units	157	114	271
%	Non-landslide slope units	93.6	6.4	100
	Landslide slope units	57.9	42.1	100

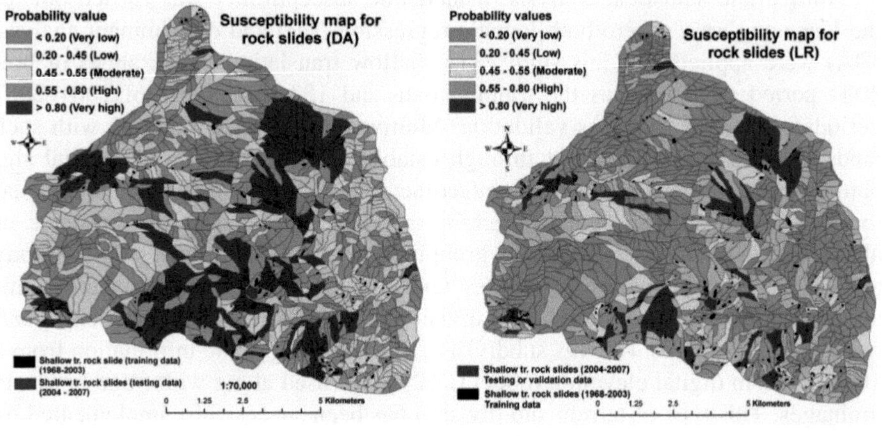

Fig. 9.12 Susceptibility maps generated using Discriminant Analysis (*left*) and logistic regression (*right*)

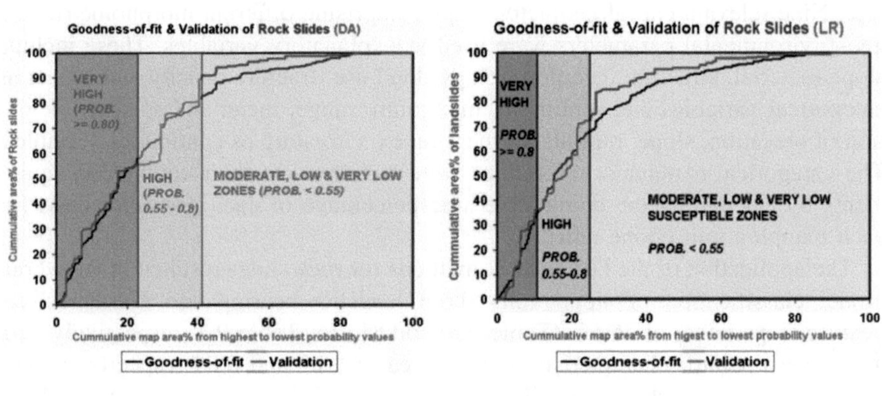

Fig. 9.13 Success and prediction rate curves. *Left* Discriminant analysis. *Right* Logistic regression

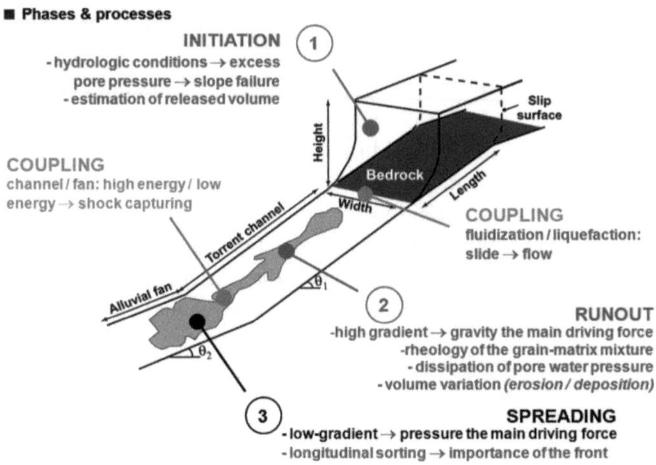

Fig. 9.14 Chronological phases and processes in shallow landslides and consequent debris flows

model. With the DA model only 23% of the landslides in the validation data set could be explained, while using the LR 72–73% of the landslides in the validation data could be explained.

This research was a joint effort with researchers from IRPI, CNR, Italy. The multivariate statistical models used were either those readily available in SPSS 4.0® or R-scripts developed at IRPI CNR Perugia (IRPI CNR 2009). The preliminary results were presented by Ghosh et al. (2009b) in the European Geosciences Union General Assembly 2009.

9.4.4 Physical modeling of landslide initiation and runout

Physically-based spatial modelling of landslide hazard necessitates the processes to be mathematically abstracted based on the known universal laws of physics (Davies 1992). The hazard area entails the three major chronological phases (Fig. 9.14): the initiation zone, transportation path and deposition fan (Chen and Lee 2004). Modelling this sequential chain of events by integrating the various governing equations into one model is complicated. This is due to the fact that the varying temporal and spatial scales of these processes will cause multi-dimensional uncertainties that cannot be quantitatively accounted. Hence, landslide initiation is modelled independent of landslide run-out; the run-out accounts for both transportation and deposition.

Physically-based models can be static or dynamic. Static models consider landslides in its stable state and seek to determine which stimuli caused the instability (Bromhead 1996). Dynamic models are capable to run forward in time, using rules of cause and effect to simulate temporal changes in the landscape

(Karssenberg 2002). A dynamic landslide hazard model addresses the spatial and temporal variation of landslide initiation (Brooks et al. 2004; van Beek and van Asch 2004) or runout (Cannon and Savage 1988; Rickenmann 2005). Even though physically-based models do not necessarily demand long term data, availability of such data can improve model calibration significantly. However, such models demand high spatial resolution and measurement precision for accurate predictions. With the increasing trend of quantitative studies in geomorphology, application of such models has substantial increased in the last two decades. Such models are valuable tools for quantitatively assessing the influence of various parameters contributing to the initiation (Kuriakose et al. 2009c) and runout of landslides.

9.4.5 The Models

There are several physically-based spatial models capable of modelling the landslide initiation and runout, though attempts to adopt them to Indian conditions are limited. This article illustrates the application of a physically-based dynamic slope hydrology coupled slope stability model [STARWARS + PROBSTAB (Storage and Redistribution of Water on Agricultural and Re-vegetated Slopes + Probability of Stability) (van Beek 2002)] and a runout model [Mass-Mov 2D (Begueria et al. 2009)] in a landslide prone region in the Western Ghats of Kerala, India. STARWARS + PROBSTAB was developed by van Beek (2002). They comprise a distributed dynamic hydrological model (STARWARS) that is coupled to a stability model (PROBSTAB). The dynamic spatial outputs of the hydrological model are the inputs for the slope stability model. An added advantage of the models is that its open architecture allows modification of the model script and thereby enables different parameterizations appropriate for the study area. Both the models are embedded in PCRaster, a GIS with an advanced Environmental Modelling Language (www.pcraster.nl).

STARWARS was originally designed to evaluate the effects of vegetation on hillslope hydrology in SE Spain. Soil hydrological properties can be assigned to specific land use types and the model originally included the processes of interception and evapotranspiration. The amount of actual evapotranspiration is scaled to the available storage and FAO crop factors (Doorenbos and Pruitt 1977). It contains a detailed description of the unsaturated zone that is present in the soil mantle over a semi-impervious lithic contact, which in this case is the Charnockites. The soil profile is subdivided into three layers that can be interpreted as the A, B and C horizons. Percolation of soil moisture is driven by gravity and depends on the unsaturated hydraulic conductivity which is prescribed by the soil water retention curve of Farrel and Larson (1972) and the unsaturated hydraulic conductivity relationship of Millington and Quirk (1959). At the lower end of the soil mantle, the percolation into the underlying bedrock is impeded and a perched water table may form. The resulting perched water table will drain laterally according to the gradient of the phreatic surface. All unsaturated fluxes are considered to be vertical only. PROBSTAB is based on the infinite slope model and as such is valid for translational slides (Skempton and DeLory 1957). This is consistent with the type of failures in the study area. PROBSTAB calculates Factor of Safety for the entire soil column (FOS), and if required the depth of failure (Z_F), based on the daily variation of water level and volumetric moisture content, which are the outputs of STARWARS. In addition, PROBSTAB uses the matric suction to calculate the unsaturated shear strength when a perched water table is absent using Fredlund's (1987) equation and it includes the mechanical effects of root reinforcement and surcharge on slope stability. Hence the calculated stability varies on a day-to-day basis with the hydrological input. The model is also capable of accounting for the probability of failure (P_F). Probability of failure was obtained using the first-order second moment (FOSM) approach (Ang and Tang 1984) which takes into account the uncertainty in the estimation of the mechanical effects of vegetation, shear strength parameters, soil depth and slope angle. FOSM method necessitates the assumption of a normal curve. The curve is implemented in the model with the first standard deviation on the positive and negative side for a given parameter. This draws from the assumption that by using the first standard deviation on either sides of the curve, 66.6% of the total possible variation of the parameter is captured. This is also a direct indicator of the sensitivity of the model to the parameter and thus can also partially address the issue of uncertainty in parameter estimation.

Owing to the data poor situation some modifications to the models and their dependency were made. For example, the model was modified to account for root-induced cohesion and surcharge that were not originally considered by van Beek (2002). Potential evapotranspiration (PET) was calculated outside the model environment using Hargreave's equation (Hargreaves and Samani 1982) which is less data demanding than the Penman's equation (Penman 1948) originally used by van Beek (2002). Interception was computed by means of "Aston's (1979)" equation (Eq. 1), and throughfall and evapotranspiration of the canopy storage was addressed outside the model environment. A detailed description of these modifications, calibration and application of the model in a larger area containing the study area can be found in Kuriakose et al. (2009c).

MassMov was developed by Begueria et al. (2009). It is linked to the PCRaster GIS package, and uses a fixed finite-differences mesh and a two-step Lax-type explicit numerical scheme with variable temporal resolution to ensure stability. The flow is modelled as a 2D continuum by using a depth-integrated form of the Navier–Stokes equations under the shallow water assumption (Saint–Venant equation). The flow behaviour is controlled by the resisting forces, for which a set of alternative rheological models can be used. For the present study Vollehmy rehology was chosen due to the known frictional behaviour of debris flows in the region. The model uses bed entrainment by defining an entrainment zone, a maximum depth of supply material and the average growth or erosion rate (McDougall and Hungr 2005).

Table 9.5 Parameters used for simulating the debris flow using MassMov 2D (Voellmy Rheology)

• Density of the DF: 2000 kg/m^3	• DEM subtracted by soil depth (SD) at the initiation zone
• Chezy roughness: 250 m/s^2	• Soil depth at the initiation zone: Initial volume—1669 m^3, Initiation area—782 m^2, Area affected by scouring—2337 m^2
• Angle of basal friction: 30°	• Soil depth along the runout and deposition zones
• Angle of internal friction: 35°	
• Density of the soil: 2000 kg/m^3	
• Scouring rate: 0.0035 m/s	
• Fluid rate: 10 m/s (Transition from solid to fluid)	
• Total time steps: 100 s	

9.4.6 The Data

The models were tested in the study area in Kerala (Fig. 9.2). The slope hydrology coupled slope stability model was applied in Aruvikkal catchment, a 9.5 km^2 sub-catchment of Tikoy River which in turn is a tributary of the Meenachil River. The run-out was tested on a specific landslide that occurred near Peringalam village in the upper reaches of Meenachil River. All necessary data for STARWARS + PROBSTAB were derived from preceding research works and associated field work (Kuriakose et al. 2009a; Thampi et al. 1998). The data available were daily rainfall, crop factors, potential evapotranspiration, MODIS NDVI 16 day composites, soil types, soil properties, contour map from topographic sheet, soil depth, root cohesion, soil depth, root cohesion, land use/land cover and a rudimentary landslide inventory containing only the date and the location of events. Calibration and validation was also carried out in a similar manner as described in Kuriakose et al. (2009a). The model validation was carried out for the year 2001 as the study area experienced six shallow landslides on 8th July 2001. The DEM and other spatially parameterized data had a spatial resolution of 10 m by 10 m.

Data such as the deposit area and depth of the Peringalam landslide necessary for calibrating and validating MassMov2D was generated during a field survey in 2007 using simple hand held Garmin GPS and information of final deposit height collected by interviews with the local inhabitants. Initial volume and scouring data was also generated in a similar manner. The pre event DEM with 1 m resolution of the area was derived from a 20 m interval contour map of the region prepared based on a survey conducted in the 1970s and interpolating the field survey points that were outside the landslide body. The parameters and the calibrated values of each of them as used in MassMov2D are provided in Table 9.5. The model was

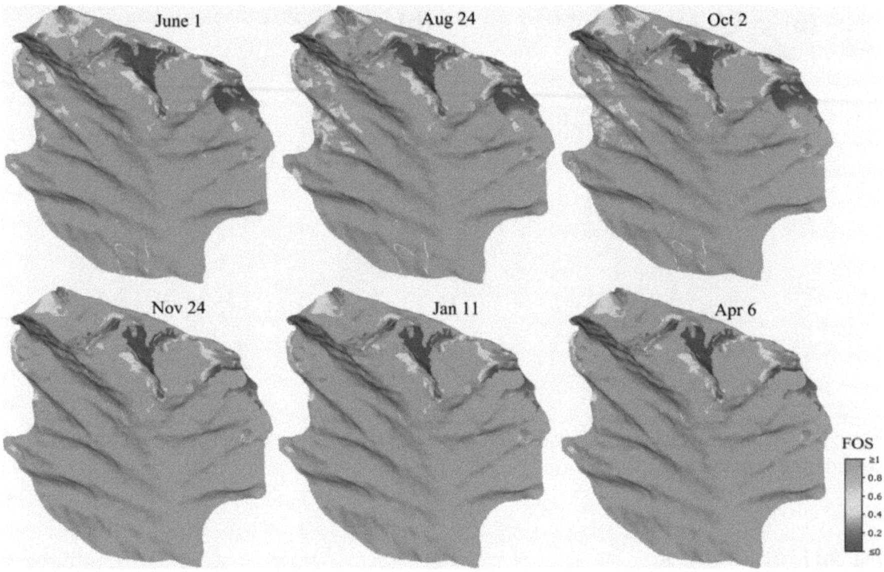

Fig. 9.15 Daily variation of factor of safety in the year 2001—Predicted by STARWARS + PROBSTAB

calibrated based on the volume of deposit and was cross validated with the observed area and depth of failure.

9.4.7 The Results

The model simulated transient hydrological and slope stability conditions on a daily time step with (Fig. 9.15) and without considering vegetation effects. A detailed investigation revealed that the hydrological effects of vegetation are crucial for the long term stability of the study area. However, its effects on slope stability during high intensity rainfall are negligible. Mechanical effect of vegetation, especially root induced cohesion was the most significant effect of vegetation on slope stability in the region. The unstable area (FOS <1) as calculated by the model, considering vegetation effects was 2.1 km^2 (Fig. 9.16a); without considering vegetation effects the unstable area was 5.5 km^2 (Fig. 9.16b). About 0.8 km^2 of this area was persistently unstable (FOS <1) which may be attributed to the inaccurate parameterization in such areas. This over estimation of unstable area was in agreement with similar studies elsewhere (Simoni et al. 2008; van Beek and van Asch 2004). Sensitivity analysis indicated that the FOS was the most sensitive to slope, angle of internal friction and soil depth (Fig. 9.17). The DEM resolution of 10 m derived from a 20 m contour map interpolation was also a significant contributor to the overestimation. All six landslide locations were predicted as failed on

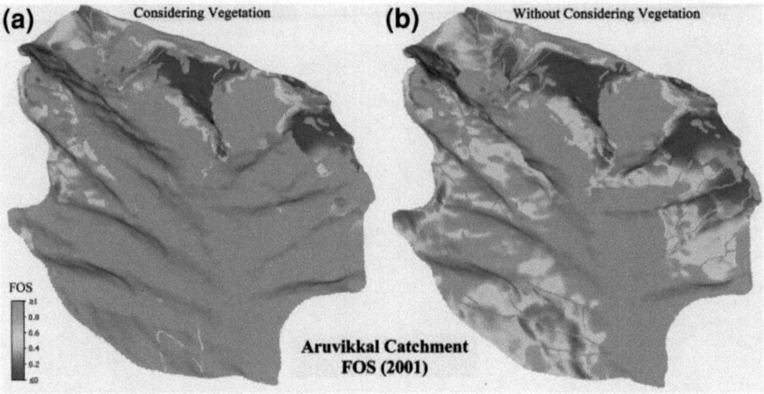

Fig. 9.16 Minimum factor of safety of every pixel (2001). **a** Considering vegetation and **b** Without considering vegetation

Fig. 9.17 Sensitivity of PROBSTAB to parameters

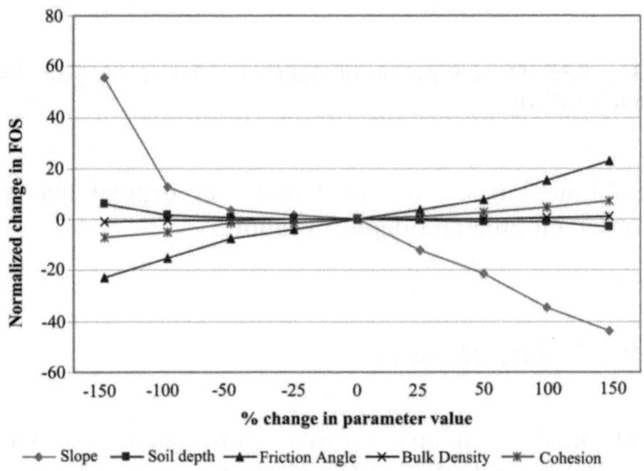

the known date of failure with P_F (FOS <1) $\geq$ 60% indicating accurate temporal predictive capability of the model.

The slope-stability modelling indicates that the antecedent moisture conditions and the persistence of high pore-water pressure for a significantly long period may have been the immediate preparatory conditions for the failures. The trigger of the events was probably an extremely high intensity rainfall which resulted in a sharp increase of pore-water pressure. This response pattern was also apparent from the instrumented monitoring of hollows in the region (Kuriakose et al. 2008). The research conclusively highlights the significance of vegetation effects on slope stability. Better input data, especially DEM and soil depth can significantly improve the predictions.

Table 9.6 shows the comparison of predicted to observed properties of the debris flow. Figure 9.18 shows the temporal evolution of the debris flow height as

Table 9.6 Comparison of observed and predicted properties of the debris flow

	Observed	Predicted by MassMov 2D
Initial volume (m^3)	443	
Entrained volume (m^3)	1226	813
Total volume (m^3)	1669	–
Total deposit volume (m^3)	1553	1253
Max. velocity (m/s)	1.5 min to reach 180 m	24
Mean max. velocity (m/s)		21
Min. deposit thickness (m)	0.1	0.1
Mean deposit thickness (m)	1.2	1.7
Max. deposit thickness (m)	6.4	3.5
Deposit area (m^2)	2686	1634
Observed deposit area overlaid by predicted deposit area (%)		61

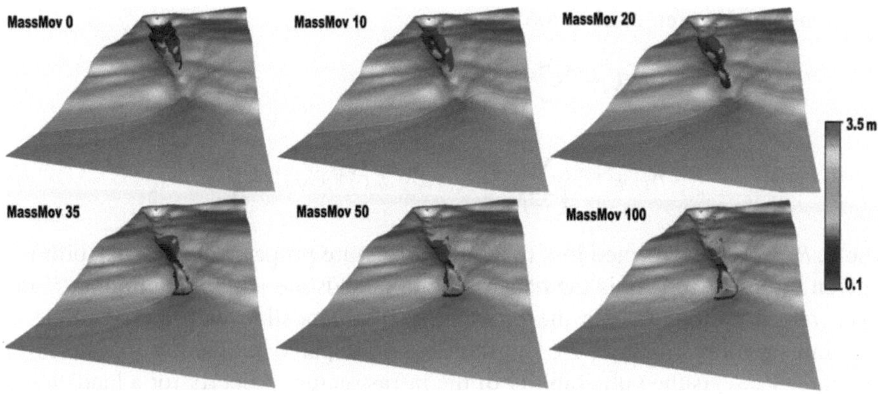

Fig. 9.18 Temporal evolution of debris flow height as predicted by MassMov2D

predicted by the model. The predicted deposit volume was underestimated; the discrepancy was not of very high orders. It has to be noted that the field survey of the slide occurred after three years causing errors in the estimation of the slide body and volume. A part of the material (the slurry) may have been eroded over time resulting in the underestimation of the observed deposit volume. The predicted deposit area was grossly underestimated. This can be attributed to the fact that in reality, the moving mass is not uniform in its rehology.

Improving the run-out model predictions necessitate high resolution DEM (>1 m). Methods for incorporating the transient rheology of the flow should be attempted. Immediate surveys after the occurrence of the events is a necessity such that accurate estimates of initial volume, scoured volume and deposit volume can be made.

The present study was an attempt to evaluate the hazard quantification potential of physically based models in data poor regions. Although both STARWARS + PROBSTAB and MassMov provided outputs that have physical meaning it was not possible to conduct an appropriate evaluation of the model performance due to the limited nature of the calibration and validation data. Thus the results are merely indicative of what can be achieved by using such models for a quantitative evaluation of landslide hazard.

It is evident from this study that physically-based spatial models are ideal to quantitatively understand the contribution of a specific parameter towards landslide hazard. As need is a motive for quality, it is certain that with more and more researchers turning their attention towards utilizing physically-based models, data quality will improve over time.

9.5 Risk Assessment

The landslide risk definition given by Varnes and IAEG Commission on Landslides and other Mass-Movements (1984) can be represented by the following equation (van Westen et al. 2006):

$$Risk = \sum \left(H \sum (VA) \right) \tag{2}$$

$$R_{prop} = \left\{ \sum_{i=1}^{k} \left[H_i x P_{T:L} x P_{S:T} x V_{prop:i} \right] x A \right\} \tag{3}$$

where, R_{prop} is the expected loss to the infrastructure property due to landsliding in a given return period, H_i is the hazard due to a landslide with a magnitude '*i*' in a given return period, $P_{T:L}$ is the probability of a landslide with a magnitude '*i*' reaching the infrastructure, $P_{S:T}$ is the temporal spatial probability of the infrastructure, $V_{prop:i}$ is the vulnerability of the infrastructure property for a landslide of magnitude '*i*', and *A* is the cost of making one kilometer of a new infrastructure. The landslide belongs to different magnitude class ranging from I to IV. The specific risk for the two important infrastructure properties i.e. the railroad (*Rs_rl*) and the highway road (*Rs_rd*) can be estimated by using Eq (3). The value of $P_{S:T}$ is taken as 1.0 as both the elements are stationary object and always remain on or in the path of the landslide. The value of $P_{T:L}$ is also taken as 1.0 as the hazard was estimated for landslides that affect the infrastructures. The assessment of vulnerability can be based on the detailed analysis of the past damage records. The degree of damage can be either monetary or physical (structural damage).

The study are in Nilgiri with the railroad and road was selected as study site for the risk assessment, as it has detailed landslide data as well as traffic information. The infrastructure vulnerability was calculated as the ratio of the total restoration cost of the damaged infrastructure due to a landslide of a given type and magnitude to the actual cost of constructing one kilometer of new railroad or tarmaced road. The total restoration cost of the railroad includes cost of removing debris of magnitude '*i*' and cost of replacing the damaged rail structure (i.e. rail, rake bar, sleeper and pebbles). Vulnerability in terms of the physical loss to the railroad and

Table 9.7 Vulnerability for elements at risk when hit by a landslide

Type of element at risk	Vulnerability in different magnitude class				Comments
	I	II	III	IV	
Physical damage to a infrastructure					Damaged caused by one landslide of a given magnitude in one kilometer
Railroad	0.01	0.08	0.25	0.60	
Road (Asphalt)	0.001	0.01	0.10	0.50	
Physical damage to a moving vehicle					Damaged calculated based on past records
Bus	0.01	0.10	0.80	1.0	
Lorry	0.01	0.10	0.80	1.0	
Car	0.10	0.50	1.0	1.0	
Motorbike	0.50	0.80	1.0	1.0	
Person in a moving vehicle					Damaged calculation partly based on past incidents and expert judgment
Bus	0.001	0.10	0.80	1.0	
Lorry	0.001	0.10	0.80	1.0	
Car	0.01	0.10	1.0	1.0	
Motorbike	0.50	1.0	1.0	1.0	

the road due to a landslide of magnitude I, II, III and IV is given in Table 9.7. Maximum vulnerability for the railroad (0.6) and for the road (0.5) was obtained for landslide of magnitude IV and the value decreases with the decrease in the size of the landslide.

The risk to a moving vehicle, i.e. a vehicle being hit by a landslide, largely depends on the temporal spatial probability of the vehicle at the time of occurrence of the landslide. This value of probability ($P_{S:T}$) can be used to calculate risk to a moving vehicle by using the following three expressions (AGS 2000):

$$R_v = \left\{ \sum_{i=1}^{k} \left[P(V) x V_{veh:j} x N_v \right] \right\} \tag{4}$$

$$P(V_i) = 1 - \left(1 - (P_{S:T})^{NR}\right) \tag{5}$$

$$P_{S:T} = (ADT\,x\,L\,x\,SL)/(24\,x\,1000\,x\,V_v) \tag{6}$$

where, R_v is the expected loss of vehicles of a given type due to landsliding in a given return period ($), $P(V_i)$ is the probability of the vehicle being hit by a landslide with a magnitude 'i', $V_{veh:i}$ is the vulnerability of the vehicle for a landslide of magnitude 'i', A is the cost of the vehicle ($), N_v is the number of vehicles of a given type in a landslide zone at any given time, $P_{S:T}$ is the temporal spatial probability of the vehicle, NR is the number of landslides of magnitude 'i' reaching the transportation line in a given return period, ADT is the average daily traffic (vehicles per day), L is the average length of the vehicle (m), SL is the length of the landslide affected zone (km), and V_v is the velocity of the vehicle (km/hr).

In this chapter, the assessment of vulnerability of different types of moving vehicles (bus, lorry, car and motorbike) and train for a landslide of magnitude '*i*' is carried out based on the historical records, experience of the local people and expert knowledge. Other variables were obtained from historical records and field calculations. The average speed for the road cruising vehicles and train was measured as 26 and 11 km/hr, respectively. The ADT for bus, lorry, car and motorbike was obtained as 137, 309, 554 and 90 vehicles per day, respectively and for train it is two per day. The length of landslide affected zone (*SL*) along the road and the railroad was calculated from landslide density, which is the ratio of the total landslide scar width to the total length of the transportation line. The total *SL* along the road and the railroad is 2.1 and 3.84 km, respectively. The average length (*L*) of a bus, lorry, car, motorbike and train was measured as 12, 8, 5, 2 and 55 m, respectively. By using the above values and Eqs (4–6) specific risk to a bus (Rs_b), lorry (Rs_l), car (Rs_c), motorbike (Rs_mb) and train (Rs_t) in terms of monetary loss ($) can be calculated for each hazard scenario.

The risk of life or the annual probability of a person losing his/her life while travelling in a vehicle depends on the probability of the vehicle being hit by a landslide and the probability of death of the person (vulnerability) given the landslide impact on the vehicle. The vulnerability of commuter to a landslide depends on the type, speed and size of the landslide, the speed and type of the vehicle, and whether the person is in the open or enclosed in a vehicle (Wilson et al. 2005). It also depends on whether the debris have directly hit the vehicle from the top or moved horizontally and hit the side of the vehicle. Even with the availability of some known incidents and damaged records the assessment of vulnerability of death still remains fuzzy. Due to such large variability of factors the assessment of vulnerability was somewhat subjective and knowledge driven. The risk of life for the commuters travelling along the road using different mode of travel such as bus, lorry, car or motorbike, and also in the train were estimated. The analysis shows that the annual probability of the person most at risk losing his/her life by driving along the road in a hazard of 3, 5, 15, 25 and 50 years return period is 1.3×10^{-6}, 2.6×10^{-6}, 5.2×10^{-6}, 6.3×10^{-6} and 7.8×10^{-6} persons/annum, respectively. Along the railroad these values are 6.8×10^{-3}, 1.0×10^{-2}, 1.7×10^{-2}, 2.0×10^{-2}, 2.4×10^{-2} persons/annum, respectively.

The total landslide risk is the summation of all the specific risks related to landslides in an area. In this study, total landslide risk in terms of the monetary loss was calculated by adding all the specific direct and indirect risks evaluated for 24 hazard scenarios as described above:

$$\begin{aligned} RT(P) &= \sum_{h=1}^{24} [RD + RI] \\ &= \sum_{h=1}^{24} [(RS\_rl + RS\_rd + RS\_b + RS\_l + RS\_c + RS\_mb + RS\_t) + RI] \end{aligned} \tag{7}$$

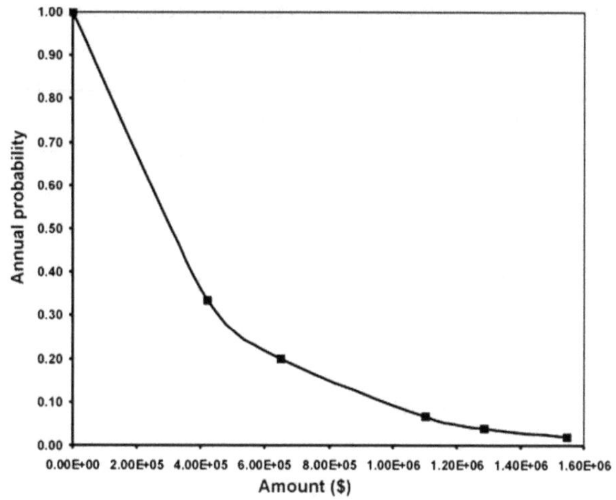

Fig. 9.19 Curve including both direct and indirect losses expressed in monetary value (in US $)

where, *RT(P)* is total risk for monetary loss expressed in US dollars. The total landslide risk for the loss of life, *RT(D)* expressed as number of people per annum was calculated by adding all specific loss of lives such as:

$$RT(D) = \sum_{h=1}^{24} [RS\_Db + RS\_Dl + RS\_Dc + RS\_Dmb + RS\_Dt] \quad (8)$$

.

The occurrence of a certain number of landslides will incur indirect loss resulting from the temporary interruption of the road (NH67) and the railroad. The indirect loss for additional fuel consumption due to the detour was calculated. The total loss for the 24 hazard scenarios amounts to $ 963470 of which the loss to Nilgiri and tourists vehicles was $ 790570 and $ 172900, respectively. In the event of the interruption of NH67 the daily commuters have to travel an additional 32 km for which they have to bear additional tickets cost for a longer journey.

Besides commuters the traffic interruption also affects the local business along NH67 and their livelihood which is totally dependent on tourists travelling along the road. The indirect loss of income for business around Katteri and Burliyar area was estimated. The total loss for the 24 hazard scenarios amounts to $ 912, $ 3009, $ 27512 for shops, commodity and liquor business respectively, located in Katteri and $ 10640 for shops in Burliyar. Besides national highway, the railroad is also prone to interruption due to landslides. The revenue loss to the railway company was estimated for all hazard scenarios, for a 50 years return period. The blockage period for the railroad is much higher than the road for a given landslide. The cumulative loss from the 24 hazard scenarios was $ 677419.

The total indirect loss resulting from the traffic interruption of the road and the railroad by landslides in six return periods amounts to $ 1801525, and in 3, 5, 15, 25, and 50 years return period it was estimated to be around $ 147580, $ 233921, $

398,283, $ 464982 and $ 556758, respectively. The total direct loss for all the hazard scenarios amounts to $ 3205711 and in 3, 5, 15, 25, and 50 years return period it was estimated to be around $ 273756, $ 416611, $ 703160, $ 822896 and $ 989286, respectively. The total risk for all hazard scenarios was estimated as $ 5007236. Similarly, human fatalities as a consequence of vehicles and trains being directly hit by landslides were estimated to be 15 persons per year. The output of the result is displayed as a risk curve, containing the relation between hazard with different annual probabilities and the corresponding total specific losses (Fig. 9.19). The result indicates that the loss.

9.6 Conclusions

Risk assessment is the final step in a chain of scientific and methodological preludes for disaster risk reduction. Thus uncertainties from all the previous steps propagate to this logical culmination of landslide hazard and risk analysis. Despite the explicit uncertainties in quantifying risk and defining vulnerability, landslide risk assessment is a formal land use planning necessity in several developed countries, especially along transportation lines. In India, Several human fatalities are reported annually as a direct consequence of landslides in the country. The sectors that are affected by continuous landslide problems are fairly well known and thus the need of the hour is to develop pragmatic remedial measures such as bio-engineering methods and structural designs capable of reducing slope instability conditions.

In January 2009 a panel discussion was held at the head office of the Geological Survey of India on some of the problems and possible solutions for a uniform and relevant method of medium-scale landslide hazard and risk assessment in India. Some of the conclusions were that:

- India should consider conducting landslide susceptibility mapping as the first step towards a comprehensive landslide hazard and risk assessment and follow this up with detailed analysis for specific areas.
- It is doubtful whether a uniform methodology can be applied to all landslide prone regions of India given the varied geological and climatic conditions. Hence, generalized susceptibility mapping approaches applicable for different landslide prone regions of the country are to be developed taking into account this variability in the prevailing environmental conditions. Also specific methods for different landslide types are required.
- Attention should on maintaining the uniformity of the output maps, which is extremely important for proper understanding by planners and other end user agencies. The scale of mapping for regional assessment could be 1:50,000. The output map may depict only three classes of susceptibility, "Low" containing <2% of all landslides, "Moderate" having ±8% of all landslides and "High" >90% of all landslides.

- The existing BIS guidelines completely ignores the necessity of a detailed landslide inventory, which is an extremely important input for a comprehensive hazard and risk assessment. A team of dedicated and trained experts should be employed to generate such a detailed landslide inventory using aerial photographs, stereo image interpretation and other ancillary datasets. Subsequently, a heuristic approach using a weighting system of geofactors may be adopted that is relevant to specific geo-environments (taking into account the landslide inventory), which could then be used to generate the susceptibility map. Statistical methods or Spatial Multi Criteria Evaluation (SMCE) could be an alternative to the heuristic weighting and rating method, depending on the availability of data and the choice of the researcher.
- The methods should be reproducible and scientifically dependable and clearly specify how the maps have been prepared and validated. For validation, the importance of using independent landslide inventory data is recommended.

GSI should work in tandem with other similar organizations in India to develop an open-source, web-based *National Landslide Database* containing information about landslide type, dimension, location, time of occurrence, topographic characteristics, etc. Though this may seem to be an enormous undertaking, given the state-of-the-art communication network and the participation and sharing of data between large numbers of interested working groups in central and state level organizations/institutions, it is quite feasible in India. It is reiterated here that an up-to-date landslide inventory is absolutely essential, without which a comprehensive landslide hazard and risk analysis is impossible.

References

AGS (2000) Landslide risk management concepts and guidelines (Prepared by Australian Geomechanics Society, Sub-committee on Landslide Risk Management). Aust Geomech 35(1):49–92

Aleotti P, Chowdhury R (1999) Landslide hazard assessment: summary review and new perspectives. Bull Eng Geol Environ 58(1):21–44

Anbalagan R (1992) Landslide hazard evaluation and zonation mapping in mountainous terrain. Eng Geol 32(4):269–277

Anbalagan R, Singh B (1996) Landslide hazard and risk assessment mapping of mountainous terrains—a case study from Kumaun Himalaya India. Eng Geol 43(4):237–246

Ang AH-S, Tang WH (1984) Probability concepts in engineering planning and design: decision, risk, and reliability (V II). Wiley, New York, 562 pp

Aston AR (1979) Rainfall interception by eight small trees. J Hydrol 42(3–4):383–396

Barlow J, Martin Y, Franklin SE (2003) Detecting translational landslide scars using segmentation of Landsat ETM + and DEM data in the northern Cascade Mountains, British Columbia. Can J Remote Sens 29(4):510–517

Basu SR, De SK (2003) Causes and consequences of landslides in Darjeeling-Sikkim Himalayas. Geogr Polonica 76(2):37–52

Begueria S, van Asch TWJ, Malet J-P, Gröndahl S (2009) A numerical simulation model of the propagation and deposition of mud and debris flows over complex terrain—Model description and modes of use. URL: http://digital.csic.es/handle/10261/11804?idioma=en. Accessed 15 May 2009

Bhattacharya A, Mishra P, Ghoshal TB, Bahuguna H, Ghatak T (1998) A geotechnical appraisal of landslides on 7th July, 1998 along National Highway No. 55, Progress report. Geological Survey of India, Government of India

BIS (1998) Preparation of landslide hazard zonation maps in mountainous terrains, IS 14496 (Part-2)— Guidelines. Bureau of Indian Standards Government of India, New Delhi

Blaschke T, Strobl T (2001) What's wrong with pixels? some recent developments interfacing remote sensing and GIS. GeoBIT/GIS, pp 12–17

Brabb E (1984) Innovative approaches to landslide hazard mapping. In:Proceeding of the IVth international symposium on landslides, Toronto, pp 307–324

Briggs RP (1974) Overdip slopes that can affect landsliding in Allegheny County, Pennsylvania. U.S. Geological Survey, Misc. Field Studies Map, MF-543

Bromhead E (1996) Slope stability modeling: an overview. In: Dikau R, Brunsden D, Schrott L, Ibsen M (eds) Landslide recognition: identification, movement and causes. Wiley, Chichester, pp 231–235

Brooks SM, Crozier MJ, Glade TW, Anderson MG (2004) Towards establishing climatic thresholds for slope instability: use of a physically-based combined soil hydrology-slope stability model. Pure Appl Geophys 161(4):881–905

Cannon SH, Savage WZ (1988) A mass change model for the estimation of debris flow runout. J Geol 96:221–227

Cardinali M, Reichenbach P, Guzzetti F, Ardizzone F, Antonini G, Galli M, Cacciano M, Castellani M (2002) A geomorphological approach to the estimation of landslide hazards and risks in Umbria, Central Italy. Nat Hazards Earth Syst Sci 2(1–2):57–72

Carrara A (1993) Potentials and pitfalls of GIS technology in assessing natural hazards. In: Guzzetti F, Reichenbach P, Carrara A (eds) Geographical information systems in assessing Natural Hazards—Abstracts. CNR, Perugia, pp 128–137

Carrara A (1999) Use of GIS technology in the prediction and monitoring of landslide hazard. Nat Hazards 20(2):117–135

Carrara A, Cardinali M, Detti R, Guzzetti F, Pasqui V, Reichenbach P (1991) GIS techniques and statistical models in evaluating landslide hazard. Earth Surf Proc Landf 16(5):427–445

Carro M, De Amicis M, Luzi L, Marzorati S (2003) The application of predictive modeling techniques to landslides induced by earthquakes: the case study of the 26 September 1997 Umbria-Marche earthquake (Italy). Eng Geol 69(1–2):139–159

Catani F, Casagli N, Ermini L, Righini G, Menduni G (2005) Landslide hazard and risk mapping at catchment scale in the Arno river basin. Landslides 2(4):329–342

Chen H, Lee CF (2004) Geohazards of slope mass movement and its prevention in Hong Kong. Eng Geol 76(1–2):3–25

Chung C-JF (2003) Validation of spatial prediction models for landslide hazard mapping. Nat Hazards 30(3):451–472

Chung CJF, Fabbri AG (1999) Probabilistic prediction models for landslide hazard mapping. Photogramm Eng Remote Sens 65(12):1389–1399

Chung C-J, Fabbri AG (2008) Predicting landslides for risk analysis—spatial models tested by a cross-validation technique. Geomorphology 94(3–4):438–452

Coe JA, Michael JA, Crovelli RA, Savage WZ, Laprade WT, Nashem WD (2004) Probabilistic assessment of precipitation-triggered landslides using historical records of landslide occurrence, Seattle, Washington. Environ Eng Geosci 10(2):103–122

Cruden D, Varnes DJ (1996) Landslide types and processes. In: Turner AK, Schuster RL (eds) Landslide: investigations and mitigation. special report 247. Transportation research Board. National Research Council. National Academy Press, Washington, pp 36–75

Davies P (1992) The mind of God—the scientific basis for a rational world. Simon & Schuster Inc, London, 256 pp

De Kok R, Schneider T, Ammer U (1999) Object based classification and applications in the Alpine forest environment fusion of sensor data knowledge sources and algorithms. ISPRS/EARSeL Workshop, Valladolid

Doorenbos J, Pruitt WO (1977) Crop water requirements. Irrigation and drainage paper 24. Food and Agriculture Organization of the United Nations, Rome

Farrel D, Larson W (1972) Modelling the pore structure of porous media. Water Resour Res 8:699–705

Fell R, Corominas J, Bonnard C, Cascini L, Leroi E, Savage WZ, Joint Technical Committee on Landslides and Engineered Slopes (JTC-1) (2008) Guidelines for landslide susceptibility, hazard and risk zoning for land use planning. Eng Geol 102(3–4):85–98

Fell R, Ho KKS, Lacasse S, Leroi E (2005) A framework for landslide risk assessment and management. In: International conference on landslide risk management, Vancouver, May 31–Jun 3 2005

Flanders D, Hall-Bayer M, Pereverzoff J (2003) Preliminary evaluation of eCognition object-based software for cut block delineation and feature extraction. Can J Remote Sens 29(4):441–452

Fredlund DG (1987) Slope stability analysis incorporating the effect of soil suction. In: Anderson MG, Richards KS (eds) Slope stability: geotechnical engineering and geomorphology. Wiley, Chichester, pp 113–144

Galli M, Ardizzone F, Cardinali M, Guzzetti F, Reichenbach P (2008) Comparing landslide inventory maps. Geomorphology 94(3–4):268–289

Ghosh S, van Westen CJ, Carranza EJM, Jetten VG (2009a) Generation of event-based landslide inventory maps in a data-scarce environment: case study around Kurseong, Darjeeling district, West Bengal, India. In: Malet JP, Bogaard TA, van Beek LPH, de Jong SM, Remaitre A (eds) Landslide processes: from geomorphologic mapping to dynamic modelling—A tribute to Prof Dr. Theo van Asch. CERG and Utrecht University, Strasbourg

Ghosh S, Reichenbach P, Rossi M, Guzzetti F, Cardinali M, van Westen CJ, Carranza EJM, (2009b) Influence of landslide types for calibrating different multivariate landslide susceptibility models. European Geosciences Union Annual General Assembly 2009, Vienna, Geophysical Research Abstracts Vol 11(EGU2009-13564), URL: http://meetingorganizer.copernicus.org/EGU2009/EGU2009-13564.pdf

Giles PT, Franklin SE (1998) An automated approach to the classification of the slope units using digital data. Geomorphology 21(3–4):251–264

Gumbel EJ (1958) Statistics of extremes. Columbia University Press, New York, 375 pp

Günther A (2005) RSS-GIS GridMap-Extension. Federal institute for geosciences and natural resources, Hannover. URL: http://www-public.tu-bs.de:8080/~aguenthe/rss-gis/. Accessed 10 May 2009

Günther A, Carstensen A, Pohl W (2004) Automated sliding susceptibility mapping of rock slopes. Nat Hazards Earth Syst Sci 4(1):95–102

Guthrie RH, Evans SG (2004) Magnitude and frequency of landslides triggered by a storm event, loughborough inlet, British Columbia. Nat Hazards Earth Syst Sci 4(3):475–483

Guzzetti F, Carrara A, Cardinali M, Reichenbach P (1999) Landslide hazard evaluation: a review of current techniques and their application in a multi-scale study central Italy. Geomorphology 31(1–4):181–216

Guzzetti F, Cardinali M, Reichenbach P, Carrara A (2000) Comparing landslide maps: a case study in the upper Tiber River Basin, central Italy. Environ Manage 25(3):247–263

Guzzetti F, Malamud BD, Turcotte DL, Reichenbach P (2002) Power-law correlations of landslide areas in central Italy. Earth Planet Sci Lett 195(3–4):169–183

Guzzetti F, Cardinali M, Reichenbach P, Cipolla F, Sebastiani C, Galli M, Salvati P (2004) Landslides triggered by the 23 November 2000 rainfall event in the Imperia Province, Western Liguria Italy. Eng Geol 73(3–4):229–245

Guzzetti F, Reichenbach P, Cardinali M, Galli M, Ardizzone F (2005) Probabilistic landslide hazard assessment at the basin scale. Geomorphology 72(1–4):272–299

Guzzetti F, Reichenbach P, Ardizzone F, Cardinali M, Galli M (2006) Estimating the quality of landslide susceptibility models. Geomorphology 81(1–2):166–184

Hansen A (1984) Landslide hazard analysis. In: Brunsden D, Prior E (eds) Slope instability. Wiley, New York, pp 523–602

Hargreaves GH, Samani ZA (1982) Estimating potential evapotranspiration. J Irrigation Drainage Eng, 108(IR3):223–230

Harp EL, Jibson RL (1996) Landslides triggered by the 1994 Northridge, California earthquake. Seismological Soc Am Bull 86:S319–S332

Hocking G (1976) A method for distinguishing between single and double plane sliding tetrahedral wedges. Int J Rock Mech Min Sci 13(7):225–226

Hoek E, Bray JW (1981) Rock slope engineering. Institute of Mining and Metallurgy, London 358 pp

IRPI CNR (2009) Tools—various software tools prepared or used by members of our team, or by scientists working with IRPI CNR. Instituto di Ricerca per la Protezione Idrogeologica, Perugia, Italy. URL: http://geomorphology.irpi.cnr.it/tools. Accessed 10 May 2009

Jaiswal P, van Westen CJ (2009) Estimating temporal probability for landslide initiation along transportation routes based on rainfall thresholds. Geomorphology. doi:10.1016/j.geomorph.2009.05.008

Karssenberg D (2002) Building dynamic spatial environmental models. PhD Thesis, University of Utrecht, The Netherlands, 222 pp

Kuriakose SL, Jetten VG, van Westen CJ, Sankar G, van Beek LPH (2008) Pore water pressure as a trigger of shallow landslides in the Western Ghats of Kerala, India: some preliminary observations from an experimental catchment. Phys Geogr 29(4):374–386

Kuriakose SL, van Beek LPH, van Westen CJ (2009a) Parameterizing a physically based shallow landslide model in a data poor region. Earth Surf Proc Land 34(6):867–881

Kuriakose SL, Sankar G, Muraleedharan C (2009b) History of landslide susceptibility and a chorology of landslide prone areas in the Western Ghats of Kerala. India Environ Geol 57(7):1153–1568

Kuriakose SL, Devkota S, Rossiter DG, Jetten VG (2009c) Prediction of soil depth using environmental variables in an anthropogenic landscape, a case study in the Western Ghats of Kerala. India Catena. doi:10.1016/j.catena.2009.05.005

Lu D, Mausel P, Brondizio E, Moran E (2004) Change detection techniques. Int J Remote Sens 25(12):2365–2407

Malamud BD, Turcotte DL, Guzzetti F, Reichenbach P (2004) Landslide inventories and their statistical properties. Earth Surf Proc Land 29(6):687–711

Mantovani F, Soeters R, van Westen CJ (1996) Remote sensing techniques for landslide studies and hazard zonation in Europe. Geomorphology 15(3–4):213–225

Mark RK, Ellen SD (1995) Statistical and simulation models for mapping debris-flow hazard. In: Carrara A, Guzzetti F (eds) Geographical information systems in assessing natural hazards. Kluwer Academic Publishers, Dordrecht, pp 93–106

McDermid GJ, Franklin SE (1994) Spectral, spatial, and geomorphometric variables for the remote sensing of slope processes. Remote Sens Environ 49(1):57–71

McDougall S, Hungr O (2005) Dynamic modelling of entrainment in rapid landslides. Can Geotech J 42(5):1437–1448

Meentemeyer RK, Moody A (2000) Automated mapping of conformity between topographic and geological surfaces. Comput Geosci 26(7):815–829

Millington RJ, Quirk JP (1959) Permeability of porous media. Nature 183:387–388

Murali Kumar B (2007) Flood Situation Report—2007 (No—137/2007, 1700 hours, LAST SITREP) disaster management division ministry of home affairs. Government of India, New Delhi

NDM (2009) The Nilgiris—District of Blue Mountains—Schemes—Disaster Management. Nilgiris District Administration, Ooty, India, Government of Tamil Nadu. URL: http://www.nilgiris.tn.gov.in/. Accessed 01 May 2009

Nichol J, Wong MS (2005) Satellite remote sensing for detailed landslide inventories using change detection and image fusion. Int J Remote Sens 26(9):1913–1926

Penman HL (1948) Natural evaporation from open water, bare soil and grass. In: Proceedings of the Royal Society of London, A(194):S120–S145

Ramasamy SM, Francis S, Neelakantan R (2003) Frequent landslides in Nilgiris, India—a phenomenon related to Pleistocene tectonism. In: Ramasamy SM (ed) Remote Sensing Geology. Rawat Publishers Pvt. Ltd., Jaipur, pp 251–255

Rib HT, Liang T (eds) (1978). Recognition and identification: landslides-analysis and control special report, 176, National Academy of Sciences. Transport Research Board, Washington, pp 34–80

Rickenmann D (2005) Runout prediction methods. In: Jacok M, Hungr O (eds) Debris-flow Hazards and Related Phenomena. Springer, Heidelberg, pp 305–321

Sarkar S, Kanungo DP, Patra AK, Kumar P (2008) GIS based spatial data analysis for landslide susceptibility mapping. J Mt Sci 5(1):52–62

Sengupta CK (1995) Detailed study of geofactors in selected hazard prone stretches along the surface communication routes in parts of Darjeeling and Sikkim Himalaya, Phase-I, Part-I (Rongtong-Kurseong road section), Annual progress report (F.S. 1993–94). Geological Survey of India, Government of India

Seshagiri DN, Badrinarayanan S, Upendran R, Lakshmikantham CB, Srinivasan V (1982) The Nilgiri Landslides: results of geotechnical and geological investigations of the G.S.I in collaboration with the state geology branch, government of Tamil Nadu. Miscellaneous Publication No. 57, Geological Survey of India, Government of India, Chennai

Simoni S, Zanotti F, Bertoldi G, Rigon R (2008) Modelling the probability of occurrence of shallow landslides and channelized debris flows using GEOtop-FS. Hydrol Process 22(4):532–545

Skempton AW, DeLory FA (1957) Stability of natural slopes in London Clay. In: 4th International Conference on Soil Mechanics and Foundation Engineering, London, pp 378–381

Soeters R, van Westen CJ (1996) Slope instability. recognition, analysis and zonation. In: Turner AK, Schuster RL (eds), Landslide: investigations and mitigation. Special report 247. Transportation research board. national research council. National Academy Press, Washington, pp 129–177

Sreekumar S (2009) Techniques for slope stability analysis: Site specific studies from Idukki district, Kerala. J Geol Soc India 73(6):813–820

Thakur VC (1996) Landslide hazard management and control in india status report. International Center for Integrated Mountain Development, Kathmandu

Thampi PK, Mathai J, Sankar G, Sidharthan S (1998) Evaluation study in terms of landslide mitigation in parts of Western Ghats, Kerala, Research report submitted to the ministry of agriculture, Government of India. Centre for Earth Science Studies, Government of Kerala, Thiruvananthapuram

van Beek LPH (2002) Assessment of the influence of changes in landuse and climate on landslide activity in a mediterranean environment. PhD Thesis, University of Utrecht, The Netherlands, 363 pp

van Beek LPH, van Asch TWJ (2004) Regional assessment of the effects of land-use change and landslide hazard by means of physically based modeling. Nat Hazards 30(3):289–304

Van Den Eeckhaut M, Reichenbach P, Guzzetti F, Rossi M, Poesen J (2009) Combined landslide inventory and susceptibility assessment based on different mapping units: an example from the Flemish Ardennes, Belgium. Nat Hazards Earth Syst Sci 9(2):507–521

van Westen CJ, Rengers N, Terlien MTJ, Soeters R (1997) Prediction of the occurrence of slope instability phenomena through GIS-based hazard zonation. Geol Rundsch 86(2):404–414

van Westen CJ, van Asch TWJ, Soeters R (2006) Landslide hazard and risk zonation: why is it still so difficult? Bull Eng Geol Environ 65(5):167–184

van Westen CJ, Castellanos E, Kuriakose SL (2008) Spatial data for landslide susceptibility, hazard, and vulnerability assessment: An overview. Eng Geol 102(3–4):112–131

Varnes DJ, IAEG Commission on Landslides and other Mass-Movements (1984) Landslide hazard zonation: review of principles and practice, natural hazards No. 3. UNESCO, Paris, 61 pp

Victor OCD (1962) Kurisumala: A socio-economic survey. St. Joseph's Apostolic Seminary, Alwaye 87 pp

Vinod Kumar K, Martha TR, Roy PS (2006) Mapping damage in the Jammu and Kashmir caused by 8 October 2005 Mw 7.3 earthquake from the Cartosat-1 and Resourcesat-1 imagery. Inter J Remote Sens 27(20):4449–4459

Wieczorek GF (1984) Preparing a detailed landslide-inventory map for hazard evaluation and reduction. Bull Assoc Eng Geol XXI(3):337–342

Wilson RA, Moon AT, Hendickx M (2005) Application of quantitative risk assessment to the Lowrence Hargrave drive project, New South Wales, Australia. In: Hungr O, Fell R, Couture R, Eberhardt E (eds) Landslide risk management. Taylor and Francis Group, London, pp 589–598

Chapter 10
Vision-Based Terrestrial Surface Monitoring

Gerhard Paar, Niko Benjamin Huber, Arnold Bauer, Michael Avian and Alexander Reiterer

Abstract The monitoring of geo-risk areas is getting more and more importance due to increasing damage caused by hazardous events such as rock slides, as a result of the environmental change. Terrestrial long-range sensing (up to several kilometres of distance between sensor and target region) is a valuable means for monitoring such sites using non-signalized targets in high resolution, which is necessary to detect regions, amount, direction and trends of motion early enough to take risk mitigation measures. The technology to realize such a sensing strategy combines various fields of research, such as sensor technology, surveying, computer vision and geological sciences. This chapter describes two vision-based sensing techniques suited for terrestrial surface monitoring (terrestrial laser scanning, and image-based tacheometers), and their sensing strategies, data processing and data exploitation issues. Examples for monitoring frameworks are given, and technical and engineering solutions are described. A set of applications from permafrost, glacier and snow cover monitoring, as well as rock fall site monitoring shows the relevance, technologic maturity and limits of existing approaches. Rock falls and other geo-hazards being the major fields of application for such systems, the chances of saving lives, protecting infrastructure and habitats and avoiding injury to field personnel are increased so that the better and more

G. Paar (✉) · N. B. Huber · A. Bauer
DIGITAL, Institute for Information and Communication Technologies,
JOANNEUM RESEARCH, Steyrergasse 17, Graz 8010, Austria
e-mail: gerhard.paar@joanneum.at

M. Avian
Institute of Remote Sensing and Photogrammetry, Graz University of Technology,
Steyrergasse 30, Graz 8010, Austria

A. Reiterer
Institute for Geodesy, GIS and Land Management, University of Technology,
Munich, Arcisstrasse 21, 80333 Munich, Germany

B. Pradhan and M. Buchroithner (eds.), *Terrigenous Mass Movements*,
DOI: 10.1007/978-3-642-25495-6_10,

accurate event can be monitored. The research and technology described in this chapter will help the surveying, photogrammetry and computer vision community fighting global warming impacts.

Keywords Terrestrial laser scanning · Image-based tacheometers · Digital surface model · Point tracking · Deformation monitoring · Rock glacier movement · Glacier change monitoring · Snow avalanche prediction · Geo-risk monitoring

10.1 Introduction and Scope

The necessity for monitoring geo-risk areas such as rock slides or avalanche areas is growing due to the increasing probability of such events caused by environmental change. For example, in Europe, rock slides cause increasing damage particularly in alpine areas. Efficient, automated, high-resolution, terrestrial, long range sensing measurement and analysis is able to monitor geo-risk (and related) objects by means of non-signalized natural target points—which is a key to such systems due to the lack of reachability and the required distance-to-the-object: commonly used systems such as locally established survey networks, on-site movement sensors or the incorporation of reflective targets fail due to their dependency on site accessibility. Even novel sensor concepts such as terrestrial laser scanning can only cover a subset of the requirements, they limit the distance of application to 1–2 km and do not provide the ability to track individual surface points in high resolution—an important feature to detect regions of motion early enough for taking measures of protection, warning inhabitants, closing infrastructure or evacuation.

Services in this area are reacting with a case-by-case strategy, using conventional technology. There is no well established market yet on this family of applications, although the occurrence of dozens of events per year indicates that a mature observation system is overdue.

Technology and research targeted to the non-contact terrestrial monitoring of such surface-changing events and processes combines methods and techniques from standard surveying, computer vision, photogrammetry, mechatronics, software engineering, and geological sciences. The main step beyond currently available techniques lies in the introduction of a novel modular suite of terrestrial visual survey sensors, be it Image Assisted Total Stations (IATS, used to perform long-range high-resolution measurements on single points), Terrestrial long-range Laser Scanners (TLS, used to survey large areas) or terrestrial radar measurements.

Sensors, their control and data processing, existing frameworks that enable expert users (geodesist/geologist) to operate the software, applications, test environments and verification procedures are still subject of dedicated research.

Rock falls and other geo-hazard events are the major fields of application for such systems, since these events generally can only be anticipated but not avoided. Chances of saving lives, protecting infrastructure and habitats and avoiding injury to field personnel increase the better and more accurate such an event can be monitored. This chapter covers some of these aspects, and together with related technologies it will help the surveying, photogrammetry and computer vision community fighting global warming impacts.

10.2 Sensing Techniques

10.2.1 Sensors

10.2.1.1 Terrestrial Laser Scanner

Scanning laser imaging has turned out to be an essential component of geotechnical disaster monitoring. The available laser devices have reached a technological fitness for this class of application over the last decade (Paar and Bauer 2001). With a maximum range of more than 2 km to naturally reflecting targets, a wide field-of-view, and a ranging accuracy of better than 2 cm remote monitoring of events like the Schwaz rock slide, Austria, Summer 1999 (Scherer 2004; Paar et al. 2000) could be accomplished.

There are quite a few systems on the market with an operating range between near-range (up to 10 m) and 300 m (hds.leica-geosystems.com 2010). However, to our knowledge the target range of more than 1 km for non-reflective targets is currently covered only by the LPM product line by Riegl Laser Measurement Systems (Horn, Austria) (www.riegl.co.at) with the current version of LPM-321 (Fig. 10.1; Table 10.1).

The distance measurement device is based on the time-of-flight method, for each single measurement a burst of several hundred laser pulses are emitted. The reflected return pulses are analyzed by a digital signal processor (DSP) to compile a single distance measurement. The distance measurement unit is mounted on a pan and tilt orientation unit motorized by step engines, similarly to a motorized theodolite the exact pan and tilt angles are read out by encoders. The device is controlled by an off-the-shelf PC via Serial- or Ethernet-interface, which handle both the device control and the data transfer from the sensor to the PC. The control software on the scanning device allows the acquisition of an (almost) rectangular regular grid of measurements, which is stored as one data file on the control PC. Each individual element within this grid consists of distance, reflectance value, the two angular measurements from the encoders of the mounting unit, and an estimated root mean square error of the distance measurement for reliability.

For the past few years, terrestrial 3D laser scanning systems have been successfully employed in the design and manufacturing industries as well as in industrial surveying (Pfeifer and Lichti 2004). Further development in terms of measurement speed, accuracy, range, field-of-view, and data sampling rate allow TLS to be applied in terrain surveying (Bauer and Paar 2004; Bauer et al. 2005) thus making this technique a very interesting instrument for measuring high mountain environments. In general several methodological, technical, and logistic

Fig. 10.1 Long range laser profiler/scanner. *Left* first generation Riegl LPM-2k. *Middle* Riegl LPM-321. *Right* Riegl LMS Z620

Table 10.1 Comparison of technical parameters of long range TLS by Riegl Laser Measurement Systems

	Riegl LPM-2K	Riegl LPM-321	Riegl LMS Z620
Market launch	1999	2007	2008
Measuring range			
good diffusely refl. targets	Up to 2,500 m	Up to 6,000 m	Up to 2,000 m
bad diffusely refl. targets	>800 m	>1,500 m	Up to 750 m
Minimum distance	10 m	10 m	2 m
Ranging accuracy	25 mm	25 mm	10 mm
Positioning accuracy	0.01 Gon	0.01 Gon	
Measurement Rate (points/s)	1–4	10–1,000	8,000–11,000
Measuring beam divergence (mrad)	1.2	0.8	0.15
Laser wavelength (μm)	0.9	0.9	0.9
Scanning range hor./vert.	400 Gon/180 Gon	400 Gon/165 Gon	400 Gon/160 Gon
Laser safety class	3B	1 M	1
Power supply (Volts DV)	11–18	12–28	12–28
Operation temperature range (°C)	−10 to +50	0 to +45	0 to +40

problems are to be encountered when establishing an integrated monitoring system in such an environment. This includes, amongst others, the stability of device control software, the automatic sensor orientation, the high number of measurements, the compensation of refraction and atmospheric influences, and the selection of reliable measurements. Many years of experience in the field of TLS (beginning in 2000) result in the development of a well-engineered and stable acquisition and analysis system which in combination with expert field work copes with all these conditions.

The integrated measurement system is capable of describing 3D motion and deformation of glacier (Avian and Bauer 2005; Avian et al. 2005b; Kellerer-Pirklbauer et al. 2005) as well as rock glacier (Bauer et al. 2003) surface within a few hours or even minutes of measurement depending on the device and the desired point density (Table 10.1). Besides the technical limitations of the measuring device maximum range also depends on the reflectivity of surface (which is favourable for snow and debris covered terrain at the wavelength of 900–1,200 nm and atmospheric visibility (best for clear visibility, bad for haze and fog). A measuring range of up to 6,000 m (Table 10.1) allows hazardous sites to be easily measured from a safe distance. Since each single measurement consists of a multitude of laser-pulses, different measurement modes (first pulse, last pulse, strongest pulse) give proper results even during bad weather conditions and on unfavourable surfaces like vegetated, moist or roughly structured terrain, that may otherwise lead to ambiguous measurements. For example, the last pulse technique allows detecting the range of the last target even if the measuring beam partially hits or penetrates other targets (like fog) before.

Monitoring systems based on a terrestrial laser scanner are described in Sect. 10.3.

10.2.1.2 Image-Based Tacheometers

Today's tacheometers measure directions with electronic sensors automatically. The axes are driven by motors controlled by a computer, automatic pointing is possible if special cooperative targets or active targets are used, and data can be captured, stored, and evaluated without human interaction. A higher degree of flexibility and automation however, will be possible, if artificial targets must no longer be fixed on the objects to be monitored. This will be possible, if the images of the telescopes visual field are used in a more flexible way.

Image-based tacheometers have a CCD/CMOS camera in its optical path. The images of the telescopes visual field are projected onto the image chip. The camera is capable of capturing mosaic panoramic images through camera rotation, if the axes of the tacheometer are driven by computer controlled motors. With appropriate calibration these images are accurately geo-referenced and oriented as the horizontal and vertical angles of rotation are continuously measured and fed into the computer. In such a system viewing angles must be addressed to image pixels inside the optical field of view. That means, especial calibration methods have to be used for the tacheometer, an autofocus unit has to be added to the optical path, and special digital image processing procedures have to be integrated.

An optical system for such a system was developed by Leica Geosystems (see Fig. 10.2) (Walser 2004). It is reduced to a two-lens system consisting of the front and the focus lens. Instead of an eyepiece a CCD sensor is placed in the intermediate focus plane of the objective lens. The image data from the CCD sensor are fed into a computer using a synchronized frame grabber. For the transformation of the measured image points into the object space the camera constant

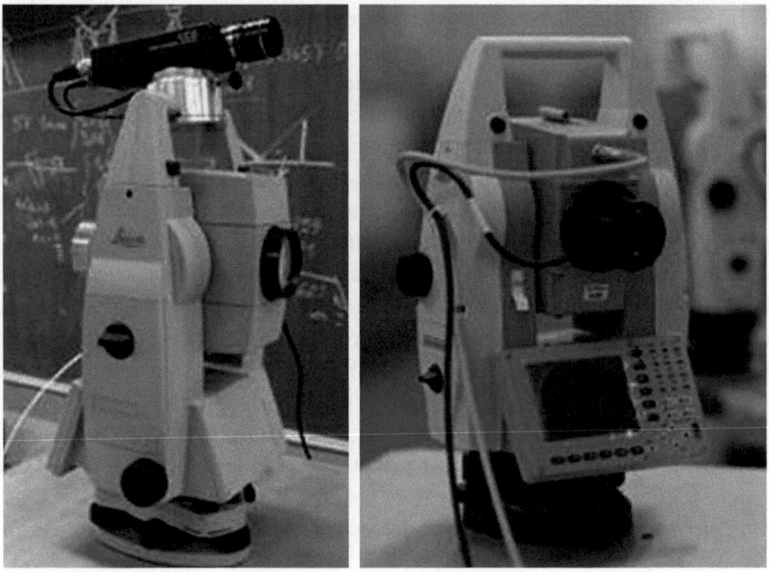

Fig. 10.2 Prototype systems developed by Leica Geosystems

must be known. In an optical system with a focus lens the camera constant, however, changes with the distance of the object.

The camera constant can be derived from the focal length. This can be performed automatically if an encoder measures the focus lens position relative to an origin, which is chosen when focusing to infinity. The optical mapping model includes not only the tacheometer axis errors and the vertical index error, but also errors resulting from a displacement of the projection centre from the intersection of the tacheometer axes and from the optical distortions for field points. Consequently calibration of an image-based tacheometer has to comprise all these errors.

Image-based tacheometers can work like a traditional theodolite without distance measuring—for the measurement and calculation of 3D point co-ordinates two or more measurement systems are used (master/slave mode); one (master) scans the object while the other one (slave) tracks it. Another approach is to combine the image-based measurement system with an integrated distance measurement unit. The system can be compared with a traditional tacheometer (total station) supported by image-based measuring. Reflectorless distance measurements can be performed until a maximal range of $\sim 1{,}000$ m.

Image-based tacheometers are currently available as prototype systems only. Such research devices have been developed by Walser (2004) and Reiterer (2004). Productive image-assisted tacheometers such as the Trimble VX (www.trimble.com) or the Topcon GPT7000i (www.topcon.eu) are using images as passive tool only and can not be used for the integration into an automated image-based geo-monitoring process.

10.2.2 Data Acquisition Strategies

10.2.2.1 Sensor Control

TLS

The laser scanner performs tasks of a predefined measurement schedule to automatically measure regions of interest (ROIs) round the clock. A ROI defines both the measurement raster, and the distance measurement parameters like integration time and mode. Several ROIs are combined to a measurement task. A measurement task list defines the measurement strategy in terms of order, priority, and point of time. On the scanner control PC a server is continuously parsing the task list and revaluating the measurement schedule. As soon as the scanner is idle, the next ROI is selected and executed. Unsuccessful measurements (i.e. due to bad environmental conditions or communication problems) are detected automatically; they are repeated and reassumed into the measurement list.

Different automatically selected strategies, like for example increasing the integration time for a single distance measurement in case of bad weather, or focusing on ROIs with top priority, allow retrieving maximum information dependent on the environmental situation.

Once a region of interest for data acquisition has been defined, the control software causes the scanner to acquire the selected rectangular region in spherical co-ordinates. A data logging scheme allows the establishment of a data base making sure that all relevant original data and their relations remain complete, unique and unchanged.

IATS

In case of the IATS, the so called system control component operates the sensors (device control, data readout, synchronization, etc.) and the algorithms (image preprocessing, point detection, etc.). The briefing about sensor unit, suitable measurement algorithms, and information about the points to be detected are located within separated subsystems.

The final output of this component is a list of coordinates (incl. additional data) of interesting points and/or regions of interest that will be stored in the common storage unit. User interaction with the system is implemented in the form of a simple graphical user interface (GUI) that allows the user to control the whole measurement process and to verify the decisions of the system.

10.2.2.2 Sensor Orientation

Sensor orientation for passive vision sensors in most cases is done by using reflective reference targets. An area around such a target is scanned with small grid width; a centroid localization algorithm (see Fig. 10.3) on the resulting image

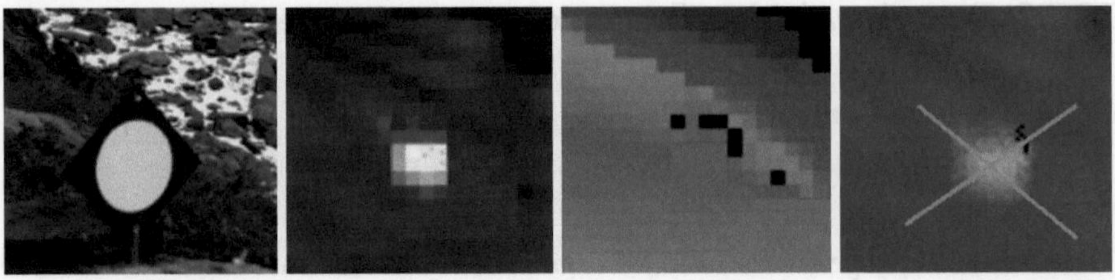

Fig. 10.3 Location of reflective target centre on sensor image. *Left* to *right* reflective target as seen by a standard camera, target reflectivity scanner image, target distance scanner image, subsampled reflectivity image and found target centroid

(subsampled, since the input measurements in general do not describe a regular grid) gains the angular components of the target co-ordinates in the scanner polar co-ordinate system. The distance is calculated as weighted average of all individual distance measurements covering the target.

Since most of the long-range TLS do not contain an electronic levelling sensor like standard theodolite it is necessary to determine all unknown position and orientation parameters using the reference targets. An optimization method is used which optionally keeps the sensor position fixed or variant. Result is a transformation vector and a rotation matrix. A minimum of 5 targets is advisable for proper sensor orientation, regardless of whether the position is fixed or not.

This step is the key to measurement accuracy; it involves several individual tasks that are still subject to improvement:

- Distribution of the orientation targets.
- Robust automatic evaluation of the usability of target measurements.
- Stable mathematical methods and algorithms for sensor orientation under various restrictions.
- Current research aims at using natural such as (non-moving) distinct landmarks (like rock formations) for continuous sensor orientation.

In addition to the determination of sensor orientation and location, the reference targets can be used to determine compensation values for atmospheric effects on the distance measurements.

The above mentioned targets forming a geodetic network can also be used by IATS for orientation purposes. If synchronized measurements of multiple devices are required for later triangulation they rely on a set of common targets for a mutual orientation.

10.2.2.3 Data Handling

All measurements are managed in a simple database that contains meta data (Sensor state, reliability, ROI data, etc.) and the measurements itself. Elements of such a database are

- Measurements in the raw file format of the sensor provider(s)
- Time period of sensing
- Calibration status of the sensor system (in simplest case the current transformation vector and rotation matrix between scanner co-ordinate system and the global co-ordinate system of the measurement project)
- Quick looks to support data handling tools
- Statistics of individual measurements to enable a quick filtering of unusable measurements
- Environment conditions during measurement such as temperature, pressure and humidity.

10.2.3 Data Evaluation

10.2.3.1 Surface Reconstruction

Pre-Processing

A classification of the raw laser measurements allows the examination of the quality and reliability of the distance measurements. Uncertain measurements are smaller weighted or even dismissed for the subsequent data evaluation.

The classification is based on an analysis of the laser beam reflectivity, the RMSE (as each single measurement consists of a multitude of laser-pulses), the structure (e.g. detected artefacts due to moisture), and external sensors (e.g. meteorological stations to estimate atmospheric influences to laser distance measurement).

Surface Model Generation

Once the measurements' reliability has been checked, all useful measurements are to be processed in order to obtain a 3D reconstruction of the current shape of the respective surface. Therefore each individual measurement (given in spherical co-ordinate system of the TLS) is transformed into a geo-referenced co-ordinate system by means of the sensor state stored with the measurement.

To represent each measurement in a reference co-ordinate system, a dense Digital Surface Model (DSM) is generated of the scene to be surveyed. A DSM is a regularly spaced grid in desired resolution on an analytical model of the local surface, in the simplest case a horizontal or vertical plane. It is used to store the elevation as a vertical distance at the grid points. We generalize the DSM to an arbitrary reference surface, to be able to represent the surface data in best resolution, since most of the potentially insecure surfaces are characterized by steep fronts. This data structure well complies with the practical requirements such as difference measuring, volume change evaluation, and various visualization tasks. Neighbourhood relations of measurement data points are directly described in the DSM structure; therefore operating on DSMs allows quick access to the surface heights in a well-defined geometry.

Direct mapping from the sensor spherical system to the DSM Cartesian co-ordinate space would result in a sparse and non-uniform elevation map, especially at large distances and flat angles to the surface. To avoid interpolation artefacts, the Laser Locus Method (Bauer and Paar 1999) for DSM (Kweon and Kanade 1992; Lieb 1991) generation proves to be a robust tool particularly with regard to glancing intersections, multiple heights, and occlusions. It supports error detection and utilization of additional confidence values provided by the range sensor. The Locus Method completely works in sensor space to compute the surface heights. For each grid point on the reference plane a hypothetical vertical line is inspected. Its intersection with the observed terrain is determined completely in sensor space utilizing regularly spaced sensor measurements. This yields a dense DSM with predefined resolution and elevation measurement uncertainty.

Since the DSMs of (temporally) different surface measurements are geo-referenced, simple differences between the DSMs reflect the changes in elevation. In consequence we can derive a full description of change in volume, spatial distribution of shape, or arbitrary profiles on the surface.

Single time-of-flight measurements are automatically combined to a measurement grid that enables the generation of a dense digital elevation model (DSM) of the object's glacier surface. Repeatable sensor orientation is performed using reflective targets fixed on stable surfaces somewhere in the spherical field of view of the sensor.

10.2.3.2 Manual Data Exploitation by Visualization

Since the DSM is given in a georefgeo-referenced co-ordinate system the visualization is mainly application driven and can be performed by standard commercial systems.

Current modes of visualization include pseudo-colour overlay on ortho image, animated 3D view, arbitrary profiles and simple numerical values within a predefined grid. Volume measurements and the output of statistical trends help to assess the further proceeding of a surface motion event. Such products include

- Interactive motion diagrams on a single ROI
- Video sequences of ROI structure
- 3D surface rendering with pseudo-colour overlay of distance change
- Lists of ROI statistics (time dependent) in different formats

10.2.3.3 Temporal Surface Comparison

The resulting DSM from each measurement campaign represents a dated state of the region covered by the sensor measurements. Since the data is georeferenced, simple differences between the DSMs reflect the changes in elevation between the campaign dates.

In a first step distance changes from the observer to selected points and ROIs can be measured and immediately statistically evaluated. Further analysis quantifies more complex movements, just as changes by debris/rock fall or accumulation. Eventually the objective is to classify the meaning of the event semantically.

The DSM difference describes only the component of the surface change perpendicular to the analytical DSM reference plane. In order to examine the complex kinematics of surface deformation the knowledge on surface motion in all three dimensions is required. 3D motion as well as structure changes like rock falls can be calculated by means of optical flow detection on the grey level images using correlation-based matching. A dedicated matching technique (Hierarchical Feature Vector Matching) (Paar and Pölzleitner 1992; Reiterer et al. 2003) can use both the surface structure (distance measurement texture) and/or the radiometric surface texture (RGB sensor or laser beam reflectivity). In such a way a single ROI behaviour can be categorized into simple classes such as *insignificant change, significant change, unusable measurement, material loss, or aggregation* (Figs. 10.36, 10.37). This enables to quickly focus the attention of experts and further automatic decision steps to potentially hazardous areas and events.

A possible hazard warning system uses the ROI tracking results and sets off an alarm if motion exceeds a critical threshold, or the structure has changed due to a recent landslide or rock-fall. Using higher-level information such as the spatial distribution of deformations on the entire hazardous site, the temporal behaviour of singular or multiple ROI motions, or a knowledge-based expert system could perform the semantic classification of the kind and relevance of the change event.

10.2.3.4 Point Tracking

In contrast to the above described comparison of DSMs, image-based systems make use of a more direct approach of motion or deformation detection. Using specific image analysis techniques it is possible to track homologue points over multiple measurement epochs and to calculate according deformation vectors.

Using a pair of synchronised image-based sensors that are capturing data of the same ROI from differing viewpoints at the same time 3D object points may be determined by spatial forward intersection provided homologues points in corresponding image sections are detected. In order to make use of these points in a geodetically deformation analysis, it is also necessary to recognise these points in multiple measurement epochs (this can also be referred to as point tracking). This result in the requirement of a dual point matching procedure: (1) inside one measurement epoch in corresponding stereo images and (2) over multiple epochs.

In the following section the method is being described that is used (with necessary adaptations) for both tasks. The hereby used process can be divided in three independent subtasks. In a first step, the single images have to be analysed for their information content. Points with a potentially high repeatability, so called *Interest Points* (IPs), are being detected and their image co-ordinates are saved. The hereby used process is called *Interest Point Detection.* As a second step every IP gets an orientation and information about the detected scale in order to being able to compare images of different size or taken from different view points. Dependent on orientation and scale the environment of the point is captured and saved in form of a descriptive feature vector. This operation is called *Interest Point Description*. The generated feature vectors can be compared and matched against a reference set generated from a corresponding image to obtain homologues pairs of points.

To track the detected points over multiple measurement epochs the same process is applied again with the adaptation that homologue points of the current epoch are matched with a saved set of descriptive feature vectors corresponding to the homologue points of the reference epoch.

A main requirement for this application is a high invariance towards changes in environmental conditions such as illumination or transformations in the image domain like differing viewpoints. Furthermore the time span used to measure and calculate one epoch is restricted by the frequency of the measurements. Ultimately the used operators must be deployable on standard field capable hardware.

Two algorithms which fulfil the above mentioned requirements are the SIFT—Scale Invariant Feature Transform (Lowe 2004) and the SURF—Speeded Up Robust Features (Bay et al. 2008) Methods.

10.2.3.5 Deformation Monitoring

The above mentioned methods are usually used to carry out a deformation monitoring, wherein any deformation of an object can be seen as the result of a physical process. Modern deformation monitoring techniques offer the possibilities to measure and analyze such a process in all details. Today geodetic deformation analysis means geodetic analysis of dynamic processes (Teskey 1985; Welsch et al. 2000). This implies time varying stresses and time varying reactions; the object is permanently in motion. Monitoring such situations requires permanent and automatic observation procedures and measurement systems. DSM differences describe only the vertical component of the surface change. In order to understand the complex kinematics of, e.g. rock glacier deformation furthermore the knowledge on surface motion in all three dimensions is required.

Kääb et al. (2003) and Kaufmann and Ladstaedter (2000) provide solutions to calculate the 3D motion by means of optical flow detection on the gray level images using correlation-based matching. This method is not applicable to the current laser scanning setup, since it cannot be assured to have similar reflectance conditions, which is a prerequisite for robust matching.

Despite this lack of textural information the tracking of objects on the surface can still be performed by the high resolution structural data provided by the DSM. State-of-art matching methods (Paar and Almer 1993) obtain dense tracking vectors only on regions where the structural surface changes are relatively small. In combination with the DSM differences mentioned above, this results in a three dimensional vector field that describes the kinematics state of object's surface between the given epochs.

To process a geodetic deformation analysis, several commercial software packages are available (e.g. GOCA, PANDA, etc.). These methods make use of 3D object coordinates which can be derived using the above described process of point tracking using input data from oriented image-based sensors. Newest trends aim to integrate numerical models or more sophisticated methods such as the application of so-called knowledge-based systems into the analysis and alerting process in order to reduce the manual input of an operator/expert (Vicovac et al. 2009). Actual hazard alerting is a topic that goes beyond the scope of most actual research projects. However, alerting strategies that are immediately derived from the obtained deformation monitoring status, such as the analysis of deformation acceleration pose a challenging field of future research.

10.3 Monitoring Frameworks

10.3.1 Geoscanner

The DIBIT Geoscanner (www.dibit-scanner.at) is a measurement framework that was developed by Joanneum Research in Graz, Austria (www.joanneum.at), to enable TLS-based 3D-measurements for geo-monitoring purposes. The main focus of the system is the automated monitoring of geo hazard zones, especially in alpine areas, with no direct access to the monitored site because of the reflectorless laserscanning approach. According to the application different laser scanner models can be used, which are installed either permanently or temporarily on the site.

The Geoscanner software (Fig. 10.4) allows to scan either individual Regions Of Interest (ROIs) or to plan and execute long-term scanning and monitoring scenarios which run automated according to a specified schedule. An integrated data interpretation module offers an immediate evaluation and visual representation of the measurement results. The Geoscanner system operates round-the-clock and automatically provides TLS measurements, 3D evaluation, and interpretation without any user interaction.

The Geoscanner framework described above is available both for scientific and commercial use. Specifically it consists of the following modules:

1. Data acquisition unit for stationary and mobile automatic scanner control and data storage.
2. Orientation unit to identify the sensor orientation within a geodetic network.
3. Visualization unit to visualize the changes obtained by the system and to integrate them into formats well known by the user community.

As mentioned before all the results are immediately available, which makes the Geoscanner monitoring a valuable tool for hazard prediction, risk evaluation and

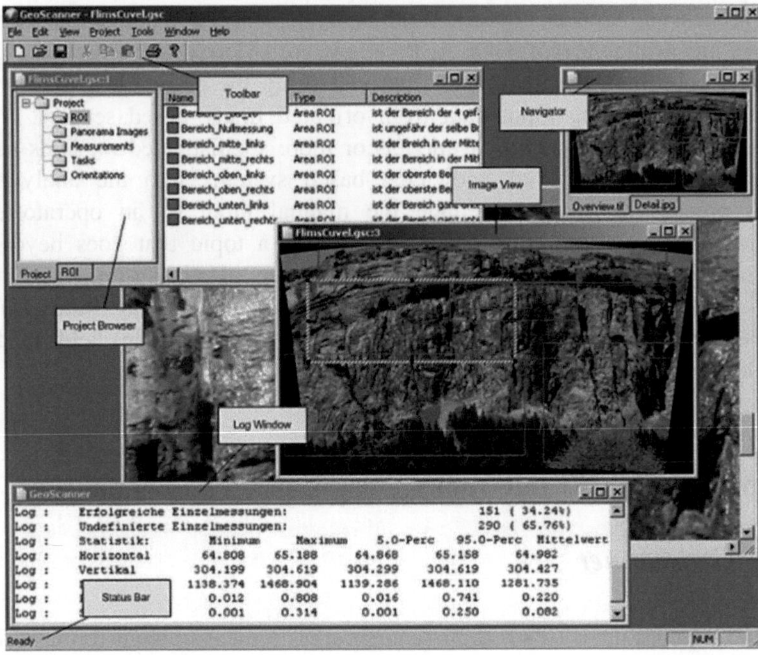

Fig. 10.4 DIBIT Geoscanner main window layout

scientific fieldwork. Such a system can be either operated non-stationary, or installed within less than two days at any location having access to power supply and mobile telephone connection. All relevant data processing and visualization can be done on-site which makes the system extremely useful for efficient research in the field.

Figure 10.5 shows a sketch of the Geoscanner framework's workflow. The laser scanning device is controlled by a server which also reads out the data. The server runs a sequence of individual measurements on ROIs (given in a task list) and stores them in a data base which is available over the internet. Certain measurements are used for continuous orientation of the sensor, also in order to store the current sensor orientation state in the data base. This is done using geodetically measured reflective targets. Auxiliary functions filter the measurements in terms of reliability, the distance measurements are corrected by compensation values gained by distance measurements to reference targets.

For further application the measurements are read from the database to generate DSMs which are aligned in the global geometrical context due to continuous knowledge of sensor orientation. The DSMs are combined to cover the entire area of interest, which serves as a data source for further display in combination with other georefgeo-referenced data like ortho image, GIS data or additional Virtual Reality (VR) objects.

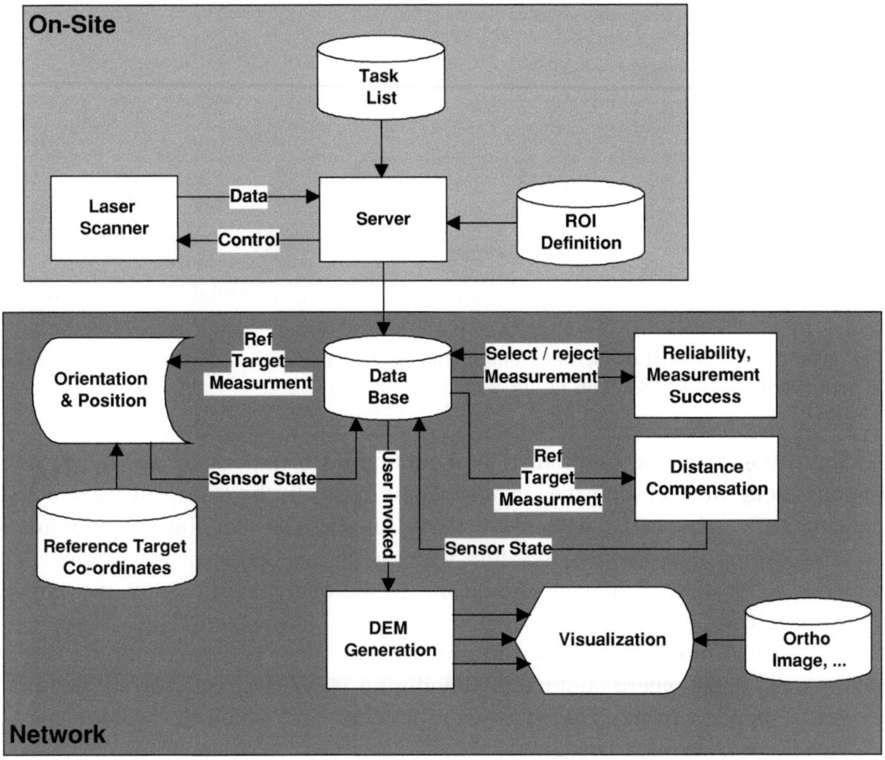

Fig. 10.5 Geoscanner system overview

10.3.2 Site Monitor

"Site Monitor" is a complete laser measurement-based system for measuring and monitoring the stability of rock faces, landslides, and buildings. It was developed by "3D Laser Mapping", a UK based company that is specialised on providing LiDAR software and hardware solutions (www.3dlasermapping.com).

The Site Monitor system uses reflectorless laser scanning (Fig. 10.6, right) to make range measurements on a pre-defined grid of points covering the area of interest, at a selected interval in order to provide comparable data for further monitoring. The reflectorless approach has the advantage of not having to apply prisms or reflector targets which would result in a much sparser grid and may not be possible in cases without physical access to the monitored site.

The influence of atmospheric conditions on the range measurements can be corrected by measuring a number of control points with known co-ordinates to compare the true range with the measured range and apply the correction to the data.

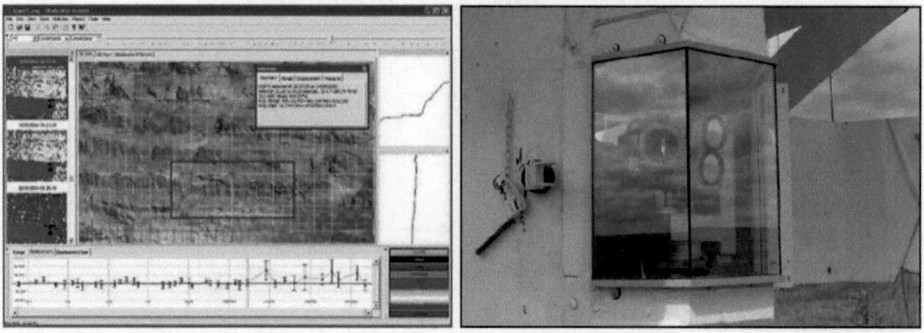

Fig. 10.6 *Left* screenshot of the SiteMonitor analysis for geotechnical monitoring. *Right* sensor in high-transmissivitiy glass housing for weather protection

Site Monitor can be used either in a continuous or a periodic setting. In continuous mode of operation the scanner would typically be set up in a protective housing on a stable pillar with fixed power supply. The scanning runs automatically according to a defined schedule without further operator intervention. Via network connection alerts can be sent out to the operator in case of movement, alarm, or error.

The manual mode can be used to switch the scanner between multiple survey pillars or in areas where continuous monitoring is not necessary or not possible. Power is supplied from a battery or from a vehicle on which the scanner can be mounted and controlled from.

Data Interpretation is carried out by a separated Analysis tool (Fig. 10.6, left) that provides 1D, 2D and 3D visualisations, horizontal and vertical cross-sections, and range, displacement and displacement rate graphs to allow an interpretation of structural integrity of the monitored site.

10.3.3 i-MeaS

Since a couple of years a strong trend towards image- and laser-based measurement systems can be seen in the field of engineering geodesy. These methods are primarily used to document an as-built state or to conduct planning and for architectural object reconstruction. In addition they are particularly well suited for ongoing deformation measurements because of the high degree of possible automation.

As mentioned above the combination of a conventional tacheometer with an imaging sensor poses a hybrid approach. By means of such a system geo-referenced images for further computation can be captured in an automated way.

Former research work was mainly focused on fundamental problems like calibration (Walser 2004; Wasmeier 2009), image pre-processing (Roic 1996), manual (Scherer 2004) and automated point detection (Mischke and Kahmen 1997;

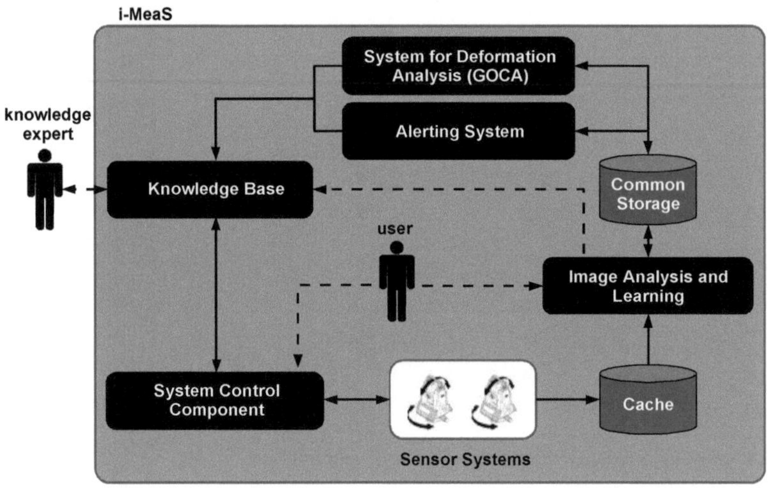

Fig. 10.7 Block scheme of information exchange between the functional system units

Reiterer et al. 2009). The aspect of repeatability which is of particular interest for the engineer geodetically deformation analysis remained untouched (examinations concerning the accuracy of manually detected points were partly carried out by Mischke and Kahmen 1997 and Wasmeier 2009). In the past couple of years intensive research on the development of a deformation measurement system based on image-based tacheometers has been carried out at the Institute of Geodesy and Geophysics at the Vienna University of Technology (Austria). A current research project (i-MeaS—An Intelligent Image-Based Measurement System for Rock Fall Monitoring) concentrates on the concrete application of geo-monitoring (rockfall, landslide, etc.). The measurement system is based on two image-based tacheometers connected with a central controlling computer running according software. Using two synchronized measurement devices, object points can be determined by spatial forward intersection thus enhancing measurement distance and accuracy.

10.3.3.1 System Layout

System layout of the i-MeaS system consists of the following components: sensors, system control component, knowledge-base, deformation analysis system, image analysis and learning unit, alerting system, cache and storage unit.

The component for system control has the task to operate the sensors (device control, data readout, synchronization, etc.) and the algorithms (image pre-processing, point detection, etc.). The briefing about sensor unit, suitable measurement algorithms (for the image-based tacheometer), and information about the points to be detected will be located within the knowledge-based subsystems. The final output of this component is a list of co-ordinates (incl. additional data) of interesting points and/or ROIs that will be stored in the common storage unit (Fig. 10.7).

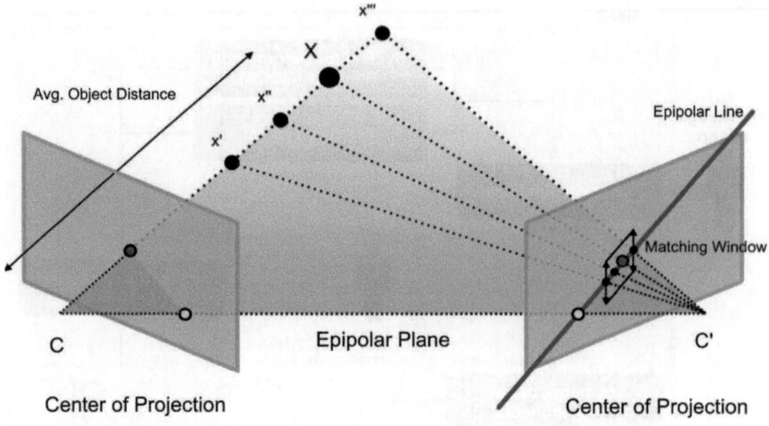

Fig. 10.8 Matching window based on epipolar-constraints

Detected points are being sorted according to their image coordinates and only compared with those that lie in a certain window in the reference data. In case of stereo image pairs this window is determined by epipolar-geometric constraints. Namely an epipolar-line is being calculated based on the known orientation of the images and further constrained using the known average object distance. The height of the matching window can be determined per parameter according to the stability of the object and the time difference between compared measurement epochs (expected deformation). Figure 10.8 illustrates the geometric constraints leading to the described matching window. The achievable speed-up can be described with the help of some example figures: two lists of feature vectors (as described above) of 25 thousand points each, require 625 million vector comparisons (quadratic runtime). Using the calculated constraints this process can be reduced to about 1,25 million comparisons (based on the assumption of an even point distribution in images of 2,560 $\times$ 1,920 pixel and a matching window of 200 $\times$ 50 pixel) resulting in a fraction of the original computation time only.

Decisions in the course of the whole processing sequence are executed by an appropriate (knowledge-based) decision system. From an architectural point of view, this component is not a monolithic system, but consists of a number of subsystems dedicated to the different subtasks. Knowledge between these subsystems is shared using a central knowledge base.

The deformation analysis system is realized by an integrated commercial system (e.g. GOCA). From the viewpoint of i-MeaS, the interaction between the deformation analysis part and the other components represents the generation of output files and the reading/importing of deformation analysis files. The output files contain the measured image point co-ordinates, the calculated 3D point co-ordinates (for more details about the calculation of 3D point co-ordinates using image-based measurement systems can be found in Reiterer et al. (2003), and the captured additional data (e.g. information from the image analysis process). All these data sets are used for deformation analysis and for the alerting system.

In contrast to conventional monitoring tools, with i-MeaS it is possible to perform a deformation interpretation. Based on the measured deformation vectors, a measurement preprocessing is performed (mainly clustering to detect areas of similar movement). On the basis of this information and additional data about velocity and orientation, some conclusions about the kind of occurring movement can be drawn. Additionally, data of different, heterogeneous sources, such as geodetic deformation measurements, geotechnical measurements, geological maps, geomorphological maps, in situ investigations, and numerical modeling methods can be included (Vicovac et al. 2009).

10.3.3.2 System Verification

Through the implementation of automated point detection and matching algorithms the necessity arises to evaluate the accuracy and reliability of the system as well as other extrinsic influencing factors. Thus the next section is dedicated to an evaluation concerning overall quality and performance of the developed i-MeaS prototype measurement system.

In the literature examinations and evaluations of the used algorithms and operators can be found, e.g. Bauer et al. (2007). A comprehensive study comparing the performance of local descriptors can be found in Mikolajczyk and Schmid (2005). According to them SIFT and SIFT-like descriptors reach their highest accuracies under viewpoint changes of up to 30°. Results in the area of textured as well as in structured scenes were significantly better then previous algorithms could achieve. The developers of the SURF operator (Bay et al. 2008) showed that their algorithm reaches a similar performance and robustness against changes in environmental conditions while the computation time was reduced by a factor of around five. These results were approved in an independent evaluation in 2007 (Bauer et al. 2007).

For the project at hand variable tests were carried out to evaluate the results of the automated point detection process. They can be divided into three main groups.

- offline-tests (simulations),
- online-tests under laboratory conditions,
- online-tests under real conditions.

Offline-Tests (simulations)

In a first step offline tests where carried out in order to verify and optimize the point detection and matching process. For the evaluation process image data from the IATS sensor system (5 MP) was used as well as pictures taken with a high resolution *Digital Single Lens Reflex* (*DSLR*) camera (12 MP).

A central issue arising from the use of a specific automated point detection process is the selection of appropriate parameters for the used algorithms. For the evaluation of the point detection and matching process a simple supporting tool was developed and implemented. Its main purpose was to help compare and evaluate the repeatability, the quality of the results and the computation time

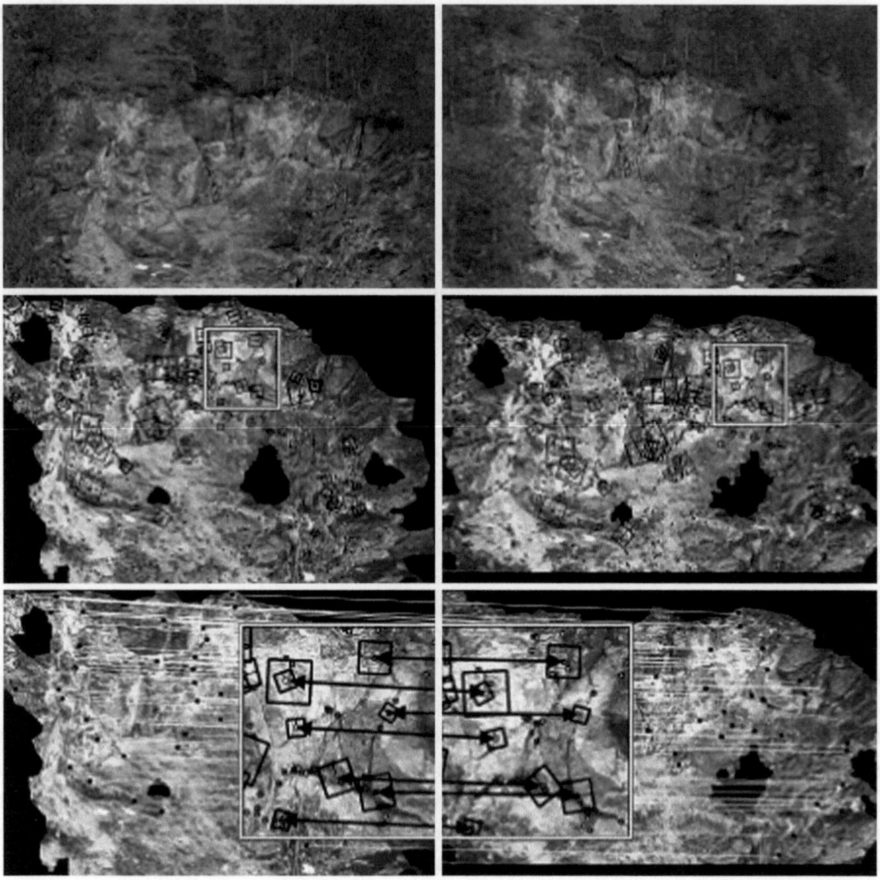

Fig. 10.9 Stereo picture pairs from *top* to *bottom* with ~50 m base and ~80 m average object distance, respectively. *Top* original high resolution (12 MP) stereo images (site in Triestingtal near Vienna). *Middle* interest points in masked area including descriptor size and orientation of the "matches" (framed sample sub-region). *Bottom* zoomed out sub-region showing homologue points (including descriptor window) connected by *arrows* (background displays all other matches)

needed by different existing operators. The separated creation of the image analysis component and the measurement system which were combined and integrated later via previously defined interfaces enabled a completely autonomous development and testing of the respective parts.

Figure 10.9 shows the application of the SURF-operator, wherein areas not suitable for deformation analysis (e.g. vegetation) were masked manually. A higher image resolution increases the amount of detected IPs (using the same set of parameters) in an almost linear manner thus extending the computation time. This means that the sensors used in the i-MeaS project pose a sound compromise for an online-system regarding resolution and computation time.

However an increased amount of available object points may still improve the quality of deformation representation (a chart illustrating this relation can be found in Fig. 10.12).

Extensive offline-tests should also determine the optimal set of parameters for the adaptation of the used operator (different image resolutions as well as the influence of changed illumination conditions were simulated). It could be demonstrated that the influence of illumination (cast shadows) could be greatly reduced by applying specific image pre-processing operations. Histogram equalization may improve the results by adapting image contrast conditions in the stereo images in a similar manner. Since these contrast values are unevenly distributed over the image, the result of such a global image enhancement is not optimal and was replaced by "adaptive histogram equalization". Therein the contrast adaptations are made within a certain predefined image neighbourhood using a sliding window technique. As a further improvement the Contrast Limited Adaptive Histogram Equalization (CLAHE) was used to threshold the maximum contrast enhancement and minimize the amplification of noise in homogeneous areas. Figure 10.10 illustrates the improvement of the results that can be achieved without image pre-processing (a), with global histogram equalization (b) and adaptive histogram equalization (c).

A standardized set of parameters controlling the SURF-algorithm proved to be not practicable—particularly in regard to the constitution of the surface (structure, colour, etc.) which requires additional adaptations.

The customisation of the parameters for the point detection algorithm to a specific test object or site poses a central challenge for an automated system. As mentioned before current development is focusing on rock surfaces. Vegetation and other non-stable or unwanted objects in the image domain have to be masked in the course of selecting a reference epoch. An automated detection of these structures may be addressed in future research but has to deal with serious problems from the field of image segmentation.

Online-Tests Under Laboratory Conditions

Primary objective of the online-tests under laboratory conditions was the confirmation of the simulation results in the controlled environment of a measurement lab. In order to carry out an automated point detection using image based tacheometers a couple of surrounding conditions have to be met. Aside from known errors of geodetic instruments like the axis error for instance, additional problems caused by the imaging sensor system have to be considered. In the present measurement system this is addressed by a highly complex calibration procedure (Walser 2004; Wasmeier 2009). Simplified this means that the mechanics' and imaging sensors' imperfections are considered so that every point on the imaging sensor can be related to its own specific angle values (horizontal Hz and vertical angle V).

The online-tests described below are focused mainly on the evaluation of the repeatability of the automated point measurements. The multi-sensor-system consisting of the tacheometer and the imaging sensor has will be considered as one single device (a separate examination will be carried out at a later date).

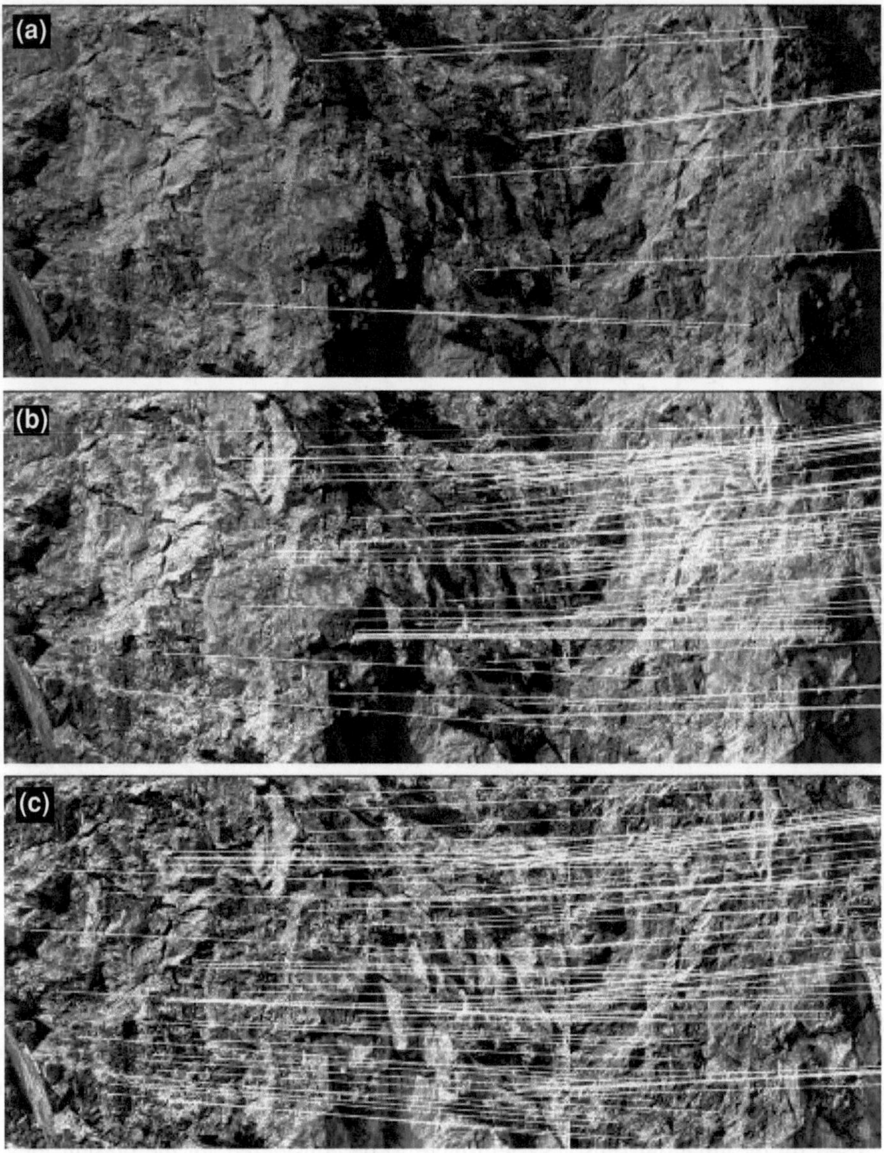

Fig. 10.10 Effect of image pre-processing on operator performance on data containing cast shadows (high threshold used to reduce the amount of detected points), **a** original data: 9 correct matches, **b** global histogram equalization: 61 correct matches, **c** contrast limited adaptive histogram equalization: 81 correct matches

In the course of the laboratory test an adequate target panel was stably mounted approximately 4 m in front of the sensors. The two imaging tacheometers were oriented and focused in the direction of the panel (the captured each covered a part of the target). Subsequently points were detected using the SURF-operator and their 3D co-ordinate in the object domain were calculated by means of spatial forward intersection. Both tacheometers were first rotated by 90° before targeting the panel again and obtaining the formerly measured 3D-points once more. This process was repeated 10 times thus creating a test series containing 10 measurement epochs. In each epoch about 250 points were detected and tracked over time. It was shown that the maximal difference between the reference- and a consecutive-epoch was 0.1 mm. This value lies within the measuring tolerance of the system. Hence it was clearly shown that the limiting factor is not the imaging chip but the tacheometer on which the sensor system was build.

In order to address the issue of changing illumination conditions and their influence on the point detection and matching process the above described 10 measurements were complemented by another set of 10 epochs. The configuration was complemented by an additional light source illuminating the target from different angles. Using appropriate image pre-processing steps and optimised image capturing (shutter speed) the system can be adapted to changed environmental conditions. The consideration of irregular illumination like cast shadows poses a much greater challenge. Simulations regarding this factor of influence have yet to be carried out. However first conclusions can be drawn based on tests executed under real conditions.

In addition these online-tests have to take the influence of temperature into consideration (thermal stability). This influence factor has been extensively covered by Wasmeier (2009) and should therefore only be referred to the existing literature for the sake of completeness.

Online-tests under real conditions

In order to test the measurement system including the used operators and algorithms under realistic field conditions a fixed installation of the sensor system was conducted over several days. The installation was made on the *Pasterze*, the largest glacier in the eastern Alps. A lower, debris covered part of the glacier as well as a geologically stable rock face was chosen as test site. Main purpose of the test was not so much the examination of accuracies but the evaluation of the point detection and matching process and the consecutive calculation of 3D co-ordinates on the object under realistic environmental (especially illumination) conditions.

The two imaging tacheometers were positioned in a mutual distance (base) of ca. 70 m and an average distance to the object of 1,000 m. A stable positioning towards the monitored object could be carried out based on local conditions. The angle of sight can be described as optimal whereas the relatively small base length compared to the large distance to the object did not represent an ideal configuration. However due to the primary focus on point detection and matching this did not pose a further problem. An overview over the region of interest (ROI) on the object is given in Fig. 10.11.

The region of interest was covered by $11 \times 23 = 253$ images with an overlap of 20% by both scanning positions. The scanning process took approximately

Fig. 10.11 Panorama of measurement site at the Pasterze glacier, including scanning position (*left*) and region of interest (ROI) on the object (area covers about 370 × 120 m)

30 min for each of the two positions. Caused by the implementation (being a prototype system) only sequential image capturing is possible, resulting in a combined scanning time of about one hour per measuring epoch.

Immediately after the image capturing the automated point detection process is started. The first measurement epoch is defined as zero or reference epoch and homologue stereo points are therefore being searched in all picture pairs and results are saved in *Descriptive Feature Vectors*. Caused by the long computation time of this operation the process was computed offline. This was made possible by the modular structure of the system: the process of measurement can be saved at certain pre-defined stages, aborted and continued later on. In consecutive epochs the feature vectors of an image pair are not mutually matched but against the respective previously saved vectors of the reference epoch. Caused by the reduced size of the saved vectors (containing only the homologue points of the reference epoch), computation time decreases to only a fraction of a full matching run. In a third step the remaining points are mutually matched to ensure a stereo relation of the tracked points in the current epoch.

The main parameter settings used were a threshold of 0.001 and 4 octaves divided into 4 intervals. This ensures a high amount of found IPs and homologue points in the reference epoch, resulting in a high amount of points being tracked over time without overly increasing computation time. It has to be mentioned that illumination conditions strongly varied both between corresponding stereo epochs (fast moving clouds) as well as between the former and consecutive epochs (time of day) this factor was compensated by a histogram equalisation. These differences in environmental conditions complicate the detection and use of only a few but very concise IPs and require a higher basic set of points.

Figure 10.12 shows a compressed analysis of a sample set of stereo- and consecutive epochs well suited to represent the region of interest. It illustrates the total amount of "interest points" found, the amount of matched homologue points in the reference epoch and the points that could be tracked in a consecutive epoch (first, fourth and fifth line from top respectively, absolute scale on horizontal axis to the left) under the influence of different threshold parameter settings (values on vertical axis). Furthermore the corresponding computation time in seconds is shown on the right.

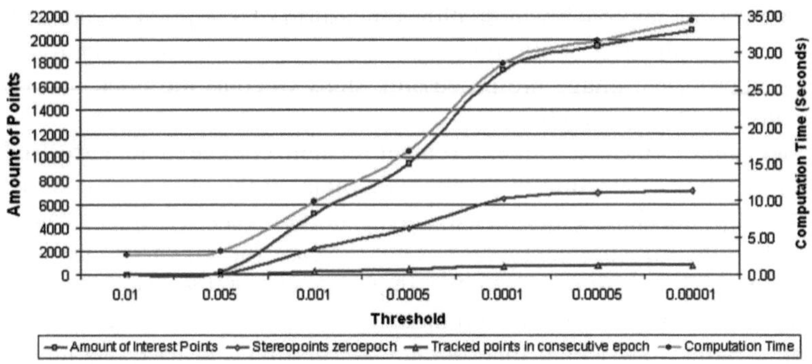

Fig. 10.12 Overview over the test series—SURF-Operator

The point quality was assessed by a manual review of the visual representation of the results to determine the amount of false matches. This was done up to about 2,500 detected homologue points resulting in a false matching rate of 5–10% in the reference epoch which could be reduced to zero for points tracked in the consecutive epoch because of the chosen multiple matching strategy. With a increasing matching rate (amount of homologue points detected) in the reference epoch more and more homologue points get lost in the course of the epoch matching.

The measurements under realistic conditions (changing illumination shadows, etc.) provide a sufficient amount of points for a representation of potential deformations. A remaining problem is the distribution of these points, which is currently only defined by mathematically defined local extrema in the image domain. A regular distribution of the points can be reached by a suitable grating of the image (the used process could be applied to each part of a regular grid with different parameter settings in order to obtain a regular distribution of detected points). The systems most urgent problem is currently the computation time of the process. The detection of IPs and the necessary matching takes about 20–30 s for one pair of images (dependent on the amount of detected IPs). For the whole region of interest this means an overall computation time of about 105 min. These values refer to the limited computational power of a conventional field notebook (2.4 GHz Intel Core 2 Duo, 2 GB Ram). This shows that the current system is only limited suitable for an online application. There is constant ongoing work on the enhancement of the prototypical implementation of the point detection, description and matching regarding computation time by reimplementation, optimisation, parallelisation, etc. Ultimate goal is a parallel measurement and evaluation of the data.

10.3.3.3 i-MeaS Prospects

The present section describes the first evaluation results of image-based deformation measurements. The used point detection and matching mechanisms have been described in detail. It could be shown that the developed system concept poses a promising approach which allows measuring large point clouds with high accuracy. The influence of illumination conditions can be strongly reduced by means of appropriate image pre-processing steps whereas the issue of cast shadows requires further research.

The advantage of such a measurement system compared to terrestrial laser scanning lies in the detection of distinctive points on the object (in contrast to an unstructured point cloud). The disadvantage is the necessity of constant illumination during the measurement which is not the case for terrestrial laser scanning.

Ongoing work concentrates on the improvement of the point detection and matching. Furthermore the measurement system will be integrated in a geo-monitoring-framework which incorporates an alerting and early-warning-system as well as an interpretation and classification of the occurred deformations.

10.4 Applications

10.4.1 Introduction

Similar relevant work on vision-based terrestrial surface monitoring started close to the end of the twentieth century. A variety of use cases have been since documented, both for commercial and scientific applications, and with different levels of maturity. Each individual case gained a set of experiences, modifications, improvements, feasibility statements and—most important in the context of this book—new research and development challenges for such systems. In the following sections a heterogeneous set of application classes and specific usage of terrestrial surface monitoring is documented.

10.4.2 Permafrost

The evaluation of rock glacier surface changes needs fast and cheap observation methods with accuracy in the range of a few centimetres. Long-range laser scanners can achieve measuring distances up to a few kilometres of range. It is shown that a system using such a device is able to successfully perform an efficient long-term change survey. We report on the sensor and software setup, the logistics and the procedure for data evaluation to perform the proposed monitoring task. An experiment was carried out at the Hinteres Langtalkar rock glacier in the Hohe Tauern range of the Austrian Alps. The obtained results enable the access to high-resolution surface deformation data in all three dimensions. Relevant parameters and advantages of the systems as well as drawbacks and ideas for further improvements are pointed out. The operational system is available for further scientific exploitation.

Surface dynamics of rock glaciers are of increasing interest due to its high relationship to thermal conditions of permafrost areas. A few rock glaciers in the

Alpine arc reacted on increasing air temperatures with extraordinary high movement rates over several years (Roer et al. 2005). Most of them now show decreasing surface velocities (Delaloye et al. 2008). The rock glacier Hinteres Langtalkar changed its behaviour most likely in 1994, moving over a prominent bedrock ridge into steeper terrain (Avian et al. 2005b). Despite the remoteness of the location, a comprehensive monitoring network has been installed to get a better understanding of present processes such as geodetic survey (annually since 1998; Kienast and Kaufmann 2004), monitoring of the near-surface and surface thermal regime (since 2006 within the project ALPCHANGE), terrestrial laser scanning (two times in 2000, two times in 2001, 2004, 2005, 2006, 2007; Bauer et al. 2003) and digital photogrammetry (1969, 1974, 1983, 1997; Kaufmann and Ladstädter 2004a) providing data of different nature and in different resolutions in time and space. Long range TLS is ought to monitor the lowest part of the rock glacier to gather information about vertical surface changes and 3D movement rates.

The automatic detection and evaluation of three-dimensional deformations, the generation of velocity fields and a full spatial high-resolution coverage of entire rock glaciers concerning these effects has only been made possible since imaging sensors in combination with computer-based data processing are available. Recent advances have been made both in the field of active remote sensing from space (Kenyi and Kaufmann 2003b) and aircraft (Baltsavias et al. 2001), aerial photogrammetry (Kaufmann and Ladstaedter 2003), and most recently with the availability of high-resolution optical remote sensing (Kääb 2002). Sharov and Gutjahr (2002) detected changes up to a height resolution of a few cm using SAR imagery.

Most of these methods rely on sensing from nearly vertical viewing angles which means that steep slopes can only be covered with strong restrictions in measurement performance, if at all. Rock glaciers are creeping ice/rock mixtures as a typical landform in permafrost areas and express themselves as highly complex in dynamics and shape. Amongst others active rock glaciers are characterized by steep fronts, causing most of the remote sensing based strategies to fail or degrade due to the above mentioned restriction. Moreover the front slopes of several rock glaciers (Roer et al. 2005) now rapidly change in shape, texture and object distribution due to rock falls, sliding processes and a general modification of the surface.

In the last years beginning with 2000 a set of experiments was started using this new technology for monitoring both glaciers and rock glaciers in the Austrian Alps. The test sites include the Pasterze glacier as well as a debris covered glacier of Gössnitzkees (glacial processes) and the rock glacier Hinteres Langtalkar (periglacial/permafrost processes), all located in the Hohe Tauern National Park.

10.4.2.1 Hinteres Langtalkar Rock Glacier

The rock glacier Hinteres Langtalkar was measured the first time in July and August 2000 and 2001, respectively (Table 10.2). The ongoing sliding process and the steepness of the rock glacier front slope prevent standard geodetic measurements as well as surface motion analysis by photogrammetric methods. A detailed

Table 10.2 Periods of data acquisition and quality parameters

Period	Total points	Used points	PR[a]	OA[b]
07/2000	27,048	26,381	0.98	0.07
08/2000	27,048	26,437	0.98	0.06
07/2001	27,048	26,274	0.97	0.02
08/2001	26,910	26,312	0.98	0.02
08/2004	20,424	19,818	0.97	0.02
08/2005	27,048	17,843	0.65	0.02
09/2006	7,836	6,048	0.77	0.02
07/2007	6,384	5,739	0.90	0.02

[a] *PR* point ratio = numbers of points used/total number of points measured, [b] *OA* orientation accuracy ± [m]

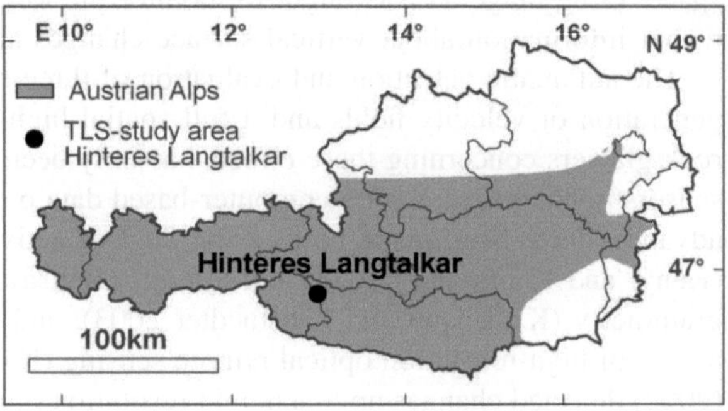

Fig. 10.13 Location of Hinteres Langtalkar within Austria

description of the situation including maps can be found in Kaufmann and Ladstaedter (2003).

Geographical Setting

The cirque-system Hinteres Langtalkar (N46°59′, E12°47′) is situated in the Schober Mountains within the Hohe Tauern Range (Central Alps, Austria) (Fig. 10.13). The cirque-system is a hanging valley at the orthographic right side of the Gössnitz Valley. The cirque covers an altitudinal range between 2,300 and 3,019 m a.s.l. The NW facing tongue shaped rock glacier itself covers the entire upper cirque floor with a lower margin at 2,455 m a.s.l., root zones beginning in appr. 2,700 m a.s.l. and a geometry of 600 × 300 m (Fig. 10.14). In general, rock glaciers are very frequent in the Schober Mountains due to favourable geological conditions leading to a total number of 77 intact rock glaciers (Lieb 1991). Furthermore the main range of the Central Alps (10 km to the N) causes pronounced continental climatic characteristics with a mean annual air temperature (MAAT) of 0°C at 2,200 m a.s.l. and precipitation of appr. 1,500 mm at 2,000 m a.s.l. Glaciations in the Schober Mountains is developed only in some cirques and decreased rapidly during the last decades due to atmospheric warming (Kellerer-Pirklbauer et al. 2005). Auer et al. (2002) report from a rising of the MAAT of 1.6°C since 1886 at the Hoher Sonnblick Meteorological station (3,106 m a.s.l., 15 km E of Hinteres Langtalkar) which is above the global average of 0.74°C (Solomon et al. 2007).

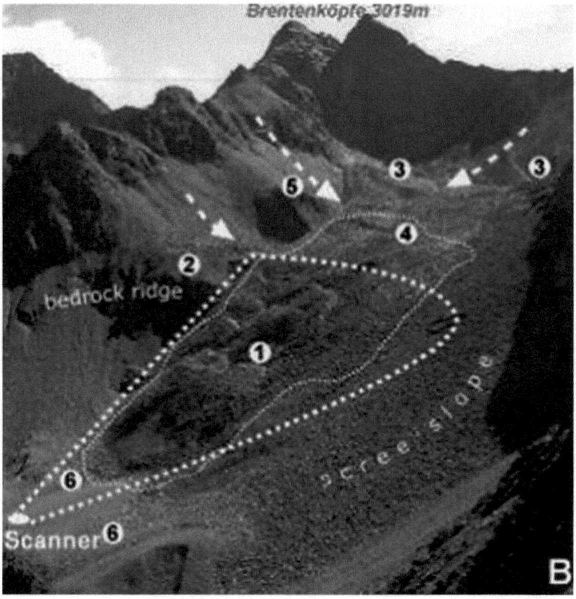

Fig. 10.14 The cirque-system Hinteres Langtalkar including the monitoring configuration. Codes in photograph: (*1*) Area of intensive rock glacier movement and disintegration, (*2*) prominent bedrock ridge partly covered by periglacially weathered debris, (*3*) latero-terminal moraine ridges dating from the Little Ice Age (~1850 AD), (*4*) crevasses on rock glacier indicating high strain rates, (*5*) meteorological station and (*6*) fresh boulders spreading over alpine meadows adjacent to the rock glacier front. The thin *dashed line* comprises the recently fast moving part of the rock glacier. Scanner position is in a distance of appr. 90 m to the rock glacier front (Photograph by Viktor Kaufmann 24.08.2003)

For sensor orientation a geodetic network of five reference points was provided by the Institute of Geodesy at TU-Graz. For each of the four individual measurements the sensor orientation was obtained independently. The sensor location was selected at a distance of about 100 m to the foot of the rock glacier front slope (Fig. 10.15). The resolution of the measurements was mainly limited by the acquisition time. A grid width of 0.5 m could be established at the centre of the front slope, corresponding to 140 × 200 single measurements.

Figure 10.16 shows the original laser measurement data and the resulting DSM. Figure 10.17, left, shows the difference DSM within a period of one year. The elevation change varies from –2.0 to +1.5 m. The bright areas indicate areas with large elevation change of the surface. Stable areas outside the rock glacier show an RMSE of ±11 cm and a systematic difference of 3 cm in height.

On Fig. 10.17, right, a debris flow event is documented. The spatial distribution of the observed mass movement can be identified and numerically evaluated. The flow has been caused by subsurface drainage after heavy rainfall. The first indications of this event can already be recognized at the diagonal vertical structure on the centre of Fig. 10.15.

Fig. 10.15 Measurement campaign of July 2000 at the Hinteres Langtalkar rock glacier in the Austrian National Park Hohe Tauern

Fig. 10.16 *Left* laser scanner distance measurement of 12-07-2000 (*grey coded black* = near, *white* = far). White areas within the DSM could not be measured due to occlusions. *Right* digital elevation model (rendered) after geocoding

Discussion

The full end-to-end chain of rock glacier monitoring using a long-range terrestrial laser scanner has been demonstrated in the experiment (Bauer et al. 2005a). Small debris falls as well as accumulation of debris and scree can be detected, the local mass movements can be evaluated down to single cubic meters. Dangerous geodetic field work in areas of difficult direct access can be completely avoided.

The result of the DSM structure based motion field analysis is shown on Fig. 10.18, combined with the vertical change obtained from the DSM differences. A horizontal motion up to 1.5 m a^{-1} and a vertical surface deformation up to 1.2 m a^{-1} could be detected. Within stable areas (e.g. upper right) no statistically

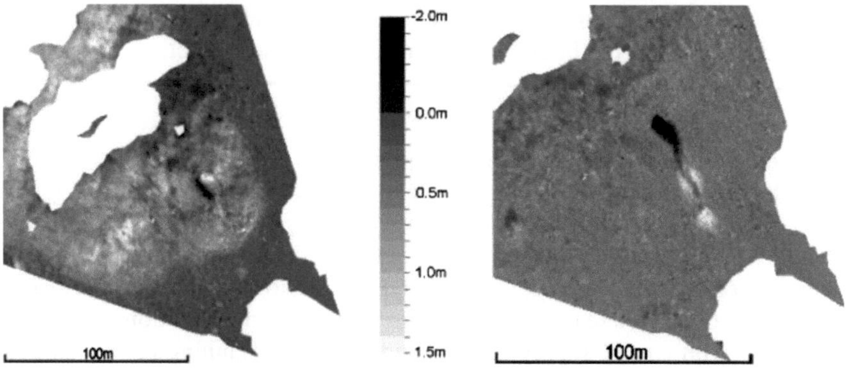

Fig. 10.17 *Left* difference DSM from August 2001 to August 2000 (*grey coded*). *Right* part of difference DSM from 21-08-2000 to the first campaign of 12-07-2000 (*grey coded*)

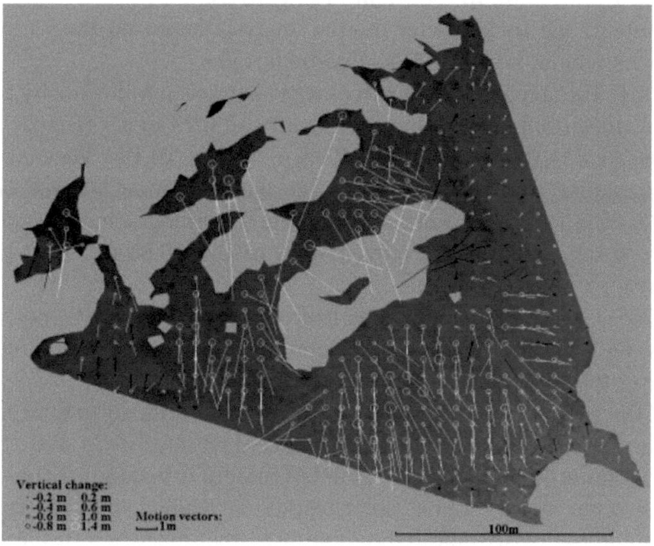

Fig. 10.18 Results of surface structure matching: change between August 2000 and August 2001. *Circles* describe vertical changes, the vector field displays the horizontal movement

significant change in all three components can be observed. The distribution of the detected motion directions is consistent with the assumption that the overall motion is mainly induced by gravity action, although local irregularities may be caused either by local sliding events or errors of the matching process.

To control the significance of the image matching data, cross checking with displacement rates from adjacent parts of the rock glacier were considered. Geodetic survey has been carried out in the middle part of the rock glacier with its lower measurement boundary at the disintegrated lower part of the rock glacier. These campaigns provide verification data for the displacement vectors derived from image matching (Table 10.2). Some limitations in interpretation have to be considered due to distances of up to 60 m between scanning area and the respective control points.

Results

To get more information about interpretability, a review of data quality is essential. In the period 2000–2005 the total number of acquired points is high with 20,000 to 30,000 per campaign. The surface topography of a rock glacier consists of boulders with sizes between a few decimetres up to some metres. This is of crucial importance in the face of a true reproduction of the real surface and subsequently to receive reasonable results in terms of surface motion patterns. Due to problems with energy supply in 2006 and 2007, the scanning increment had to be reduced during these campaigns leading to total point numbers below 10,000. This low number is unfavourable for interpolation and obtaining a DSM for an area wide precise motion analysis. Orientation accuracy is satisfying for all periods with ranging from ±0.02 to ±0.07 m (Table 10.2).

The results of the rock glacier motion analysis based on image matching are given in mean annual horizontal displacement rates.

2000–2001: The stepped lowest part differentiates clearly in velocity patterns. The ridges show mean annual displacement rates of 0.78 m a^{-1} (max. 1.38 m a^{-1}) (Fig. 10.19a; zone 1), 0.95 m a^{1} (max. 1.73 m a^{-1}) (Fig. 10.19a; zone 2), 1.38 m a^{-1} (max. 1.64 m a^{-1}) (Fig. 10.19a; zone 3), and 1.35 m a^{-1} (max. 1.75 m a^{-1}) (Fig. 10.19a; zone 4). The adjacent scree slope at the orthographic left side of the rock glacier shows very constant movement rates of 0.02–0.06 m a^{-1} over the entire scanning area.

2004–2005: Quality of velocity data from this period tend to be not satisfying in some areas. Distinct differentiation between obvious moving and akinetic areas at the margin of the rock glacier is not possible.

2005–2006: Flow velocities and patterns are more difficult to interpret although the ridges are detectable due to higher displacement rates (0.75 m a^{-1}). The main front shows rates around 0.15 m a^{-1}, the left margin is distinguishable towards the non moving scree slope. Coarse point resolution leading to small spots of areal data does not allow reasonable interpretation of the upper part of the rock glacier tongue (uppermost scanning area in Fig. 10.14).

Direction of movement at the terminus zone of the rock glacier shows outwards movement of displacement vectors as expected. The ongoing development of the imbrications and the distinct shifting of detritus is visible in abrupt changes of the magnitude and the direction of velocity vectors (Fig. 10.19a–c, 1–4).

High deviation values in a cross check with data acquired in a geodetic survey, at some areas of the right margin of the rock glacier, the bedrock, and vegetated areas are results due to unfavourable scanning geometry and therefore not taken into account as the exact rock glacier extent is known from fieldwork.

In terms of the accuracy of the measurements a crucial factor is the availability of independent measurements to validate the acquired data. Furthermore stable bedrock areas—forming “natural targets”—are rare in the scanning sector, but will

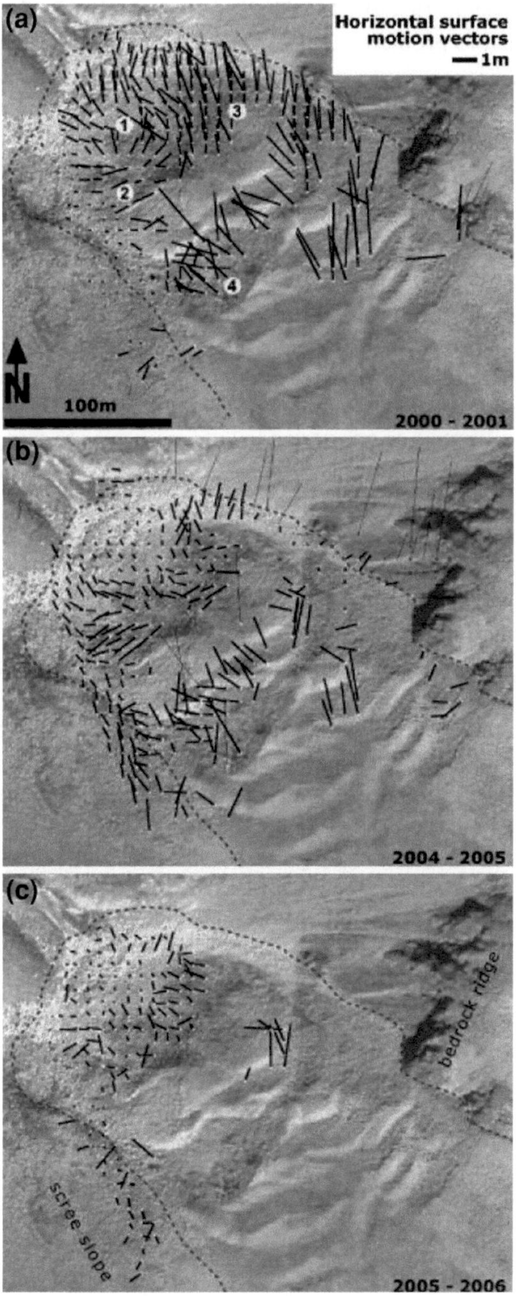

Fig. 10.19 Horizontal displacement rates [m a^{-1}] for three annual periods: **a** 2000/01, **b** 2004/05 and **c** 2005/06. Motion vectors are obtained from DSM differences, DSMs were derived from TLS data. Numbers (*1–4*) indicate zones of different behaviour in terms of surface velocity, for details refer to text (Orthophotograph © Nationalpark Hohe Tauern 1998)

be integrated in future measurements. An airborne LiDAR survey was carried out in late summer 2008. These data will be compared with TLS data to validate results and accuracy. The next monitoring campaigns will be carried out with a new sensor system (Riegl LMS-Z620 see Table 10.1) which allows faster data acquisition to avoid problems concerning sensor levelling during the measurements.

10.4.3 Glaciers

Monitoring of glacier behaviour is an important task in environmental research. For detailed detection of glacier surface and volumetric changes terrestrial laser scanning is a very effective and cheap observation method, due to the ability to acquire high-resolution 3D data. Advancing or retreating glaciers have a complex impact on their neighbouring environment, as e.g. on permafrost, geomorphic processes (thus natural hazards) or vegetation. In Austria, monitoring of glacier behaviour is an important task in environmental research. Since many decades the Austrian Alpine Association (OeAV) co-ordinates annual glacier measurement campaigns. Of the 925 Austrian glaciers—according to the first Austrian glacier inventory from the year 1969 (Patzelt 1980)—107 glaciers are currently monitored on an annual basis by this program (Patzelt 2005); the majority of them only in a very general way (e.g. change of glacier terminus). In addition to these activities, annual mass balance measurements are carried out at a much smaller number of Austrian glaciers within the framework of the World Glacier Monitoring Service (WGMS). Currently, a new Austrian glacier inventory is in progress. For this purpose, 49 glaciers of different types have been surveyed between 1995 and 2003 with the aid of ground penetrating radar (Würländer and Kuhn 2000; Fischer and Span 2005; Lambrecht et al. 2005). Knowledge of glaciation changes is essential for the interpretation of glacier-climate interaction and glacier-climate modelling studies. Furthermore, the total volume of a glacier body is an interesting parameter for water resource management, in particular in areas where glacier water is essentially used for irrigation during the summer period (e.g. Hunza valley, Pakistan) or used for hydropower production (e.g. Kaprun, Austria). Numerical requirements on the quality of glaciological data are high in order to allow reliable predictions for all these issues. A number of different methods on a local, regional and global scale for glacier monitoring are available. For detection of glacier surface and thus volumetric changes in a very high spatial resolution terrestrial laser scanning is a very effective observation method, due to the ability to acquire high-resolution 3D data in a very short period of time. The automatic detection and evaluation of three-dimensional (3D) deformations, the generation of velocity fields and a full spatial high-resolution coverage of entire glaciers concerning these effects have only been made possible since imaging sensors in combination with computer-based data processing are available.

10.4.3.1 Pasterze Glacier

This section discusses the results of annual terrestrial laser scanning campaigns beginning in 2001 within the monitoring framework at the Pasterze glacier tongue.

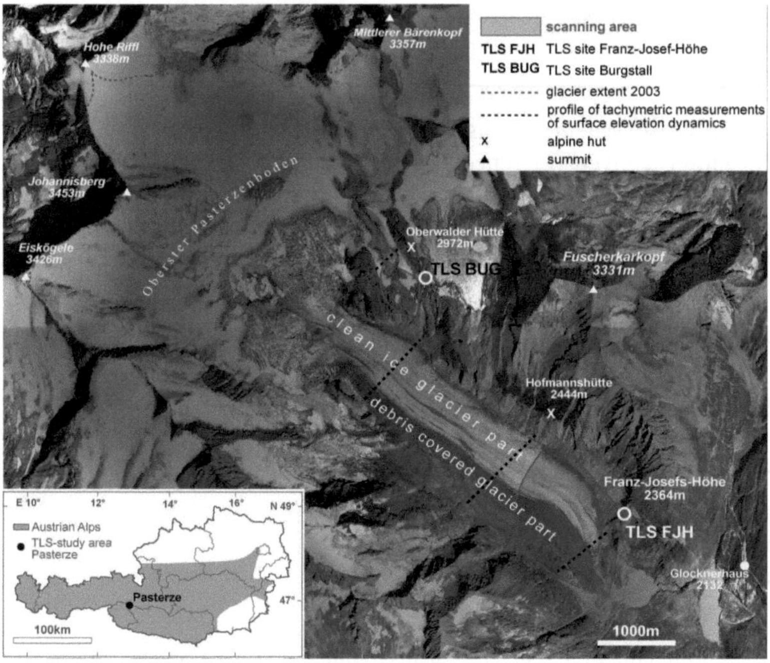

Fig. 10.20 Location of the Pasterze glacier, glacier extent and position of the TLS-sites as well as profiles of tachymetric measurements (Orthophotograph 1998, © National Park Hohe Tauern 2006)

Geographical Setting

The Glockner Mountains comprise the longest glacier in the Eastern Alps with the main attraction Pasterze Glacier (Central Alps, Fig. 10.20, 12°44′E, 47°04′N). The glacier's dimension changed from 1852 to 2002 as follows: length 11.4 to 8.4 km (−26%) and area 26.5 to 18.5 km^2 (−30%). The dramatic loss of ice in the last decades is documented in a comprehensive modification of this alpine environment on the glacier itself as well as in the paraglacial area. Quantification of these changes such as spatial measurements of glacier surface changes are a primary task in the last 20 years with new remote sensing techniques using photogrammetric analyses (Kaufmann and Ladstädter 2004c), Airborne laserscanning (Geist et al. 2003; Kellerer-Pirklbauer et al. 2005) and radar-interferometry (Kaufmann et al. 2005). The selection of terrestrial laser scanning on this part of the Pasterze glacier is due to following reasons providing several advantages:

- Perfect accessibility keeps costs low (mountain road "Großglockner-Hochalpenstraße" to end point "Franz-Josefs-Höhe").
- The lowest part of the glacier, i.e. the glacier terminus, is object of an intense retreat which goes along with a massive modification of its proglacial landscape. In the upper part of the scanning area (at the "Seeland-Linie" profile beneath the

Hofmanns-Hütte) the glacier lost 38% in width, 61% of thickness, and 75% of the area concerning the vertical cross-section (Wakonigg and Lieb 1996; Krobath 2003).

Terrestrial laser scanning measures the position of a theoretical stable point on the surface within a given reference-system. Resulting elevation differences do not reflect mass-losses at this particular point because surface velocity has not been taken into account yet. To ensure to probability to analyse surface velocity distribution, measurements have to be carried out with higher resolution e.g. 0.02 Gon instead of 0.2 Gon, exploiting the abilities of newest sensor technology.

Scanning Campaigns

Surface lowering and ongoing deglaciation of terrain were successfully measured four times resulting in three data-sets of surface elevation changing rates (xy-resolution 1 m, z-resolution 5 cm) at following dates:

- Campaign 1: 19./20.10.2001
- Campaign 2: 21./22.9.2002
- Campaign 3: 14./15./16.8.2003
- Campaign 4: 20./21.9.2004

2001–2002: This first year of measurements is mainly focused on the set up and instrumentation than taking into account glacial questions. The main question is to figure out the best increment for the different aspects in monitoring glaciers regarding time and resolution. As glaciers retreat recently deglaciated areas are characterized by high rates of sediment delivery (e.g. kame terraces) and deposit (e.g. sandur). This paraglacial area at the Pasterze is subject to conditions of comprehensive modifications at a mean magnitude from −1 to +1 m in vertical surface elevation changes (max. up to −3.4 m, Fig. 10.21 left, a).

The debris covered, right part of the glacier shows a very interesting linear succession of surface lowering. Beginning at its right margin rates increase from around −1.5 to −4.8 m at the supra-glacial melt water channel marking the border between debris covered and bare ice part of the Pasterze glacier (Fig. 10.21 left, b). This left part of the glacier is subject of dramatic modifications. Area wide surface elevation changes range from −6.2 to −7.2 m and is very homogenous in terms of spatial distribution at the entire bar-ice glacier terminus. These rates do not include areas showing the beginning of intensive collapses of the ice body that mainly occurs at the left margin of the bare ice glacier surface. The lowest event in terms of absolute elevation is already visible due to developing crevasses near the fenced tourist areas (Fig. 10.21 left, c, Fig. 10.24 right), large areas show sinking rates over −12 m, maximum rates reach up to −19.2 m. Following the left glacier margin 550 m upwards an area of extraordinary sinking rates is detectable in the results (mean −7.8 m, max. −10.2 m). The magnitude of the sinking rates within this area exceeds surrounding rates by ∼2.0 m. The occurrence of debris cover in some parts of the bare ice left part is visible in significant smaller sinking rates. Albedo of clean ice is about 40% compared to shallow debris covered ice where

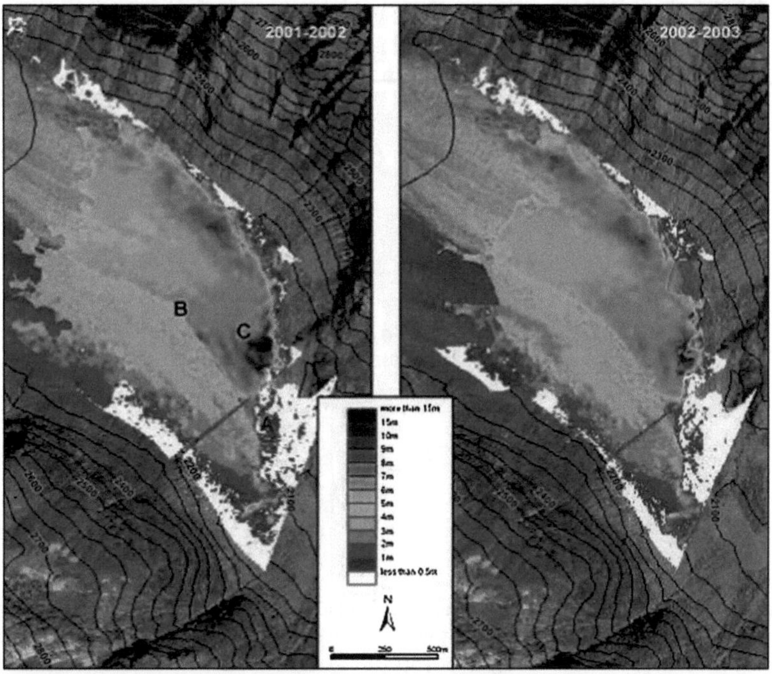

Fig. 10.21 Spatial distribution of surface elevation changes on the Pasterze glacier in the period 2001/2002 and 2002/2003. Orthophotograph 1998, © Nationalpark Hohe Tauern

albedo reaches values in the range of 10–15%. Less energy is available for ablation on snow surfaces than on debris covered ice. Highest ablation rates occur at a thickness of debris cover of 0.3 cm; if debris cover increases to 5 cm, ablation rates on bare ice and debris covered parts are equal. Debris cover with a thickness of at least 7 cm is a significant protection against incoming short wave radiation (Benn and Evans 1998; Nakawo et al. 2000). The thickness of debris cover was not analysed particularly but rough estimations on the glacier during fieldwork provided results of around 10–15 cm.

2002–2003: Dynamics of modification of paraglacial areas follows the retreating glacier terminus. Surface elevation changes decrease depending on the distance to the glacier terminus (as expected). Surface lowering probably connected to melting of dead ice bodies is also visible as in the previous period. The two mentioned collapsing areas still remain in extraordinary spatial dynamics; especially the upper one increases sinking rates (mean −8.3 m, max. −10.8 m; Fig. 10.21 right; Fig. 10.24 right). A new zone of massive surface lowering is developing another 400 m upwards on the left glacier margin (max. −9.5 m, mean −8.5 m, surrounding: −7.3 m). Ongoing increasing surface lowering towards the glacier terminus of the entire bare ice part is visible. This process is strongly influenced by a decrease of ice-supply due to reduced glacier surface velocities in the lower part since 1983 (Wakonigg and Lieb 1996). Surface elevation changes at

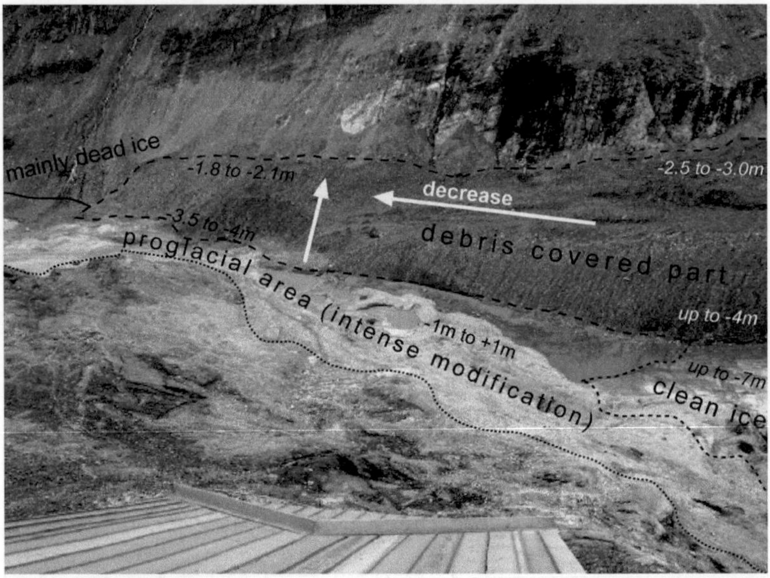

Fig. 10.22 Lowest part of Pasterze glacier and proglacial area from the scanning position. Note the decrease of surface elevation changes towards to glacier terminus (debris covered), and to the *right margin* of the debris covered glacier part (Image 25.09.2006)

the debris covered part decrease mainly at the upper part, increase at the middle section and decrease to the lowest part. The mass loss on the debris covered glacier tongue is detectable at a magnitude of ~ -3.5 m at lower edge decreasing to ~ -2.1 m at the upper edge (Fig. 10.22).

2003–2004: In 2004 we planned to increase the temporal resolution with measurements in the summer period at four epochs (mid of June, July, August, and September) to get a better picture of the inter-annual ablation dynamics. Complications due to unstable weather conditions and problems in sensor orientation inhibited the first three campaigns completely. Results were only carried out in September, where only 33% of the area of 2001, 2002, and 2003 could be used for reasonable comparisons.

The most obvious indication is a beginning of a comprehensive collapse of the entire left part of the bare ice glacier tongue. Gentle signs of this process are already visible in some parts since 2001, massive and homogenous surface lowering (over -9 m) as well as newly developed crevasses supports this assumption (Figs. 10.23, 10.24). Mean sinking rates arise from -9.8 to -10.5 m with maximums in collapsed areas up to -20.5 m. The sub-glacial melt water channel is visible a few meters (near the tourist-area) beginning to isolate several ice-bodies. The entire lowest part of the bare ice glacier part shows surface lowering with more than -7.7 m (Fig. 10.23).

Adjacent slopes to the Franz-Josefs-Höhe also show surface lowering due to processes such as melting of dead ice and different types of mass movements. This

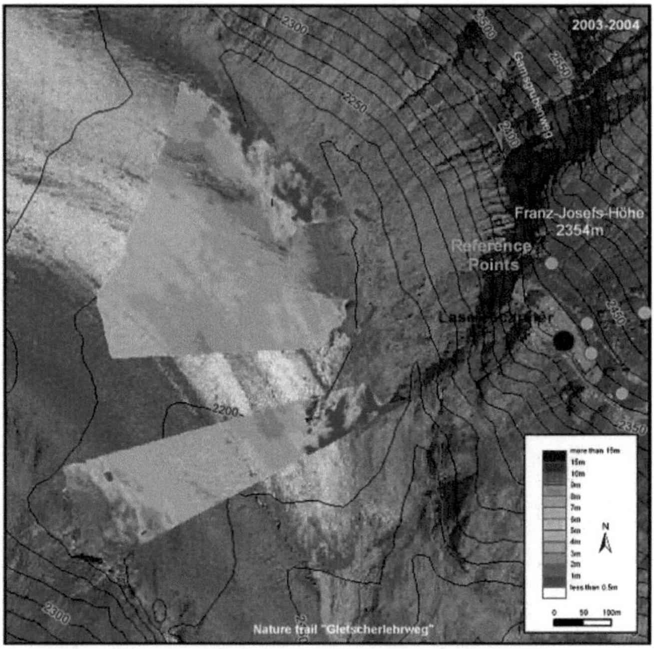

Fig. 10.23 Spatial distribution of surface elevation changes on the Pasterze glacier in the period 2003–2004 with scanner position, reference points and tourist infrastructure. Orthophotograph 1998, © National Park Hohe Tauern

part has not been measured consequently in the last periods but visual observations as well as the well known succession following glacier retreat support this assumption. Due to the lack of data on the debris covered part statements are not reasonable. Area wide information is not available, but the scanned sections show similar behaviour than in previous periods.

In order to get comparable data sets, a prior frequency distribution of all data sets was carried out to provide sub-zones on the clean ice part named collapse area, ice margin and main ice part. Surface elevation changes on the glacier tongue express the continuation and intensification of ice disintegration processes in the entire monitoring period of TLS. Surface lowering increases over the entire glacier tongue area with a growth of 38% since 2001/2002 (Table 10.3).

All periods show a constant lowering of the glacier surface with a noticeable acceleration of the process since 2003. This is again important to underline since collapsing areas—or the different behaviour of the ice margin—have no influence on the areal statistics. The period 2004/2005 demonstrates the tremendous magnitude of landscape modification in the Pasterze area with mean sinking rates of −7.45 m on the clean ice part (max. −16.85 m, 55% quartile −7.85 m, median −7.45 m, 25% quartile −7.10 m, min. 0.25 m) and −4.65 m on the debris cover part (max. −16.85 m, 75% quartile −5.00 m, median −4.65 m, 25% quartile −4.35 m, min. −0.15 m) on the debris cover part (Fig. 10.25). As a consequence

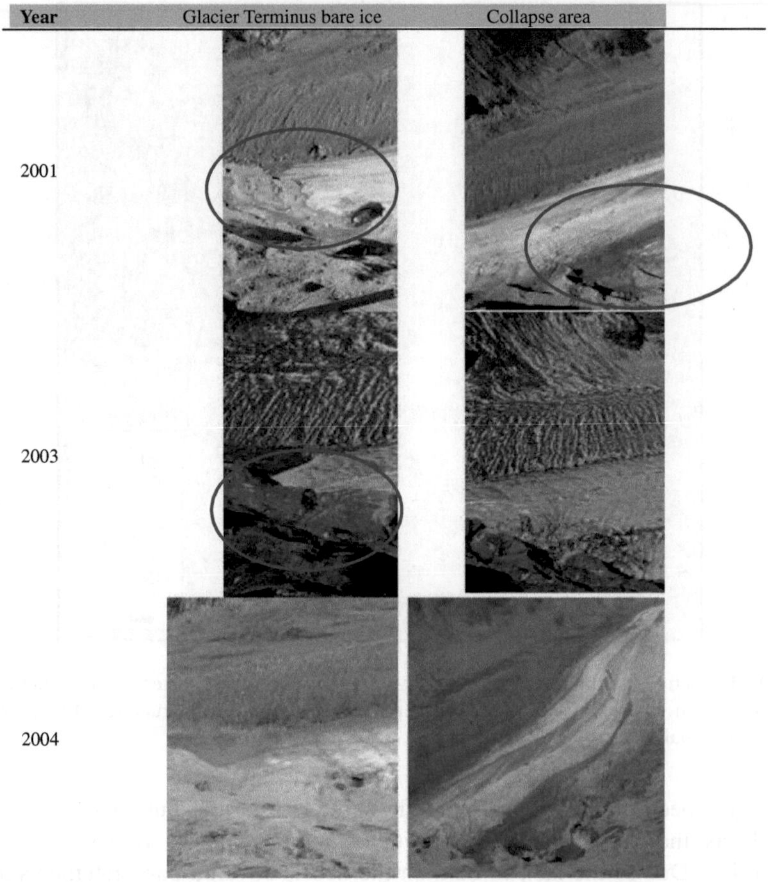

Fig. 10.24 Progression of glacier retreat at the lowest part of Pasterze glacier. Images: Avian Michael (15.09.2002, 20.09.2004); Kaltenböck Alexander (19.10.2001), Kellerer-Pirklbauer Andreas (18.09.2003)

of these accompanying morphological processes technical problems occur due to disturbed terrain characteristics like shadowing effects or malfunctions in pulse reception due to increasing melt water discharge.

The clean ice part retreats rapidly in the last decade (mean $-6.50\ \mathrm{m\ a}^{-1}$, 2001–2005) exposing wide proglacial areas consisting of large debris deposits (e.g. kame terraces), dead ice bodies and melt water lakes. This micro-scale topography tends to be an excellent area for methodological tests for TLS. Several dead ice holes have been developed in the last years visible in Fig. 10.26. Annual mean surface elevation differences are within the range of 2.10–2.50 m at these three examples. The erosion of kame terraces marks a very frequent process in this very young and instable deposit landscape. Melting interior ice bodies and lateral erosion of the glacier creek caused a total loss of $\sim 1{,}600\ \mathrm{m}^3$ and very stable surface elevation losses of mean -5.20 m within one year (2004–2005).

Table 10.3 Surface elevation changes, statistical parameters and scanning areas for the periods 2001–2005

Zone		Median of surface lowering (m) (Xmed), Standard deviation (SD)				Max. surface lowering (m) Area (ha)			
Period		01/02	02/03	03/04	04/05	01/02	02/03	03/04	04/05
Clean ice	x_{med}	−3.65	−3.05	−4.20	−4.65	−13.85	−8.05	−10.00	−10.95
Ice margin	SD	1.94	1.71	2.50	1.63	0.41	3.10	0.92	0.97
Clean ice	x_{med}	−8.70	−7.20	−4.95	−8.55	−18.20	−17.70	−17.50	−16.85
Collapse	SD	2.47	2.24	3.60	1.53	0.21	2.12	1.92	3.28
Clean ice	x_{med}	−5.95	−6.05	−7.15	−7.10	−19.30	−11.20	−13.50	−11.15
Main ice part	SD	0.73	0.67	0.78	0.58	42.84	38.87	38.16	23.61
Clean ice	x_{med}	−5.95	−6.00	−7.15	−7.45	−19.30	−11.20	−17.50	−16.85
	SD	1.40	1.26	1.48	1.20	43.46	44.09	41.00	27.86
Debris cover	x_{med}	−3.75	−3.45	−3.80	−4.65	−10.20	−9.15	−8.90	−9.35
	SD	1.07	1.04	0.70	0.74	30.48	29.72	22.79	9.28
Glacier tongue	x_{med}	−5.25	−5.35	−6.95	−7.25	−19.30	−17.70	−17.50	−16.85
	SD	1.75	1.76	1.87	1.64	73.94	73.81	63.79	37.14

The differentiation clean ice margin, clean ice collapse, and clean ice main ice part aggregates in the term clean ice; clean ice and debris cover aggregates in the term glacier tongue respectively

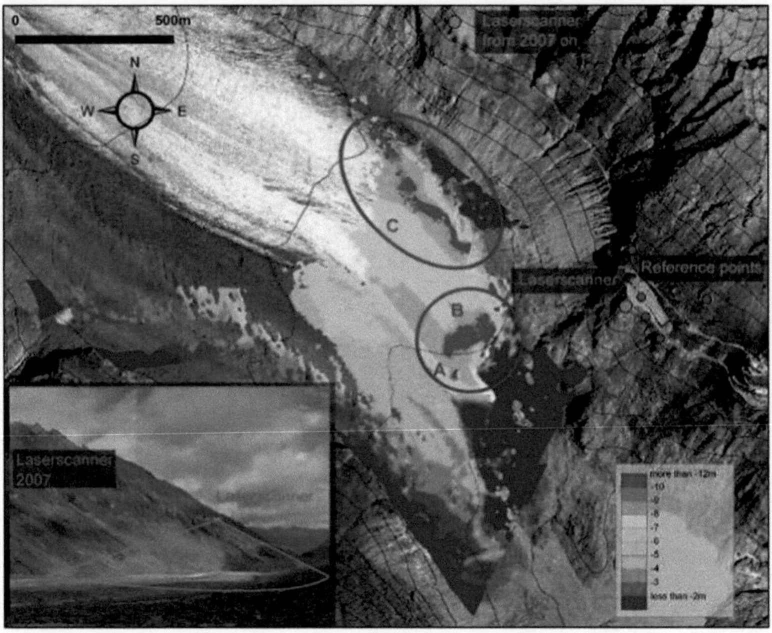

Fig. 10.25 Monitoring configuration of TLS with additional scanner position for 2007. Surface elevation changes 2004–2005 with ice disintegration zones and areas of above-average surface lowering. The current position of the laser-scanner is getting unfavourable due to the fast retreat of the Pasterze glacier. The glacier terminus reached the scanning centreline in 2006, so only 50% of the collected data represent clean glacier ice. We defined a second scanning position located on the trail "Gamsgrubenweg" in the vicinity of the former Hofmannshütte (elevation: 2,456 m) in September 2006 to ensure high quality data and an area wide analysis of the glacier variation, exact positioning and first measurements will be carried out in June 2007 (Photographs by A. Kellerer-Pirklbauer, 23.09.2006, Orthophotograph 12.09.1997, © BEV Vienna)

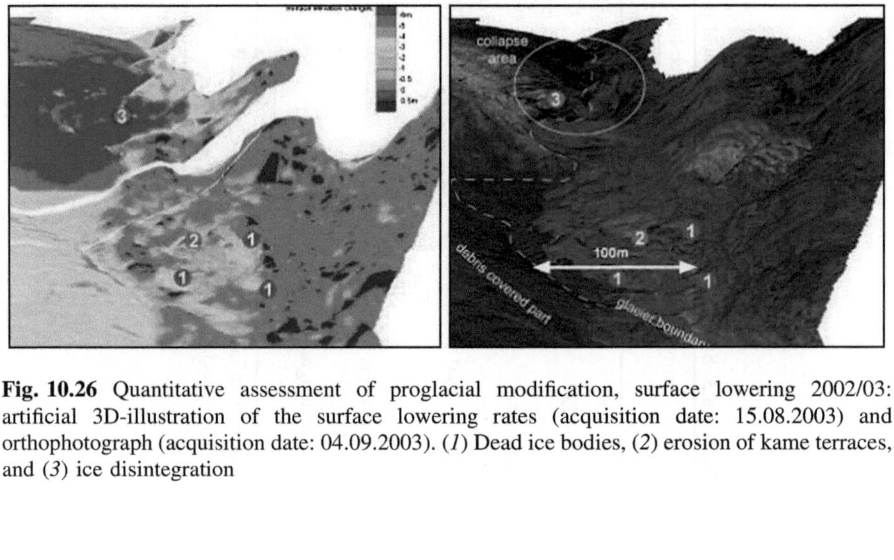

Fig. 10.26 Quantitative assessment of proglacial modification, surface lowering 2002/03: artificial 3D-illustration of the surface lowering rates (acquisition date: 15.08.2003) and orthophotograph (acquisition date: 04.09.2003). (*1*) Dead ice bodies, (*2*) erosion of kame terraces, and (*3*) ice disintegration

Fig. 10.27 Features: (*1*) Sandur with water channels, (*2*) kame terraces, (*3*) lateral slope of terraces, and (*4* and *5*) melt water lake (image by A. Kellerer-Pirklbauer 19.09.2006)

At the glacier itself ice disintegration is in progress due to several reasons. The glacier creek—subglacial until 2005—eroded the lowest part of the glacier tongue leading to a collapse of the ice body showing first signs of large crevasses in 2001. Now it isolates large ice bodies from the main glacier on the left rim of the glacier. In this area erosive processes are more dominant than ablation causing major problem especially in the tourist area. Even with TLS a point density of 1 m^{-1} is possible leading to an excellent geometric resolution for these proglacial areas in reasonable time providing an adequate data set for geomorphic interpretation (Fig. 10.27).

Conclusion

As expected from simultaneous tacheometric measurements all data sets show a clear trend in spatial variations of glacier retreat. There is a significant distribution of mass loss from the SW to the NE which also increases towards the glacier terminus on the bare ice glacier part. Here we can observe a clear leap in mass loss difference between the two obviously different glacier parts from the upper part to the lowest part. Debris cover with a thickness of at least 7 cm proofs to be a significant protection against incoming short wave radiation and therefore ablation (Kellerer-Pirklbauer 2008). Another aspect in terms of different ablation rates of glacier parts is less potential radiation in the summer period due to the shadowing effect of the south-east facing, adjacent Großglockner ridge.

The left part of the entire bare ice glacier part is beginning to collapse comprehensively. The lowermost event is already visible in the last years with a dramatic landscape modification near the fenced off tourist area. The dynamics of the lowest event is already decelerating. Following the glacier upwards two further small "basins" flanked by circular to semi-circular crevasses are developing with a distance of 500 m from the lower one which seem to coalesce in 2004. All indications on the lowest bare ice part of the Pasterze glacier including the behaviour of the melt water channel lead to the prediction that the entire foot slope is about to collapse. The lower part of the adjacent slope facing to the "Gamsgrubenweg" is also getting increasingly unstable which is observable in mass losses already above dead ice. This part has not been measured consequently in the last years, no comparable rates are calculated.

The lowest part of the study area comprises the paraglacial area where a consequent decelerating of processes depending on the distance to the glacier as well as the outcrop of dead ice bodies is observable.

Terrestrial laser scanning proved to be a promising tool in monitoring local events with dimensions of 2–2.5 km. Single day's campaigns of one person with consecutively short post-processing keep costs low and allow measures at short notice to react on dynamic processes.

10.4.3.2 Gößnitzkees Glacier

Since summer 2000 this method is applied at the Gößnitzkees glacier, a small debris-covered glacier located in central Austria (12°45′E, 46°58′N; size c.0.75 km^2. Note: "*Kees*" is a regional term for glacier). More than 60% of the glacier is covered by a prominent debris mantle. Measurements with TLS allow the comparison of three different time scales (intermonthly, interannual, four years). The results demonstrate that accumulation and ablation (snow/firn/ice) can be monitored very accurately. The debris cover reduces net ablation at the glacier surface by up to 75% whereas the amount of incoming solar radiation is less important. A highly active feature is the retreating steep ice wall at the glacier terminus. It is shown that by use of this method it is easily possible to detect small changes on a glacier surface (clean and debris-covered) relevant for glacier-climate modelling but also for aspects in hydrology and natural hazard management.

Study Area

Gößnitzkees is located to the south of the main crest of the Hohe Tauern range in the central part of the Schober group at the valley head of the Gößnitz valley and is thus in the inner zone of the Hohe Tauern National Park at 12°45′E and 46°58′N (Fig. 10.28). Gößnitzkees was included into the network of the mentioned annual glacier measurements of the OeAV in 1982. Due to the unsuitable topographic (steep rock faces, narrow crests, lack of flat surfaces at high elevations above the regional ELA) and climatic conditions (continental climate: low precipitation—c.1,500 mm at 2,000 m a.s.l., 0°C at 2,300 m a.s.l.) of the Schober group, the glaciation is limited to a few positions at the foot of rock faces in northern expositions. The mean size of the glaciers does not exceed 0.18 km^2 (N = 29) and making Gößnitzkees with its c.0.75 km^2 in 1997 the largest glacier of this mountain group (Lieb 2000; Kaufmann and Ladstaedter 2003). The general exposure of the glacier is NW with high crests and mountain tops to the S. The accumulation area of the glacier is very small. A high amount of snow accumulation originates from avalanches, in particular at the western head of the glacier at the foot of some pronounced couloirs (Fig. 10.29). More than 60% of the glacier is covered by a prominent debris mantle with variable thickness. The overall appearance of the Gößnitzkees indicates a very inactive glacier also expressed by low mean annual flow velocities of 30–60 cm a^{-1} (Kaufmann and Ladstädter 2004b).

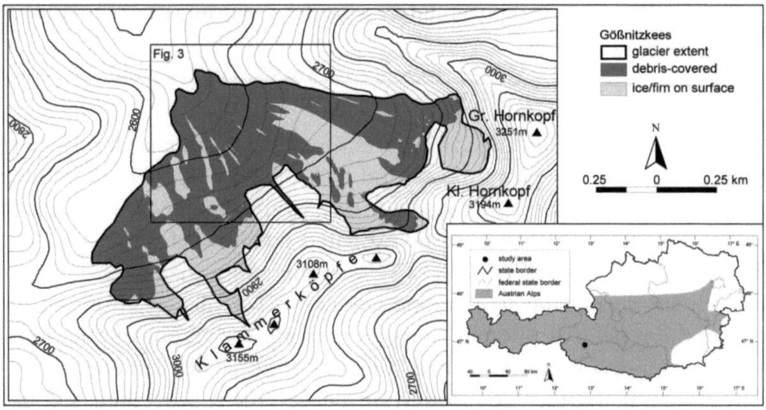

Fig. 10.28 Location and setting of the study area Gößnitzkees. The delineation of the glacier and the distribution of debris-covered and clean surfaces are based on the aerial photographs from the year 1998. The square in the main map indicates the location of the maps shown in Fig. 10.30

Fig. 10.29 Terrestrial overview of the study area. Note the steep ice front of the mostly debris-covered glacier adjacent to the proglacial lake and the group of people standing next to the location of the laser scanner; view towards SE (Photo: Kellerer-Pirklbauer 20-08-2004)

Due to the described topographic and climatic situation in the Schober group the entire area favours permafrost conditions. In NW-, N- and NE-aspects discontinuous permafrost can be expected above 2,600 m with frequent occurrence of creeping permafrost features, i.e. active rock glaciers. The existence of the 77 intact rock glaciers is further enhanced by the mentioned lithologies; some of them may also contain some glacier ice (Lieb 1991; Krainer and Mostler 2000; Lieb et al. 2004; Avian et al. 2005a). Summing up, relief and geoecological elements of the

Schober group are highly representative for the Central Alps and thus for the National Park Hohe Tauern.

Methods and Experimental Setup

Between July 2000 and August 2009 ten terrestrial laser scanning campaigns have been carried out covering the central part of the Gößnitzkees including the glacier terminus. This chapter focuses on the results of the monitoring term of 2000–2004. The area that has been investigated during each laser scanning campaign covers 0.09–0.13 km^2 or 11.8–17.1% of the entire glacier surface gives the spatial coverage during the five campaigns as well as the locations of the laser scanner and the reflective reference targets relevant for sensor orientation. DTM-differences were performed for three different time scales (intermonthly, interannual, four years). Further description of measurement procedure and data processing is found in Bauer et al. (2003).

Results

Based on the measured terrain data glaciation changes over three different time scales have been calculated: two times intermonthly—07 to 08-2000, 07 to 08-2001, two times interannual—07-2000 to 07-2001, 08-2000 to 08-2001, and once over four years—08-2000 to 08-2004. The results are presented in Fig. 10.30. Table 10.4 gives a numerical overview of the calculated results and an estimation of volume changes for the total glacier for each time interval. The scanned area covers a profile sector from the cirque headwall to the glacier terminus and is representative for the entire area. Thus, it is assumed that the detected changes are not only valid for the measured sector but also can be—more or less—extrapolated on the entire glacier.

Discussion

The results shown in Fig. 10.30 clearly demonstrate non-uniform and peculiar retreat behaviour of the Gößnitzkees during all three studied periods. The distribution and characteristics of the debris cover plays a crucial role in the behaviour of the Gößnitzkees. During all three time scales, glacier surface changes are greatly influenced by the presence or absence of a debris mantle (Figs. 10.30, 10.31). In some areas the debris cover reduced net ablation at the glacier surface by up to 75% compared to clean ice surfaces in close neighbourhood. Glaciers mantled by a pronounced debris-cover behave differently to normal or 'clean' glaciers. An important aspect in this context is the increasing input of debris on glaciers; lower glaciation causes a higher input of debris on the remaining glacier body due to pressure release and paraglacial instabilities on adjacent slopes and the cirque headwall. The results measured and calculated for the entire monitoring period give a mean elevation change of about c.-5.8 m (-1.45 m a^{-1}) and an estimated total volume loss of more than 4.3 Mio m^3 (1.08 Mio m^3 a^{-1} for the whole glacier) strongly indicating once more a retreating glacier. As an example of utilizing high resolution laser scanner data only a few glaciological aspects are pointed out in this paper. Further analysis and interpretations are in progress.

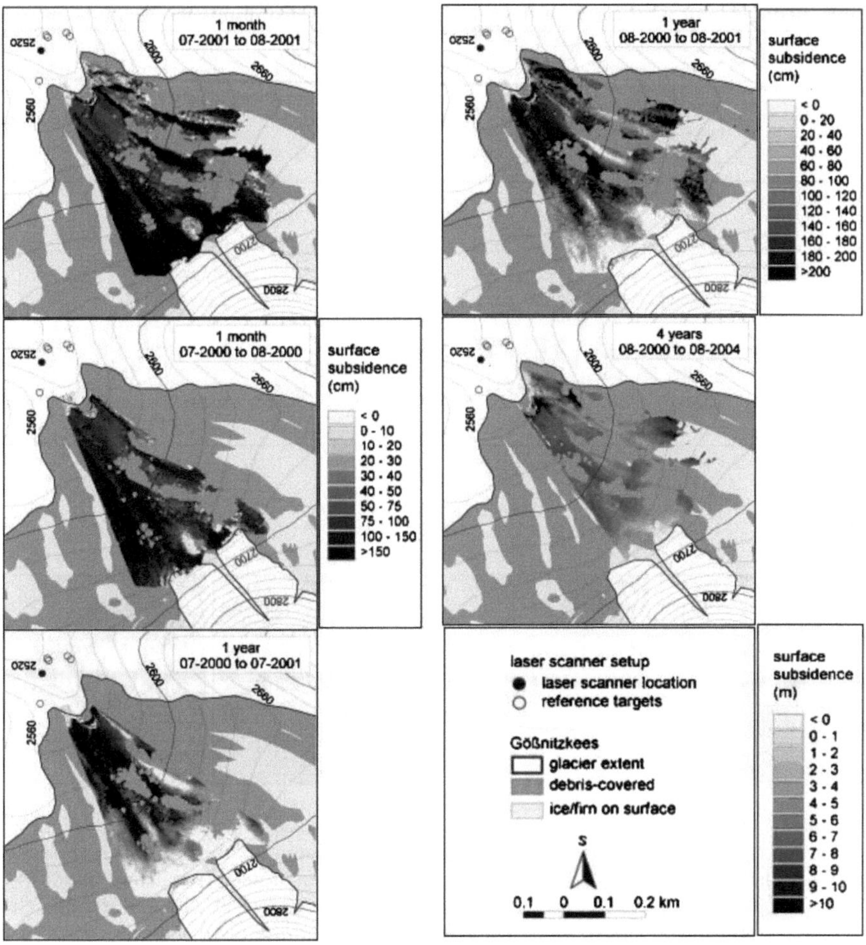

Fig. 10.30 Glacier surface changes during three different time scales (intermonthly, interannual, four years). Positive values indicate areas of thinning and <0 values indicate areas of thickening respectively (i.e. accumulation of snow/firn). Note the increase in thickness in some areas at the 1 year time scales due to a favourable glaciological year 2000/2001

Technical and glaciological conclusions

- Monitoring of glacier surface changes with high temporal and spatial resolution in alpine terrain is feasible.
- Operational system is available as mobile or stationary unit.
- Results are available immediately after measurement.
- Accuracy depends on viewing geometry and footprint size of the laser beam.
- 3D high resolution surface change data (accumulation of snow and debris; ablation of snow/firn/ice) is obtained by DSM analysis.

Table 10.4 Numerical overview of the calculated and estimated results for each time interval

Difference calculations	Area considered in the difference calculations				Area considered on the entire glacier[a] (%)	Mean change in eleva-tion (m)	Measured volume change at the monitored glacier section (m^3)	Estimated volume change for the entire glacier[a] (m^3)
	Total [a] (m^2)	On the glacier [b] (m^2)	On the glacier [b] (%)	δ of a and b (%)				
1 month: 07 to 08-2000	87,763	85,262	97.2	2.8	11.4	−0.971	−82,789	−728,250
1 month: 07 to 08-2001	120,241	118,175	98.3	1.7	15.8	−1.113	−131,529	−834,750
1 year: 07-2000 to 07-2001	86,262	84,224	97.6	2.4	11.2	−0.829	−69,822	−621,750
1 year: 08-2000 to 08-2001	119,962	117,896	98.3	1.7	15.7	−1.179	−138,999	−884,250
4 years: 08-2000 to 08-2004	63,083	62,894	99.7	0.3	8.4	−5.755	−361,955	−4,316,250

[a] Glacier size 0.75 km^2

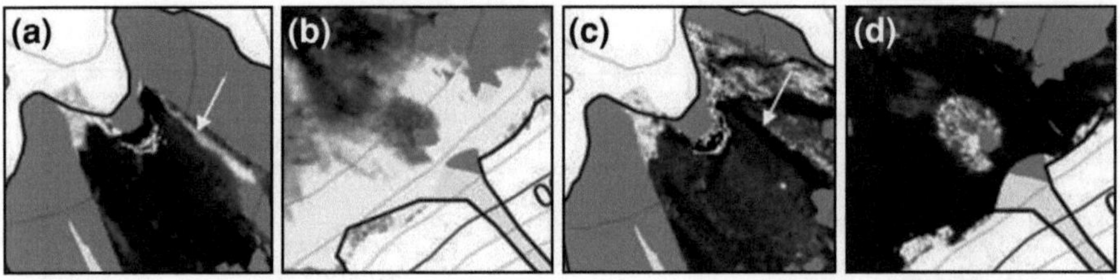

Fig. 10.31 Section examples of Fig. 10.30 showing surface changes—both increase and decrease—at the glacier at different time scales and at different locations; (**a** and **b**): sections of 1-year change maps (07.2000 to 07.2001), bright areas indicate accumulation of snow—**a**: in the supraglacial melt water channel (*white arrow*), **b**: at the foot of cirque headwall; (**c** and **d**): sections of 1-month change maps (07.2001 to 08.2001); *dark areas* indicate enhanced ice melting—**c**: at the supraglacial melt water channel (*white arrow*) at the glacier terminus and in areas with a minor debris cover, **d**: melting of winter snow at the foot of the cirque headwall

- Both large scale (e.g. estimation of total glacier volume change) and small scale (e.g. quantification of supraglacial melt water channel incision) glaciological features can be quantified.
- PSWR during summer plays a minor role in different ablation and accumulation behaviour on a heavily debris-covered glacier.
- An existing debris cover causes a striking difference in net ablation at areas mantled by debris relative to clean ice surfaces within close distance (up to 75%), in particular at the debris-free vertical glacier terminus. A detailed analysis of the debris cover on the glacier (e.g. thickness of layer, spatial distribution, clast size) is planned in the near future.

- Obviously, the principal factor causing an uneven distribution of surface changes within a short horizontal distance between two compared time periods is the specific accumulation/ablation history.
- It is clearly shown that multi-temporal terrestrial laser scanning analyses provide a high potential for mass balance estimates and thus glacier dynamics studies.

10.4.4 Snow

Although the data generated by the measurement devices can in principle be directly used for further visualization and measurement, several methodological, technical and logistic problems are to be encountered when establishing a fully automatic monitoring system. Stability of device control software, automatic sensor orientation, high number of measurements, compensation of weather influences such as fog or precipitation, and a selection of reliable measurements are some of them. One of the limitations in accuracy is a highly heterogeneous surface in terms of material (rock, vegetation, and humidity in general) and structure. In the case of snow cover measurement surface structure and surface material are more cooperative in this respect.

10.4.4.1 Monitoring for Snow Avalanche Prediction

During winters 1999/2000 and 2000/2001, the SAMPLE study (SAMPLE) dealing with feasibility of snow avalanche prognosis used a sensor framework containing a laser scanner as key component. It resulted in the necessary knowledge in terms of sensor technology, control software and data handling for that purpose. Beside valuable hints in the possiblities and limits of such technology for the purpose of avalanche prognosis and warning, already some explicit cases for usability could be demonstrated. Figure 10.32 gives an idea of such a use case: one important indicator of growing avalanche risk at a certain slope are cracks on top of the hazardous region, particularly the speed of width increase of such cracks. The installation at St. Anton (see next section) allowed to evaluate several of such dynamic processes.

10.4.4.2 The St. Anton Campaign: Touristic Visualization of Arlberg Snow Cover

Starting in winter 2000/2001 together with the St. Anton Snow Avalanche Commission as an experienced target user, a prototype system for snow cover monitoring on the Valluga Mountain at St. Anton, midst a large alpine skiing area, was established.

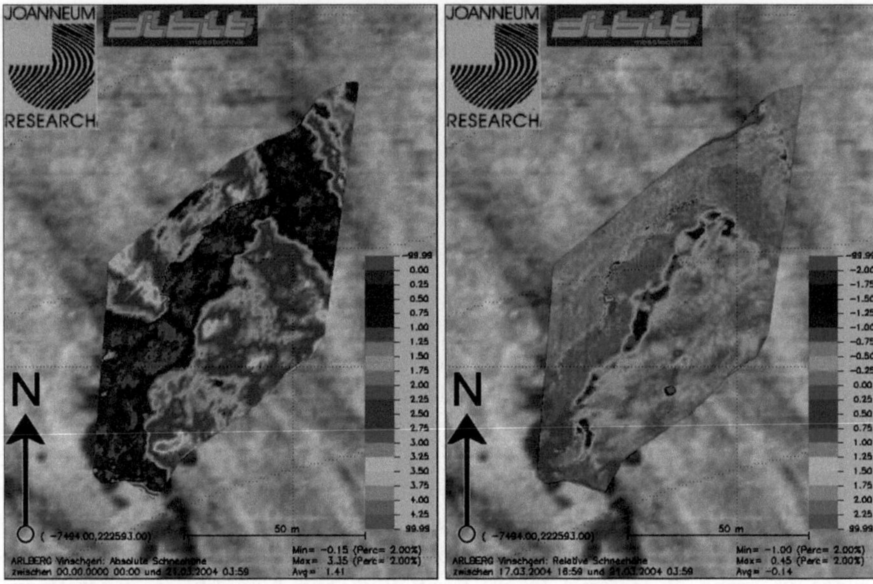

Fig. 10.32 Glide cracks. *Left* absolute snow cover. *Right* snow cover change (3 days)

To get measurements without snow cover for later determination of absolute snow height, in September 2000 a set of six ROIs on different slopes were defined. In addition, five ROIs were used for the reflective reference targets. All the ROIs were scanned within stable temperature conditions during night time. The reference targets were measured using standard geodetic methods which enabled further geocoding of the scanner data. Standard resolution on the slope ROIs was 0.2 Gon (with two of them having 0.1 Gon in one direction due to flat view), the reference targets were scanned with 0.02 Gon resolution.

For continuous monitoring the scanner was placed within a weatherproof heated housing. To facilitate maintenance during this first field application a placement near the upper station of the Valluga cable car at an altitude of about 2,600 m above sea level was selected. The housing is mounted on a 2 m high concrete pile (Fig. 10.33) with a wire channel inside which enables a direct connection to power supply and PC in the cable car station.

The sensor orientation was determined four times within the entire measurement period. At the beginning of the campaign a significant movement of the sensor could be detected, which was probably caused by beginning freezing of the ground (hence causing slight movements of the pile with respect to the rocks underneath). Each orientation process resulted in an average inconsistency of about 2 cm for the five reflective targets, which was mainly caused by atmospheric effects that changed the distance measurement.

In January 2001 the operational phase of data acquisition started. Each ROI was measured 2–3 times a day, giving access to one pseudo-coloured snow cover map

Fig. 10.33 Laser scanner LPM-2k with weatherproof housing, near Valluga cable car station

(snow height as difference to zero measurements in September 2000), and one pseudo-coloured snow cover change map (difference to one of the prior measurements). Figure 10.34 shows an example for a snow cover map which was the primary result of the system passed to the avalanche commission.

10.4.5 Disaster Monitoring Application Case Studies

The evaluation and classification of instable surfaces need fast and cheap automatic sensing methods with accuracy in the range of a few centimetres. A terrestrial laser scanning system is able to successfully perform an efficient change survey and is capable to automatically detect changes and motion on the surface of an active rockslide area.

A landslide is the movement of a mass of rock, debris or earth down a slope. A rockslide involves a downward, usually sudden and rapid movement of newly detached segments of bedrock over an inclined surface. A rock fall is the fastest moving landslide. A newly detached segment of bedrock of any size suddenly falls down from a very steep slope.

Once a landslide is triggered along a plane of weakness, material is transported by various mechanisms including sliding, flowing or falling. After falling or precipitously moving, the mass of materials deposits at the base of the slope. The moving mass is greatly deformed and usually breaks up into many smaller slides. Rockslides can vary in size from a single boulder in a rock fall or topple to tens of millions of cubic meters of material in a debris avalanche.

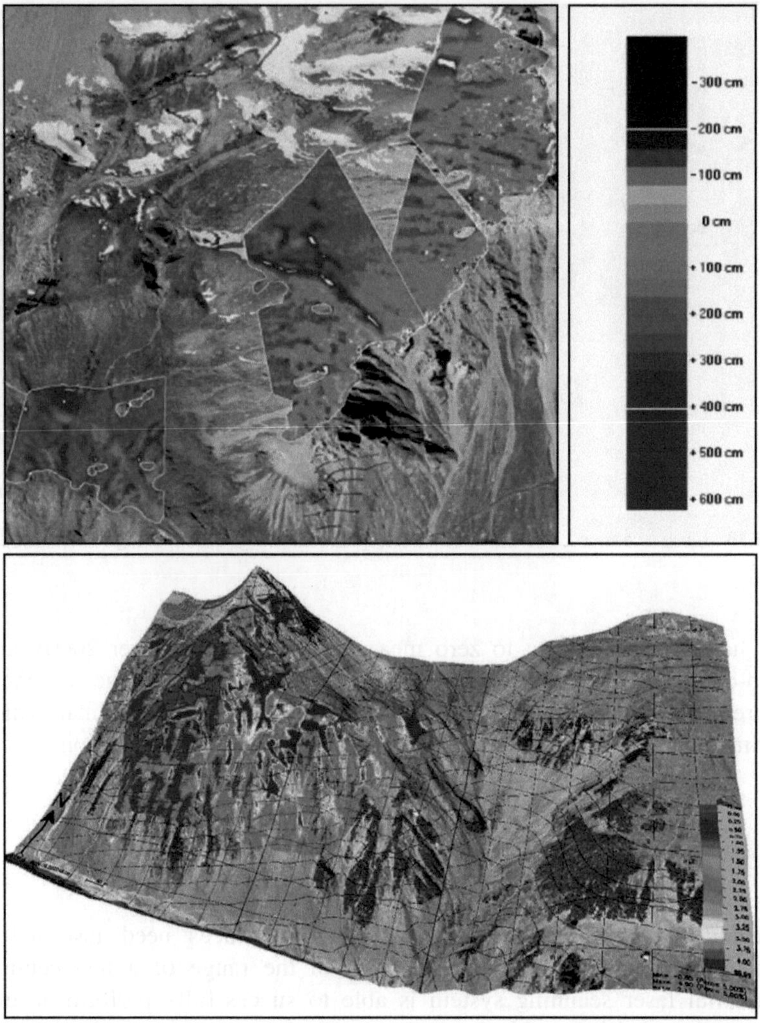

Fig. 10.34 *Top* snow cover map (27.2.2001) overlaid on ortho image (4 ROIs). Area about ½ km^2: large snow height in valleys is clearly visible. In the *lower left* ROI a flat ski track passes which causes lower snow height. Small areas with negative snow height are caused by noise at highly structured rocks with insufficient DSM representation. *Bottom* 3D visualization of a similar area

Rockslides can be triggered by natural causes or by human activity. Natural causes include extreme weather conditions like heavy rainfalls, saturation of slope material from rainfall or seepage, vibrations caused by earthquakes, or undercutting of banks by rivers. Also rockslides frequently occur in high mountain areas during spring and autumn when there is repeated freezing and thawing. Another reason is the climate change observed in the last decades. In particular the global warming increases the permafrost level and material formerly frozen gets instable.

Human activities may include the removal of vegetation like the removal of protection forest, interference with or changes to natural drainage, the modification of slopes by construction of trails, roads, railways or buildings, housing sprawl, mining activities, vibrations from heavy traffic or blasting, and excavation or displacement of rocks.

Rockslides are extremely hazardous. They involve a rapid sliding of large masses of fractured rock and regolith. They can move millions of tons of rocks in a short time. Rockslides may endanger or damage buildings, roads, railways, pipelines, protection forest, agricultural land and crops.

The risk assessment of landslides requires an accurate evaluation of the geology, hydrogeology, landform, and interrelated factors such as environmental conditions and human activities. It is of particular importance for engineers and geologists to assess slope stability and landslides in order to take appropriate, effective, and timely measures. Potential measures may vary from road closure, and excavation of buildings within the area of risk, to investments in protection buildings like protective barriers, retaining walls, retention capacity, and rock anchors. Disaster alert plans have to be developed. Above all engineering and geotechnical investigations have to define the landslide hazard and risk.

Potential indicators of active landslides include slope cracks, curved tree trunks, tilted poles, tilted walls, and the presence of wet or seepage areas. Before a rock fall, in most cases a slight shear continuous distortion over a specific period can be observed.

Rock fall models can be useful tools to predict the risk posed by individual falling rocks (Dorren 2003). One essential part of the management strategy for slope instability risk mitigation is a remote monitoring system for a continuous observation of the mass movement, which is often performed over long periods of time (Jaboyedoff et al. 2004). Remote monitoring of slope movement of unstable or potentially unstable slopes normally is a multidisciplinary approach incorporating several sensors. For example, movements and deformation can be measured with inclinometers, tilt-meters, extensometers, time-domain reflectometry, radar, and GPS. Water levels can be observed using vibrating wire piezometers (Kane and beck 2000).

In the standard case the area of a rockslide event is not achievable for the application of standard geodetic targets. Therefore remote techniques must be used that are continuously available, which excludes air-based or satellite remote sensing (Kenyi and Kaufmann 2003a) from the list of candidate techniques.

Terrestrial scanning laser imaging has turned out to be an essential component of geo-technical disaster monitoring, since it provides high resolution, a wide field of view, medium accuracy and high availability over long periods of time with comparably low cost.

10.4.5.1 Schwaz

On July, 10th, 1999 an unexpected large rockfall occurred at the Eiblschrofen near Schwaz (Tyrol, Austria). Several 1,000 m^3 of material were falling towards the valley, endangering the city of Schwaz. About 50 buildings and ten companies were evacuated, and about 250 inhabitants had to leave their homes immediately.

Due to the imminent threat for the inhabitants, immediately after the rockfall, geodetic, geological, geophysical, and geotechnical monitoring started at the Eiblschrofen area in order to monitor the actual situation, to find out the long-term trend of the movement, and to guarantee the safety of the workers at the protection dams. The rapid alert system (Scheikl et al. 2000a) was amongst others based on the following sensors: video surveillance, tachymeter, GPS, extensometer, high-precision levelling, infrared measurements, radio interferometry, acoustic monitoring, and terrestrial laser scanning.

Since long-range TLS systems (up to 2 km to naturally reflecting targets) were first available in 1998, this rockfall was the first practical test and major challenge to prove that ground-based remote monitoring using TLS offers the possibility for detailed real-time monitoring of surface motion with high temporal resolution.

The TLS monitoring system applied at Eiblschrofen (Scheikl et al. 2000b) was established in a successful co-operation between the scanner manufacturer (Riegl Laser Measurement Systems GmbH—Horn, Austria, www.riegl.co.at), software development (Joanneum Research—Graz, Austria, www.joanneum.at) and the main contractor (ILF—Innsbruck, Austria, www.ilf.com) who performed the integration into the geodetic framework as well as the operation. The acquired and processed data (with an accuracy of few centimetres for displacement measurements) were reviewed to be essential as a decision-making basis for the emergency response team.

10.4.5.2 Gries

The prototype of the Dibit Geoscanner monitoring system was field-tested at a rock fall near Gries, Austria. In June 2003 pieces of cliff as big as humans had dislodged due to mass movements in terms of a rockslide. Safety measures including the evacuation of several houses and road closures in the affected residential area were quickly initiated.

To assess the risk of succeeding rock falls, a laser-scanner monitoring system was installed which measured potentially instable regions over a period of more than two months.

Figure 10.35 shows an overview of the Gries rock fall area. In collaboration with geologists, a set of 11 ROIs were identified to be measured in high resolution round the clock. Additionally two ROIs, covering almost the whole instable area, were defined to be measured in lower resolution once a day. Five ROIs were used for the reflective reference targets to compensate atmospheric influences, and another five ROIs in the close-up range of the scanner position were used for sensor orientation. Each reference target was measured once using standard geodetic methods, which enabled further geocoding and comparability of the scanner data. Standard resolution on the slope ROIs was 0.05 Gon, and the reference targets were scanned with 0.02 Gon resolution.

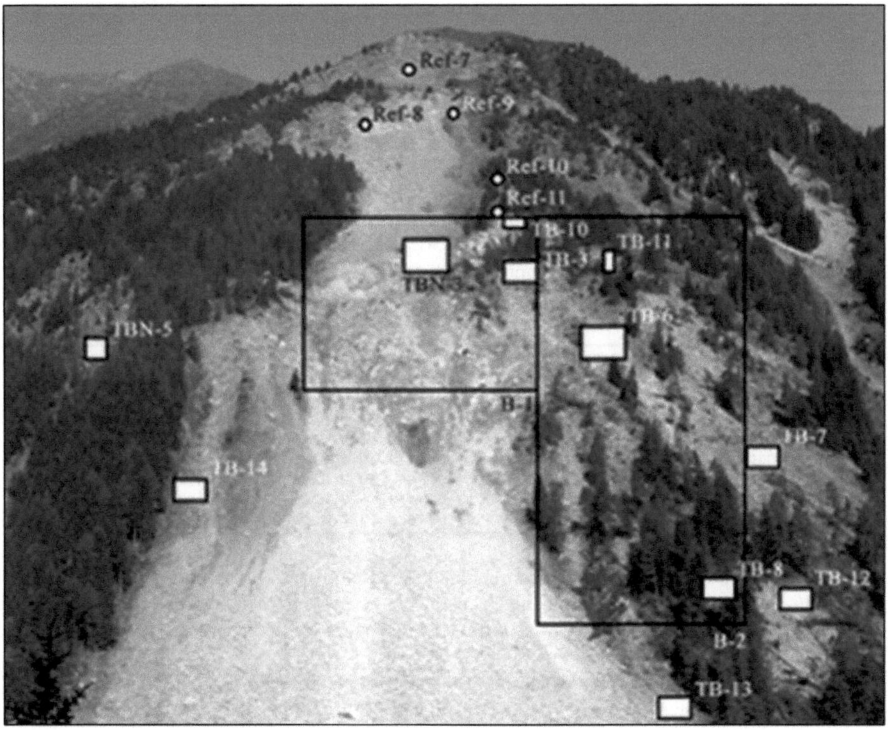

Fig. 10.35 Overview of the Gries rock fall area, Austria. A set of 13 monitoring ROIs were scanned, together with five reflective targets

For continuous monitoring the scanner was mounted on a stable console on the opposite slope with an average distance to the target area of 800 m.

The sensor orientation was determined four times a day to compensate slight movements and misalignments of the scanner. Before each slope measurement, a reference target was measured to compensate for atmospheric effects that have an influence on the scanner distance measurements.

Figure 10.36 shows an example of subsequent distance measurements of one ROI over a period of two days. The accuracy obtained is within a range of 5 cm in all three co-ordinate axes. Involving image processing and classification allows deriving simple deformation categories as shown in Fig. 10.37.

Figure 10.38 illustrates the 3D deformation of the whole instable area, and Fig. 10.39 shows an example of the deformation of one ROI as a result from region tracking.

Currently the measurement results are verified using geodetic measurements and ground truth information. Future research will emphasize the application dependent knowledge-based systems for ROI tracking, detection and correction of measurement outliers, accuracy investigations, an optimised sensing strategy and

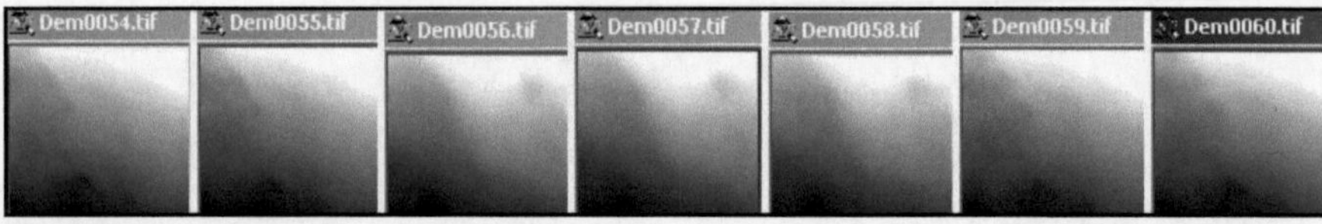

Fig. 10.36 *Grey-level coded* distance images of a ROI (10 × 7 m) in the Gries rock fall area measured over a period of two days (data acquisition every 6 h). *Bright* 635 m, *dark* 620 m distance. The motion between frames 55 and 56 was considerably slow, whereas a rock fall occurred between frames 58 and 59 (note that the structure changes completely)

```
078 TB8_20030721_054935 TB8_20030720_234921 0.86 0.13 0.0013 ROI_M_IrrelevantDisplacement
079 TB8_20030720_234921 TB8_20030720_174926 0.10 0.24 0.1860 ROI_M_StructureChanged
080 TB8_20030720_174926 TB8_20030720_114934 0.61 0.19 0.0200 ROI_M_LargeDisplacement
081 TB8_20030720_114934 TB8_20030720_054929 0.64 0.20 0.0023 ROI_M_IrrelevantDisplacement
082 TB8_20030720_054929 TB8_20030719_234942 0.81 0.16 0.0019 ROI_M_IrrelevantDisplacement
083 TB8_20030719_234942 TB8_20030719_174933 0.64 0.20 0.0188 ROI_M_LargeDisplacement
084 TB8_20030719_174933 TB8_20030719_114919 0.64 0.20 0.0194 ROI_M_LargeDisplacement
085 TB8_20030719_114919 TB8_20030718_235057 0.76 0.18 0.0006 ROI_M_IrrelevantDisplacement
086 TB8_20030718_235057 TB8_20030719_054929 0.76 0.17 0.0014 ROI_M_IrrelevantDisplacement
```

Fig. 10.37 Output of a simple deformation categorization process based on the data depicted in Fig. 10.36. In the rightmost column the decision of the system is displayed (structure change caused by a rock fall, irrelevant motion of the terrain surface, or considerable side motion of the terrain between subsequent measurements). The decision is based on statistics of matching results between subsequent distance images (some of the statistical parameters—matching area covered, reliability, and angular displacements—are displayed in the output list as separate columns)

additional sensors (Reiterer 2004) as well as the automatic selection of regions of interest.

The results show the usability of this approach in various related application areas such as landslides, glaciers and the monitoring of large edifices. All the results are available immediately, which makes the terrestrial laser scanner monitoring a valuable tool for risk evaluation and prediction.

10.4.6 Geo-Risk Monitoring Case Studies and Systems

Scanning laser imaging has turned out to be an essential component of geotechnical disaster monitoring. The following section will give a short overview over existing applications and case studies conducted in the field of geo-hazard monitoring.

Two TLS based measurement frameworks focused on the monitoring of geo-hazard zones are the “DIBIT Geoscanner” (Sect. 10.3.1), developed by Joanneum Research and DMG and the “Site Monitor” (Sect. 10.3.2) developed by the UK based company “3D Laser Mapping”. Both systems combine TLS data acquisition, visualisation and evaluation and offer an automated monitoring process.

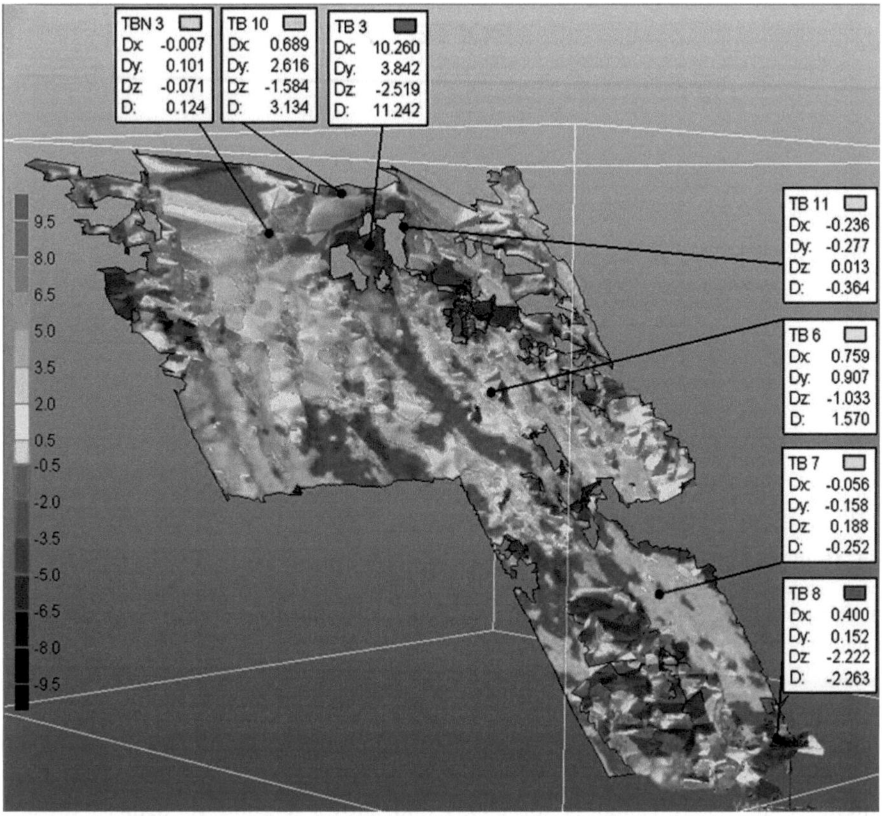

Fig. 10.38 3D deformation model of the whole instable area (ROI B-1 and B-2 in Fig. 10.35.) over a period of three weeks. The deformation is *grey level coded* with *white colour* indicating almost no changes regarding to the reference measurement at the beginning of the monitoring period, and *black colour* indicating up to 9 m of break-off or accumulation. For significant terrain regions the deformation amount is displayed in all three dimensions

Beside these commercial attempts, some case studies have reached a level of maturity in very specific aspects (see Table 10.5 for an overview): applications for geo-hazard monitoring were tested all around the world and several specific measurement frameworks have been developed for the scientific community (Lehmann et al. 2009) as well as for commercial applications (e.g. mine monitoring www.3dlasermapping.com, see above), or both (www.dibit-scanner.at). For various applications different TLS systems exist on the market with an operating distance between near-range (up to 10 m), midrange ($\sim$300 m) and long range (up to 6,000 m). The ability of TLS to acquire high-resolution 3D data of surface structures makes this technique a very interesting instrument for measuring hazardous or non accessible sites such as high mountain environments and extensive or susceptible man made structures. Another approach is the so called Ground Based Interferometric Synthetic Aperture Radar (GBInSAR) technique. It allows

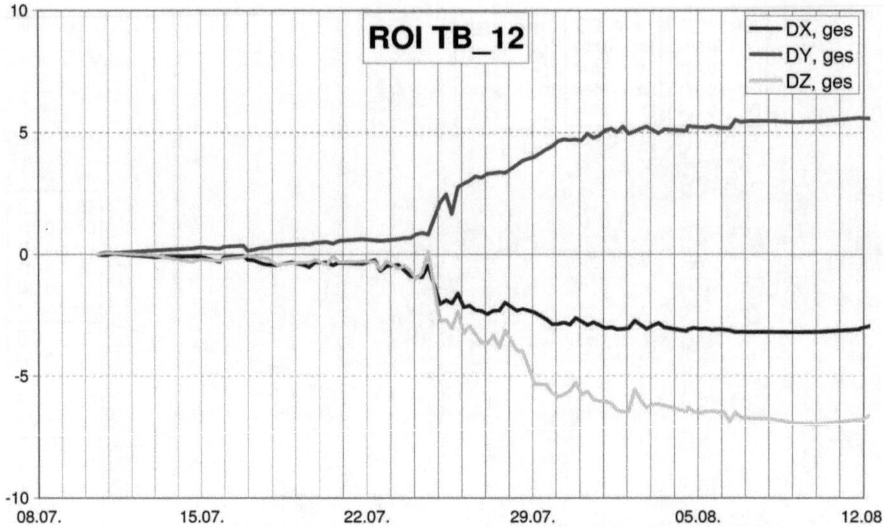

Fig. 10.39 Deformation of a significant area over a period of one month. Starting from July 11th the average position was tracked. The area moved slightly until July 24th. Mainly due to a heavy rainfall period the region slid up to 5 m until July 31st. Subsequently the region movement stabilized

to measure considerably higher accuracies (deviations in the range of millimeters (Alba et al. 2008) compared to centimeter using TLS (e.g. Abellán et al. 2009; Bitelli et al. 2004; Hsiao et al. 2003) and shows a good robustness towards fog and rain.

On the other hand these techniques have several shortcomings. Compared to IATS based measurement systems (as introduced in Lehmann et al. 2009) TLS offer a limited range and less reliable measurement accuracies whereas GBInSAR is more accurate but requires heavy stationary equipment (Alba et al. 2008) which confines possible measurement sites.

10.5 Conclusions and Prospects

The major advantages of the terrestrial remote sensing of surface motion lie in the accessibility of unreachable (steep slope, hazardous) areas, together with the ability to identify and evaluate both global and local surface motion and change effects. Furthermore, fix installed systems allow a frequent geo-coded measurement of the same area for long periods of time.

High resolution DSMs derived particularly from long range terrestrial laser scanning are a good data basis for monitoring processes related to geo-hazards. Derivates like surface elevation changes and displacement vectors (in case of the

Table 10.5 Geo-risk monitoring case studies and systems

Approach/results				
Comparison of TLS, tachymetry, probing (Prokop et al. 2007)	Davos, Switzerland	Riegl LPM-800HA Leica TCRP1201	20–450 m	800 m 300 m
Snowpack measurement via tachymeter, TLS and simple probing. Comparison with tachymeter as reference: offset ±8 cm. Std.dev. 1σ: 2 cm-TLS, 10 cm-probing. TLS showed to be most efficient				
Landslide change detection via LiDAR (Hsiao et al. 2003)	Hsin-chu and Jiu-fen-er, Taiwan	Optech ILRIS-3D	<100 m	1,500 m
Measurements by total stations, GPS, air and ground based LIDAR. Landslide topography change can be detected (cm-m scale), LIDAR methods not enough for erosion evaluation (requires mm accuracy)				
Rockslide monitoring using TLS (Oppikofer et al. 2009)	Åknes western Norway	Optech ILRIS-3D	300–400 m	1,500 m
Four scans, unwanted obj. manually removed. Co-registration by iterative closest point algorithm. Geo-referenced through ALS point cloud of assumed stable rock part to measure displacements				
TLS and photogrammetry landslide monitoring (Bitelli et al. 2004)	Cà di Malta, near Bologna, Italy	Riegl LMS-Z210 Riegl LMS-Z420i	<300 m	350 m 800 m
Aerial photographs and two TLS surveys. Displacement detection, method comparison. TLS: effective rapid solution for small/medium landslides in high frequency/accuracy. Aerial photographs: more expensive, DTM more difficult to construct. However: no masked regions; can cover larger areas				
Detecting millimetric deformation via TLS (Abellán et al. 2009)	Catalonia, Spain	Optech ILRIS-3D	50–190 m	1,500 m
The experimental setup (mm deformation assessment) and evaluation of basalt cliff TLS scans before and after rock-fall. Comparison suggests that cliff underwent deformation prior to the actual rock-fall				
GBInSAR dam deformat. measurement (Alba et al. 2008)	Cancano dam, Alta Valtellina, Italy	IBIS-L microwave interferometer	400 m	4,000 m
Novel GBInSAR sensor used to measure daily deformations of arch-gravity dam (due to water basin and temperature change). Validation comparison with coordinatometer shows a good agreement				
Landslide analysis via GBInSAR (Leva et al. 2003)	Schwaz, Austria	LISA SAR System	1,000 m	n/a
Interferometer used to retrieve landslide velocity field. Deformation map agrees with measurements by GPS. Results show that GBInSAR is a cost-effective solution for the monitoring of landslides				
SiteMonitor (www.3dlasermapping.com)	Potgietersrust Mine, South Africa	Riegl LPM-2K	525 m	2,500 m
TLS based framework for mine monitoring. Data acquisition, visualisation and evaluation (Sect. 10.3.2)				
Dibit GeoScanner (www.dibit-scanner.at)	Pasterze Glacier, Austria	Riegl LPM-2 K	<2,000 m	2,500 m
TLS based framework for geo-monitoring. Data acquisition, visualisation and evaluation (Sect. 10.3.1)				
i-MeaS (Lehmann et al. 2009)	Pasterze Glacier, Austria	2x Leica TCA 1201 based IATS	1,000 m	5,000 m
Geo-hazard monitoring system based on two image-based tacheometers, automatic point detection and determination by spatial forward intersection (enhanced distance and accuracy) (Sect. 10.3.3)				

Table 10.6 Key parameters for a long-range terrestrial geo-risk monitoring system

Quantitative objective	Target value
Measurement range	>5 km
Accuracy of single 3D point detection (for 1,000 m distance)	<3 cm (x, y, z)
Necessary measurement grid width on surface	<1 m
Accuracy of full area cover change detection	<5 cm (distance)
Operation effort	1 person operating, 2 h establishment
Application maturity	Similar to tacheometers: ruggedized, weatherproof → field capable
Measurement speed (single point)	>3 points/s
Measurement speed (full area cover 10 * 10 degrees)	<1 h

availability of multi-temporal high resolution DSMs) provide useful information about 3D surface dynamics. However, the quality of data sets and concepts for calculating morphometric parameters have to be assessed critically. Point density is a crucial factor (Bodin et al. 2008) as well as the quality of the measurement itself (high point ratio and total point number). Remoteness (sufficient energy supply) and atmospheric conditions (e.g. air humidity) are crucial limiting factors. Furthermore, exemplary studies (e.g. those conducted for rock glaciers) demonstrate the importance of independent control data for evaluation and the problems in acquisition of the latter. High landscape dynamics such as frequent shifting of material and block falls inhibit terrestrial surveys. Methodological problems in automatic data interpretation (e.g. inadequate texture for photogrammetry) complicate remote sensing approaches.

In the following, a vision of a productive system to cope with the requirements of the various deformation monitoring cases is given. Table 10.6 lists key parameters to be fulfilled, up-to-date sensor concepts appear mature enough to deliver proper data.

For highly mature terrestrial monitoring systems as sketched in this section, the **Measurement and Data evaluation** will combine several methods and data:

(1) A suitable **calibration method**. Due to the complex system design calibration of the whole multi-sensor system is necessary. This process includes instrument calibration as well as the calibration of image-related parts. Main goal is to be able to collect any point of the observed surface within an accuracy of 0.3 mGon, which enables a 3D point localization in the range of $\sim$25 mm from a distance of 1 km.

(2) **Selection of robust targets for monitoring**: Vegetation needs to be separated from soil, rock or debris. It is targeted to be able to automatically determine at least 80 percent of the usable—non vegetated—area.

(3) **Decision on adequate target selection**: Due to a relatively low measurement speed, point measurement devices such as IATS can only cover a

limited set of measurement points within a repetitive period. It is therefore essential to decide upon the relevant target before the measurement campaign starts.

(4) A **3D point detection and matching system** which is nearly independent towards changing illumination conditions (Cast shadows, direct sunlight, ambient light). This is the key to accurate 3D measurement in time-series.

(5) **Filtering of measurements to evaluate their reliability**, taking into account various conditions: Meteorological (temperature distribution of atmosphere, haze, rainfall, etc.), surface (e.g. moisture), and illumination (sun incidence angle, ambient, white balance, etc.) influences need to be compensated in order to be able to compare measurements from different instances of time.

(6) Any vision-based long-range measurements are strongly exposed to the atmospheric conditions. One of the main problems when working with image-based sensor systems therefore is the consideration of the **refraction of light**, due to different densities of the atmosphere: The influence of the refraction depends on the measured azimuth (vertical angle) and the meteorological conditions. For a distance of 1,000 m and comfort meteorological conditions (20°C, 1,013 hPa) this influence can be in the order of 1–2 cm. For long-range measurements also the earth curvature has to be taken into account. To consider this influence will be a challenging task.

(7) **Combine 2D and 3D information** for target recognition and matching: It is not always possible to use just image texture for matching, particularly if different epochs with different illumination conditions are involved. Such matching is robust, however, when the 3D structure (which does not change locally in the narrow field-of-view of the IATS cameras) is used, which can be a by-product from target selection. To remain below the influence of refraction is the requirement here assuming it cannot be feasibly modeled.

(8) A major challenge is the **independence/robustness** with regard to illumination conditions[1] (see previous point). A robust system can only be implemented regarding changes in light and shadow during point capturing. This can be considered on the one hand by recognizing similar illumination conditions (in the simplest case selection the same time of day for comparative measurements), on the other hand by computer vision methods that are robust to illumination conditions. The selection of proper targets on the basis of 3D information (see above) is another means to partly solve this problem, also here keeping accuracy better than the refraction influence is the target.

(9) For data interpretation, **adequate data exploitation** is required: The problem of automated and semi-automated data interpretation is the hugeness of a 3D point detection system. Efficient data interpretation can only be processed on a properly preprocessed data set. This preprocessing can be

[1] Of course an active system such as Radar Interferometry is not prone to such restrictions.

done in form of data clustering (e.g. points belonging to a unique form of movement are summarized to one cluster and in further processing steps represented by this cluster). Outlier detection and removal is the most important measure of filtering, a rate close to zero will be the target for a commercial system in order to keep the rate of unmotivated sensible measures (e.g. evacuation due to false alarm) as low as possible.

(10) For implementing and combining all techniques into a **productive prototype system**, the use of an **open architecture and the connection with existing program modules such as sensor control**, **Graphical User Interface (GUI)**, **and a sensing data base** is essential.

Acknowledgments The content of this chapter was produced in multiple research projects which deserve further mentioning. Namely the i-MeaS—"An Intelligent Image-Based Measurement system for Geo-Hazard Monitoring" project (info.tuwien.ac.at/ingeo/research/imeas) which is funded by the Austrian Science Fund (Fond zur Förderung der wissenschaftlichen Forschung Österreich, FWF) (project number: L514), the project ALPCHANGE (www.alpchange.at) also funded by the FWF (project number P18304-N10) and the "K plus program" of "K plus Competence Center Advanced Computer Vision" together with FWF Project P14664. Furthermore we kindly acknowledge the help of Viktor Kaufmann (Institute of Remote Sensing and Photogrammetry, Graz University of Technology) for providing data from geodetic surveys and kindly reviewing this chapter. Viktor Kaufmann, Gerhard Karl Lieb, Andreas Kellerer-Pirklbauer-Eulenstein and Herwig Proske provided valuable source material from related publications, this is very much appreciated. We also thank students of the Institute of Geography and Regional Science, University of Graz and the Institute of Remote Sensing and Photogrammetry, Graz University of Technology, Austria as well as several volunteers of the National Park Hohe Tauern for their support during field campaigns. Last but not least we thank our important research partners and funding sources Joanneum Research (www.joanneum.at) and Dibit Messtechnik GmbH (www.dibit-scanner.at).

References

Abellán A, Jaboyedoff M, Oppikofer T, Vilaplana JM (2009) Detection of millimetric deformation using a terrestrial laser scanner: experiment and application to a rockfall event. Nat Hazards Earth Syst Sci 9:365–372

Alba M, Bernardini G, Giussani A, Ricci PP, Roncoroni F, Scaioni M, Valgoi P, Zhang K (2008) Measurement of dam deformations by terrestrial interferometric techniques. Int Arch Photogramm Remote Sens Spat Inf Sci XXXVII (part B1):133–139

Auer I, Böhm R, Leymüller M, Schöner W (2002) Das Klima des Sonnblicks—Klimaatlas and Klimageographie der GAW-Station Sonnblick einschließlich der umgebenden Gebirgsregion. Österreichische Beiträge zur Meteorologie und Geophysik 28:305

Avian M, Bauer A (2005) The use of long range laser scanners in terrestrial monitoring of glacier dynamics, Pasterze glacier (Hohe Tauern, Austria). Geophys Res Abstr 7:06779 (European Geosciences Union General Assembly, Vienna, Austria. 24–29 Apr 2005)

Avian M, Kaufmann V, Lieb GK (2005a) Recent and Holocene dynamics of a rock glacier system: the example of Langtalkar (Central Alps, Austria). Nor J Geogr 59:149–156

Avian M, Bauer A, Lieb GK (2005b) Monitoring modification of alpine environments: new techniques and perspectives. In: 3rd symposium of the Hohe Tauern National Park for research in protected areas, Kaprun, Salzburg, 15–17 Sept 2005

Baltsavias EP, Favey E, Bauder A, Boesch H, Pateraki M (2001) Digital surface modelling by airborne laser scanning and digital photogrammetry for glacier monitoring. Photogramm Rec 17(98):243–273

Bauer A, Paar G (1999) Elevation modeling in real time using active 3D sensors. In: Proceedings of the 23rd workshop of the Austrian association for pattern recognition, AAPR, Robust vision for industrial applications 1999, Steyr, Austria, 27–28 May 1999, Schriftenreihe der Österreichischen Computer Gesellschaft, vol 128, pp 89–98

Bauer A, Paar G (2004) Monitoring von Schneehöhen mittels terrestrischem Laserscanner zur Risikoanalyse von Lawinen. In: Proceedings of the 14th international course on engineering surveying, Zurich, Switzerland, 15–19 Mar 2004

Bauer A, Paar G, Kaufmann V (2003) Terrestrial laser scanning for rock glacier monitoring. In: Phillips M, Springman SM, Arenson LU (eds) Proceedings of the 8th international permafrost conference, Zurich, pp 55–60

Bauer A, Kellerer-Pirklbauer A, Avian M, Kaufmann V (2005a) Five years of monitoring the front slope of the highly active Hinteres Langtal rock glacier using terrestrial laser scanning: a case study in the Central Alps, Austria, Terra Nostra. In: 2nd European conference on Permafrost, vol 91, Potsdam

Bauer A, Paar G, Kaltenböck A (2005b) Mass movement monitoring using terrestrial laser scanner for rock fall management. In: Proceedings of the 1st international symposium on geo-information for disaster management, Delft, The Netherlands. Springer, Berlin, pp 393–406

Bauer J, Sünderhauf N, Protzel P (2007) Comparing several implementations of two recently published feature detectors. In: Proceedings of the international conference on intelligent and autonomous systems, IAV, Toulouse

Bay H, Ess A, Tuytelaars T, Van Gool L (2008) SURF: speeded up robust features. Comput Vis Image Underst 110(3):346–359

Benn DI, Evans DJA (1998) Glaciers and glaciation. Arnold, London, p 734

Bitelli G, Dubbini M, Zanutta A (2004) Terrestrial laser scanning and digital photogrammetry techniques to monitor landslide bodies. In: Proceedings of the XXth ISPRS congress, vol XXXV, part B5, Istanbul, pp 246–251

Bodin X, Schoeneich P, Jaillet S (2008) High resolution DSM extraction from terrestrial LIDAR topometry and surface kinematics of the creeping alpine permafrost: the Laurichard Rockglacier case study (French Southern Alps). In: Kane DL, Hinkel KM (eds) Ninth international conference on permafrost, Institute of Northern Engineering, University of Alaska at Fairbanks, vol 1, pp 137–142

Delaloye R, Perruchoud E, Avian M, Kaufmann V, Bodin X, Ikeda A, Hausmann H, Kääb A, Kellerer-Pirklbauer A, Krainer K, Lambiel C, Mihajlovic D, Staub B, Roer I, Thibert E (2008) Recent interannual variations of Rockglaciers creep in the European Alps. In: 9th international conference on permafrost, Fairbanks, Alaska, 29 June–03 July 2008, pp 343–348

Dorren L (2003) A review of rock fall mechanics and modelling approaches. Prog Phys Geogr 27(1):69–87

Fischer A, Span N (2005) A volume inventory of glaciers in the Austrian Alps. European Geosciences Union (EGU), second assembly, Vienna, 24–29 Apr 2005, CD-ROM

Geist T, Lutz E, Stötter J (2003) Airborne laser scanning technology and its potential for applications in glaciology. In: Proceedings of the ISPRS workshop on 3-D reconstruction from airborne laserscanner and INSAR data, Dresden, pp 101–106

hds.leica-geosystems.com (2010) Official web-site of Leica Geosystems. Accessed 29 Jan 2010

Hsiao KH, Yu MF, Liu JK, Tseng YH (2003) Change detection of landslide terrains using ground-based lidar data. In: Proceedings of 2003 annual symposium of the society of Chinese association of geographic information

Jaboyedoff M, Ornstein P, Rouiller JD (2004) Design of a geodetic database and associated tools for monitoring rock-slope movements: the example of the top of Randa rock fall scar. Nat Hazards Earth Syst Sci 4:187–196

Kääb A (2002) Monitoring high-mountain terrain deformation from digital aerial imagery and ASTER data. ISPRS J Photogramm Remote Sens 57:39–52 (1–2 Novemb 2002)

Kääb A, Kaufmann V, Ladstädter R, Eiken T (2003) Rock glacier dynamics: implications from high-resolution measurements of surface velocity fields. In: Proceedings of the eighth international conference on permafrost, vol 1, 21–25 July 2003, Zurich, Switzerland, pp 501–506

Kane WF, Beck TJ (2000) Instrumentation practice for slope monitoring. In: Engineering geology practice in Northern California. association of engineering geologists Sacramento and San Francisco sections

Kaufmann V, Ladstaedter R (2000) Spatio-temporal analysis of the dynamic behaviour of the Hochebenkar rock glaciers (Oetztal Alps, Austria) by means of digital photogrammetric methods. In: Proceedings of the 6th international symposium on high mountain remote sensing cartography, Grazer Schriften der Geographie und Raumforschung, Band 37, Institute of Geography and Regional Sciences, University of Graz, pp 119–139

Kaufmann V, Ladstaedter R (2003) Quantitative analysis of rock glacier creep by means of digital photogrammetry using multi-temporal aerial photographs: two case studies in the Austrian Alps. In: Proceedings of the eighth international conference on permafrost, vol 1, 21–25 July, Zurich, Switzerland. Balkema Publishers, Rotterdam, pp 525–530

Kaufmann V, Ladstädter R (2004a) Documentation of the movement of the Hinteres Langtalkar rock glacier. In: Proceedings of the 20th congress of the international society for photogrammetry and remote sensing, vol 35, part B7, Istanbul, Turkey, 12–23 July 2004, IAPRS, pp 893–898

Kaufmann V, Ladstädter R (2004b) Terrestrisch-photogrammetrische Dokumentation des Gletscherrückgangs am Gößnitzkees (Schobergruppe, Nationalpark Hohe Tauern). Pangeo 2004, Graz, 24–26 Sept 2004, Beitrags-Kurzfassungen, Erdwissenschaften und Öffentlichkeit, vol 9, pp 240–242

Kaufmann V, Ladstädter R (2004c) Documentation of the retreat of a small debris-covered cirque glacier Goessnitzkees, Austrian Alps by menas of terrestrial photogrammetry. In: Proceedings of the 4th ICA mountain cartography workshop, Vall de Nuria, Catalonia, Spain, 30 Sept–02 Oct 2004, pp 65–76

Kaufmann V, Kenyi LW, Avian M (2005) Messung der Fließgeschwindigkeit von Gletschern mittels satellitengestützter Radar-Interferometrie in der Schobergruppe (Nationalpark Hohe Tauern, Kärnten). Endbericht zum Forschungsprojekt (Projektleiter V. Kaufmann) des Kärntner Nationalparkfonds, Institut für Fernerkundung und Photogrammetrie, TU Graz, p 59

Kellerer-Pirklbauer A (2008) The supraglacial debris system at the Pasterze glacier, Austria: spatial distribution, characteristics and transport of Debris. Z Geomorph NF 52(Suppl 1):3–25

Kellerer-Pirklbauer A, Bauer A, Proske H (2005) Terrestrial laser scanning for glacier monitoring: Glaciation changes of the Gößnitzkees glacier (Schober group, Austria) between 2000 and 2004. Third symposion of the Hohe Tauern national park for research in protected areas, Kaprun, Austria, 15–17 Sept 2005, pp 97–106

Kenyi LW, Kaufmann V (2003a) Estimation of rock glacier surface deformation using SAR interferometry data. IEEE Trans Geosci Remote 41(6):1512–1515

Kenyi LW, Kaufmann V (2003b) Measuring rock glacier surface deformation using SAR interferometry. In: Proceedings of the 8th international permafrost conference, vol 1, Zurich, Switzerland, 21–25 July. Balkema Publishers, Lisse, pp 537–541

Kienast G, Kaufmann V (2004) Geodetic measurements on glaciers and rock glaciers in the Hohe Tauern National park (Austria). In: Proceedings of the 4th ICA mountain cartography workshop, Vall de Núria, Catalonia, Spain, 30 Sept–2 Oct 2004, Monografies tècniques 8, Institut Cartogràfic de Catalunya, Barcelona, pp 101–108

Krainer K, Mostler W (2000) Reichenkar rock glacier: a glacier derived debris-ice-system in the Western Stubai Alps, Austria. Permafr Periglac Proc 11:267–275

Krobath M (2003) Gletscherschwund—Wasserland Steiermark 3:18–23

Kweon IS, Kanade T (1992) High-resolution terrain map from multiple sensor data. IEEE Trans Pattern Anal Mach Intell 14(2):278–292

Lambrecht A, Würländer R, Kuhn M (2005) The new Austrian glacier inventory: a tool for the analysis of modern glacier change. European Geosciences Union (EGU), second assembly, Vienna, 24–29 Apr 2005, CD-ROM

Lehmann M, Reiterer A, Huber NB, Bauer A (2009) An automated optical rockfall monitoring system. In: 9th conference on optical 3-D measurement techniques, vol 1, Vienna, 2009, pp 91–101

Leva D, Nico G, Tarchi D, Fortuny-Guasch J, Sieber AJ (2003) Temporal analysis of a landslide by means of a ground-based SAR interferometer. GeoRS 41(4):745–752 (Apr 2003)

Lieb GK (1991) Die horizontale und vertikale Verbreitung von Blockgletschern in den Hohen Tauern (Österreich). Zeitschrift für Geomorphologie NF 35(3):345–365

Lieb GK (2000) Die Flächenänderung von Gößnitz- und Hornkees (Schobergruppe, Hohe Tauern) von 1850 bis 1997. Festschrift für Heinz Slupetzky zum 60. Geburtstag, Salzburger Geographische Arbeiten 36:83–96

Lieb GK, Kaufmann V, Avian M (2004) Das Hintere Langtalkar (Schobergruppe, Nationalpark Hohe Tauern)—ein Beispiel für die komplexe Morphodynamik in der Hochgebirgsstufe der Zentralalpen. Mitt d Österr Geogr 146:147–164 (Gesellschaft, Wien)

Lowe DG (2004) Distinctive image features from scale-invariant keypoints. Int J Comput Vis 60:91–110

Mikolajczyk K, Schmid C (2005) A performance evaluation of local descriptors. IEEE Trans Pattern Anal Mach Intell 27(10):1615–1630

Mischke A, Kahmen H, (1997) A new kind of measurement robot system for surveying of non signalized targets. In: Optical 3-D measurement techniques, vol IV. Herbert Wichmann, Karlsruhe

Nakawo M, Raymond CF, Fountain A (eds) (2000) Debris-covered glaciers. In: Proceedings of an international workshop held at the University of Washington in Seattle, vol 264, Washington, 13–15 Sept 2000. IAHS publication, Wallingford, p 288

Oppikofer T, Jaboyedoff M, Blikra L, Derron M-H, Metzger R (2009) Characterization and monitoring of the Åknes rockslide using terrestrial laser scanning. Nat Hazards Earth Syst Sci 9:1003–1019

Paar G, Almer A (1993) Fast hierarchical stereo reconstruction. In: Proceedings of the 2nd conference on optical 3-D measurement techniques, Zurich, Switzerland, pp 460–466

Paar G, Bauer A (2001) Terrestrial long range laser scanning for high density snow cover measurement. In: Proceedings of the 5th conference on optical 3D measurement techniques. Vienna, pp 33–40

Paar G, Pölzleitner W (1992) Robust disparity estimation in terrain modelling for spacecraft navigation. In: Proceedings 11th ICPR, international association for pattern recognition, pp 738–741

Paar G, Nauschnegg B, Ullrich A (2000) Laser scanner monitoring—technical concepts, possibilities and limits. Workshop on advances techniques for the assessment of natural hazards in mountain areas, Igls, Austria, 4–6 June

Patzelt G (1980) The Austrian glacier inventory: status and first results. In: Workshop proceedings, vol 126. IAHS, Riederalp, pp 267–280

Patzelt G (2005) Gletscherbericht 2003/2004: Sammelbericht über die Gletschermessungen des Österreichischen Alpenvereins im Jahre 2004. Mitteilungen des Österreichischen Alpenvereins 60(2):24–31

Pfeifer N, Lichti D (2004) Terrestrial laser scanning: developments, applications and challenges. GIM International 18(12):50–53 (Dec 2004)

Prokop A, Schirmer M, Rub M, Lehning M, Stocker M (2007) A comparison of measurement methods: terrestrial laser scanning, tachymetry and snow probing, for the determination of the spatial snow depth distribution on slopes. In: International symposium of snow science. Moscow

Reiterer A (2004) Knowledge-based decision system for an on-line videotheodolite-based multisensor system. PhD thesis, Vienna University of Technology

Reiterer A, Kahmen H, Egly U, Eiter T (2003) 3D-Messverfahren mit Videotheodoliten und automatisierte Zielpunkterfassung mit Hilfe von Interest Operatoren. Allgemeine Vermessungs-Nachrichten 110:150–156

Reiterer A, Lehmann M, Miljanovic M, Ali H, Paar G, Egly U, Eiter T, Kahmen H (2009) A 3D optical deformation measurement system supported by knowledge-based and learning techniques. J Appl Geodesy 3:1–13

Roer I, Avian M, Delaloye R, Lambiel C, Dousse JP, Bodin X, Thibert E, Kääb A, Kaufmann V, Damm B, Langer M (2005) Rock glacier "speed-up" throughout European Alps—a climatic signal? In: Proceedings of the second european conference on permafrost, Potsdam, Germany, 12–16 June 2005, pp 101–102

Roic M (1996) Erfassung von nicht signalisierten 3D-Strukturen mit Videotheodoliten. Dissertation, TU-Wien

Scheikl M, Angerer H, Dölzlmüller J, Poisel R, Poscher G (2000a) Multidisciplinary monitoring demonstrated in the case study of the Eiblschrofen rock fall. Felsbau 18(1):24–29

Scheikl M, Poscher G, Grafinger H (2000b) Application of the new automatic laser remote monitoring system (ALARM) for the continuous observation of the mass movement at the Eiblschrofen rockfall area—Tyrol. Workshop on advances techniques for the assessment of natural hazards in mountain areas, Igls, Austria, 4–6 June 2000

Scherer M (2004) Intelligent scanning with robot-tacheometer and image processing—a low cost alternative to 3D laser scanning? In: FIG working week 2004, Athens, Greece, 22–27 May, pp 1–12

Sharov A, Gutjahr K (2002) Some methodological enhancements to INSAR surveying of polar ice caps. In: Begni G (ed) Observing our environment from space. Proceedings of the 21st EARSeL symposium in Paris, 14–16 May 2001. Balkema, Lisse, pp 65–72

Solomon S, Qin D, Manning M, Chen Z, Marquis M, Averyt KB, Tignor M, Miller HL (2007) IPCC: climate change 2007: the physical science basis. Contribution of working group I to the fourth assessment report of the intergovernmental panel on climate change. Cambridge University Press, Cambridge, pp 235–336

Steffan H, Bauer A, Schaffhauser H, Randeu W (2001) SAMPLE—Snow avalanche monitoring and prognosis by laser equipment. Final report. EU target area II regional support funded, Styrian Government ref. AAW 11 L 6 97/5

Teskey WF (1985) Determining deformation by combining measurement data with structural data. In: Teskey WF, Gruendig L (eds) Papers for the precise engineering and deformation surveys workshop, Calgary Alberta

Vicovac T, Reiterer A, Egly U, Eiter T, Rieke-Zapp D (2009) First development steps for an automated knowledge-based deformation interpretation system. In: Grün A, Kahmen H (eds) Optical 3-D measurement techniques IX, vol 1, Zurich, Switzerland, pp 61–90

Wakonigg H, Lieb GK (1996) Die Pasterze und ihre Erforschung im Rahmen der Gletschermessungen. Kärntner Nationalparkschriften 8, Großkirchheim, pp 99–115

Walser B (2004) Development and calibration of an image assisted total station. Dissertation, ETH-Zürich

Wasmeier P (2009) Grundlagen der Deformationsbestimmung mit Messdaten bildgebender Tachymeter. Dissertation, TU-München

Welsch W, Heunecke O, Kuhlmann H (2000) Auswertung geodätischer Überwachungsmessungen. Wichmann, Heidelberg

Würländer R, Kuhn M (2000) Zur Erstellung und Anwendung der Produkte des neuen Österreichischen Gletscherkatasters. Festschrift für Heinz Slupetzky zum 60. Geburtstag, Salzburger Geographische Arbeiten 36:57–67

www.3dlasermapping.com (as from 8 Dec 2011)

www.dibit-scanner.at: official web-site of DIBIT GeoScanner (as from 8 Dec 2011)

www.ilf.com: official web-site of ILF Consulting Engineers (as from 8 Dec 2011)

www.joanneum.at: official web-site of JOANNEUM RESEARCH (as from 8 Dec 2011)

www.riegl.co.at: official web-site of Riegl Laser Measurement Systems (as from 8 Dec 2011)

www.topcon.eu: official web-site of Topcon Europe Positioning B.V. (as from 8 Dec 2011)

www.trimble.com: official web-site of Trimble Measurement Systems (as from 8 Dec 2011)

Chapter 11
LaSIRF: Landslide Safe Intelligent Route Finder for Mountainous Terrain in GIS Environment

M. K. Arora, A. K. Saha, P. Gupta and R. P. Gupta

Abstract Route planning in hilly areas is a compound job as it involves consideration of a number of factors. The conventional route planning practice is time consuming and does not consider factors related to geo-hazards such as landslide hazard zones, geological faults etc., thereby resulting in increased cost of road design, maintenance etc. The aim of this chapter is to develop a Geographic Information System (GIS) based software for planning a road route that passes through landslide safe areas. A number of thematic cost factors have been integrated in GIS. Dijkstra's least-cost finding algorithm together with improved neighbourhood analysis to compute the neighbourhood movement cost has been used to find landslide safe route. Working examples have been presented to demonstrate the utility of the software for route planning in highly landslide prone area in the Himalayas.

Keywords Least-cost route · Dijkstra · Landslides · GIS

M. K. Arora (✉) · P. Gupta
Department of Civil Engineering, Indian Institute of Technology Roorkee, Roorkee 247667, India
e-mail: manoj.arora@gmail.com

A. K. Saha
Department of Geography, University of Delhi, Delhi 110007, India

R. P. Gupta
Department of Earth Sciences, Indian Institute of Technology Roorkee, Roorkee 247667, India

B. Pradhan and M. Buchroithner (eds.), *Terrigenous Mass Movements*,
DOI: 10.1007/978-3-642-25495-6_11,

11.1 Introduction

In mountainous terrain, roads generally constitute the only mode of transport to link various habitat areas, which are often very distant from each other. In a rugged and tectonically active area, such as the Himalayas, occurrence of frequent landslides, particularly in the monsoon season, happen to be a major problem causing extensive damage to life, property and communication every year. It has been observed that in general, most of the roads are constructed disregarding the distribution of Landslide Hazard Zones and geological factors such as thrusts, shear zones *etc.* in the Himalayan region. Therefore, a huge amount of money and manpower are spent on the maintenance of roads throughout the year. Moreover, several roads usually get closed due to landslides in rainy season, thereby disconnecting many villages and towns in the area causing acute local civic miseries. Hence, there is a tremendous need for proper planning of roads that considers efficient engineering design as well as geological factors in terms of slope stability and safety measures.

Route planning and development in hilly areas has always been a difficult task as the nature of terrain plays an important role. A road in the mountainous region has to traverse a longer path with numerous curves and turnings to meet the elevation requirements in hilly areas, in contrast to the roads in the plain areas, which are usually straight and traverse the shortest distance. Therefore, for hill road planning, a number of factors ought to be taken into consideration (Khanna and Justo 1987). These include:

a. Distance from source to destination
b. Geological safety and soundness
c. Stability of slopes
d. Landslide hazards
e. Drainage crossings and extent of waterways
f. Necessity of passing through obligatory points, *e.g.*, villages, *etc.*
g. Topography and gradient
h. Need for special structures, *e.g.*, bridge, tunnel, retaining wall, *etc.*
i. Value of land
j. Construction cost, earthworks, *etc.*
k. Availability of construction material

Although, these factors are expected to be considered for efficient and accurate route planning, the hill roads are generally planned only on the basis of the topography and the length of the road. Other factors such as the geological hazards are often ignored.

The conventional route planning practice adopted in India is typically based on the manual method. A reconnaissance survey for selection of alternative route alignments, based on the Survey of India toposheets (scale commonly 1:50,000 and contour interval: ~50 m), is first carried out, which are followed with preliminary and detail surveys for complete route planning. On the base (contour) map, traces are drawn with a divider to demarcate the possible routes at a specified gradient.

In this way, various alternative routes are drawn and the same are comparatively evaluated in terms of length and cost of the route to identify the best route. Hence, the process is largely manual, tedious and time consuming, and based only on the topographic data obtained from the SOI maps. Therefore, there is always a possibility that all the possible routes may not be considered and a route, which may not be the best, gets selected.

Introduction of the aerial photogrammetric techniques in 1960s resulted in overcoming some of the limitations of the conventional manual route planning approach. The aerial photographs provide a synoptic overview as well as the three-dimensional perspective of the terrain. These also help the experts to work in the laboratory and reduces the necessity of cumbersome field surveys.

However, in either manual or photogrammetric route planning approach, geological aspects, particularly landslide distribution, which is a key factor for cost escalation in hilly areas, are generally not taken into consideration; this may lead to recurring problems of landslide activities all along the route. Besides, in the manual approach, it is also nearly impossible to integrate all the factors as mentioned earlier.

Hence, there is a great demand to utilise computer-assisted methodologies for route planning that may consider various factors including landslide hazards. The availability of digital remote sensing data and Geographic Information System (GIS) based computing techniques, has given further impetus on computer-assisted methodologies by virtue of their numerous advantages in route planning, as it requires efficient processing, interpretation and analysis of a large amount of spatial data corresponding to various factors.

The GIS-based methodology for route planning is based on integration of raster and vector spatial data thematic layers. Thus, the success of a GIS-based methodology depends on the efficient design of network patterns and consideration of various factors in the form of thematic layers. For example, the G-Route, a raster GIS tool developed by ITC, Netherlands (Ellis 1990) for route planning is based on a 3×3 cells neighbourhood analysis approach. This approach, however, results into the creation of zig-zag pattern of predicted least-cost route and also high and abnormal direction dependent gradient for normal vehicle movement. A few other commercial software containing appropriate route planning modules to generate least cost path have also been developed. These include PATHDISTANCE in ArcGIS, VARCOST and PATHWAY in IDRISI software. However, in all these route planning modules, the neighbourhood search is limited to 3×3 cells only. In a 3×3 cells kernel, a maximum of 8 neighbourhood directions, and in a 5×5 cells matrix, 16 neighbourhood directions are possible. The use of these neighbourhoods may, however, result in rather sharply zigzag and unrealistic paths.

Yu et al. (2003) presented a GIS based algorithm for route planning based on a 5×5 cells neighbourhood pattern establishing an analogy with the game of chess. Only, two factors, namely landuse and topography in the form of Digital Elevation Model (DEM) were used for route planning.

In this chapter, the work Yu et al. (2003) has been extended further to propose a GIS based approach for route planning in landslide hazardous areas by considering a number of other factors, and also two new neighbourhood patterns based on 7 × 7 cells kernels for the generation of least cost route that permits gentler gradients. Similar to other studies (*e.g.*, Yu et al. 2003; Rees 2004), the least cost route has been determined based on Dijkstra's algorithm in the proposed GIS based approach. An indigenous software acronymed as **La**ndslide **S**afe **I**ntelligent **R**oute **F**inder (LaSIRF) has been developed. The conceptual background and detailed description of the proposed approach can be found in Saha et al. (2005b). In this chapter, focus is placed on the design and description of the LaSIRF software, its usage and implementation to some typical route planning problems in the Himalayas.

11.2 Methodology

In this section, a brief description of the methodology adopted in the route planning approach has been provided. It consists of four major steps:

11.2.1 Generation of Thematic Cost Map

The thematic cost map has been defined as a raster map, where attribute of each cell indicates the estimated relative cost of route development and maintenance. The cost is cumulative, having possible inputs from various thematic data layers pertaining to the factors considered, namely, landslide distribution, landslide hazard zonation, landuse/landcover, lithology and drainage order *etc.* Additional relevant thematic data layers may also be considered as per the requirements of the application. Remote sensing and GIS tools have been used to generate various thematic data layers. These include landslide distribution (size), landslide hazard zonation, drainage order, landuse/landcover and lithology. The attributes of each data layers are of diverse data types (*e.g.*, categorical, ordinal and ratio/interval data) at different measurement scales and units. Therefore, an ordinal weighting-rating scheme based on the experience of route planning projects or experts' opinion has been applied on each thematic data layer individually, which have then been integrated to produce the cumulative data layer. The weights range from 0 to 9, with 0 signifying the minimum cost and 9 implying the highest cost. For example, a higher cost at a cell may be related to higher cost of bridge construction, road maintenance due to landslides, land acquisition, blasting, *etc.*

11.2.2 Selection of Connected Neighbours

In raster or grid based models, such as those used in the proposed approach, each cell can be regarded as a node. The term neighbourhood may be defined as the location within proximity of some starting point or grid cell. Neighbourhood

	32	23		24	25	
31	33	15	34	16	35	26
22	14	8	1	5	9	17
	40	4	S	2	36	
21	13	7	3	6	10	18
30	39	12	38	11	37	27
	29	20		19	28	

	Possible movement from source **S** to pixel number
Rook's Pattern:	1,2,3,4
Bishop's Pattern:	5,6,7,8
Knight's Pattern:	9,10,11,12,13,14,15,16
Knight31's Pattern:	17,18,19,20,21,22,23,24
Knight32's Pattern:	25,26,27,28,29,30,31,32

Note: Pixel numbers 33-40 are not treated as connected neighbours but used in Table 11.1 for thematic cost calculation.

Fig. 11.1 Various possible neighbourhood pattern in a 7 × 7 cell kernel

analysis is important in route planning to find various possibilities of movement from a source node to its immediate neighbour, such that the cost associated with that connection may be calculated. In this study, a 7 × 7 cells kernel has been considered and 32 unique neighbourhood patterns have been defined (Fig. 11.1).

For example, considering 'S' as the source node, its unique neighbours may be defined as 1 to 32. The connections from 'S' to either of 1 to 32 may be considered similar to the moves in the game of chess. The vertical and horizontal movements (1 to 4) are similar to the moves of Rook and the diagonal movements (5 to 8) are similar to the moves Bishop (Yu et al. 2003; Xu and Lathrop 1994). The movements from 9 to 16 can be considered as analogous to the movements of Knight, which moves two-step straight and one step left or right of the direction of movement, to occupy the new position. Saha et al. (2005b) considered two additional neighbourhood patterns, which do not exist in the game of chess, these originate from the concept of Knight's movement, and have therefore been named as Knight31 and Knight32. The Knight31 first moves 3 nodes straight and then one node towards either left or right (17 to 24), whereas Knight32 moves 3 nodes straight and then 2 nodes towards either left or right (25 to 32). It may thus be noted that in these new neighbourhood patters, since the location of neighbours is far from the central node, the identified route may be smoother with gentler gradients. Each node has been given a unique identification number and a mathematical relationship has been used to find out the neighbourhood pattern.

11.2.3 Computation of Neighbourhood Movement Cost (NM-cost)

Once the connected neighbours have been selected, the next step is to calculate the cost of moving from a source node to the connected neighbour. This has been termed here as NM-cost and is computed from three input data: cost related to neighbour-distance, cost related to gradient and the thematic cost.

In a real-world situation, the topography is uneven, particularly in a terrain such as the Himalayas where slopes vary in different directions. Hence, the NM-cost must also consider this direction-dependency (anisotropy) (Yu et al. 2003). For an anisotropic surface, the NM-cost may be given as,

$$\text{NM-cost} = \text{Distance} \times (\text{p} \times \text{Thematic cost} + \text{q} \times \text{Gradient cost}) \quad (11.1)$$

where, p and q are weights to normalize the thematic and gradient costs.

The actual distance between successive neighbours in a three-dimensional space can be calculated using Euclidean distance, which requires spatial dimension (*i.e.*, size) of each cell and elevation difference between the connected neighbours. In case of raster data, the mid points of various cells may be used to determine the spatial extents for distance computations.

Gradient or slope is also a key factor to be considered in route planning in a terrain, such as the Himalayas. The gradient can be defined as the rate of rise or fall along the length of the road with respect to the horizontal. It may be noted that this gradient angle is direction dependent. A vehicle may move to its connected neighbour according to the permissible gradient. The gradient values may be categorized and a weight may be assigned to each as per the difficulty in the movement of a vehicle on a slope.

Thus, based on values of p and q, the NM-cost for various neighbourhood patterns can be formulated and are given in Table 11.1.

11.2.4 Selection of Least-Cost Route

Dijkstra's algorithm (Dijkstra 1959) has been widely used for finding least-cost path. Here also, the classical Dijkstra's algorithm has been designed for identifying the shortest path in a network with nodes connected by weighted links. The algorithm works only in the case where all the associated costs of neighbourhood connections are positive. It generates the least-cost route from the source node to all the nodes in a raster network. To increase the computational efficiency, the algorithm stops when the pre-defined destination node is reached.

11.3 Description of the Software

In a standard raster GIS route planning model, each pixel in an image is usually associated with single attribute information along with its geographic location. In the proposed approach, a maximum of 7 × 7 cells kernel has been considered to incorporate 32 unique neighbourhood movements from a cell, there is a need to store 32 NM-cost values for each cell in addition to the direction of the movement. Therefore, a higher level of data structuring models is required, which may not be

Table 11.1 NM-cost functions for various neighbourhood patterns

Neighbourhood pattern	Formulation for neighbourhood movement cost (NM-cost)
Rook's pattern	$\text{NM-cost}_{(\text{S},\text{P}_i)} = \sqrt{\mu^2 + (H_{P_i} - H_S)^2}\left(p \times \frac{C_S + C_{P_i}}{2} + q \times f\left(\tan^{-1}\left(\frac{H_{P_i} - H_S}{\mu}\right)\right)\right)$ i = 1 to 4
Bishop's pattern	$\text{NM-cost}_{(\text{S},\text{P}_i)} = \sqrt{2\mu^2 + (H_{P_i} - H_S)^2}\left(p \times \frac{C_S + C_{P_i}}{2} + q \times f\left(\tan^{-1}\left(\frac{H_{P_i} - H_S}{\sqrt{2}\mu}\right)\right)\right)$ $i = 5$ to 8
Knight's pattern	$\text{NM-cost}_{(\text{S},\text{P}_i)} = \sqrt{5\mu^2 + (H_{P_i} - H_S)^2}\left(p \times \frac{C_S + C_{Pa} + C_{Pb} + C_{P_i}}{4} + q \times f\left(\tan^{-1}\left(\frac{H_{P_i} - H_S}{\sqrt{5}\mu}\right)\right)\right)$ $i = 8$ to 16; for $i = 9$, $a = 2$ and $b = 5$ and so on
Knight31's pattern	$\text{NM-cost}_{(\text{S},\text{P}_i)} = \sqrt{10\mu^2 + (H_{P_i} - H_S)^2}\left(p \times \frac{C_S + C_{Pa} + C_{Pb} + C_{Pc} + C_{Pd} + C_{P_i}}{6} + q \times f\left(\tan^{-1}\left(\frac{H_{P_i} - H_S}{\sqrt{10}\mu}\right)\right)\right)$ $i = 17$ to 24; for $i = 17$, $a = 2$, $b = 5$, $c = 9$ and $d = 36$ and so on
Knight32's pattern	$\text{NM-cost}_{(S,P_i)} = \sqrt{13\mu^2 + (H_{P_i} - H_S)^2}\left(p \times \frac{C_S + C_{Pa} + C_{Pb} + C_{Pc} + C_{Pd} + C_{P_i}}{6} + q \times f\left(\tan^{-1}\left(\frac{H_{P_i} - H_S}{\sqrt{13}\mu}\right)\right)\right)$ $i = 25$ to 32; for $i = 25$, $a = 1$, $b = 5$, $c = 16$ and $d = 35$ and so on (see Fig. 11.1)

(Definition of terms: NM-cost is the neighbourhood movement cost to move from S (source pixel) to pixel *Pi* (connected neighbour); μ is the pixel size; H_S, H_{Pi} are the elevation of S (source pixel) and the elevation of *Pi* (connected neighbour); *C* is the thematic cost to pass through the pixel under consideration; *p*, *q* are the weights; *f()* is a gradient cost function; and *i, a, b, c, d* are the neighbourhood pixel numbers (*a, b, c, d* depend on i, see Fig. 11.1))

easily interfaced with existing raster-based GIS. Moreover, Dijkstra's algorithm consists of a number of complex steps, which require storage of a large volume of temporary data and pointers. Therefore, a customised software, named as LaSIRF has been written in C++ and interfaced with a raster based GIS to implement the route planning approach as described earlier.

The output from the LaSIRF will be a safe route that avoids potential landslide-prone areas in a hilly terrain. The software can be compiled and executed using a Windows based Microsoft Visual Studio compiler. The process is computationally intensive and hence at least dual core processor with a minimum of 1GB RAM hardware is recommended.

The DEM, and the thematic cost map generated via a number of spatial thematic data layers are the major inputs to the LaSIRF to generate the least-cost route between a user-defined source and destination nodes. The software has four basic modules:

a) DATEXP (for creation of input data file)
b) NMCOST (for computation of neighbourhood movement cost)
c) DIJKSTRA (for finding out least-cost route)
d) INTERFACE (for interfacing with ILWIS).

a) Creation of Input Data Files

The DEM and thematic cost map in ILWIS raster format (**.mpr* and **.mp#*) have to be exported to ASCII (**.asc* format for ILWIS ASCII export) for their direct input to LaSIRF. ILWIS ASCII export option was used for this purpose (Fig. 11.2). The DATEXP module reads the ASCII data, converts row/column location of each cell into a node number and stores the thematic cost and the elevation data for each node into a data file (*cost_elev.dat*). The structure of the output data file is shown in Fig. 11.2.

b) NM-cost Calculation

NMCOST module computes the neighbourhood movement cost for each node. The module requires '*cost_elev.dat*' file, generated by DATEXP as input, as node wise processing has been carried out. For example, it reads the node number of a pixel's 32 neighbours (in 7×7 pixel window) and calculates the distance, gradient and thematic cost according to the pattern of connection (*viz.*, Rook, Bishop, Knight, *etc.*). The gradient value is classified and a cost value is assigned to each class. The distance, gradient cost and thematic cost are integrated for each node to calculate the NM-cost for various patterns. The output NM-cost values are stored in a data file (*nmcost.dat*). If there is no possible connection between any two nodes or the NM-cost associated is too high, '-1' is assigned to such connection. As Dijkstra's algorithm is based only on 'non-negative' values, these connections are automatically eliminated from the network. Fig. 11.3 shows the procedural layout of the NMCOST module and the associated output data file structure.

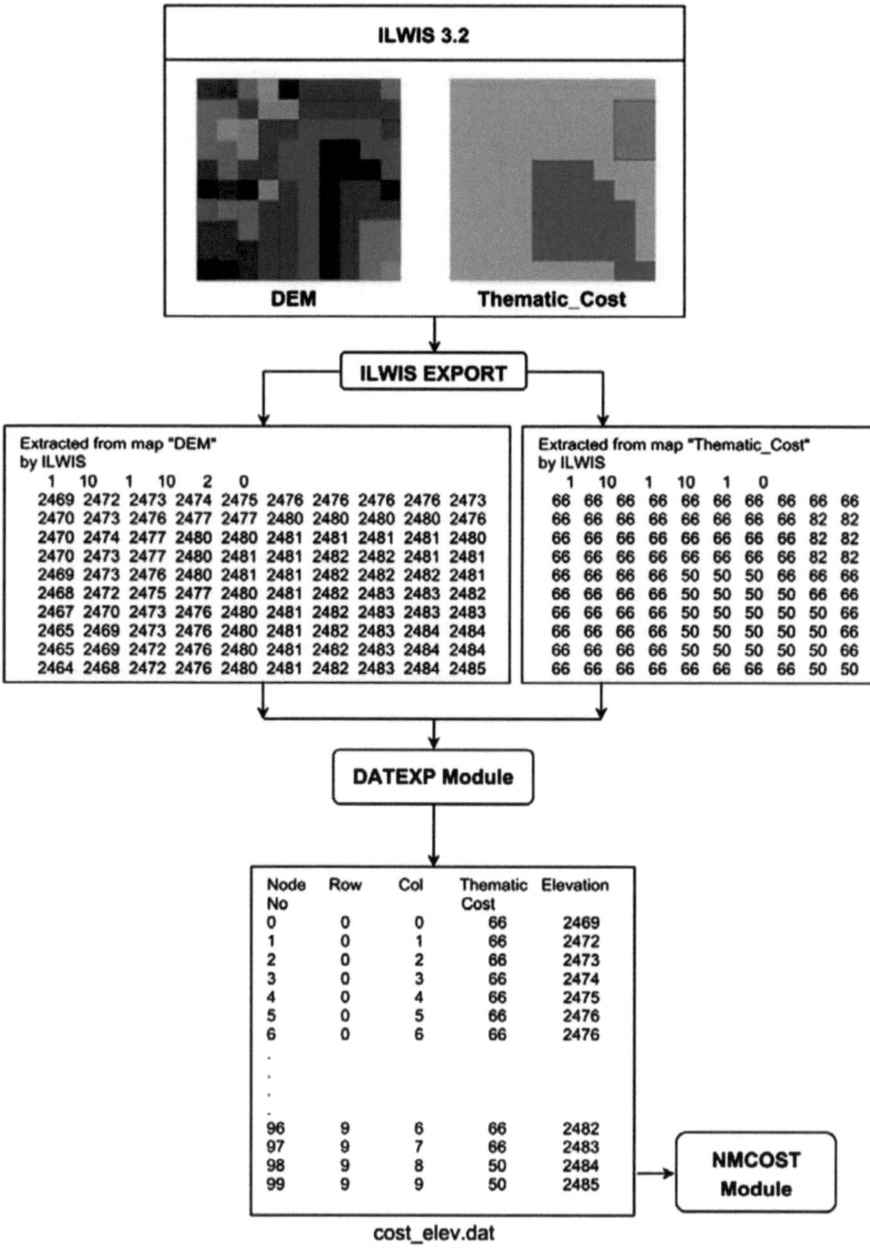

Fig. 11.2 Flow diagram showing the steps involved in input data preparation with the help of DATEXP module of LaSIRF. The output from this module '*cost_elev.dat*' is used as the input for neighbourhood movement cost calculation in the next step

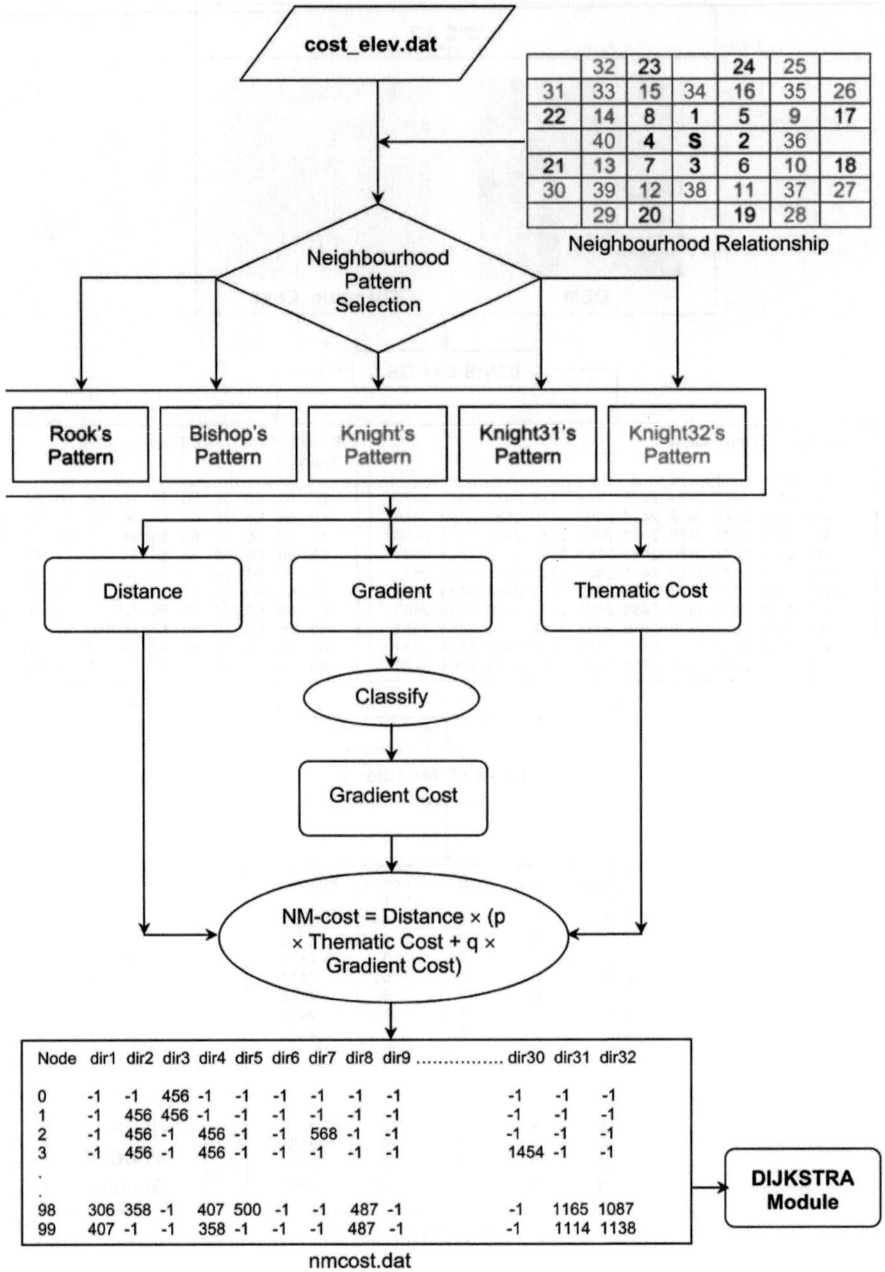

Fig. 11.3 Flow diagram showing the steps of NMCOST module for calculating the neighbourhood movement cost for each cell/node. The output from this module is used as input to DIJKSTRA module for least-cost route selection. ('dir 1', 'dir 2' denote the neighbourhood position from the source

Table 11.2 Functions created in DIJKSTRA module and their purpose

C ++ functions	Purpose
get_nmcostmatrix()	- This function reads the 'nmcost.dat' file as a two-dimensional matrix - Store the data in RAM
nmcost_between()	- This function checks if the nodes are connected - If there is a connection, it supplies the NM-cost value to leastcost_path() else '–1' i.e. no connection
leastcost_path()	- Iteratively checks the next node to be on the path as per the algorithm - When destination node is reached, it prints the nodes on least-cost path in 'result.dat' file

c) Route Finding Using Dijkstra's Algorithm

DIJKSTRA module is the most important component of the software LaSIRF to find the least-cost route between user-defined source and destination nodes. DIJKSTRA can read the source and destination information in the form of node numbers. A function ROWCOL2NODE supplements the DIJKSTRA to determine the node number of a pixel if the pixel location information is in the form of row and column coordinates. In this case, the size of the map (i.e., total number of rows and columns under consideration) is to be provided. DIJKSTRA reads the '*nmcost.dat*' file, generated by NMCOST and allocates it in its Random Access Memory (RAM) for fast retrieval and computation. It calculates the least-cost route using Dijkstra's algorithm, as described in Rees (2004). A number of functions associated with this module and their utility are given in Table 11.2. The least-cost route and the cumulative cost for the route are stored in a data file (*result.dat*). The processing steps for this module are shown in Fig. 11.4.

d) Interface of LaSIRF with ILWIS

The INTERFACE module converts the least-cost route stored in the form of nodes in '*result.dat*' file, generated by DIJKSTRA module, into ILWIS readable segment file in ASCII (**.smt*). The user has to enter cell size, number of rows and columns and the projection information for the map being created. These data are required since the thematic cost and elevation ASCII files are in the form of row and column matrix. The projection information can be readily obtained from the properties of raster maps (in ILWIS). INTERFACE module converts the '*result.dat*' into '*least_cost_path.smt*' file, which can be imported directly through import option of ILWIS and laid over elevation or thematic map to visualise the least-cost route computed (Fig. 11.5.)

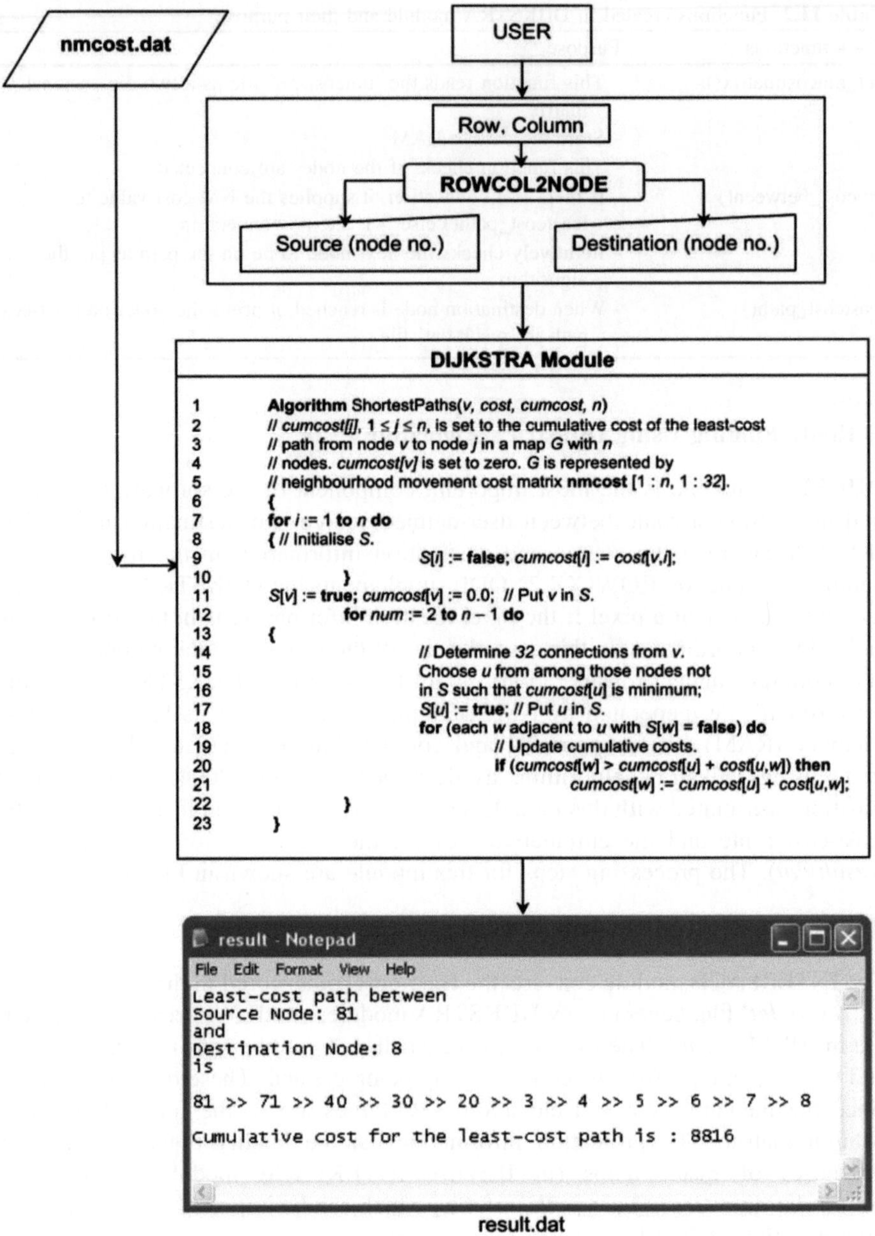

Fig. 11.4 Flow diagram showing the steps of DIJKSTRA module of LaSIRF software for calculating the least-cost path

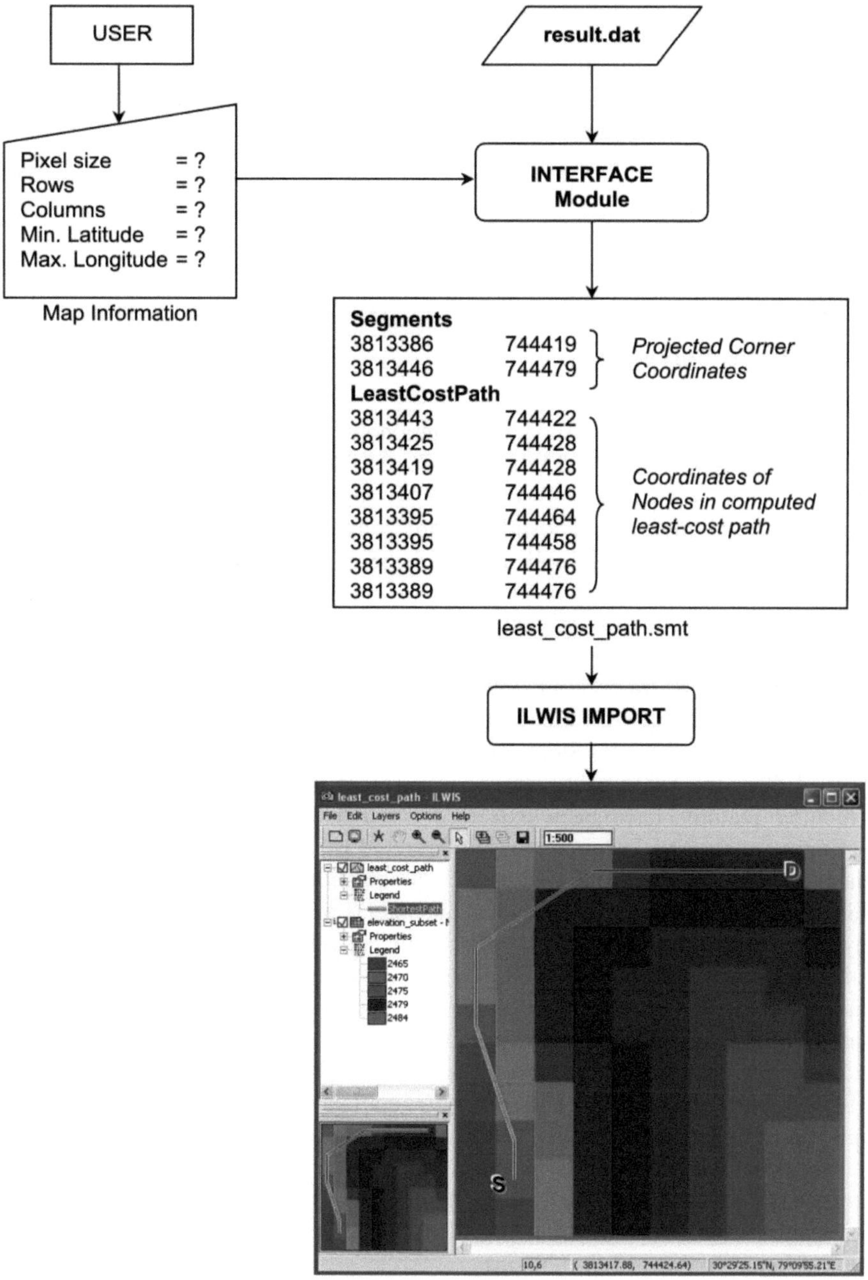

Fig. 11.5 Flow diagram showing the INTERFACE module of LaSIRF software. The least-cost path computed by DIJKSTRA is displayed on DEM of the area in ILWIS GIS

The salient features of LaSIRF can thus be summarised as,

- The software is fast with reasonable computational efficiency.
- The software can be applicable to a study area of any dimension, though it also depends on the hardware specification.
- The software can be easily interfaced with ILWIS GIS.

11.4 Working Examples to Demonstrate Implementation of the Software

A landslide prone highly rugged mountainous area in the Indian Himalayas (Latitude 30°20′-30°34′N and Longitude 79°05′-79°22′E) is selected. A number of thematic data layers *viz.*, DEM, lithology, landuse/landcover, drainage order, landslide distribution and landslide hazard zonation (LHZ) have been generated using remote sensing - GIS tools. IRS 1C LISS-III multispectral and PAN images, topographic and geological maps, and road design parameters are used as data sources to generate various thematic data layers. Field data were also collected for verification purposes.

Contours at 40 m contour interval in Survey of India topographic map (1:50,000) are digitised, interpolated and resampled to generate the Digital Elevation Model (DEM). Slope, aspect and relative relief data layers are then derived from the DEM using standard processes in raster GIS. Lithological and structural features have been extracted through digitisation of features in the geological map. The lineaments are interpreted by applying a 3×3 pixels edge-enhancement filter on LISS-III image in association with the geological map. At this stage, the structural features and lineaments are merged in a single thematic layer and a distance lineament buffer data layer has been generated, which is suitably categorised. The drainage features have been digitised from topographic map and classified according to Strahler's ordering of streams. A drainage density data layer is also generated. Landuse/landcover data layer has been produced from multi-source image classification of an integrated dataset consisting of IRS-1C LISS-III image, Normalized Difference Vegetation Index (NDVI) and DEM, using maximum likelihood classifier. Nine landuse/landcover classes that may have impact on landslide activities in the region are considered. These include dense forest, sparse vegetation, agriculture, fallow land, barren land, settlements, fresh sediments, water body and snow. Further description on multisource classification can be found in Saha et al. (2005a).

The PAN-sharpened LISS-III image, augmented with substantial field observations have been used to produce an existing landslide distribution map, has been used as a reference data source for development and validation of statistical model for the generation of landslide hazard zonation (LHZ) data layer. For the preparation of the LHZ data layer, a new statistical model, Landslide Nominal

Table 11.3 Comparative study of efficiency of the LaSIRF software

LaSIRF modules	Test area size in cells	
	250 × 250	500 × 500
DATEXP	14 s.	1 min.
NMCOST	7 min.	28 min.
DIJKSTRA	5 min. 11 s.	1 h. 45 min.
INTERFACE	2 s.	5 s.

Susceptibility Factor (LNSF) model has been developed. The inputs to this model are attribute information from thematic data layers, namely, slope, aspect, relative relief, lithology, structure-buffers, drainage density and landuse/landcover. Further details on this statistical approach for LHZ can be found in Saha et al. (2005c).

For computing the cost of road development and maintenance, following thematic data layers have been considered:

a) existing landslide distribution map classified according to size
b) landslide hazard zones
c) higher order drainage (to consider bridge construction cost)
d) landuse/landcover
e) lithology

The data layers have been integrated using an ordinal scale weighting-rating method. The thematic layers are arranged in a hierarchical fashion, in ascending order of cost and a weight number (from 0 to 9) has been given to each layer. Similarly, each class within a thematic data layer has been given an ordinal rating. These weighted layers are aggregated to generate a thematic cost map. The attribute value of each cell in this map implies the thematic cost to move through that cell.

The thematic cost map and the DEM are used as the inputs to the LaSIRF software for determining least-cost route between a source node and a destination node.

11.4.1 Efficiency of the Software

The efficiency of the software has been tested using a sample test areas of 250 × 250 cells and 500 × 500 cells on an Intel P4 2.8GHz and 512MB RAM workstation. It has been found that the LaSIRF software takes 13 min and 2 h 30 min respectively to find a route between the diagonally opposite corners for the two selected tests areas. The computational time for individual module has been listed in Table 11.3. It may be noted that for a specific study area, the computation of Neighbourhood Movement Cost (NM-cost) is only a single time activity. Once the

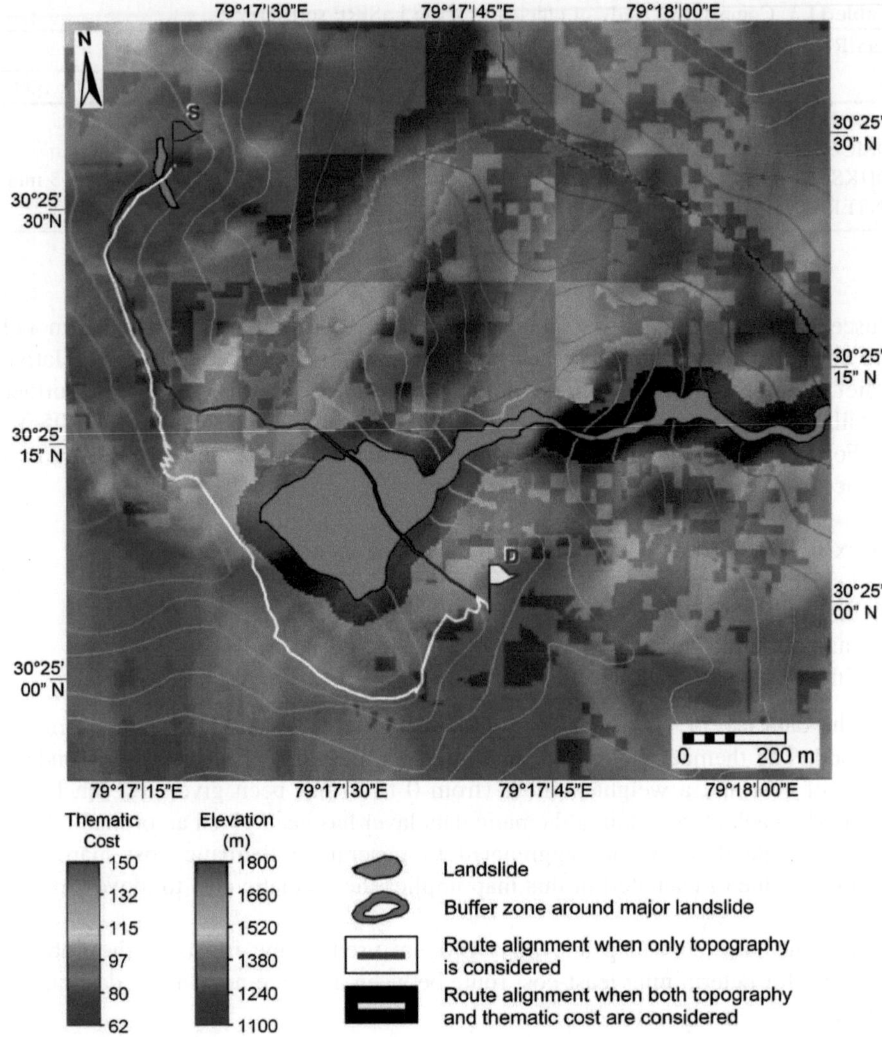

Fig. 11.6 Given the task that the northeast corner has to be joined with the southwest corner, this example shows alternative route alignments. When only topography is considered, the alignment passes through higher thematic cost zones and landslide zones; when both the topography and thematic cost are considered, the route passes through mainly lower thematic cost zones and avoids the landslide areas. (S–Source, D–Destination)

NM-cost file ('*nmcost.dat*') is prepared for the study area, the same file can be used for determination of multiple routes in that study area. This may result in elimination of time factor required for generation of NM-cost and, hence, in improving the efficiency of the software while determinations of multiple routes in the same study area.

It may also be desirable to use the software on a high-end workstation having high computational power (such as dual or quad core CPUs) and large Random Access Memory (RAM) size to boost the computation speed of the NM-cost and Dijkstra's algorithm.

11.4.2 Working Examples

In order to examine the implementation of the concept and the methodology via LaSIRF, for route planning in a landslide-prone rugged terrain, a few test areas of 1.5 km × 1.5 km size, with different combinations of terrain conditions and landslide susceptibility have been selected. The results of these test areas are discussed in the following.

11.4.2.1 Working Example 1

This example considers the case of occurrence of major and minor landslides in adjacent areas such that the source and destination points are located on opposite sides of a set of landslides. Figure 11.6 shows the test area, which is situated west of the Gopeshwar town (latitude 30°25′ N, longitude 79°19′E). There is a major landslide along with a debris flow channel that carries the debris to the river valley in the east.

Assuming that the entire area covered by the landslide and the associated debris flow track is unsuitable for route location, the landslides must be avoided. For such a situation, a buffer zone can be created around the landslide and debris flow track and the entire zone can be assigned a high value in the thematic cost layer. The source and destination points have been selected at nearly same elevation (approximately 1600 m) on either side of the set of major and minor landslides (Fig. 11.6). First, a route has been generated by considering only topography. This route normally follows the contour line but crosses the minor and major landslides right through the middle. However, when both topography and thematic cost are used in the proposed software, the identified route passes through the minor landslides but avoids the major landslide and its buffer zone, though it takes a longer path to reach the destination point.

11.4.2.2 Working Example 2

This example shows the suitability of the route planning methodology for cases involving no-trespassing zones. The no-trespassing zone may include, for example, archaeological-cultural heritage sites, strategic installations, exceptionally high land costs or reserved parks. Thus, the condition is that the route may touch the boundary of the no-trespassing area, but can not pass through the area. For this

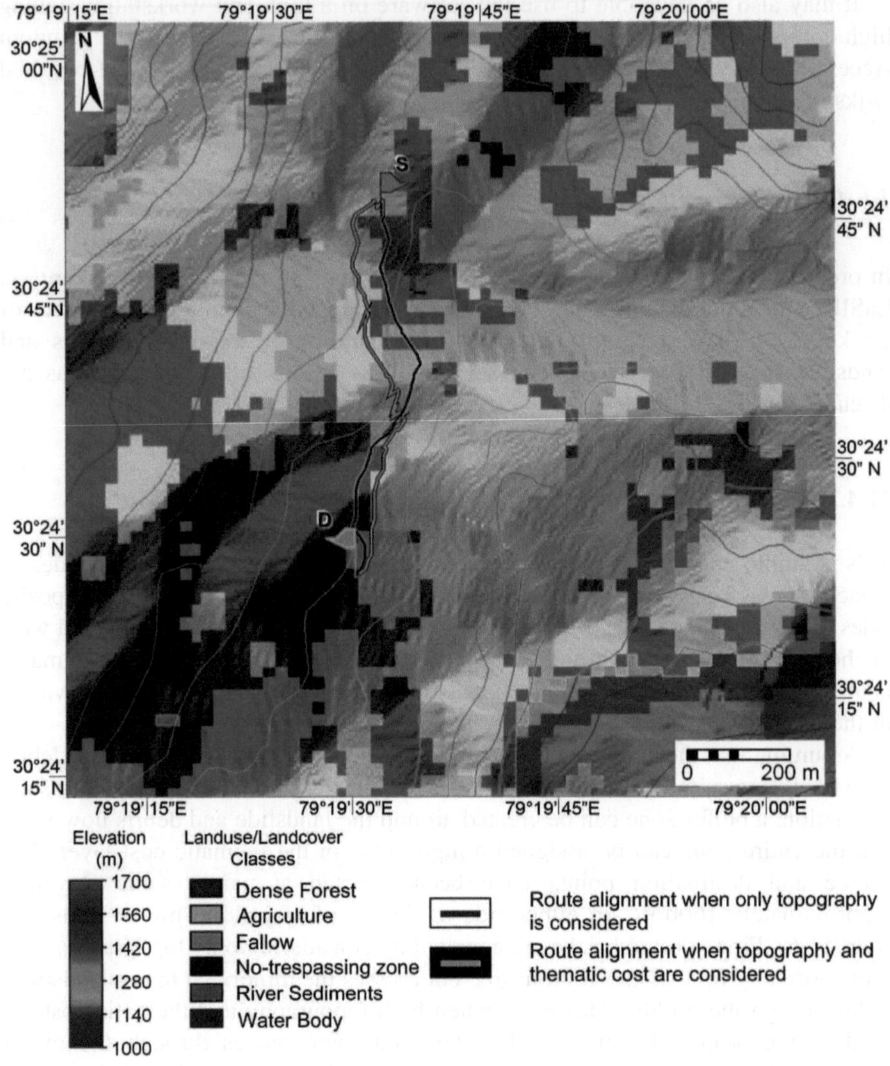

Fig. 11.7 Route alignment for connecting no-trespassing zones. (S–Source, D–Destination)

case, a landuse/landcover map has been draped over shaded relief model such that the no-trespassing area is depicted in red colour (very high attribute value of the cell) (Fig. 11.7).

The source and destination nodes have been selected at the corners of two no-trespassing zones. The route computed from thematic cost, gradient cost and distance is shown in yellow, as determined through LaSIRF, passes outside the periphery of the no-trespassing zones.

These two examples sufficiently demonstrate the versatility of the methodology and the software LaSIRF developed in this study.

11.5 Summary and Conclusions

In this chapter, particulars of an indigenous software named as LaSIRF developed exclusively for route planning in landslide prone hilly areas in GIS environment were discussed. A methodology based on Dijkstra's Algorithm for finding the least-cost path was implemented in the software.

The software is built on four basic modules one each for creation of input data files, computation of neighbourhood movement cost, finding of least-cost route and interfacing with commercial GIS software ILWIS respectively. The LaSIRF software is intelligent, fast and efficient and considers all possible combinations of routes between the source and the destination points. The least-cost path generated by the software provides the best option with greater degree of confidence.

The efficacy of the software was successfully examined through working examples in a landslide-prone hilly terrain. It was observed that the routes identified by the proposed software in the high-altitude, rugged Himalayan terrain with markedly different landslide susceptibility conditions passed through relatively safe areas avoiding major landslides.

The route planning through this software was also found to be very efficient in comparison to the conventional manual practice. The working examples showed that for a test area of 1.5 km x 1.5 km, a CPU time of only 13 min was consumed, a task which would have been completed in many days by the manual approach even with all the data in hand. Nevertheless, the efficiency of the proposed algorithm can further be increased with efficient data structuring techniques.

Moreover, the LaSIRF software considers gradient for various direction-dependent connected neighbours, thematic cost and distance in a three-dimensional space. The path gradient can be adjusted interactively in the software as per requirements, depending upon the terrain conditions, thus permitting design of quite a realistic route in an automated way by merely changing some of the parameters in the software. A GUI based interface of the software is under development.

References

Dijkstra EW (1959) A note on two problems in connection with graphs. Numer Math 1:269–271

Ellis MC (1990) Analytical free-range route planning "G-Route" GIS simulation model. Technical Note Internal Publication ITC, Enschede

Khanna SK, Justo CEG (1987) Highway engineering, 6th edn. Nem Chand and Bros, Roorkee, p 908

Rees WG (2004) Least-cost paths in mountainous terrain. Comput Geosci 30:203–209

Saha AK, Arora MK, Csaplovics E, Gupta RP (2005a) Land cover classification using IRS LISS III imagery and DEM in a rugged terrain: a case study in Himalaya. GeoCarto 20(2):33–40

Saha AK, Arora MK, Gupta RP, Virdi ML, Csaplovics E (2005b) GIS-based route planning in landslide prone areas. Int J Geog Inf Sci 19(10):1149–1175

Saha AK, Gupta RP, Sarkar I, Arora MK, Csaplovics E (2005c) An approach for GIS-based statistical landslide susceptibility zonation–with a case study in the Himalayas. Landslides 2(1):61–69

Xu J, Lathrop RG (1994) Improving cost-path tracing in a raster data format. Comput Geosci 20(10):1455–1465

Yu C, Lee J, Munro-Stasiuk MJ (2003) Extensions to least-cost path algorithms for roadway planning. Int J Geog Inf Sci 17(4):361–376

Chapter 12
Identification of Potentially Dangerous Glacial Lakes in the Northern Tian Shan

Tobias Bolch, Juliane Peters, Alexandr Yegorov, Biswajeet Pradhan, Manfred Buchroithner and Victor Blagoveshchensky

Abstract Like in many other parts of the world, the glaciers in northern Tian Shan are receding, and the permafrost is thawing. Concomitantly, glacial lakes are developing. Historically, outbursts of these glacial lakes have resulted in severe hazards for infrastructures and livelihood. Multi-temporal space imageries are an ideal means to study and monitor glaciers and glacial lakes over large areas. Geomorphometric analysis and modeling allows to estimate the potential danger for glacial lake outburst floods (GLOFs). This paper presents a comprehensive approach by coupling of remote sensing, geomorphometric analyses aided with GIS modelling for the identification of potentially dangerous glacial lakes. We suggest a classification scheme based on an additive ratio scale in order to prioritise sites for detailed investigations. The identification and monitoring of glacial lakes was carried out semi-automatically using band ratioing and the normalised difference water index (NDWI) based on multi-temporal space imagery from the

Slightly improved version of an earlier article published in Natural Hazards 59(3): 1691-1714. doi: 10.1007/s11069-011-9860-2. For citation please refer to this initial paper.

T. Bolch (✉)
Geographisches Institut, Universität Zürich, 8059 Zürich, Switzerland
e-mail: tobias.bolch@geo.uzh.ch

T. Bolch · J. Peters · M. Buchroithner
Institut für Kartographie, Technische Universität Dresden, 01062 Dresden, Germany

A. Yegorov · V. Blagoveshchensky
Institute of Geography of Ministry of Education and Sciences
of the Republic of Kazakhstan, Almaty, Kazakhstan

B. Pradhan
Institute of Advanced Technology, University Putra Malaysia (UPM),
43400 Serdang, Malaysia
e-mail: biswajeet24@gmail.com; biswajeet@lycos.com

B. Pradhan and M. Buchroithner (eds.), *Terrigenous Mass Movements*,
DOI: 10.1007/978-3-642-25495-6_12,

years 1971–2008 using Corona, ASTER and Landsat data. The results were manually edited when required. The probability of the growth of a glacial lake was estimated by analysing glacier changes, glacier motion and slope analysis. A permafrost model was developed based on geomorphometric parameters, solar radiation and regionalised temperature conditions which permitted to assess the influence of potential permafrost thawing. Finally, a GIS-based model was applied to simulate the possibly affected area of lake outbursts. The findings of this study indicate an increasing number and area of glacial lakes in the northern Tian Shan region. We identified several lakes with a medium to high potential for an outburst after classification according to their outburst probability and their downstream impact. These lakes should be investigated more in detail.

Keywords Glacial lakes · GLOF · Debris-flow · Remote sensing · Geomatics · GIS · Modelling · Hazard assessment · Tian Shan

12.1 Introduction

Climate change and concomitant glacier recession has caused the development and expansion of glacial lakes in mountain areas of the world which leads to an increasing risk of lake outbursts. Outbursts of glacial lakes represent a serious hazard especially for populated regions in the mountains all over the world (Clague and Evans 2000; Huggel et al. 2003, 2005; Iwata et al. 2002; Ma et al. 2004; Popov 1988; Richardson and Reynolds 2000). These glacial lake outburst floods (GLOF, also called jökullhlaup) can cause extremely high water discharges as well as large mudflow events. Triggering events for an outburst can be moraine failures induced by an earthquake, by the degradation of permafrost and increased water pressure, or a rock, snow, or ice avalanche into the lake causing a flood wave with a subsequent outburst (Buchroithner et al. 1982; Fujita et al. 2008; Ives 1986; Vuichard and Zimmermann 1987). The potential downstream path has to be taken into account in order to asses the potential effect of a GLOF event on the infrastructure and human population. Previous studies showed that the susceptibility of lake development is highest where the glaciers have a low surface slope angle and a low flow velocity or are stagnant (Bolch et al. 2008; Frey et al. 2010a; Quincey et al. 2007; Reynolds 2000).

Glacial lakes that develop in remote mountainous areas are often difficult to access and field studies are laborious and cost-intensive. Therefore, remote sensing data and GIS are ideal tools for studying and monitoring glacial lakes and assessing their hazard potential (Buchroithner 1996; Huggel et al. 2003; Schneider 2004; Kääb et al. 2005; Quincey et al. 2005; Bolch et al. 2008).

Recently, several studies demonstrated the suitability of optical remote sensing data for detection of glacial lakes in an automated way (Huggel et al. 2002; Quincey et al. 2005; Bolch et al. 2008; Frey et al. 2010a). The aforementioned literature review indicates that one of the main drawbacks in the automated methods is the difficulty in differentiating the lakes with turbid water and the fact that the areas with cast shadow are usually misclassified. While the latter can be addressed by applying a shadow mask using a precise digital elevation model (Huggel et al. 2002), so far no real promising method for turbid water is existing in the literature. Hence, manual checking and editing is still essential. Manual digitizing is also required for panchromatic data such as aerial imagery and declassified intelligence data such as Corona.

A digital elevation model (DEM) of the study area is essential to obtain the geomorphometric data of the glaciers, glacial lakes and its surroundings and especially for modeling of the probable outburst path. The freely available near-global void-filled SRTM3 DEM and the ASTER GDEM are a good choice if no other detailed local DEM is available (Frey et al. 2010b). The SRTM3 DEM and ASTER derived DEMs were shown to be suitable with the limitation that the elevation and characteristics of smaller features such as the lateral moraines and deep gorges may not be accurately and precisely depicted (Fujita et al. 2008; Huggel et al. 2003; Kamp et al. 2005). Additional errors occur especially on steep slopes due to low contrast in areas with cast shadow in the utilised imagery (GDEM) and layover and foreshortening of the radar data (SRTM DEM, Kocak et al. 2004). Reported RSME values are 15 m for mountainous terrain (Berry et al. 2007; Falorni et al. 2005).

The northern Tian Shan is prone to natural hazards due to gravitative processes such as avalanches, landslides, debris flows and flash floods (Havenith et al. 2003; Passmore et al. 2008; Severskiy and Zichu 2000; Storm and Korup 2006; Yegorov 2007). Several catastrophic mudflows have been documented during the last 100 years and before (Gorbunov and Severskiy 2001; Blagoveshchenskiy and Yegorov 2009, Table 12.1). It has been shown that about 11% of the catastrophic mudflows were triggered by GLOFs (Popov 1988; Medeuov and Nurlanov 1996; Medeuov et al. 1993, Yegorov 2007).

The aim of this study is to investigate the suitability of a comprehensive geomatics-based approach to detect and monitor potential dangerous glacial lakes (PDGL) in the northern Tian Shan and classify the glacial lakes according to their hazard potential. Hence, we address level 1 (basic detection of glacial lakes) and level 2 (assessment of hazard potential) based on the multi-level strategy as suggested by Huggel et al. (2002). A further aim is to recommend which lakes should be further investigated using high resolution imagery and in the field (level 3). Addressing the risk for the society of a GLOF event is beyond the scope of this study. The utilised approach should be based on standardised criteria, sophisticated yet rather simple and suitable for assessing a large number of glaciers simultaneously. The conditioning parameters which influence the potential danger of a glacial lake are widely published in some of the aforementioned papers on GLOFs. The interpretation, however, is mainly based on description or subjective classification. The analysis presented here uses the above mentioned conditioning parameters and is more objective.

Table 12.1 Examples of past GLOF events in northern Tian Shan and their triggering factors

Date	Location	Elevation (m)	Discharge volume (m^3)	Transported volume (m^3)	Cause/trigger	Reference
15 July 1973	Proglacial lake, Kishi Almaty	3,370	~180,000	~4 mio	Overfilling of the lake kettle; high surface discharge over the dam, dam erosion	Baimoldayev and Vinohodov (2007) Plehanov et al. (1975)
03 August 1977	Khumbel river basin right inflow of Ulken Almaty	3,400	~200,000	~4 mio	Superficial discharge initiated by lake overfilling	Popov (1984a)
21 June 1979	Srednij Talgar	3,400	~80,000	~113,000	Subglacial tunnel	Shusharin and Popov (1981)
23 July 1980	Kaskelen river basin	3,500	~200,000	~2 mio	Tunnel in buried ice	Popov (1984b)

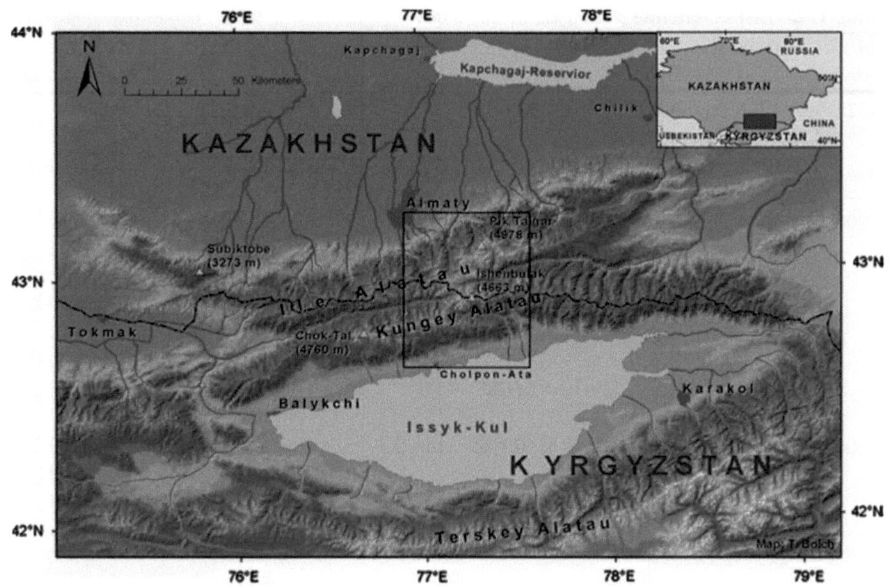

Fig. 12.1 Location of the study area

12.2 Study Area

The mountain ranges Ile Alatau and Kungöj Ala–Too (also: Zailijskij and Kungej Alatau, 42°30′–43°30′N, 75°–79°E) of the northern Tian Shan are located in Central Asia at the border between Kazakhstan and Kyrgyzstan (Fig. 12.1). These ranges rise from the Kazakh Steppe at an elevation of about 800 m asl to nearly 5,000 m asl. The southern edge is an intramountainous basin filled by Lake Issyk–Kul (also: Isyk–Köl, 1,608 m asl). Many villages, the million habitant city of Almaty and important tourist destinations such as Cholpon–Ata at lake Issyk–Kul as well as major roads are situated directly at the foothills of these mountains.

The mountain chain of the northern Tian Shan originate from the Caledonian orogenesis but are still affected by compression and are still slightly uplifting. The area is situated within the Chilik–Kemin Seismic Zone (Chedija 1986; Utirov 1978) with several predominant WSW–ENE-striking faults.

The appearance of the seismically active mountain ranges is mainly formed by neotectonic activity (Chedija 1986). Several major earthquakes have occurred since the end of the 19th century and thousands of smaller seismic events have been recorded (Lukk et al. 1995). The major earthquake of the year 1887 (Ms = 7.3) affected the large town Almaty (called Vernyi at that time) situated at the edge of Ile Alatau (Yadav and Kulieshhius 1992). Another major earthquake (Kemin earthquake, Ms = 8.2) occurred in 1911 and caused numerous landslides and rock avalanches in northern Tian Shan (Delevaux et al. 2001). Moreover, evidence of several prehistoric earthquakes also exists in this region (Korjenkov et al. 2004).

Table 12.2 List of satellite data used in this study

Time period	Date	Satellite and sensor	Resolution (m)	Source	RMSx,y	Spectral bands
1971/1972	17 September 1971	Corona KH-4B	~5	USGS	56 m	PAN
	07 September 1972	Landsat MSS	60 (res.)	GLCF	47 m	VIS, NIR
~2000	08 August 1999	Landsat ETM+	15/30	USGS	Reference	VIS, NIR, SWIR
	13 October 2000	Terra ASTER	15	USGS	64 m	VIS, NIR, SWIR
	05 September 2001	Terra ASTER	15	USGS	40	VIS, NIR, SWIR
	30 September 2001	Terra ASTER	15	USGS	41	VIS, NIR, SWIR
~2007	14 August 2007	Landsat SLCoff	15/30	USGS	21	VIS, NIR, SWIR
	13 June 2008	Landsat SLCoff	15/30	USGS	11	VIS, NIR, SWIR

Due to the topography, the overall continental climate is characterised by distinct local variability. Precipitation at altitudes about 3,000 m asl ranges from more than 1,000 mm/a on windward northern slopes to less than 800 mm in a leeward valley south of the main mountain ridges (Bolch 2007). The minimum precipitation occurs in the study area during winter due to the Siberian anticyclone and the maximum occurs in early summer due to both cyclonic activity and convective precipitation (Böhner 1996). Mean annual air temperature (MAAT) recorded at Tuyuksu glacier station (3,434 m asl) is about −4°C. The zero degree isotherm is situated just above 2,700 m asl. The steady-state equilibrium line altitude of glaciers is situated at about 3,800 m asl on northern slopes and between 3,900 and 4,000 m asl on southern slopes (Bolch 2007). A characteristic feature of the northern Tian Shan is its pronounced periglacial zone with many large and active rock glaciers. This zone is characterised by frequent diurnal freeze–thaw cycles (Marchenko 1999). Permafrost is sporadic at about 2,700–3,200 m asl, discontinuous at 3,200–3,500 m asl, and continuous above 3,500 m asl (Gorbunov et al. 1996). Hence, the drained lakes discussed above were situated in the discontinuous permafrost zone.

12.3 Data and Methods

12.3.1 Data

Important data used in this study are remote sensing imagery from different time periods (Table 12.2). The earliest available remote sensing data are Corona and Landsat MSS from the 1970s while the recent data are ASTER and Landsat ETM+

Table 12.3 Overview of the utilised NDWI thresholds for different sensors

	Landsat MSS	ASTER	Landsat ETM+
Threshold	0.45–0.9	0.3–0.7	0.3–0.9

SLCoff (with data gaps due to a scan-line error). Fortunately, larger parts of the study region are within the central part of the SLCoff scenes and are thus not or only partly affected by the gaps.

Most of the utilised scenes show little seasonal snow cover and are therefore suitable for lake and glacier identification. Only the ASTER scene from 13 September 2001 showed the presence of larger snow cover. Nevertheless, it was used as supplementary information as most of the lakes are identifiable. Unfortunately, no suitable scene from the 1990s was available for this study. Soviet topographical maps with the scale, 1:100,000 from the 1980s and GPS points obtained during multiple fieldworks in the years 2001–2004 were used for orthorectification of the master image (Landsat ETM+, 1999). All other images were co-registered to this image. The void-filled SRTM3 DEM, vers. 4 from CGIAR was used (srtm.csi.cgiar.org/) for rectification. The geometry of the Corona imagery is complex with the least distortion in the centre of the image (Dashora et al. 2007). Fortunately, the area of interest is located around this centre and hence, the distortion is found to be low. For co-registration purposes, a projective transformation followed by rubber sheeting was used in ERDAS Imagine 9.1. The latter was necessary to improve the accuracy which is caused by panoramic distortion. The overall RMSx,y error is less than two Landsat pixels ($\sim$56 m, Table 12.2) which is acceptable considering the complex image geometry and the mountainous terrain. The RMSx,y of the other imagery was lower, mostly in the range of one Landsat TM pixel ($\sim$30 m, Table 12.2). The void-filled SRTM3 DEM was also used for the modelling (e.g. mass movements, probable outburst path). The reported accuracy for the SRTM DEM proved true for the study area (Bolch et al. 2005; Bolch 2008).

12.3.2 Glacial Lake Identification

Water reflects mainly within the visible spectrum with a maximum value in the green band. This typical characteristic enables to identify the clean water using multi-spectral imagery. We tested several methods like ratioing and the normalised differenced water index (NDWI) with different band combinations (e.g. Blue, Green, NIR, SWIR). The main aim was to obtain the most precise delineation of lakes with some ice on the water and turbid lakes with minimum misclassification error. The best results were obtained using the NDWI approach by employing NIR and Blue or NIR and Green bands. The water index using Blue performed better in the shaded areas while the index with Green had lesser problems with ice on the

Fig. 12.2 Procedure showing automated lake identification

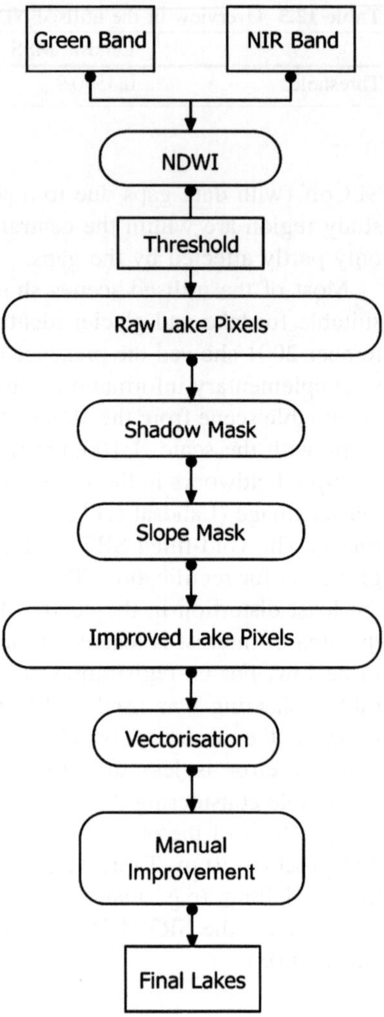

water bodies. Finally, the NDWI was used by employing NIR and Green bands (Green − NIR/Green + NIR). The utilised thresholds were adapted for each of the individual scene (Table 12.3). The misclassified shadow areas were eliminated using a shadow mask (Huggel et al. 2002). Hence, few lakes in shadow had to be digitized manually. The procedure for the lake identification is shown by Fig. 12.2.

The delineated water bodies with the coarser resolution of MSS were manually improved using the Corona data. An image stretching algorithm was used in order to facilitate the visual detection of water bodies in the panchromatic Corona images. The position of some lakes did not match perfectly with MSS images due to the distortion of the Corona images. In this case, the lakes were completely digitised from the Corona imagery and manually shifted in order to match with the correct position in the MSS data. Not all glacial lakes could be identified on each image. This largely depended on the image quality. Presence of shadow and especially snow cover hampers the identification of high altitude lakes. On the other hand, some lakes might have disappeared due to slow lake drainage or a sudden outburst.

Apart from the quality of the scene, sources of possible errors are attributed to the accuracy of the lake delineation and the error during co-registration. The highest RMSEx,y in co-registration is about 50 m (Table 12.2). However, as we compared the absolute change in area and not the exact location, therefore the impact of this on the derived data is relatively negligible. The remaining errors were estimated based on a buffer method using similar approach as suggested by Granshaw and Fountain (2006). Half a pixel was chosen as the buffer size as this is supposed to represent the worst possible case in which all margin pixels were wrongly classified. This method also includes relatively higher error of smaller polygons as those may have a higher number of edge pixels.

12.3.3 Hazard Assessment

Several factors need to be taken into account when assessing the hazard potential of a glacial lake outburst. In this study, the methodology of Huggel et al. (2002) and Bolch et al. (2008) was adapted and expanded as we introduce a higher number of variables for hazard assessment. The hazard assessment can be summarised by four major parameter groups: lake characteristics, characteristics of the lake surroundings, characteristics of the adjacent glaciers, impact on downstream areas. Each of those groups consists of several variables. Table 12.4 gives an overview of the addressed variables and its applicability using remote sensing and references for this task.

12.3.3.1 Lake Characteristics

One of the most important variables for analysing the potential danger of a GLOF is the change of the glacial lake. We address the issue of changes based on space imagery from three different time periods (Table 12.4). The growth of a supraglacial or proglacial lake depends primarily on the glacier characteristics and retreat. We expanded the analysis on glacier shrinkage based on Bolch (2007, 2008) and include the glacier area of 2007 for those lakes adjacent to a glacier. The volume of the lakes is addressed based on the empirical formula (Eq. 12.1) suggested by Huggel et al. (2002) which is primarily based on 15 lakes with existing depth measurements.

$$V = 0.104\,A^{1.42} \tag{12.1}$$

Table 12.4 Key factors contributing to the hazard risk of a glacial lake and its investigation using remote sensing data

Characteristics group	Factor	Remote sensing data source and applicable techniques	Suitable for automatisation	References
Lake characteristics	Lake area and volume	Detection using multitemporal multi-spectral (MS) satellite data	Yes	Wessels et al. (2002), Huggel et al. (2002)
	Rate of lake formation and growth	Change detection using multi-temporal (MT) and MS satellite data	Yes	Bolch et al. (2008)
Glacier characteristics	Fluctuations of the glacier	Investigation of area and volume change of the glacier based on MS and MT satellite data, MT digital elevation models (DEMs)	Yes	Bolch et al. (2010, 2011), Paul et al. (2002), Aizen et al. (2006)
	Activity of the glacier	Derive glacier velocity using feature tracking or DInSAR based on MT optical or radar data	Yes	Kääb (2005), Luckman et al. (2007), Scherler et al. (2008), Bolch et al. (2008)
	Geomorphometric characteristics of the glacier	Geomorphometric DEM analysis, slope classification	Yes	Bolch et al. (2007), Quincey et al. (2005)
Characteristics of the lake surrounding	Freeboard between lake and crest of moraine ridge	Geomorphometric DEM analysis	Partly	
	Width and height of the moraine dam	Geomorphometric DEM analysis	Partly	
	Stability of the moraine dam/ presence of dead ice in the moraine dam	Investigation of surface deformation based on MT DEM analysis, permafrost modelling	Partly	Fujita et al. (2008)
	Possibility of mass movements into the lakes	Mapping of ice cover and geology using MS data, Geomorphometric DEM analysis of the surrounding catchment areas, flow modelling	Yes	Huggel et al. (2003), Salzmann et al. (2004), Allen et al. (2009)
Impact of a GLOF to downstream areas	Affected area	Flow modelling	Yes	Huggel et al. (2003), Mergili et al. (2011)
	Infrastucture down-valley	Detection of human infrastructure based on MS satellite data analysis.	Partly	

(based on Richardson and Reynolds 2000; Huggel et al. 2002; Quincey et al. 2005; Bolch et al. 2008)

However, it is important to highlight here that this scaling formula may serve as a first estimation only as the lake volume depends on several variables. By applying this formula, (Eq. 12.1) for the lakes in Kishi Almaty valley with existing measurements of the bathymetry (Kasatkin and Kapista 2009; Tokmagambetov 2009) reveals an overestimation of the volume of up to 20% (Lake no. 1: modelled value: 33,040 m^3, measured value 27,618 m^3, lake no. 9: 21,563, 17,165 m^3). Similar uncertainty was also mentioned by Huggel et al. (2002).

12.3.3.2 Characteristics of Adjacent Glaciers

The recent glaciers were delineated based on the 2007 and 2008 Landsat ETM+ data (Table 12.2) using band ratioing (Band 4/Band 5) approach in order to be consistent with the method used by Bolch (2007) for the 1999 ETM+ Scene. The glacier velocity was estimated from multi-temporal optical imagery based on feature tracking using cross-correlation techniques (Berthier et al. 2005; Bolch et al. 2008; Kääb 2005). Unfortunately, no suitable ASTER data were available. Hence, we chose Landsat scenes from 1999, 2007 and 2008 and used the near infrared band due to the better contrast than in the higher resolution panchromatic band. The open source software 'Cosi–Corr' (Leprince et al. 2007) was applied for the automated estimation of the velocity. This software proved to be well suitable for mountain glaciers (Scherler et al. 2008). Here, it should be noted that Cosi–Corr was developed and applied for imagery from push broom scanners such as SPOT and ASTER with known image geometry while Landsat TM/ETM+ data has some inaccuracies in this respect. However, this affects the obtained results only marginally as the main interest here is to know about the activity and not the absolute values of the glacier movement. We estimate a glacier to be stagnant at the snout if the calculated velocity is below the uncertainty of one pixel (30 m).

The slope of the glacier surface gives a hint where glacial lakes can develop or an existing lake can extend in the near future. A threshold of 2° for supraglacial lake formation on debris-covered glaciers in the Himalaya (Quincey et al. 2007; Reynolds 2000; cf Bolch et al. 2008) or 5° for the formation of proglacial lakes in overdeepenings of debrisfree glaciers in the Alps (Frey et al. 2010b) have been suggested. We applied a slope threshold of 5° as only very few glaciers in the study area have larger portions of debris cover.

12.3.3.3 Characteristics of Lake Surroundings

Mass movements like rock fall or ice avalanches into a lake are important triggering mechanisms for an outburst. Hence, an analysis of the surrounding topography is highly sought after. We applied the modified single-flow model (MSF, Huggel et al. 2003) which was developed to model mass movements like debris flows and ice avalanches based on the surrounding topography. The model is a modified D8 flow direction algorithm and calculates the likelihood that a raster

Table 12.5 Parameters and their thresholds used for modelling of the probability of mass movements

Parameter	Rock avalanche (Kaibori et al. 1988) (°)	Ice avalanche (Alean 1985) (°)	Debris flow (Haeberli 1983, Huggel et al. 2002) (°)	Flood wave (Allen et al. 2009) (°)
Minimum slope at the detachment/starting zone	30	25	0	0
Angle of friction (average incline)	20	17	11	3

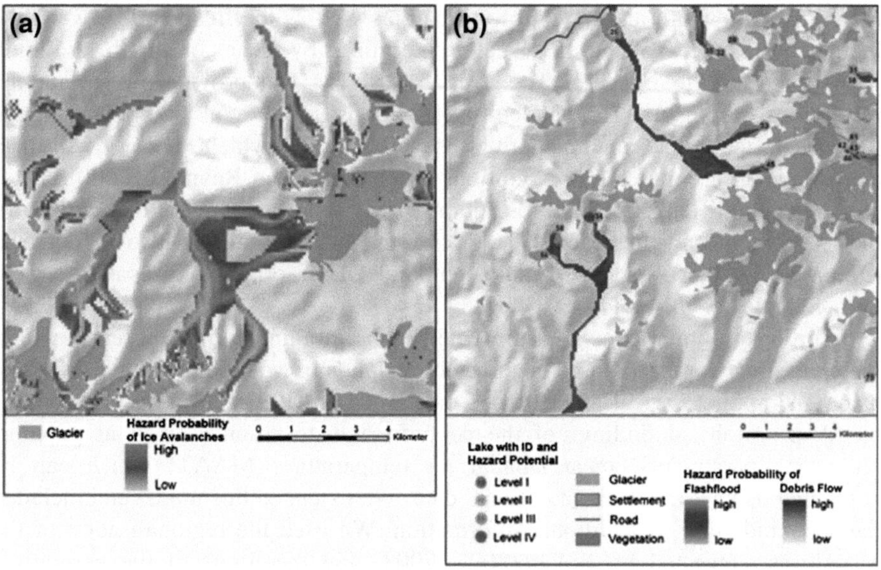

Fig. 12.3 Examples from the flow modelling. **a** Probability of an area affected by ice avalanches, **b** probability of an area affected by flash floods and mudflows

cell will be affected by such a mass movement. Similar methods to model rock and ice avalanches were applied by Allen et al. (2009) and Salzmann et al. (2004). Unfortunately, no detailed ground information is available for the study area. Hence, we chose to model the probability of a rock fall based on Kaibori et al. (1988) who presented detailed statistics for the slope at the detachment zone and for the angle of friction. The angle of friction defines the average slope between the starting and end points of the mass movement (Hsü 1975). In contrast to the results of Kaibori et al. (1988) who chose average values, we estimated the minimum values so that 90% of all occurred events are included. The threshold values are 30° for the slope at the detachment zone and 20° for the angle of friction (Table 12.5). We modelled ice avalanches in a similar way but chose the thresholds to be 25° and 17° based on the empirical work of Alean (1985, Fig. 12.3a). Although the utilised values are based on studies in the Alps and high mountains in Japan, and may be slightly different for the northern Tian Shan these values seem to be reasonable for estimation as they represent a worst case scenario. For example, Alean (1985) suggested a threshold of 45° for the slope of the detachment zone for cold glaciers and 25° for warm glaciers. Most of the glaciers of the study area are, however, either cold or polythermal. Van der Woerd et al. (2004) estimated a slope of about 45° for the origin of ice avalanches for cold glaciers in the Central Asian Kunlun Shan.

The probability of a dam failure mainly depends on the characteristics of the lake dam itself. In the study area, most of the lake dams comprise morainic materials. The width and height of the dam as well as the freeboard between the lake level and the crest can be visually determined by means of a high resolution DEM and satellite imagery. The available SRTM can provide a hint but the resolution is too coarse for detailed investigations (Fujita et al. 2008). A dam can become unstable if it contains permafrost or buried ice which thaws or will thaw due to changing temperature conditions (Richardson and Reynolds 2000). Comparison of multi-temporal high resolution DEMs can give a hint at the thawing of the ice content and the extent of the lowering of the dam can also be detected (Fujita et al. 2008). However, multi-temporal DEMs of suitable resolution were not available for this study.

In order to obtain some measure whether the moraine dam is currently within the permafrost zone and could be affected by thawing, we modelled the permafrost using a simple empirical model based on Permakart (Keller 1992). This model is based on empirical findings of the permafrost distribution as well as geomorphometric parameters, mean annual air temperature (MAAT) which can be computed using a DEM and additional data. We extended this model and included the solar radiation as additional information. We used the regionalisation of the MAAT as suggested by Bolch (2007, 2008) and the limits of the permafrost distribution [sporadic: 2,700 m asl, discontinuous: 3,200, and continuous: 3,500 m asl, Gorbunov et al. (1996)]. The physical model by Marchenko (2001) developed for a small subset of the study area (Kishi and Ulken Almaty valley) served as an evaluation dataset. Overall, the model showed a good agreement with the results obtained by Marchenko (2001), but small-scale variability (e.g. caused by the land cover) could not be captured. Climate change had also an impact on the permafrost distribution and permafrost area diminished during the last 130 years (Marchenko et al. 2007). We consider that a dam may become unstable if it is outside the continuous permafrost area. Although being a rough estimation, especially when taking into account that the blocky morainic material itself may retard thawing (Gorbunov et al. 2004) this approach provides a relatively quick estimation about the possible current existence and condition of permafrost in the moraine dam (Fig. 12.4).

12.3.3.4 Impact of Glacial Lake Outburst Floods

A GLOF presents a risk if human life and infrastructures would be affected. Therefore, the probability was calculated to which the downstream area would be affected by a GLOF using the previously mentioned MSF model introduced by Huggel et al. (2003). Flashfloods in the Tian Shan and elsewhere often lead to debris flows. Besides the presence of loose sediments, a certain velocity of the water is needed to transport the debris. This again depends mainly on the steepness of slope. In this research, the values for a debris flow were adopted based on the findings of Haeberli (1983) and Huggel et al. (2002) who suggested that the debris flow ends if an average incline of 11° is reached. We estimated that flash flood would occur when if the angle of friction is lower than this threshold and stopped

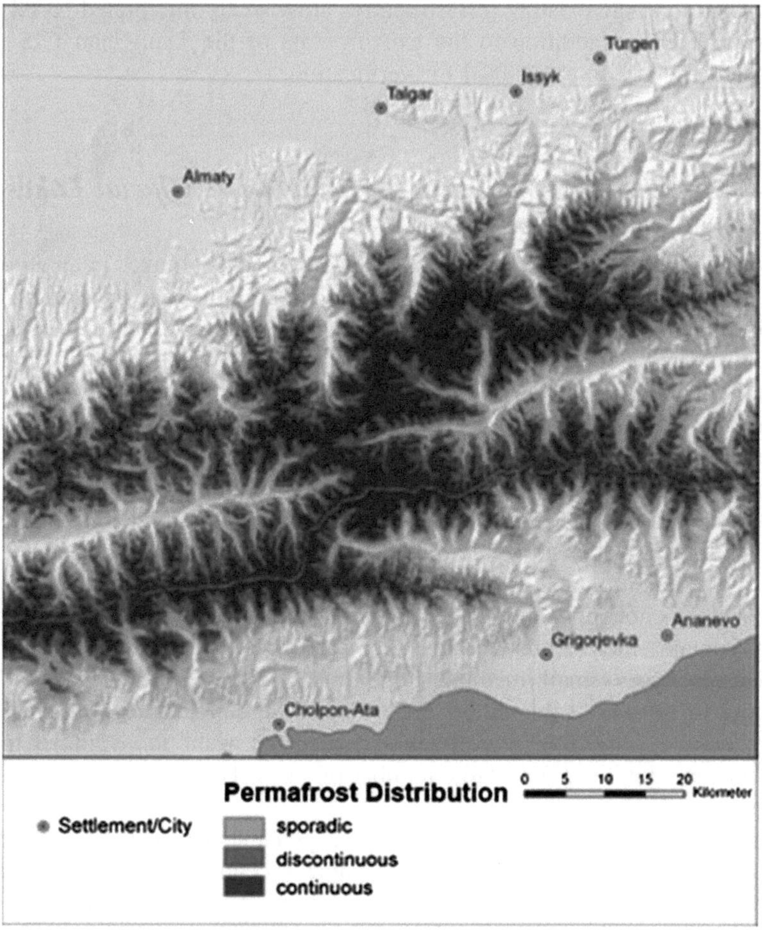

Fig. 12.4 Modelled permafrost extend

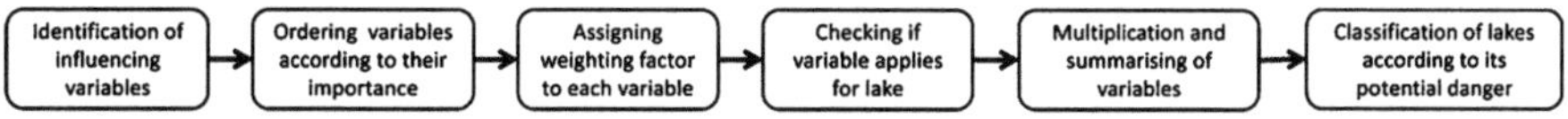

Fig. 12.5 General workflow for the classification of potential dangerous glacial lakes

the calculation at an angle of friction of less than 3°_ (Allen et al. 2009, Table 12.5). However, these thresholds are rough estimations and in reality transitions exist with different flow types occurring in the same event.

As a result, the relative probabilities of an affected downstream area can be calculated (Fig. 12.3b). A major decisive parameter is the quality of the applied

DEM. Therefore, the existing infrastructures are visually interpreted based on the satellite imagery in addition to the existing data of the Tian–Shan–GIS [major roads, settlements, Bolch (2008)] for verification.

12.3.4 Identification of Potentially Dangerous Glacial Lakes

In order to be able to identify potentially dangerous glacial lakes in an automated and a most objective way, it is necessary to combine the above mentioned conditioning parameters. We suggest a numerical approach on the basis of additive ratio scales similar to those utilised in business studies (Kahle 1998). The general workflow is presented in Fig. 12.5. The aim in this context is to have an efficient tool to help to make decisions (e.g. to find the right location(s) for the new business centre). In this case, this approach helps to find out which lakes are potentially of high danger and should be further investigated. For this, each introduced variable has to be tested if it applies to the investigated lake. If so (e.g. if a potential ice avalanche would reach a lake or a lake is in direct contact to the glacier), a value of one (1) is assigned to the lake otherwise a zero (0). However, this approach is not applicable for the lake area and lake growth. A larger lake area usually contains more water and can therefore cause higher damage. We introduced three classes (small, medium, large) and assign 0.5, 1 and 1.5 to each lake according to its area (Table 12.6). We do not differentiate further as this would usually require subjective interpretation which we want to avoid so that the approach can be utilised for a large number of lakes. The lake growth was treated in a similar way (Table 12.6).

Ideally, the utilised variables should be independent. This is the case with most of the variables we choose but there are some exceptions: For example, a flash flood will always occur if a mudflow is modelled. Therefore we introduced the precondition factor that a flash flood is only considered if a mudflow did not occur. Also, the increase of lake area depends at least partly on the glacier retreat if a glacier is in direct contact to the glacier. Glacier flow velocity and the variable slope below 5°_ is also not independent. We considered these two issues while assigning the weighting factors to each variable. A weighting scheme is also needed in order to account for the different impacts on the potential danger of the investigated lakes. However, the weighting is often subjective and depends also on the special situation in the study area.

We suggest a weighting scheme after a sequential order of the parameters as this is most objective and each variable is treated separately. The first and crucial step for the suggested scheme is the ordering of the variables after the estimated hazard potential from the highest to the lowest. We considered the knowledge from literature and past GLOF events (Table 12.1) for this step. Then, the weights are linearly distributed while the 2nd lowest weight is two times the lowest weight, the 3rd lowest is the sum of the 2nd lowest plus the lowest weight and so on. The sum of the weighting factor is set to 1 by default (Table 12.6). The variables which

Table 12.6 Weight factors of the selected variables

Variable	Weight factor	
Lake area change	0.1661	0: Shrinkage or no significant growth 0.5: growth < 50% of the initial area 1: growth < 100% of the initial area 1.5: growth < 150% of the initial area 2: growth > 150% of the initial area
Risk of ice avalanche	0.1510	1: Modelled deposits hit lake 0: Modelled deposits do not hit lake
Risk of rock fall/avalanche	0.1359	1: Modelled deposits hit lake 0: Modelled deposits do not hit lake
Instable dam[a]	0.1208	1: Dam is within discontinuous permafrost 0: Dam is outside discontinuous permafrost
Debris flow	0.1057	1: Debris flow would occur if an outburst would happen 0: Debris flow would not occur if an outburst would happen
Flash flood	0.0906	1: Flash flood would occur if an outburst would happen 0: Flash flood would not occur if an outburst would happen
Contact to glacier	0.0755	1: Lake is in direct contact with glacier 0: Lake is not in direct contact with glacier
Lake area	0.0604	0.5: Small (size < 50,000 m^2) 1.0: Medium (> 50,000 and < 100,000 m^2) 1.5: Large (> 100,000 m^2)
Glacier shrinkage	0.0453	1: Significant glacier shrinkage 0: No significant glacier shrinkage
Glacier slope <5° at the terminus	0.0302	1: Glacier has slope angels below 5° adjacent to the lake 0: Glacier has slope angels above 5° adjacent to the lake
Stagnant ice at the terminus	0.0151	1: No significant glacier velocity was detected at terminus 0: Significant glacier velocity was detected at the terminus
Sum of the weights	1.000	

[a] Although an unstable dam is one of the most important parameter for the potentially danger lake, we consider a lower weighting as we can only address roughly the thawing of permafrost based on the permafrost model

were applied for each lake are then multiplied with the weighting factor and subsequently added up. Thereafter, a total of nine remote sensing data derived parameters were included and modelled in GIS environment for the current situation, and two additional parameters are also included which indicate whether the glacial lake may continue to grow in the near future.

The characteristics of the moraine dams (width, height, freeboard) could only be addressed visually while other critical measures of the dam stability such as material composition or piping can hardly be addressed from remote sensing. The importance of those is included in the discussion section.

The final classification was established by the definition of qualitative threshold values ranging from very low potential danger to a high danger. The calculated values for the lakes range between 0.03 and 0.88. A very low hazard potential

Table 12.7 Number, area and the estimated volume and its changes of the glacial lakes in the study area for 1972, ~1999 and ~2007

	1972	2000	2007	1972–2000		2000–2007		1972–2007	
				Δ	Δ/a	Δ	Δ/a	Δ	Δ/a
Number	66	127	132	+61	+2.26	+5	+0.5	+65	+1.86
Area (km^2)	2.56 ± 0.14	3.28 ± 0.17	3.44 ± 0.18	+0.72 ± 0.22	+0.027 ± 0.008	+0.12 ± 0.24	+0.015 ± 0.03	+0.88 ± 0.23	+0.025 ± 0.007
Volume[a] ($\times 10^6$ m^3)	42.4 ± 10.6	47.2 ± 11.8	50.1 ± 12.5	+4.8 ± 15.9	+0.17 ± 0.57	+2.9 ± 17.2	+0.41 ± 2.46	+7.7 ± 16.4	+0.22 ± 0.47

[a] We included an additional uncertainty of 20% in the lake volume calculation (see text)

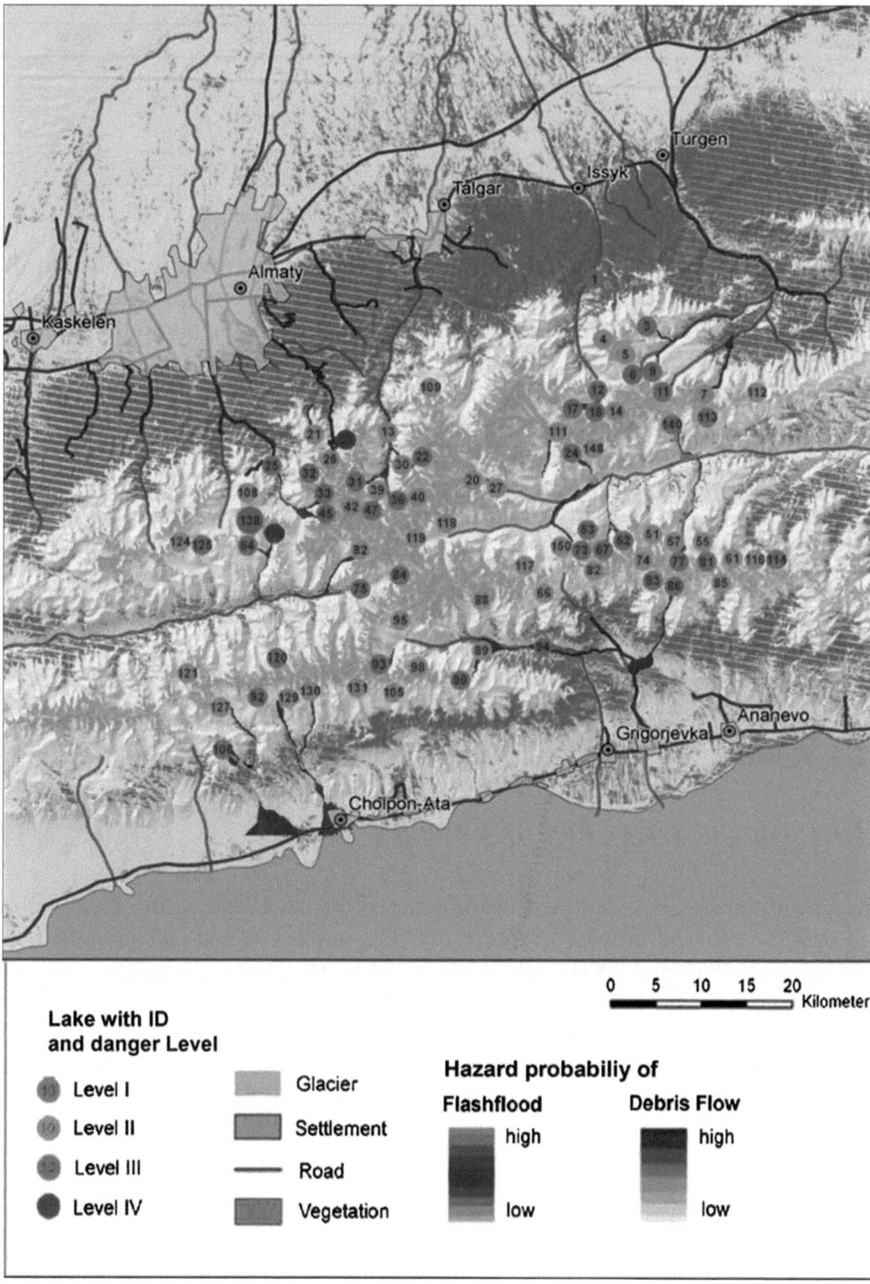

Fig. 12.6 The location of potentially dangerous glacial lakes in the study area of northern Tian Shan

Table 12.8 Numbers of glacial lakes in each class

Class	Risk index	No. of lakes
I (very low danger)	≤0.1	6
II (low danger)	0.1–0.325	79
III (medium danger)	0.325–0.574	45
IV (high danger)	>0.574	2

should have a lake only if no or only one factor with low weight applies to the lake. Hence, we choose 0.1 as the first threshold. We consider a lake can be potentially of high danger if the four most important factors apply to the lake or a combination of several factors reaching the sum of the weights of the four most important factors (in our case 0.574). The threshold between the low and medium potentially dangerous glacial lakes should be the mean value between class 1 and 4 (0.325, Table 12.8). We evaluated the weighting and classification scheme based on visual interpretation of the morphometric variables and the satellite imagery of selected case studies, previous GLOF events and knowledge from field visits from the authors and their colleagues from the Institute of Geography, Almaty and the State Agency for Mudflow Protection of the Ministry of Emergency Situations of the Republic of Kazakhstan (Kazselezashchita). We focused on the identified lakes with a high danger and its surrounding lakes.

12.4 Results

12.4.1 Glacial Lakes and its Changes

Overall, 66 lakes were identified in the imagery of the 1970s, while this number had increased to 132 in 2007 (Table 12.7). The number of the lakes almost doubled between 1972 and 2000 while it remained nearly constant between 2000 and 2007. Twelve of the lakes showed no significant changes over the investigated period. The two largest lakes are Bolshoje Almatinskoje lake and lake Dzhazil–Köl. The former, which is close to Almaty, developed after a rock avalanche and is now dammed by an artificial dam. The latter, situated at the end of Chon–Kemin valley, is dammed by two joining rock glaciers. About 60% of the identified lakes are in direct contact with the glacier ice. The overall area of the lakes increased from about 2.56 km^2 to about 3.44 km^2 and the estimated volume from $\sim 42.4 \times 10^6\ m^3$ to $\sim 50.1 \times 10^6\ m^3$ (1972–2007). Both the absolute lake area and the change rate increased from the periods 1972–2000 to 2000–2007. However, during this time it is observed that several lakes also lost surface area or disappeared completely. This is likely to be caused by lake drainages some of which may have occurred rapidly causing a GLOF, as e.g. in Kishi Almaty (Malaya Almatinka) valley in 1973 (Popov 1988).

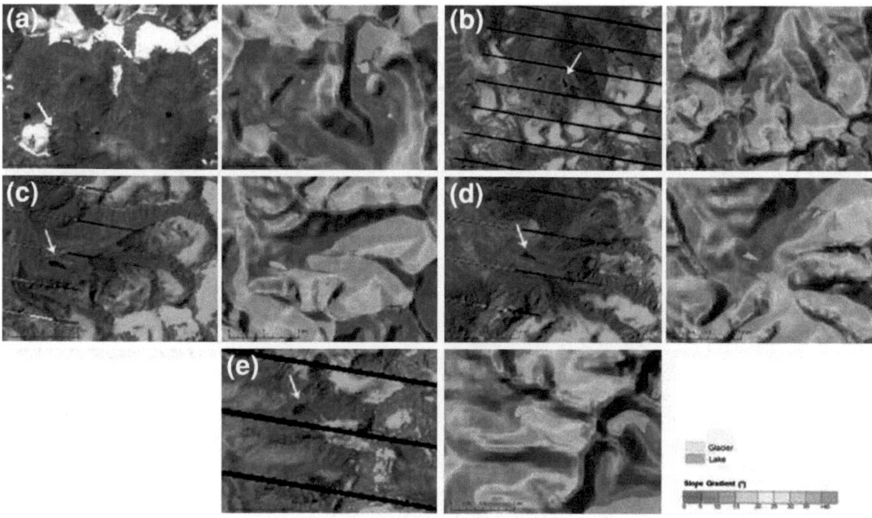

Fig. 12.7 Visualisation of the situation of selected glacial lakes. For each example, we show a satellite image and the slope of the topography. The *arrow* indicates a lake

12.4.2 Potentially Dangerous Glacial Lakes (PDGL)

The results of the hazard assessment showed that the majority of the lakes are classified as 'low danger' (Table 12.8; Fig. 12.6). These lakes are usually small and did not change significantly over the time. Overall, 47 out of 132 lakes are mapped in 2007 and are falling within the high and medium danger category. There is a medium to high possibility of an outburst of these lakes that can affect infrastructures or human beings. Therefore, these lakes should be studied and monitored more in detail. In the following section, we describe the situation of identified two highly dangerous lakes and some further lakes as case studies.

The lake identified as the highest danger is Lake No. 23 [named Lake No. 6 in the local literature, e.g. Kasatkin and Kapista (2009)] and situated in Kishi Almaty (Malaya Almatinka) valley which is close to the Million City of Almaty and was identified as highly dangerous (Figs. 12.7e, 12.8a). It is in close contact to a steep glacier and grew significantly since the 1970s. We modelled a high danger of ice avalanches or rock falls into the lake. Permafrost is likely to be in the dam and, in addition, the outburst modelling shows a high probability of a mudflow. And in fact this lake was also identified by the State Agency for Mudflow Protection of the Ministry of Emergency Situations of the Republic of Kazakhstan (Kazselezashchita) as a highly dangerous lake (Popov 1988; Kasatkin and Kapista 2009; Tokmagambetov 2009). This lake has formed in 1959 and increased concomitantly with the shrinkage of the adjacent Manshuk Mametova Glacier at the altitude of 3,600 m. Its tongue retreated approximately 250 m since 1958. By the end of the 1990s, the lake had a length of 230 m, a width of 150 m, and the area reached 24,000 m^2 with a volume close to 250,000 m^3. It is estimated that an outburst would

Fig. 12.8 Two examples of glacial lakes in the northern Tian Shan. *Top* The highly dangerous lake No. 23 (Nr. 6 of the valley) in Kishi Almaty (Malaya Almatinka) valley after the surface lowering by deepening of the outflow channel in 2010 (**a** view from dam to glacier, **b** view to dam, photos: Blagoveshchenskiy), *bottom* Lake No. 93 in Chon Aksu valley in 1994 (**c** photo: D. Wagner) and 2003 (**d** photo: T. Bolch)

cause a mudflow with a volume of more than 2 million m^3. Kazselezashchita built a channel in the dam to decrease the mudflow danger in 1997. Therefore, the lake volume decreased up to 150,000 m^3. However, by the year 2010 because of melting of ice, the average depth of the lake increased again and reached a maximum depth of 22 m, and the volume again reached 250,000 m^3.

Therefore, Kazselezashchita has deepened the channel which has lowered a water level on 6.6 m to a volume slightly higher than 100,000 m^3. The second highly dangerous lake, Lake No. 122, is situated in a side valley of the Chon–Kemin river (Fig. 12.7a). The lake is in direct contact to a glacier which has not only retreated but has also a very flat tongue which makes a further growth of the lake very likely. In addition, the lake has adjacent steep slopes and is exposed to both ice and rock avalanches. It is anticipated that no settlements would be affected by an outburst. However, there is a possibility that the gravel road that connects the Issyk–Kul with Almaty and which is frequently used by tourists and trekkers might be destroyed. Similarly, the neighbouring lake has a medium risk. It has similar conditions but is much smaller and the lake growth is slower. An example for a medium dangerous lake is No. 67 situated at south of the Shilek (Chilik) river

(Fig. 12.7b). The parent glacier shows characteristics for increasing the outburst hazard of the glacial lake, such as strong recession and flat snout. In addition, the lake area has increased by about 40% between 1999 and 2007. However, the affected area of an outburst will be probably small as the overall slope is gentle.

The characteristics of Lake No. 36 which was classified as medium dangerous and situated at the end of Levij Talgar valley are similar to those of Lake No. 67 (Fig. 12.7c). In addition, it is comparatively large. Permafrost thaw might destabilise the moraine dam. However, the dam is relatively wide. A new lake has developed up-valley which can, in case of an outburst, trigger the overflow of Lake No. 36. Hence, this lake should be continuously monitored also because Levij Talgar valley is frequently visited by tourists and an outburst in this valley can destroy infrastructures at the foothills such as in the case of the 2003 outburst event (Bolch 2008). Another larger lake (No. 22) in this valley is classified as a medium risk (Fig. 12.7d). This seems reasonable as it is not adjacent to a glacier and mass movements will not reach the lake. Another example of medium risk is Lake No. 93 in Chon Aksu valley (Fig. 12.8c), although the lake did not grow significantly over the observed period. The dam could be affected by permafrost thaw and an outburst would trigger a flash flood.

12.5 Discussion

12.5.1 Glacial Lakes in Northern Tian Shan

Previous studies have shown that the number and the area of glacial lakes have increased concomitantly in terms of glacier shrinkage in many regions (Bajracharya and Mool 2009; Gardelle et al. 2011). The susceptibility of a lake outburst increases when the thawing of permafrost destabilises the moraine dam or if the glacial lake is growing (Richardson and Reynolds 2000). Temperatures in northern Tian Shan have increased about 0.8 K/100a between the period 1900–2000 and about 2.0 K/100a on average for the second half of the last century (Giese et al. 2007; Bolch 2007). The increase was found to be less pronounced in the mountainous areas, but is still obvious. No significant trend was found with the precipitation (Aizen et al. 1997; Bolch 2007). The glaciers in the study area are shrinking concomitantly with respect to the climate change. The glacier shrinkage at the northern edge of the Tian Shan is more pronounced than the central parts with its continental mountain ranges (Aizen et al. 2006; Bolch 2007; Kutuzov and Shahgedanova 2009; Narama et al. 2010b; Solomina et al. 2004). The lower boundary of permafrost has shifted upwardly about 150–200 m since the beginning of the 20th century while the area of permafrost has decreased by approximately 18% (Marchenko et al. 2007). Hence, it could be expected that the number and area of potentially dangerous lakes increases in the study area, a fact which is confirmed by this study. Other studies on glacial lakes in the same and adjacent mountain ranges of the Tian Shan show similar results

(Erohin and Cerny 2009; Janský et al. 2011; Narama et al. 2009, 2010a; Popov 1988). Further glacier shrinkage will probably lead to higher danger of a GLOF. Also in the past, several glacial lake outburst events were caused by rapid glacier shrinkage followed by an overflow of the moraine dam.

The national scientists and the responsible authorities in Kazakhstan and Kyrgyzstan have been well aware of the glacial lake hazards for several decades. Different hazard mitigation activities such as the activities for the lowering of the lake levels (e.g. Kasatkin and Kapista 2009; Popov 1988), mudflow protection walls and dams (such as in Kishi and Ulken Almatinka valleys, Yegorov 2007) reduced the risk significantly. However, GLOFs and mudflows still occur and can endanger infrastructures. This is especially true for areas above the protection walls and dams.

12.5.2 Methodology

The case studies show that our suggested approach which is based on remote sensing analysis and modelling can be successfully applied to identify potentially dangerous glacial lakes. It is hence, suitable for a first comprehensive assessment of PDGLs for a larger area and addresses the levels 1 and 2 of the suggested approach by Huggel et al. (2002). We combined the manifold conditioning parameters which had not been addressed previously in this comprehensive way. The approach is easily reproducible as it is based on welldeveloped methods such as the detection of water bodies using multispectral imagery (Huggel et al. (2002), automated detection of glaciers (Bolch and Kamp 2006; Paul et al. 2002), and their velocities (Bolch et al. 2008; Kääb 2005) and simple but robust models such as the modelling of an outburst path or rock/ice avalanches (Huggel et al. 2003), or a permafrost model (Keller 1992). Older panchromatic imagery such as Corona proved to be suitable to extend the analysis back in time which was already shown for glaciers (Narama et al. 2010b; Bolch et al. 2010). It could also be demonstrated that it is possible to detect PDGLs for a larger area such as mountain ranges covering a whole Landsat scene ($\sim$ 185 by 185 km) within a short period of time. However, the results need to be carefully evaluated and the weighting scheme possibly adjusted for the special situation in the respective study region. Data from former outburst events are valuable sources for calibration. In case of the current study area, they confirmed the importance of lake growth and the possibility of existing ice in the moraine dam.

The three main limitations of the geomatics-based approach are that (1) the dam characteristics and the probability of a dam failure can only be addressed marginally, (2) lakes in shaded areas and turbid lakes are difficult to identify automatically, (3) the modelling was based on the SRTM3 DEM with a spatial resolution of 30°. The result of the modelling of the outburst path is based on the SRTM DEM which highlights the most endangered areas. Comparisons with historical outbursts show that the path length and, hence, the affected areas are underestimated. For instance, a flash flood caused by a lake outburst which happened in 2003 in Leviy Talgar valley after heavy rainfalls destroyed a bridge at the foothills but the modelled flow path stopped shortly before this point. Similar uncertainty exists in the modelling of the mass movements based on the SRTM. Hence, the SRTM3 DEM proved to be suitable for a first assessment but has inaccuracies which were also concluded by Frey et al. (2010b). A more accurate DEM should be used if available.

The major limitation of a remote sensed based study is that the characteristics and stability of the moraine/lake dam cannot or only roughly taken into account. The used permafrost model is coarse and does not consider the material composition which can strongly alter the thermal conditions within the materials and also retard the thawing of the ice. However, this model can be easily applied to other mountainous regions in order to provide a first hint about the existence of permafrost. In addition, this model can be applied for estimation of the future permafrost extent and to address the danger of increased mass movements due to permafrost thaw. Other critical measures of the stability of the dam are the types of drainage. While the drainage over the dam could be detected at least with using high resolution imagery outflows under or through the dam or piping cannot be addressed. A further limitation is that the water volume of the lakes can only be roughly calculated based on the area. Field investigations would be necessary to measure the lake depth and to address the grain size distribution of the moraine.

The applied methodology for the categorisation of the glacial lakes produced reasonable results. The utilised weighting according to the importance of the variables, however, should be carefully chosen by an expert and adjusted to the study region.

12.6 Conclusion

The presented geomatics-based approach successfully detects potentially dangerous glacial lakes across a larger area and presents decision criteria where time-and cost-intensive field studies should be performed. The availability of data from previous outburst events helps to adjust the weighting scheme for the respective study region. The main drawback is that the stability of the moraine can currently only roughly be addressed by remote sensing and modelling. The nearly globally available SRTM DEM is a good choice for the first assessment but a more accurate DEM of higher resolution should be utilised for the analysis if available.

We detected two highly dangerous lakes which should be continuously monitored. For these lakes and lakes of medium danger which could affect infrastructures and endanger human lives (e.g. at the northern slope of Ile Alatau and the southern slope of Kungey Alatau and in the mountains along trekking routes) visual checking and manual interpretation are needed. Continuous climate warming and the resultant permafrost thaw and glacier recession will increase the potential danger of lake outbursts.

Acknowledgments The authors would like to thank D. Quincey and C. Huggel for their thorough comments which significantly improved the quality of the manuscript. The logistic support by I. Severskiy, I. Shesterova and A. Kokarev (Institut for Geography, Almaty) is appreciated.

References

Aizen VB, Aizen EM, Melack JM, Dozier J (1997) Climate and hydrologic changes in the Tien Shan, central Asia. J Clim 10:1393–1404

Aizen VB, Kuzmichenok VA, Surazakov AB, Aizen EM (2006) Glacier changes in the central and northern Tien Shan during the last 140 years based on surface and remote-sensing data. Ann Glaciol 43:202–213

Alean J (1985) Ice avalanches: some empirical information about their formation and reach. J Glaciol 31:324–333

Allen SK, Schneider D, Owens IF (2009) First approaches towards modelling glacial hazards in the Mount Cook region of New Zealand's southern Alps. Nat Hazard Earth Syst Sci 9(2):481–499

Baimoldayev T, Vinohodov V (2007) Kazselezashchita–operativnye mery do I posle stihii–Izd. Bastau, Almaty, p 284

Bajracharya SR, Mool PK (2009) Glaciers, glacial lakes and glacial lake outburst floods in the Mount Everest region, Nepal. Ann Glaciol 50(53):81–86

Berry PA, Garlick JD, Smith RG (2007) Near-global validation of the SRTM DEM using satellite radar altimetry. Remote Sens Environ 106(1):17–27

Berthier E, Vadon H, Baratoux D, Arnaud Y, Vincent C, Feigl KL, Rémy F, Legrésy B (2005) Surface motion of mountain glaciers derived from satellite optical imagery. Remote Sens Environ 95(1):14–28

Blagoveshchenskiy VP, Yegorov AB (2009) The Ile Alatau range natural hazards. In: Materials of the international conference on mitigation of natural hazards in mountain areas, 15–18 Sept 2009. Bishkek, pp 153–157

Böhner J (1996) Säkulare Klimaschwankungen und rezente Klimatrends Zentral-und Hochasiens. Göttinger Geographische Abhandlungen, vol 101, pp 166

Bolch T (2007) Climate change and glacier retreat in northern Tien Shan (Kazakhstan/Kyrgyzstan) using remote sensing data. Glob Planet Change 56:1–12

Bolch T (2008) Klima-und Gletscheränderungen im nõrdlichen Tien Shan (Kasachstan/Kyrgyzstan) mit einem Vergleich zur Bernina-Gruppe/Alpen (in German with English and Russian abstract and captions). VDM Verlag Dr. Müller, p 264, ISBN: 3-639-03742-1. (= Bolch T (2006) Dissertation, Universität Erlangen-Nürnberg, p 210. http://www.opus.ub.uni-erlangen.de/opus/volltexte/2006/447/)

Bolch T, Kamp U (2006) Glacier mapping in high mountains using DEMs, landsat and ASTER data, Grazer Schriften der Geographie und Raumforschung. In: Proceedings 8th international symposium on high mountain remote sensing cartography, vol 41, 20–27 March 2005. La Paz, Bolivia, pp 13–24

Bolch T, Kamp U, Olsenholler J (2005) Using ASTER and SRTM DEMs for studying geomorphology and glaciation in high mountain areas. In: Oluic M (ed) New strategies for European remote sensing. Millpress, Rotterdam, pp 119–127

Bolch T, Buchroithner MF, Kunert A, Kamp U (2007) Automated delineation of debris-covered glaciers based on ASTER data. In: Gomarasca MA (ed) GeoInformation in Europe (= Proceedings of 27th EARSeL-symposium, 4–7 June 2007, Bozen, Italy). Millpress, Netherlands, pp 403–410

Bolch T, Buchroithner MF, Bajracharya SR, Peters J, Baessler M (2008) Identification of glacier motion and potentially dangerous glacier lakes at Mt. Everest area/Nepal using spaceborne imagery. Nat Hazard Earth Syst Sci 8(6):1329–1340

Bolch T, Yao T, Kang S, Buchroithner MF, Scherer D, Maussion F, Huintjes E, Schneider C (2010) A glacier inventory for the western Nyainqentanglha range and Nam Co Basin, Tibet, and glacier changes 1976–2009. Cryosphere 4:419–433

Bolch T, Pieczonka T, Benn DI (2011) Multi-decadal mass loss of glaciers in the Everest area (Nepal, Himalaya) derived from stereo imagery. Cryosphere 5:349–358

Buchroithner M (1996) Jökulhlaup mapping in the Himalaya by means of remote sensing. Kartographische Bausteine 12:75–86

Buchroithner M, Jentsch G, Wanivenhaus B (1982) Monitoring of recent geological events in the Khumbu area (Himalaya, Nepal) by digital processing of Landsat MSS data. Rock Mech 15:181–197

Chedija OK (1986) Morfostruktury i novejzhij Tektogenez Tjan-Shanja, Ilm, Frunse

Clague JJ, Evans SG (2000) A review of catastrophic drainage of moraine-dammed lakes in British Columbia. Quat Sci Rev 19:1763–1783

Dashora A, Lohani B, Malik JN (2007) A repository of earth resource information—CORONA satellite programme. Curr Sci 92(7):926–932

Delevaux D, Abdrakhmatov KE, Lemzin IN, Strom L (2001) Opolszni i Razpyvy Keminskogo Zemletrjacenija 1911 g. s Ms 8, 2 v Kirgizii. Geologija i Geofisika 42(10): 1667–1677

Erohin S, Cerny M (2009) Monitoring of out bursting lakes of Kyrgyzstan. In: Material of the international conference on mitigation of natural hazards in mountain areas, 15–18 Sept 2009, Bishkek, pp 30–34

Falorni G, Teles V, Vivoni ER, Bras RL, Amartunga KS (2005) Analysis and characterization of the vertical accuracy of digital elevation models from the shuttle radar topography mission. J Geophys Res 110 (F02005). doi: 10.1029/2003JF000113

Frey H, Huggel C, Paul F, Haeberli W (2010a) Automated detection of glacier lakes based on remote sensing in view of assessing associated hazard potentials. In: Kaufmann V, Sulzer W (eds) Proceedings of 10th international symposium on high mountain remote sensing cartography. Grazer Schriften der Geographie und Raumforschung, vol 45, pp 261–272

Frey H, Haeberli W, Linsbauer A, Huggel C, Paul F (2010b) A multi-level strategy for anticipating future glacier lake formation and associated hazard potentials. Nat Hazard Earth Syst Sci 10(2):339–352

Fujita K, Suzuki R, Nuimura T, Sakai A (2008) Performance of ASTER and SRTM DEMs, and their potential for assessing glacial lakes in the Lunana region, Bhutan Himalaya. J Glaciol 54(185):220–228

Gardelle J, Arnaud Y, Berthier E (2011) Contrasted evolution of glacial lakes along the Hindu Kush Himalaya mountain range between 1990 and 2009. Glob Planet Chang 75:47–55

Giese E, Moßig I, Rybski D, Bunde A (2007) Long-term analysis of air temperature trends in central Asia. Erdkunde 61(2):186–202

Gorbunov AP, Severskiy EV (2001) Seli okrestnostej Almaty. Almaty

Gorbunov AP, Severskiy EV, Titkov SN (1996) Geokriologichesije Uslovija Tjan-Shanja iPamira, Institut Merzlotovedenija, Rassiyskaja Akademija Nauk, Irkutsk

Gorbunov AP, Marchenko SS, Severskiy EV (2004) The thermal environment of blocky materials in the mountains of central Asia. Permafr Periglac Process 15(1):95–98

Granshaw FD, Fountain AD (2006) Glacier change (1958–1998) in the north Cascades national park complex, Washington, USA. J Glaciol 52(177):251–256

Haeberli W (1983) Frequency and characteristics of glacier floods in the Swiss Alps. Ann Glaciol 4:85–90

Havenith HB, Strom A, Jongmans D, Abdrakhmatov A, Delvaux D, Tréfois P (2003) Seismic triggering of landslides, part A: field evidence from the northern Tien Shan. Nat Hazard Earth Syst Sci 3(1/2):135

Hsü KJ (1975) Catastrophic debris streams (Sulzstroms) generated by rockfalls. Geol Soc Am Bull 86:129–140

Huggel C, Kääb A, Haeberli W, Teysseire P, Paul F (2002) Remote sensing based assessment of hazards from glacier lake outbursts: a case study in the Swiss Alps. Can Geotech J 39: 316–330

Huggel C, Kääb A, Haeberli W, Krummenacher B (2003) Regional-scale GIS-models for assessment of hazards from glacier lake outbursts: evaluation and application in the Swiss Alps. Nat Hazard Earth Syst Sci 3:647–662

Huggel C, Zgraggen-Oswald S, Haeberli W, Kääb A, Polkvoj A, Galushkin I, Evans SG (2005) The 2002 rock/ice avalanche at Kolka/Karmadon, Russian Caucasus: assessment of extraordinary avalanche formation and mobility, and application of QuickBird satellite imagery. Nat Hazard Earth Syst Sci 5:173–187

Ives J (1986) Glacial lake outburst floods and risk engineering in the Himalaya. ICIMOD, Kathmandu

Iwata S, Ageta Y, Naito N, Sakai A, Narama C, Karma (2002) Glacial lakes and their outburst flood assessment in the Bhutan Himalaya. Glob Environ Res 6(1):3–17

Janský B, Sobr M, Engel Z (2011) Outburst flood hazard: case studies from the Tien-Shan mountains, Kyrgyzstan. Limnologica—ecology and management of inland waters. doi: 10.1016/j.limno.2009.11.013

Kääb A (2005) Combination of SRTM3 and repeat ASTER data for deriving alpine glacier flow velocities in the Bhutan Himalaya. Remote Sens Environ 94:463–474

Kääb A, Huggel C, Fischer L, Guex S, Paul F, Roer I, Salzmann N, Schlaefli S, Schmutz K, Schneider D, Strozzi T, Weidmann Y (2005) Remote sensing of glacier-and permafrost-related hazards in high mountains: an overview. Nat Hazard Earth Syst Sci 5:527–554

Kahle E (1998) Betriebswirtschalftliche Entscheidungen: Lehrbuch zur Einführung betreibswirtschaftlicher Entscheidungstheorie. München

Kaibori M, Sassa K, Tochiki S (1988) Betrachtung über die Bewegung von Absturzmaterialien. International symposium of INTERPRAEVENT 1988/07, vol 2, pp 227–242

Kamp U, Bolch T, Olsenholler J (2005) Geomorphometry of Cerro Sillajhuay, Chile/Bolivia: comparison of DEMs derived from ASTER remote sensing data and contour maps. Geocarto Int 20(1):23–34

Kasatkin NE, Kapista VP (2009) The morainic lakes dynamics in the Ile Alatau. In: Materials of the international conference on mitigation of natural hazards in mountain areas, 15–18 Sept 2009. Bishkek, pp 55–58

Keller F (1992) Automated mapping of mountain permafrost using the program PERMAKART within the geographical information system ARC/INFO. Permafr Periglac Process 3(2): 133–138

Kocak G, Büyüksalih G, Jacobsen K (2004) Analysis of digital elevation models determined by high resolution space images. Int Arch Photogramm Remote 35(B4):636–641

Korjenkov AM, Kovalenko VA, Usmanov SF (2004) Long-term preservation of paleoseismic deformation as a tool for revealing traces of ancient seismic catastrophes (example of the Chon-Kemin Valley, Kyrgyzstan). Kartographische Bausteine 28:137–154

Kutuzov S, Shahgedanova M (2009) Glacier retreat and climatic variability in the eastern Terskey-Alatoo, inner Tien Shan between the middle of the nineteenth century and beginning of the twenty first century. Glob Planet Change 69(1–2):59–70

Leprince S, Barbot S, Ayoub F (2007) Automatic and precise orthorectification, coregistration, and subpixel correlation of satellite images, application to ground deformation measurements. IEEE Trans Geosci Remote Sens 45(6):1529–1558

Luckman A, Quincey D, Bevan S (2007) The potential of satellite radar interferometry and feature tracking for monitoring flow rates of Himalayan glaciers. Remote Sens Environ 111:172–181

Lukk AA, Yunga S, Shevchenko VI, Hamburger MW (1995) Earthquake focal mechanisms, deformation state and seismotektoniks of the Pamir-Tien Shan region, central Asia. J Geophys Res 100(B100):20321–20343

Ma D, Tu J, Cui P, Lu R (2004) Approach to mountain hazards in Tibet, China. J Mt Sci 1(2):143–154

Marchenko SS (1999) Sovremennoje i ozhidaemje v 21. V. izmenenije merzlotno-klimaticheskikh uslovnij b severnom Tjan-Shanje. Criosfera Semli 3(2):13–21

Marchenko SS (2001) A model of permafrost formation and occurrences in the intracontinental mountains. Norsk Geograf Tidsskrift 55(4):230–234

Marchenko SS, Gorbunov AP, Romanovsky VE (2007) Permafrost warming in the Tien Shan mountains, central Asia. Glob Planet Change 56(3–4):311–327

Medeuov AR, Nurlanov MT (1996) Selevye Javlenija seismoaktivnykh Territorij Kazakhstana. Akademiya Nauk, Almaty

Medeuov A, Kolotilin NF, Keremkulov VA (1993) Seli Kazakhstana. Gylym, Almaty, p 160

Mergili M, Schneider D, Worni R, Schneider J (2011) Glacial lake outburst floods in the Pamir of Tajikistan: challenges in prediction and modelling. In: 5th international conference on debris-flow hazards mitigation: mechanics, prediction and assessment, University of Padova, Italy, 14–17 June 2011, accepted

Narama C, Severskiy I, Yegorov A (2009) Current state of glacier changes, glacial lakes, and outburst floods in the Ile Ala-Tau and Kungoy Ala-Too ranges, northern Tien Shan mountains. Ann Hokkaido Geogr 84:22–32

Narama C, Duishonakunov M, Kääb A, Daiyrov M, Abdrakhmatov K (2010a) The 24 July 2008 outburst flood at the western Zyndan glacier lake and recent regional changes in glacier lakes of the Teskey Ala-Too range, Tien Shan, Kyrgyzstan. Nat Hazard Earth Syst Sci 10:647–659

Narama C, Kääb A, Duishonakunov M, Abdrakhmatov K (2010b) Spatial variability of recent glacier area changes in the Tien Shan mountains, central Asia, using Corona (1970), Landsat (2000), and ALOS (2007) satellite data. Glob Planet Change 71(1–2):42–54

Passmore DG, Harrison S, Winchester V, Rae A, Severskiy IV, Pimankina NV (2008) Late Holocene debris flows and valley floor development in the northern Zailiiskiy Alatau, Tien Shan mountains, Kazakhstan. Arctic Antarct Alp Res 40(3):548–560

Paul F, Kääb A, Maisch M, Kellenberger T, Haeberli W (2002) The new remote sensing derived swiss glacier inventory: I. Methods. Ann Glaciol 34:355–361

Plehanov PA, Sudakov PA, Tokmagambetov GA (1975) Prichiny obrazovaniya I processy formirovaniya selevogo potoka na reke Malaya Almatinka 15 iyulya 1973 g. Vestnik AN KazSSR 4:24–25

Popov NV (1984a) O seleopasnyh ozerah glacialnoi zony Zailiyskogo Alatau. Problemy protivoselevyh meropriy. Alma-Ata: 96–105

Popov NV (1984b) O glyacialnom selevom potoke 23 Iyulya, 1980 goda v basseine reki Kaskelen Zailiyskogo Alatau. Problemy protivoselevyh meropriyatiy. Alma-Ata: 222–230

Popov NV (1988) Die Kontrolle gefährlicher Gletscherseen im nördlichen Tienschan. In: Proceedings of the international symposium interpraevent, vol 4, pp 29–41

Quincey DJ, Lucas RM, Richardson SD, Glasser NF, Hambrey MJ, Reynolds JM (2005) Optical remote sensing techniques in high-mountain environments: application to glacial hazards. Prog Phys Geogr 29(4):475–505

Quincey DJ, Richardson SD, Luckman A, Lucas RM, Reynolds JM, Hambrey MJ, Glasser NF (2007) Early recognition of glacial lake hazards in the Himalaya using remote sensing datasets. Glob Planet Change 56(1–2):137–152

Reynolds JM (2000) On the formation of supraglacial lakes on debris-covered glaciers. IAHS publication, vol 264 (= Debris-covered Glaciers), pp 153–161

Richardson SD, Reynolds JM (2000) An overview of glacial hazards in the Himalayas. Quat Int 65/66(1):31–47

Salzmann N, Kääb A, Huggel C, Allgöwer B, Haeberli W (2004) Assessment of the hazard potential of ice avalanches using remote sensing and GIS-modelling. Norsk Geograf Tidsskrift 58:74–84

Scherler D, Leprince S, Strecker MR (2008) Glacier-surface velocities in alpine terrain from optical satellite imagery—accuracy improvement and quality assessment. Remote Sens Environ 112(10):3806–3819

Schneider JF (2004) Risk assessment of remote geohazards in western Pamir, GBAO, Tajikistan. In: Proceedings of the international conference on high mountain hazard prevention, Vladikavkaz Moscow, 23–26 June 2004, pp 252–255

Severskiy IV, Zichu X (eds) (2000) Snow cover and avalanches in the Tien Shan mountains. VAC Publishing House, Almaty

Shusharin VI, Popov NV (1981) Razvitie selevogo potoka v basseine reki Sredniy Talgar—problemy protivoselevyh meropriyatiy. Alma-Ata: 153–157

Solomina ON, Barry R, Bodnya M (2004) The retreat of Tien Shan glaciers (Kygryzstan) since the little ice age estimated from aerial photographs, lichenometric and historical data. Geograf Ann Ser A 86(2):205–216

Storm AL, Korup O (2006) Extremly large rockslides and rock avalaches in the Tien Shan mountains, Kyrgyzstan. Landslides 3:125–136

Tokmagambetov TG (2009) The moraine-dammed glacial lakes current state in the Iliy Alatau. In: Materials of the international conference on mitigation of natural hazards in mountain areas, 15–18 Sept 2009, Bishkek, pp 82–83

Utirov CU (1978) Seismic deformations. Geological principles of the seismic zoning of the Issuk-Kul depression. Ilm, Frunze, pp 91–111

Vuichard D, Zimmermann M (1987) The 1985 catastrophic drainage of a moraine-dammed lake, Khumbu Himal, Nepal: cause and consequences. Mt Res Dev 7:91–110

Wessels R, Kargel J, Kieffer H (2002) ASTER measurement of supraglacial lakes in the mount Everest region of the Himalaya. Ann Glaciol 34:399–408

Woerd J, van der Owen LA, Tapponnier P, Xu X, Kervyn F, Finkel RC, Barnard PL (2004) Giant, _M8 earthquake triggered ice avalanches in the eastern Kunlun Shan, northern Tibet: characteristics, nature and dynamics. Geol Soc Am Bull 116(3):394–406

Yadav RR, Kulieshhius P (1992) Dating of Earthquakes: tree ring responses to the catastrophic earth quake of 1887 in Alma Ata, Kasakhstan. Geograph J 158:295–299

Yegorov A (2007) Analyse und Bewertung von Naturgefahren als Grundlage für ein nachhaltiges Naturgefahrenmanagment in Gebirgssystemen SO-Kasachstans am Beispiel des Ile-Alatau. Dissertation, LMU München

MIX
Papier aus verantwortungsvollen Quellen
Paper from responsible sources
FSC® C105338

If you have any concerns about our products,
you can contact us on
ProductSafety@springernature.com

In case Publisher is established outside the EU,
the EU authorized representative is:
Springer Nature Customer Service Center GmbH
Europaplatz 3, 69115 Heidelberg, Germany

Printed by Libri Plureos GmbH
in Hamburg, Germany